T0181995

Lecture Notes in Computer Science   13969

Founding Editors

Gerhard Goos
Juris Hartmanis

The series Lecture Notes in Computer Science (LNCS), including its subseries Lecture Notes in Artificial Intelligence (LNAI) and Lecture Notes in Bioinformatics (LNBI), has established itself as a medium for the publication of new developments in computer science and information technology research, teaching, and education.

LNCS enjoys close cooperation with the computer science R & D community, the series counts many renowned academics among its volume editors and paper authors, and collaborates with prestigious societies. Its mission is to serve this international community by providing an invaluable service, mainly focused on the publication of conference and workshop proceedings and postproceedings. LNCS commenced publication in 1973.

Ying Tan · Yuhui Shi · Wenjian Luo
Editors

Advances in Swarm Intelligence

14th International Conference, ICSI 2023
Shenzhen, China, July 14–18, 2023
Proceedings, Part II

 Springer

Editors
Ying Tan 🄳
Peking University
Beijing, China

Yuhui Shi
Southern University of Science
and Technology
Shenzhen, China

Wenjian Luo
Harbin Institute of Technology
Shenzhen, China

ISSN 0302-9743   ISSN 1611-3349 (electronic)
Lecture Notes in Computer Science
ISBN 978-3-031-36624-6   ISBN 978-3-031-36625-3 (eBook)
https://doi.org/10.1007/978-3-031-36625-3

This Springer imprint is published by the registered company Springer Nature Switzerland AG
The registered company address is: Gewerbestrasse 11, 6330 Cham, Switzerland

Preface

This book and its companion volume, LNCS vols. 13968 and 13969, constitute the proceedings of the Fourteenth International Conference on Swarm Intelligence (ICSI 2023) held during July 14–18, 2023 in Shenzhen, China, both onsite and online.

The theme of ICSI 2023 was "Serving Life with Swarm Intelligence." ICSI 2023 provided an excellent opportunity for academics and practitioners to present and discuss the latest scientific results and methods, innovative ideas, and advances in theories, technologies, and applications in swarm intelligence. The technical program covered a number of aspects of swarm intelligence and its related areas. ICSI 2023 was the fourteenth international gathering for academics and researchers working on most aspects of swarm intelligence, following successful events in Xi'an (ICSI 2022) virtually, Qingdao (ICSI 2021), Serbia (ICSI 2020) virtually, Chiang Mai (ICSI 2019), Shanghai (ICSI 2018), Fukuoka (ICSI 2017), Bali (ICSI 2016), Beijing (ICSI-CCI 2015), Hefei (ICSI 2014), Harbin (ICSI 2013), Shenzhen (ICSI 2012), Chongqing (ICSI 2011), and Beijing (ICSI 2010), which provided a high-level academic forum for participants to disseminate their new research findings and discuss emerging areas of research. The conference also created a stimulating environment for participants to interact and exchange information on future challenges and opportunities in the field of swarm intelligence research.

Due to the continuing global COVID-19 pandemic, ICSI 2023 provided both online and offline presentations. On one hand, ICSI 2023 was normally held in Shenzhen, China. On the other hand, the ICSI 2023 technical team provided the ability for the authors of accepted papers who had restrictions on overseas travel to present their work through an interactive online platform or video replay. The presentations by accepted authors were made available to all registered attendees onsite and online.

The host city of ICSI 2023 was Shenzhen, China, which, located on the southern coast of China, adjacent to Hong Kong, is an international modern city in Guangdong Province. It has developed rapidly due to Reform and Opening-up. It is currently the third largest city in mainland China in terms of economic aggregate. Shenzhen is an important high-tech R&D and manufacturing base in southern China, and is often referred to as the "Silicon Valley of China". It is the most suitable dynamic city for talents at home and abroad to expand their business.

ICSI 2023 received a total of 170 submissions and invited submissions from about 452 authors in 15 countries and regions (Armenia, Brazil, Canada, Chile, China, France, India, Japan, Pakistan, Russia, South Africa, Taiwan(China), Thailand, UK, and USA) across 5 continents (Asia, Europe, North America, South America, and Africa). Each submission was reviewed by at least 2 reviewers, and on average 2.8 reviewers. Based on rigorous reviews by the program committee members and reviewers, 81 high-quality papers were selected for publication in this proceedings volume with an acceptance rate of 47.65%. The papers are organized into 12 cohesive sections covering major topics of swarm intelligence research and its development and applications along with a competition session entitled "Competition on Single-Objective Bounded Optimization Problems (ICSI-OC'2023)".

On behalf of the Organizing Committee of ICSI 2023, we would like to express our sincere thanks to the International Association of Swarm and Evolutionary Intelligence (IASEI) (iasei.org), which is the premier international scholarly society devoted to advancing the theories, algorithms, real-world applications, and developments of swarm intelligence and evolutionary intelligence. We would also like to thank Harbin Institute of Technology Shenzhen, Peking University, and Southern University of Science and Technology for their co-sponsorships, and Computational Intelligence Laboratory of Peking University and IEEE Beijing Chapter for their technical co-sponsorships, Nanjing Kanbo iHealth Academy for its technical and financial co-sponsorship, as well as to our supporters of International Neural Network Society, World Federation on SoftComputing, MDPI's journal Entropy Beijing Xinghui Hi-Tech Co., and Springer Nature.

We would also like to thank the members of the Advisory Committee for their guidance, the members of the International Program Committee and additional reviewers for reviewing the papers, and the members of the Publication Committee for checking the accepted papers in a short period of time. We are particularly grateful to the proceedings publisher Springer for publishing the proceedings in the prestigious series of Lecture Notes in Computer Science. Moreover, we wish to express our heartfelt appreciation to the plenary speakers, session chairs, and student helpers. In addition, there are still many more colleagues, associates, friends, and supporters who helped us in immeasurable ways; we express our sincere gratitude to them all. Last but not least, we would like to thank all the speakers, authors, and participants for their great contributions that made ICSI 2023 successful and all the hard work worthwhile.

May 2023

Ying Tan
Yuhui Shi
Wenjian Luo

Organization

Honorary Co-chairs

Russell C. Eberhart   IUPUI, USA
Yan Jia   Harbin University of Technology, Shenzhen, China

General Co-chairs

Ying Tan   Peking University, China
Wenjian Luo   Harbin University of Technology, Shenzhen, China

Programme Committee Chair

Yuhui Shi   Southern University of Science and Technology, China

Advisory Committee Chairs

Gary G. Yen   Oklahoma State University, USA
Xin Yao   Southern University of Science and Technology of China, China
Yaochu Jin   Bielefeld University, Germany
Xuan Wang   Harbin University of Technology, Shenzhen, China

Technical Committee Co-chairs

Kay Chen Tan   Hong Kong Polytechnic University, China
Qingfu Zhang   City University of Hong Kong, China
Haibo He   University of Rhode Island Kingston, USA
Martin Middendorf   University of Leipzig, Germany
Xiaodong Li   RMIT University, Australia
Hideyuki Takagi   Kyushu University, Japan
Ponnuthurai Nagaratnam Suganthan   Nanyang Technological University, Singapore
Mengjie Zhang   Victoria University of Wellington, New Zealand

Nikola Kasabov	Aukland University of Technology, New Zealand
Jinliang Ding	Northeast University, China
Shengxiang Yang	De Montfort University, UK
Yew-Soon Ong	Nanyang Technological University, Singapore
Andreas Engelbrecht	University of Pretoria, South Africa
Yun Li	Shenzhen Higher Academy, Electronic Technology University, Shenzhen, China
Honggui Han	Beijing University of Technology, China
Ling Wang	Tsinghua University, China
Haibin Duan	Beihang University, China

Plenary Session Co-chairs

Zhexuan Zhu	Shenzhen University, China
Andres Iglesias	University of Cantabria, Spain
Chaoming Luo	University of Mississippi, USA

Invited Session Co-chairs

Zhihui Zhan	South China University of Technology, China
Zhun Fan	Shantou University, China
Maoguo Gong	Xidian University, China

Special Sessions Co-chairs

Ben Niu	Shenzhen University, China
Yan Pei	University of Aizu, Japan
Ming Jiang	Xiamen University, China

Tutorial Co-chairs

Jiahai Wang	Sun Yatsan University, China
Junqi Zhang	Tongji University, China
Han Huang	South China University of Technology, China

Publications Co-chairs

| Swagatam Das | Indian Statistical Institute, India |
| Radu-Emil Precup | Politehnica University of Timisoara, Romania |

Publicity Co-chairs

Fernando Buarque	Universidade of Pernambuco, Brazil
Eugene Semenkin	Siberian Aerospace University, Russia
Jing Liu	Xidian University, Guangzhou Institute of Technology, China
Hongwei Mo	Harbin Engineering University, China
Liangjun Ke	Xi'an Jiaotong University, China
Shenli Xie	South China University of Technology, China
Qiuzhen	Shenzhen University, China
Mario F. Pavone	University of Catania, Italy

Finance and Registration Chairs

Andreas Janecek	University of Vienna, Austria
Suicheng Gu	Google Corporation, USA

Local Arrangement Chairs

Qing Liao	Harbin University of Technology, Shenzhen, China
Shuhan Qi	Harbin University of Technology, Shenzhen, China
Zhaoguo Wang	Harbin University of Technology, Shenzhen, China

Conference Secretariat

Yifan Liu	Peking University, China

Program Committee

Ashik Ahmed	Islamic University of Technology, Bangladesh
Shakhnaz Akhmedova	Robert Koch Institute, Germany
Abdelmalek Amine	Tahar Moulay University of Saida, Algeria
Sz Apotecas	.
Sabri Arik	Istanbul University, Turkey
Nebojsa Bacanin	Singidunum University, Serbia

Liangjun Ke	Xi'an Jiaotong Univ, China
Waqas Haider Khan	Kohsar University Murree, Pakistan
Feifei Kou	.
Jakub Kudela	Brno University of Technology, Czech Republic
Germano Lambert-Torres	PS Solutions, Brazil
Tingjun Lei	Mississippi State University, USA
Xj Lei	Northwest A&F University, China
Bin Li	University of Science and Technology of China, China
Lixiang Li	Beijing University of Posts and Telecommunications, China
Xiaobo Li	Lishui University, China
Xuelong Li	Chinese Academy of Sciences, China
Ya Li	Southwest University, China
Jing Liang	Zhengzhou University, China
Peng Lin	Capital University of Economics and Business, China
Xin Lin	Nanjing University of Information Science and Technology, China
Jianhua Liu	Fujian University of Technology, China
Jing Liu	University of New South Wales, Australia
Ju Liu	Shandong University, China
Qunfeng Liu	Dongguan University of Technology, China
Wenlian Lu	Fudan University, China
Chaomin Luo	Mississippi State University, USA
Dingsheng Luo	Peking University, China
Wenjian Luo	Harbin Institute of Technology, China
Lianbo Ma	Northeastern University, China
Chengying Mao	School Jiangxi University of Finance and Economics, China
Michalis Mavrovouniotis	University of Cyprus, Cyprus
Yi Mei	Victoria University of Wellington, New Zealand
Carsten Mueller	Baden-Wuerttemberg Cooperative State University, Germany
Sreeja N. K.	PSG College of Technology, India
Qingjian Ni	Southeast University, China
Ben Niu	Shenzhen University, China
Lie Meng Pang	Southern University of Science and Technology, China
Endre Pap	Singidunum University, Serbia
Mario Pavone	University of Catania, Italy
Yan Pei	University of Aizu, Japan
Thomas Potok	ORNL, USA

Mukesh Prasad	University of Technology Sydney, USA
Radu-Emil Precup	Politehnica University Timisoara, Romania
Robert Reynolds	Wayne State University, USA
Yuji Sato	Hosei University, Japan
Carlos Segura	Centro de Investigación en Matemáticas, A.C. (CIMAT), Mexico
Timothy Sellers	Mississippi State University, USA
Kevin Seppi	Brigham Young University, USA
Ke Shang	Southern University of Science and Technology, China
Zhongzhi Shi	Institute of Computing Technology Chinese Academy of Sciences, China
Suwin Sleesongsom	KMITL, Thailand
Joao Soares	GECAD/ISEP, Portugal
Wei Song	North China University of Technology, China
Jianyong Sun	University of Essex, UK
Meng Sun	Peking University, China
Yifei Sun	Shaanxi Normal University, China
Ying Tan	Peking University, China
Qirong Tang	Tongji University, China
Anastasiia Timofeeva	Novosibirsk State Technical University, Russia
Eva Tuba	University of Belgrade, Serbia
Mladen Veinović	Singidunum University, Serbia
Gai-Ge Wang	Ocean University of China, China
Guoyin Wang	Chongqing University of Posts and Telecommunications, China
Hong Wang	Shenzhen University, China
Zhen Wang	Northwestern Polytechnical University, China
Wenjie Yi	University of Nottingham, UK
Ka-Chun Wong	City University of Hong Kong, China
Ning Xiong	Mälardalen University, Sweden
Benlian Xu	.
Rui Xu	Hohai University, China
Yu Xue	Nanjing University of Information Science & Technology, China
Wu Yali	Xi'an University of Technology, China
Yingjie Yang	De Montfort University, UK
Guo Yi-Nan	China University of Mining and Technology, China
Peng-Yeng Yin	National Chi Nan University, Taiwan
Jun Yu	Niigata University, Japan
Ling Yu	Jinan University, China

Zhengfei Yu	National University of Defense Technology, China
Zhi-Hui Zhan	South China University of Technology, China
Fangfang Zhang	Victoria University of Wellington, New Zealand
Jie Zhang	Newcastle University, UK
Jiwei Zhang	Beijing University of Posts and Telecommunications, China
Junqi Zhang	Tongji University, China
Lifeng Zhang	Renmin University of China, China
Tao Zhang	Tianjin University, China
Xiangyin Zhang	Beijing University of Technology, China
Xiaoming Zhang	Anhui Agriculture University, Hefei
Xingyi Zhang	.
Zili Zhang	Deakin University, Australia
Xinchao Zhao	Beijing University of Posts and Telecommunications, China
Zheng PKU	.
Yujun Zheng	Zhejiang University of Technology, China
Mengchu Zhou	New Jersey Institute of Technology, USA
Yun Zhou	National University of Defense Technology, China
Tao Zhu	University of South China, China
Mingyin Zou	National University of Defense Technology, China

Additional Reviewers

Cai, Gaocheng
Chen, Wei
Cheng, Dongdong
Elbakri, Idris
Jia, Zhan Xiao
Jiang, Yi
Khan, Muhammad Saqib Nawaz

Li, Xinxin
Liu, Jianhua
Qiu, Haiyun
Wang, Yixin
Yang, Ming
Zivkovic, Miodrag

Contents – Part II

Routing and Scheduling Problems

Stock Prediction and Portfolio Optimization

ICSI-Optimization Competition

Contents – Part I

Particle Swarm Optimization Algorithms

Genetic Algorithms

Optimization Computing Algorithms

Neural Network Search and Large-Scale Optimization

Multi-objective Optimization

Swarm Robotics and UAV

A Blockchain-Based Service-Oriented Framework to Enable Cooperation of Swarm Robots

Chung-Yu Huang, Jhih-Yi Li, Jhih-Yuan Huang, and Wei-Po Lee[✉]

Department of Information Management, National Sun Yat-sen University, Kaohsiung, Taiwan
wplee@mail.nsysu.edu.tw

Abstract. Swarm robotics has been proposed to achieve tasks by the cooperation and coordination among distributed simple robots. To overcome the practical issues of deploying a swarm robotic system, we present a blockchain-based service-oriented framework to ensure the secured cooperation among heterogeneous swarm robots. Our work adopts smart contracts working on the blockchain to design service functions for swarm robots. The major service functions for the robot nodes include requesting the services, selecting suitable services based on the quality and creating composite services to achieve complex tasks. Considering the market-oriented cooperation condition of the real world, we design a service trading mechanism to take into account utilities of service requesters and providers. We conduct a series of experiments to evaluate the framework, and the results show the promising of the presented approach.

Keywords: Swarm Robotics · Blockchain · Collective Intelligence · Smart Contract · Service Composition · Service Trading

1 Introduction

Swarm robotics has been proposed to establish a system of simple robots with limited perceptions to emerge complex group behaviors [1]. Swarm robotics is considered to have the potential of implementing a highly fault-tolerant, scalable, and flexible system to achieve complex tasks [1, 2]. Moreover, the rise of blockchain technology provides a secure channel for robot communication. In this work, we develop a blockchain-enabled service-oriented framework to address some issues in swarm robotics to enrich its functions.

The decentralized and autonomous characteristics make swarm robots more robust and scalable than a single powerful centralized robot. However, before a swarm robotic system can be practically deployed, there are still some difficulties to be overcome [1, 3]. The first issue is to provide a safe communication channel among robots. Some recent studies have proposed to adopt the blockchain technology [4, 5]. Besides the encryption algorithms, the development of smart contracts allows the blockchain to further formulate protocols, extending blockchain transactions to the stage of inputting information, executing the corresponding programs, and trigger events with the obtained

Y. Tan et al. (Eds.): ICSI 2023, LNCS 13969, pp. 3–15, 2023.
https://doi.org/10.1007/978-3-031-36625-3_1

outputs. The programmable control of smart contracts enables a variety of applications to be combined with the blockchain. For example, Fernandes presented RobotChain [6], a private blockchain enabling robots to obtain global information and shared knowledge through the blockchain. Strobel *et al.* used blockchain to detect malicious robots, in which the contract was used to conduct consensus voting on the color of the floor [3]. Ferrer *et al.* used blockchain and smart contract to develop communication strategies to make secured group decisions [5].

To achieve complex tasks by swarm robots, it is critical to provide a task planning mechanism that involves the composition of various simple services carried by homogeneous swarm robots. Service composition can be solved by workflow-based or AI planning methods. Hierarchical Task Network (HTN) planners are typical examples [7, 8]. Such kind of planners indicates the trend of combining automatic planners with human efforts to generate composite services; that is, under the guidance of a pre-defined task ontology. To solve complex tasks, in this work, we adopt the AI planning-based service composition mechanism previously developed for a cloud robot to our current blockchain-based swarm robotic framework. As the blockchain is a shared ledger that can broadcast information to all members of the network, a robot node here needs to regularly interact with smart contracts to update its status and position on the blockchain. In this way, when a robot initiates a task that requires assistance, the smart contract can find suitable robots to provide services. This study aims to overcome the related challenges and exploit the benefits of using blockchain to perform service composition for heterogeneous swarm robots.

After identifying the required services, a robot needs to request these services from others. However, services are usually not freely offered. This work thus investigates how to obtain these services through a trading mechanism. Some studies have studied the resource auction problem among robots [9, 10]. The auctions were combined with blockchain to construct a secure, decentralized and verifiable trading environment. For example, De Biase *et al.* created a model based on the reputation and price of the robots, in which the transaction information was added to the blockchain to ensure the information traceable and difficult to be modified [11]. Though many studies discussed the integration of transaction and blockchain, most of them stayed in the conceptual stage. This work implements transactions using smart contracts on the blockchain and explores the results. We adopt a more efficient multi-resource dual auction algorithm to solve the transactions involving multiple buyers and sellers. The system utilities are pursued under the constraints of all parties [12, 13].

To enrich the functions and enhance the performance of a swarm robotic system, this work presents a blockchain-based service-oriented framework for secured cooperation among heterogeneous swarm robots. Our framework inherits the advantages from both blockchain and swarm robotics technologies, and the framework further uses smart contracts to design service functions and service trading from practical perspectives. To evaluate the proposed approach, we conduct a series of experiments and the results show the promise of the presented approach.

2   A Blockchain-Based Framework for Swarm Robots

2.1   Framework

To establish a blockchain-based system for the cooperation of heterogeneous swarm robots, we design and implement a service-oriented system framework to integrate robot functions and smart contracts on the blockchain. Figure 1 illustrates the framework that integrates blockchain and service computing to achieve the service scalability, information transparency, trade traceability and node management. The proposed framework includes two major mechanisms: task planning (service composition) and collaborative task execution. The task planning means to decompose a complex target task into a set of simple tasks by certain robot services, based on the predefined task ontology and the available services. Then, the service set is listed in the smart contract and posted on the Ethereum blockchain to call for assistances from other robots. Here, the smart contract is responsible for the management of available services and can decide which services to perform to achieve the target task. Through the distributed ledger technology, the smart contract is synchronized with the nodes (i.e., robot with services) registered to the blockchain. There could be many robots providing the same service, so they have to verify their functional abilities to complete for performing the task and obtain the reward. In this work, we adopt a simplified strategy of selecting the nodes based on their registration time; more comprehensive methods can be developed to obtain better performance.

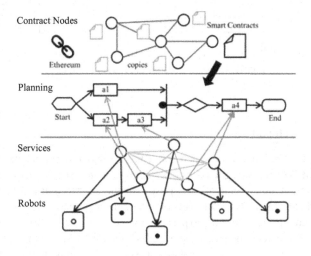

Fig. 1. Proposed framework of integrating blockchain and service computing.

2.2   Service Registration and State Update

In this framework, a robot needs to register as a blockchain node to the smart contract for service management, with the services it can provide and the initial states of these services. The contract maintains a list for each type of services. In this way, the contract can thus know which nodes (i.e., robot services) are suitable for the requests by other nodes. The blockchain nodes use their addresses (for example, "0x4e..") to represent the identities, and the registered services are maintained accordingly by the smart contract. After that, the robot nodes can directly perform the operations (functions) provided by the contract, with a single identity of blockchain node to prevent the malicious identify forgery. Figure 2(a) illustrates a registration example.

Blockchain is a distributed ledger that synchronizes information with all nodes, and the smart contract is not authorized to update the information of each node. Therefore, to maintain the most up-to-date information in a timely manner, the robot has to regularly update the current states of the services it has registered in the smart contract. Based on the correct information, the smart contract can choose the most suitable services for the task needing helps. For example, for an object moving task, the robot has to indicate the weight it can carry; or for a charging service, the robot has to provide the amount of electricity it can offer. Also, the robot has to update its profile information (such as it is in a busy state or its current position) for the smart contract to make correct service decision. Figure 2(b) illustrates an example.

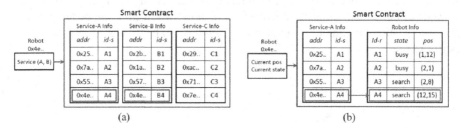

Fig. 2. (a) robot registration; (b) state update.

2.3 Service Search and Task Tracking

To participate task collaboration, each robot has to register its services to the smart contract, and afterwards it regularly checks if there is any service request from other nodes (the states of other related tasks posted in the smart contract) to find the task it can assist. If the robot decides to accept the request, it can retrieve the task-related information from the smart contract to perform its service. Figure 3 (a) shows an example in which the robot checks the smart contract and finds the task it can perform from the list, and then the robot accepts and provides its service to complete the task.

On the other hand, when a robot cannot complete a task alone, it seeks for assistances from other robots through the smart contract. The task-initiated robot has to provide all task-related information to the smart contract, allowing it to find a suitable service (robot) for assistance. After initializing a task, the robot needs to regularly track the task states to check if the required services have been provided.

As mentioned, there could be many services fulfill the requirement specified by the task initiator. To ensure the best selection, the smart contract calculates a score for each candidate by a predefined strategy and then chooses the best robot accordingly. Our current implementation adopts a simple strategy of weighted sum over all requirements to determine the score. Alternatively, the task initiator can define its own strategy under specific considerations to choose the suitable one. If any robot is available to provide the required service at the time of request, the service (with the information of the task initiator, task requirements, and the selected robot) is added to the list of request for the task to be completed; otherwise a failure message is returned. The smart contract then provides task-related information to the selected robot and asks its assistance. The task initiator can regularly check the state of each service afterwards to see if any robot for the selected service is available.

It's worth noted that even for the same task, different initiators may propose different task requirements due to the consideration on the robot ability it needs (such as sensing or acting). Therefore, the service provided by the smart contract may be different. For example, if the task initiator needs to find the target object and move it, and this task requires two kinds of robots to complete (one detecting robot and one carrying robot). The initiator has to provide the task-related Information to the smart contract so the smart contract can determine the suitable services accordingly.

Figure 3 (b) illustrates an example in which the task initiator has to keep tracking if its task has been accepted and completed, due to some uncertain situations during the collaboration process (such as a robot node accepted the task but unable to complete it). If the task is not accepted or completed for a certain period of time, the initiator has to reinitiate the task again to ask for other assistance.

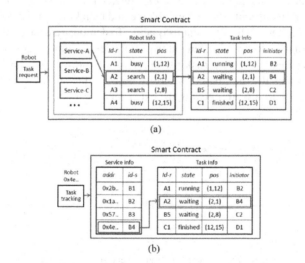

Fig. 3. Examples of service initiate and tracking.

2.4 Service Composition

As indicated in the above section, a robot linked to the blockchain can initiate a task request to seek service assistance, and the service search procedure can find a service to meet the requirement. However, for complex task that cannot be achieved by a single service, more advanced mechanism needs to be developed. In this work, we design a function of service composition to enable robot collaboration. The overall workflow is illustrated in Fig. 4. As shown, this function includes two phases. The first phase is to adopt AI planning techniques to perform service composition. In the task planning domain, HTN planning has been shown to be a well-designed approach suitable to work for the service composition. A required task is decomposed according to the pre-defined task ontology to break the task hierarchically into several subtasks, and then the planner solves the subtasks in the reverse order. We choose to employ this planning approach to achieve service composition and utilize the task ontology we developed previously for a home service robot to guide the decomposition. We used OWL-S to implement the ontology (including the service profile with profile elements of Input, Output, Precondition, and Effect), and a HTN planner can thus efficiently deal with the hierarchically structured OWL-S process model. The obtained subtasks are then sent to the smart contract which retrieves the possible services for each subtask from the repository based on a set of matching rules.

The second phase is to select suitable services, similar to the service selection procedure described on the above section. For each satisfied service, the smart contract examines if the task demand has been satisfied and inspects if there is a robot available to provide this service at the time of request. If available, this service is then considered candidates to constitute the composite service. In the above algorithm, an quality-of-service scheme can be defined by several attributes, such as cost, response time, and reputation to select suitable services. For example, service reputation is an objective and easy-to-measure attribute. It can be implemented by taking the average rating of all users for a specific service to represent its quality. To demonstrate the quality-based service selection, in this work we implement two strategies: a simple first-come-first-select strategy that decides the service rank by its arrival time (Sect. 3.2), and a cost-based strategy that decides the rank based on utility between the service requester/provider (Sect. 3.3). The details are presented in the experiment section.

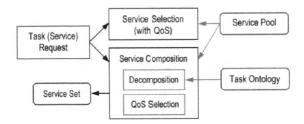

Fig. 4. Flow of the function of service composition.

2.5 Service Trading

As the service supports are not free by the providers in reality, the service requesters need to purchase the services with certain cost. Therefore, our system includes a trading mechanism that allows the service requesters and providers to perform trading after the useful services are selected. We develop a game theory based pricing strategy for the robots to purchase or sell the services.

In this work, the trading activities with multiple requesters and providers are market-oriented. The service requesters and providers aim to obtain maximal utilities. The robots compete with each other to obtain the services they need from the providers, and they thus do not reveal their trading information (i.e., the service quantities and prices) to others. Each robot engaging in the trading market knows the minimal and maximal prices given by the system and knows that the value price distributes uniformly in the interval. Under such circumstances, we develop an approach (called PCDA, Priority-based Combinatorial Double Auction) that adopts the static non-cooperative game theory to find the optimal solution. That is, to find the Nash equilibrium for the auction, in which any robot cannot change its strategy to enlarge the overall system utility [12, 13].

We assume the requester and the provider employ linear functions to be their pricing strategies as following:

$$P_{i,s}^r\left(V_{i,s}^r\right) = a_{i,s}^r + b_{i,s}^r \times V_{i,s}^r \tag{1}$$

$$P_{j,i,s}^p\left(V_{j,i,s}^p\right) = a_{j,s}^p + b_{j,s}^p \times V_{j,i,s}^p \tag{2}$$

In which $P_{i,s}^r$ is the price of requester i bidding for service s; $V_{i,s}^r$ is the value price of requester i bidding for service s; $P_{j,i,s}^p$ is the price of provider j offering service s to requester i; $V_{j,i,s}^p$ is the value price of provider j offering service s to requester I; and $P_{i,s}^r < V_{i,s}^r$, $P_{j,i,s}^p < V_{j,i,s}^p$. In addition, $a_{i,s}^r$, $b_{i,s}^r$, $a_{i,s}^p$, $b_{i,s}^p$ are the coefficients of the above linear equations. The optimal price is determined by the prices offered by both sides, shown in Eq. (3). The utilities for the requester and the provider are defined as (4) and (5) respectively. As mentioned above, they both aim to maximize their utilities, and under such constraints, we can formulate the objective functions of requester and the provider, which are (6) and (7).

$$Pdeal_{i,j}^s = (P_{i,s}^{r*} + P_{j,i,s}^{p*})/2 \tag{3}$$

$$U_i^r = \sum_{s=1}^{k}[V_{i,s}^r - Pdeal_{i,j}^s] \times C_{i,s}^r \tag{4}$$

$$U_j^p = \sum_{s=1}^{k}[Pdeal_{i,j}^s - V_{j,i,s}^p] \times C_{i,s}^r \tag{5}$$

$$\max_{P_{i,s}^r}\{[V_{i,s}^p - \frac{P_{i,s}^r + E[P_{j,i,s}^p(V_{j,i,s}^p)|P_{j,i,s}^r \geq P_{j,i,s}^p(V_{j,i,s}^p)]}{2}] \times P[P_{i,s}^r \geq P_{j,i,s}^p(V_{j,i,s}^p)]\} \tag{6}$$

$$\max_{P_{j,i,s}^p}\left\{\left[\frac{E\left[P_{i,s}^r\left(V_{i,s}^r\right)|P_{i,s}^r\left(V_{i,s}^r\right) \geq P_{j,i,s}^p\right] + P_{j,i,s}^p}{2} - V_{j,i,s}^p\right] \times P\left[P_{i,s}^r \geq P_{j,i,s}^p\left(V_{j,i,s}^p\right)\right]\right\} \tag{7}$$

In the above equations, $E[P_{j,i,s}^p(V_{j,i,s}^p)|P_{j,i,s}^r \geq P_{j,i,s}^p(V_{j,i,s}^p)]$ is the price the requester expects the providers will offer in the situation of a successful trading; $P[P_{i,s}^r \geq P_{j,i,s}^p(V_{j,i,s}^p)]$ is the probability of such a situation; $E[P_{i,s}^r(V_{i,s}^r)|P_{i,s}^r(V_{i,s}^r) \geq P_{j,i,s}^p]$ is the price the provider expects the requester will offer in a successful trading; and $P[P_{i,s}^r(V_{i,s}^r) \geq V_{j,i,s}^p]$ is the probability of such a situation. We can then apply (1) and (2) to (6) and (7) to derive the expected prices and the probabilities for requester and provider. The results can be further taken back to the objective functions to obtain the optimal solutions for the requester and the provider: $P_{i,s}^{r*}$ and $P_{j,i,s}^{p*}$ shown in (8) and (9).

$$P_{i,s}^{r*} = \frac{Pmax_s}{12} + \frac{Pmin_r}{4} + \frac{2V_{i,s}^r}{3} \tag{8}$$

$$P_{j,i,s}^{p*} = \frac{Pmax_s}{4} + \frac{Pmin_r}{12} + \frac{2V_{j,i,s}^p}{3} \tag{9}$$

In the implementation for service trading, the above prices are adopted for service requesters and providers. They then attempt to compete for the services they need accordingly. With the above prices sent from the requesters and providers, the smart contract continues to process the relevant information of both sides, such as the reputation of the robots involved, the priorities of different services for each robot, substitutability of each requested service. The smart contract can then use the information to link both sides more appropriately to perform trading. As this work means to provide a fundamental mechanism along with the function of service composition presented in the above sections, rather than to develop an optimal matching mechanism; therefore, we adopt a simplified way of considering only the service price. More comprehensive strategy can be developed to further enhance the trading performance.

3   Experiments and Results

3.1   Experimental Setup

To evaluate the proposed approach for robot collaboration, we adopted a swarm robot simulator and conducted several sets with it. We adopted the ARGoS simulator [14] to establish a swarm robotic environment, in which each robot is a container formed by the Ethereum image file which was used to simulate the blockchain node. Also, to simulate the actual working situations of swarm robots, we restricted the resources on the containers: 1.5 cores (CPU) and 1.8 GB memory (RAM).

3.2   Evaluation of Service Planning

In the experiments, the first set of experiments was to evaluate the service composition module described in Sect. 2.4. Here, we took a popular swarm foraging task that included several subtasks (services) and required cooperation of other types of robots to complete. The task was that the robots needed to search the target area, found and recognized objects, and then cleaned the red (heavy) objects and moved the black (light) objects to home.

There were three types of robots in the environment. FB_D: a small robot able to move freely and detected objects; FB_C: can move freely, recognize red/black objects, and move red objects; and FB_R: It can move but only on request, and then it cleans black objects and returns to a fixed location. As specified, each type of robots can perform part of the overall task. They need to collaborate with each other to complete the task defined above. That is, FB_D move around and when it detects a task, it seeks for the assistance of FB_C. FB_C performs object detection when it reaches the task location. If the object is red, the robot carries it back to home; if it is a black object, the robot seeks for the assistance of FB_R to clean the object. The FB_D robot is referred to the E-puck robot [15], and FB_C and FB_R, the CoroBot Explorer robots. The energy consumption rates of the three types of robots are defined based on real robots: the ratio of their energy consumption per unit time is 1:6:6.

To complete the overall task, the framework first decomposed it into several subtasks according to the task ontology. Figure 5 presents the results. As shown, the original was decomposed to four types of subtasks (explore/move, recognize, collect and clean) that can be performed by different types of robots (with the services they provide). The small robots FB_D moved around in the environment to search the target objects. As this type of robots cannot complete the task alone, they asked for assistance (services) from others through our blockchain-based service-oriented framework. The robots that can provide required services then responded to the framework for collaboration to complete the original foraging task. Figure 6 illustrates two examples of task workflows summarized from the robot collaboration in the simulated environment. Figure 6(a) shows that a FB_D robot explored and detected a target object at time t_1. It initiated a request at time t_2 for the subtask of recognizing and a FB_C robot accepted the request. Then, the FB_C robot moved to the target position and it arrived at time t_3 to perform the service. Based on the recognized result (object with red color), the robot collected the object back to the home area at time t_4. Figure 6(b) illustrates another example in which the recognized object (object with red color) needed to be cleaned. The FB_C robot initiated a new request at time t_4 and a FB_R robot accepted this request. The FB_R robot moved to the target position and arrived at time t_5, and then provided its service and completed the clean task at time t_6.

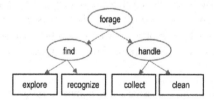

Fig. 5. Result of service composition for the target task.

We also conducted a set of quantitative experiments to evaluate the performance of swarm robot under the blockchain-based environment. We fixed the number of robots and varied the number of tasks to investigate the efficiency of collaboration. In the experiments, 6 robots were used, including 3 FB_D, 2 FB_C and 1 FB_R. The reason for selecting this combination was that due to the hardware limitation (for stably running

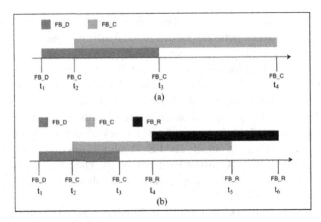

Fig. 6. Examples of summarized task workflows: (a) collect; and (b) clean.

Ethereum container nodes), the maximal number of robot was 10, and at the same time we followed the cost principle (small number for expensive robot) to determine their numbers (i.e., FB_D ≥ FB_C ≥ FB_R). Also, the number of tasks varied from 7 to 15, defined to be roughly 1 to 1.5 times of the number of robots. This choice was to distribute objects in the environment with certain distance while avoiding the situation of spending a long time in waiting for required services.

In the experiments, we measured the response time (the averaged time between a task was initiated and completed) to evaluate the collaboration performance. As the execution of smart contract on the blockchain network was effected by certain randomness, the experimental trials were thus performed five times and the averaged results are provided. Table 1 presents the results, in which the number of tasks varied from 7 to 15 as indicated. For each case, the numbers of two different types of tasks (i.e., move_red-object_ to_home and clean_black-object) are also listed. The overall tasks can be achieved successfully with the developed service composition mechanism. In detail, the response time has not obvious correlation to the number of tasks. This is because the tasks were initiated and performed dynamically by the robots only at the time when requested. Also, there were other factors occurring in the interval could take additional time, such as moving to a target position. Thus, the averaged time spend on the tasks was not necessarily proportional to the number of tasks. Similarly, the time spent on the individual tasks is not correlated to the number of task.

Table 1. Response time (in seconds) of using a fixed number of robots, in which "b" is "move_red-object_to_home", and "r" is "clean_black-object" tasks.

num of task	7 (5b 2r)	8 (6b 2r)	9 (7b 2r)	10 (7b 3r)	11 (8b 3r)	15 (12b 3r)
Time (all)	993.6	1022.4	967.8	986.8	922.1	1101.5
Time (b)	601.1	605.4	603.2	602.8	712.3	690.5
Time (r)	1371.6	1390.2	1311.7	1322.4	1123.5	1521.2

3.3   Evaluation of Service Trading

After finding the requested services from the service composition and selection procedures, in this section we presented the evaluation of trading performance in obtaining the services. Due to the hardware limitation for running the blockchain-swarm simulation, in this set of experiments, the maximal number of robots (including service requesters and providers) was 13 and the number of requester/provider was between 1 and 10. The metric of utility defined in Sect. 2.5 was used for performance evaluation with the price range of 0.012 and 0.1046 (based on the setting in the reference work [13]). For simplicity, the multi-service trading was performed in the one-to-one manner. That is, for each trading, the requests by a single requester were regarded a set (non-separable) and only the providers met all requests can participate the trading.

The first set of experiments was to observe the performance change under the condition of using a fixed number of robots to ask for different types of services. Here, eight requesters and five providers (more requesters to compete the resources) were used, and different types (from 4 to 10) of service types were offered. The requested and the offered services for each robot were randomly assigned (Table 2 illustrates an example). Two trading strategies were performed for each assignment, including the proposed game-based strategy and a traditional strategy (two methods). The results (utilities) for requesters and providers under the cases of different numbers of service types are presented in Fig. 7. As presented, the utilities for both types of robots (i.e., requester and provider) increased along with the increase of services. It shows that in general more types of services can better satisfy the needs of requests. The results also show that the proposed game theory-based trading strategy outperformed traditional methods, CDAPA (Combinatorial Double Auction Resource Allocation [16]) and GCDA (Game-based Combinatorial Double Auction [17]). This confirms the effectiveness and usefulness of the trading mechanism in the proposed framework.

In the second set of experiments, we investigated the effect of the number of service requesters on the trading utilities (for requesters and providers respectively). In this set of experiments, the number of service types was fixed to a medium number of 8, and the number of requesters was increased. The results are shown in Fig. 8. As shown, for a fixed number of service types, more requesters generated a higher service need, and thus increased the utilities of both service requesters and providers. Also, we can observe that the proposed PCDA strategy is more effective than the traditional CDAPA method. Again, these results show the promise of the proposed approach.

Table 2. An example of service settings: eight requesters (r1 to r8), five providers (p1 to p5) and four types of services (s1 to s4).

requester	provider
r1: [2, 0, 0, 1]	p1: [0, 0, 4, 3]
r2: [3, 0, 0, 0]	p2: [2, 8, 5, 0]
r3: [0, 2, 1, 0]	p3: [3, 3, 5, 1]
r4: [2, 1, 0, 0]	p4: [1, 8, 0, 0]
r5: [0, 0, 2, 1]	p5: [1, 1, 7, 2]
r6: [1, 2, 0, 0]	
r7: [2, 0, 0, 0]	
r8: [0, 1, 2, 0]	

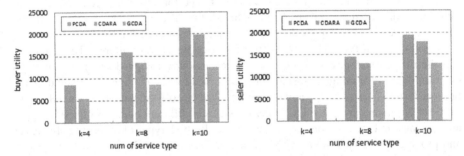

Fig. 7. Total utilities of seller (left) and buyers (right) with different numbers of service types.

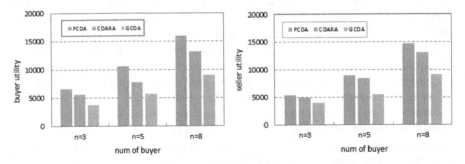

Fig. 8. Total utilities of sellers (left) and buyers (right) under different numbers of buyers.

4   Conclusion

In this work, we presented a blockchain-based service-oriented framework to ensure the secured cooperation among heterogeneous swarm robots. Our framework used smart contracts on the blockchain to design service functions, including service registration and tracking, service selection, and service composition. Moreover, we developed a service trading mechanism to consider the service utility between service requesters and providers. To evaluate the proposed approach, we conduct a series of experiments and the results show the usefulness and effectiveness of the presented approach. Based on the presented framework, we are now developing more comprehensive composition and trading methods to further enhance the performance.

References

1. Brambilla, M., Ferrante, E., Birattari, M., Dorigo, M.: Swarm robotics: a review from the swarm engineering perspective. Swarm Intell. 7(1), 1–41 (2013)
2. Dorigo, M., Floreano, D., Gambardella, L.M., et al.: Swarmanoid: a novel concept for the study of heterogeneous robotic swarms. IEEE Robot. Autom. Mag. 20(4), 60–71 (2013)
3. Strobel, V., Ferrer, E.C.: Managing byzantine robots via blockchain technology in a swarm robotics collective decision making scenario. In: Proceedings of 17th International Conference on Autonomous Agents and MultiAgent Systems, pp. 541–549 (2018)
4. Alsamhi, S.H., Shvetsov, A.V., Shvetsova, S.V., et al.: Blockchain-empowered security and energy efficiency of drone swarm consensus for environment exploration. IEEE Trans. Green Commun. Netw. 7(1), 328–338 (2023)
5. Ferrer, E.C., Hardjono, T., Pentland, A., Dorigo, M.: Secure and secret cooperation in robot swarms. Sci. Robot. 6(56) (2021)
6. Fernandes, M., Alexandre, L.A.: Robotchain: using Tezos technology for robot event management. Ledger 4, 32–41 (2019)
7. Yang, T.-H., Lee, W.-P.: A service-oriented framework for the development of home robots. Int. J. Adv. Robot. Syst. 10 (2013). Article no. 122
8. Canal, G., Alenyà, G., Torras, C.: Adapting robot task planning to user preferences: an assistive shoe dressing example. Auton. Robot. 43(6), 1343–1356 (2019)
9. Wang, L., Liu, M., Meng, Q.-H.: Hierarchical auction-based mechanism for real-time resource retrieval in cloud mobile robotic system. In: Proceedings of IEEE International Conference on Robotics and Automation (2014)
10. Lee, D.-H.: Resource-based task allocation for multi-robot systems. Robot. Auton. Syst. 103, 151–161 (2018)
11. De Biase, L.C., Calcina-Ccori, P.C., Fedrecheski, G., et al.: Swarm economy: a model for transactions in a distributed and organic IoT platform. IEEE Internet Things J. 6(3), 4561–4572 (2018)
12. Kim, M.-H., Kim, S.-P., Lee, S.: Social-welfare based task allocation for multi-robot systems with resource constraints. Comput. Ind. Eng. 63, 994–1002 (2012)
13. Baranwal, G., Vidyarthi, D.P.: A fair multi-attribute combinatorial double auction moel for resource allocation in cloud computing. J. Syst. Softw. 108 (2015)
14. Pinciroli, C., Trianni, V., O'Grady, R., et al.: ARGoS: a modular, parallel, multi-engine simulator for multi-robot systems. Swarm Intell. 6(4), 271–295 (2012)
15. Mondada, F., Bonani, M., Raemy, X., et al.: The e-puck, a robot designed for education in engineering. In: Proceedings of the 9th Conference on Autonomous Robot Systems and Competitions, pp. 59–65 (2009)
16. Samimi, P., Teimouri, Y., Mukhtar, M.J.I.: A combinatorial double auction resource allocation model in cloud computing. Inf. Sci. 357, 201–216 (2016)
17. Li, Q., Huang, C., Bao, H., Fu, B., et al.: A game-based combinatorial double auction model for cloud resource allocation. In: Proceedings of the 28th International Conference on Computer Communication and Networks, pp. 1–8 (2019)

Collective Behavior for Swarm Robots with Distributed Learning

Junqiao Zhang, Qiang Qu, and Xue-Bo Chen$^{(\boxtimes)}$

School of Electronics and Information Engineering, University of Science and Technology
Liaoning, Anshan 114051, China
xuebochen@126.com

Abstract. The swarm robot system has great application potential in surveillance and reconnaissance tasks. In this paper, we apply a swarm robot system, which integrates distributed learning and flocking control, allowing robots to find task targets in a distributed learning way. Then, a new control scheme based on potential field is adopted to drive the robot to move towards these task targets without collision. Secondly, the convergence and applicability of the system are proved. Finally, the effectiveness of the system is verified by simulation and experiment.

Keywords: Collective Behavior · Swarm Robots · Flocking Control · Distributed Learning

1 Introduction

As one of the most common phenomena in nature, collective behavior widely exists in various complex systems, including animal groups, human populations, bacterial colonies, and traffic flows. Collective behavior is a macro collective behavior presented by continuous and orderly individuals, and its research has attracted the attention of many fields [1]. The research of behavior science shows that the formation of complex collective behavior originates from the interaction between individuals, and the appearance of this interaction will spontaneously form the collective behavior. In the study of biology, the collective behavior of animals or microorganisms through swarm intelligent decision-making and information exchange is considered to be a remarkable manifestation of evolution [2]. In the research of physics, the swarm is regarded as a collection of particles, and the interaction and movement of the swarm are described by the swarm dynamics equation [3]. Researchers of computer science regard the recognition and simulation of collective behavior as an important research branch in the field of computer vision and computer graphics [4]. Therefore, groups are often able to obtain relatively certain results from random individual behaviors. The individuals in a swarm only interact with the environment or other members through limited perception ability, and eventually generate the corresponding swarm Intelligence and Emergent Intelligence. Inspired by this, Swarm Robotics, which uses swarm intelligence to simulate collective behavior, has attracted the attention of many researchers in recent years, and is an important research direction of multi-robot and cooperative robotics.

© The Author(s), under exclusive license to Springer Nature Switzerland AG 2023
Y. Tan et al. (Eds.): ICSI 2023, LNCS 13969, pp. 16–25, 2023.
https://doi.org/10.1007/978-3-031-36625-3_2

A system of multiple robots can cooperate to accomplish tasks that cannot be done by a single robot. Swarm robots have the remarkable characteristics of spatial distribution and temporal distribution, which has a wide range of application prospects. In the industrial field, unmanned systems composed of a large number of low-cost robots can be used for the inspection, handling and collection of specific items [5]. In the military field, a large number of member robots can be deployed in sensitive areas to monitor and sense changes in the enemy situation in a wide range of time and space. Large-scale Uavs, unmanned ships and robots can replace the army to carry out reconnaissance and combat, which can minimize casualties [6]. On land and in the air, it can be used for detection, target tracking, surveillance and disaster rescue. Underwater, it can be used in environmental monitoring, mine countermeasure, archaeology, pollution source location, search for ship wreckage and crashed aircraft and other tasks. In the civil field, it can be used in indoor and outdoor disaster search and rescue, fire protection, construction, transportation and even medical care and other tasks [7]. However, for the search task of a specific target, the current typical method is to use a small number of robots or specific equipment to search the relevant area repeatedly. The disadvantages of such methods are: long search time, poor economy, low coverage and so on. In terms of the current level of robot technology, the ability of single robot in information acquisition, processing and coverage is limited, especially for complex search tasks and changing working environments, the ability is even more insufficient. On the contrary, the swarm robot system has the typical characteristics of decentralization, with the advantages of low complexity and low cost. In addition, the basic reason for studying the multi-robot system is that the efficiency of the group working at the same time is higher than that of the individual, and it can overcome the task that the individual is unable to undertake, which is consistent with the cooperation law between insects and humans in nature.

The research on swarm robot system is inspired by the emerging intelligent behaviors of a large number of social animals, such as the nesting behavior of termites, the formation of fish and birds, etc. [8]. Exploring how to reproduce Swarm intelligence (SI) shown by social animals into swarm robot system is an active topic in the research of control science. In general, swarm robot is a special implementation method of multi-robot system, which is obviously different from other methods. First, swarm robot systems are highly autonomous and decentralized. The robot only relies on limited environment perception ability, interacts with the environment and neighboring robots in a certain range, and then realizes decision-making and executes corresponding actions. It has higher flexibility than other centralized control multi-robot systems. Secondly, the swarm robot members are isomorphic. In general, there are fewer members in the swarm whose member robots have physically significant differences, which has strong interchangeability. In addition, the decentralized structure has small amount of calculation, fast reaction speed and no dependence on other special resources. Third, swarm robots have good cooperative characteristics. Swarm robots have low individual ability. For example, ants are very weak and can capture prey much larger than themselves through collaborative force. Swarm robots themselves only have low complexity, but through cooperation, they can complete tasks that are difficult for a single high-level robot. Fourth, swarm robots have good robustness and scalability. The coordination mechanism of the system does not change with the size of the group, and has good scalability. In addition, swarm robots are special multi-robot systems composed of many indifferent autonomous robots with typical distributed system characteristics. Swarm intelligence is achieved through limited individual robots through interaction, coordination and control to collaborate on relatively complex prescribed tasks [9]. The research direction has broad development prospects, and a large number of research results have emerged in recent years.

For the cooperative control of swarm robots studied in this paper, much work has focused on heuristic search. A lot of work applies the corresponding heuristic optimization algorithm directly to the cooperative search problems of swarm robots, with little or no consideration given to the kinematic constraints, random motion, potential field, communication constraints and overlapping positions of the robots. Therefore, the introduction of distributed learning and optimal collaborative control model is a problem that must be solved when applying this method to a real swarm robot system. This research is a cross topic in the field of robotics, control science and applied mathematics. The task allocation of multi-robot system through distributed learning is studied to meet the self-organization requirements in multi-robot cooperative control [10]. The aim is to achieve distributed, parallel and coordinated work of multiple robots in various locations of the workspace.

The rest of this article is organized as follows. Section 2 introduces flocking control algorithms. Section 3 analyses the convergence of the proposed learning algorithm. Section 4 gives the simulation and experimental results. Finally, Sect. 5 gives the conclusions of this paper and further work.

2   Flocking Control Algorithm

2.1   Collective Behavior

At different locations, a group of robots collects information samples to search for the target range of interest. The swarm motion of a robot is determined by an artificial potential function, which provides obstacle avoidance attributes for the robot. Potential functions of attraction and exclusion can generate collective behavior [11]. We need to introduce a smooth collective potential function that enables a set of robots to ensure a class of algebraic constraints $\|q_i - q_j\| = d$ for all $j \in \mathcal{N}(i, q)$.

$$
\begin{aligned}
U_1(q) := & \sum_i \sum_{j \in \mathcal{N}(i,q), j \neq i} U_{ij}\left(q_i - q_j^2\right) \\
= & \sum_i \sum_{j \in \mathcal{N}(i,q), j \neq i} U_{ij}(r_{ij})
\end{aligned}
\tag{1}
$$

where $r_{ij} = \|q_i - q_j\|^2$. The attraction and repulsion potential functions $U_{ij}(r_{ij})$ can be described as

$$
U_{ij}(r_{ij}) := \frac{1}{2}\left(\log(\alpha + r_{ij}) + \frac{\alpha + d^2}{\alpha + r_{ij}}\right), \; if \; r_{ij} < d_0^2,
\tag{2}
$$

where $\alpha, d \in \mathbb{R}_{>0}$ and $d < d_0$. The gradient of the potential function of robot i with respect to q_i can be described as:

$$\nabla U_1(q_i) = \frac{\partial U_1(q)}{\partial q_i} = \sum_{j\neq i} \frac{\partial U_{ij}(r)}{\partial r}\Big|_{r=r_{ij}} 2(q_i - q_j)$$

$$= \begin{cases} \sum_{j\neq i} \frac{(r_{ij}-d^2)(q_i-q_j)}{(\alpha+r_{ij})^2} & \text{if } r_{ij} < d_0^2 \\ \sum_{j\neq i} \rho\left(\frac{\sqrt{r_{ij}}-d_0}{|d_1-d_0|}\right) \frac{d_0^2-d^2}{(\alpha+d_0^2)^2}(q_i - q_j) & \text{if } r_{ij} \geq d_0^2 \end{cases} \tag{3}$$

where $\rho: \mathrm{R} \geq 0 \to [0, 1]$ is the bump function

$$\rho(z) := \begin{cases} 1, & z \in [0, h); \\ \frac{1}{2}\left[1 + \cos\left(\pi \frac{(z-h)}{(1-h)}\right)\right], & z \in [h, 1]; \\ 0, & \text{otherwise.} \end{cases} \tag{4}$$

where, ρ Smooth change between 1 and 0. Form (1)–(4) can be continuously differentiable between two random robots. Model parameter α, d, d_0, and d_1 form an artificial potential function. To avoid the divergence of reaction forces, we also add a non-zero gain factor α. The continuously differentiable collective potential function is constructed between random two robots. In addition, we need to introduce the potential field U2 to simulate the simulation environment. U2 is able to keep each robot in a monitoring area M within a certain range. It also prevents the robot from colliding with obstacles [12].

$$U(q) := k_1 U_1(q) + k_2 U_2(q) \tag{5}$$

where k1, k2 > 0 are weighting factors.

2.2 Models for Mobile Robots

Let n be the number of robots distributed in the region $\mathcal{M} \subset \mathbb{R}^2$, which can be assumed to be a convex and compact set. $\ell = \{1, 2, ..., N_s\}$ represents the intrinsic properties of each robot. The time $t \in M$ is defined as the position of the ith robot, and q denotes the configuration of the multi-robot system. The high-order dynamic model is described as [13]:

$$q_i(t + \Delta_t) = q_i(t) + \Delta_t v_i(t) \tag{6}$$

Here $q_i(t) \in \mathbb{R}^2$ is the position of the robot i at time $t \in \mathbb{R}_{\geq 0}$. $v_i(t) \in \mathbb{R}^2$ is the control input. $\Delta t \in \mathbb{R}_{>0}$ indicates the iteration step size. Assume that the measured value $y(q_i(t))$ of the ith robot is as follows:

$$y(q_i(t)) := \mu(q_i(t)) + w(t) \tag{7}$$

where $\mu(q_i(t))$ is an unknown field. $w(t)$ is sensing robot noise. t is sampling time. The basic principle of the algorithm is that each robot uses information interaction to search the task target (unknown area). Distributed learning is used to control multiple robots to find the maximum value in the unknown area.

2.3 Distributed Learning

Each robot needs to use distributed learning algorithm to search unknown task target area $\mu:M \rightarrow [0, \mu : max]$. Let $\mu(v)$ be the task target area is defined as [12]:

$$\mu(v): = \sum_{j=1}^{m} \phi_j(v)\theta_j = \phi^{\mathrm{T}}(v)\Theta \tag{8}$$

where $\varphi^T(v)$ and θ are defined as respectively:

$$\phi^{\mathrm{T}}(v): = \left[\phi_1(v)\ \phi_2(v)\ \cdots\ \phi_m(v) \right],$$
$$\Theta: = \left[\theta_1\ \theta_2\ \cdots\ \theta_m \right]^{\mathrm{T}} \in \Theta \tag{9}$$

where $\theta \subset R^m$ is a compact set. $\left\{ \varphi_j(v) \right\}$ is Gaussian radial basis functions

$$\phi_j(v): = \frac{1}{\Gamma_j} \exp\left(\frac{-v - \kappa_j^2}{2\sigma_j^2} \right) \tag{10}$$

where, the width of the Gaussian radial basis is σ_j. a standardizing constant is $\daleth_j$.

Based on the observation data and the regression coefficient $\{(y(v_k),\ \phi(v_k))\}_{k=1}^{n}$. The goal is to search for the minimum least-square error Θ:

$$\sum_{k=1}^{n} \left| y(v_k) - \phi^T(v_k)\hat{\Theta} \right|^2 \tag{11}$$

2.4 Cooperative Control

First, update the gradient by using a recursive algorithm. Then, according to the gradient state of the algorithm update, the collaborative control is determined. Therefore, the new recursive algorithm is defined as [12]:

$$K_i(t + 1) = P_i(t)\Phi_{*i}^{\mathrm{T}}\left(I_s + \Phi_{*i}P_i(t)\Phi_{*i}^{\mathrm{T}}\right)^{-1}$$
$$P_i(t + 1) = (I_m - K_i(t + 1)\Phi_{*i})P_i(t)$$
$$\hat{\Theta}_i(t + 1) = \hat{\Theta}_i(t) + K_i(t + 1)\left[Y_{*i} - \Phi_{*i}\hat{\Theta}_i(t) \right]$$
$$\nabla\hat{\mu}_i(t, q_i(t)) = \phi'^{\mathrm{T}}(q_i(t))\hat{\Theta}_i(t + 1) \tag{12}$$

Here $\nabla\hat{\mu}_i(t, v) : \mathbb{Z}_{\geq 0} \times \mathcal{M} \rightarrow \mathbb{R}^2$ is the gradient of the estimated field. Y_{*i} is the sum number of measurement data. According to (2), for all $j \in \mathcal{N}(i, q(t)) \cup \{i\}$, we have

$$Y_{*i} = \Phi_{*i}\Theta + \begin{bmatrix} \vdots \\ w_j(k) \\ \vdots \end{bmatrix} =: \Phi_{*i}\Theta + w_{*i}(t), \tag{13}$$

where t represents the exponent of any sampling time, in the measurement period between t and $t + 1$. w_j (k) represents the measurement noise information of robot j, and is uniformly and independently distributed on $j \in \ell$. According to the latest update of field gradient, distributed control v_i $(t + 1)$ is defined as follows:

$$v_i(t + 1) := \frac{\gamma(t + 1)}{\Delta_t}\left[\frac{\Delta_t}{\gamma(t)}v_i(t) + \gamma(t)u_i(t)\right] \tag{14}$$

When the parameters in (14) change, the dynamics of robot i is shown as follows

$$\begin{cases} q_i(t + 1) = q_i(t) + \gamma(t)p_i(t), \\ p_i(t + 1) = p_i(t) + \gamma(t)u_i(t), \end{cases} \tag{15}$$

Here, the new notations is applied to (1). $\Delta_t v_i(t)$ is replaced by $\gamma(t)p_i(t)$, $t + \Delta_t \in \mathbb{R}_{\geq 0}$ is replaced by $t + 1 \in \mathbb{Z}_{\geq 0}$, $t \in \mathbb{R}_{\geq 0}$ is replaced by by $t \in \mathbb{Z}_{\geq 0}$.

3 Convergence Analysis

In order to analyze the convergence of the algorithm, the ordinary differential equation (ODE) method of Ljung is used to prove it. The analysis method of recursive random algorithm is as follows [14]:

$$x(t) = x(t - 1) + \gamma(t)Q(t; x(t - 1), \varphi(t)) \tag{16}$$

and the observation process

$$\varphi(t) = g(t; x(t - 1), \varphi(t - 1), e(t)) \tag{17}$$

The ODE way needs to satisfy the regularity condition in the nonlinear observation process. Suppose DR is a subset of x space, where the regularity condition can be satisfied.

C1: $\|(x, \varphi, e)\| < C$ for all ϕ, e for all $x \in D_R$.

C2: For $x \in D_R$, the function $Q(t, x, \varphi)$ is continuously differentiable relative to x and φ. For fixed x and ϕ, the derivatives are bounded in t.

C3: $g(t; x, \varphi, e)$ is continuously differentiable relative to $x \in D_R$.

C4: Define $\overline{\varphi}(t, \overline{x})$ as $\overline{\varphi}(t, \overline{x}) = g(t; \overline{x}, \varphi(t - 1, \overline{x}), e(t)), \overline{\varphi}(0, \overline{x}) = 0$, (38) and suppose that $g(\cdot)$ has the following properties: $\|\overline{\varphi}(t, \overline{x}) - \varphi(t)\| < C \max_{n \leq k \leq t} \|\overline{x} - x(k)\|$, if $\overline{\varphi}(n, \overline{x}) = \varphi(n)$. This shows that a small change in x in (37) will not be increased to the higher amplitude of the observed φ.

C5: let $\overline{\varphi}_1(t, \overline{x})$ and $\overline{\varphi}_2(t, \overline{x})$ be the solution of (38), where $\overline{\varphi}_1(s, \overline{x}) := \overline{\varphi}_1^0$ and $\overline{\varphi}_1(s, \overline{x}) := \overline{\varphi}_2^0$. D_S can be defined as a set of all $\overline{x}$ as follows:

$$\|\overline{\varphi}_1(t, \overline{x}) - \overline{\varphi}_2(t, \overline{x})\| < C(\overline{\varphi}_1^0, \overline{\varphi}_2^0)\lambda^{t-s}(\overline{x}),$$

where $t > s$ and $\lambda(\overline{x}) < 1$. This is the exponential stable area of (37).

C6: $\lim_{t \to \infty} EQ(t, \overline{x}, \overline{\varphi}(t, \overline{x}))$ exists for $\overline{x} \in D_R$ and is denoted by $f(\overline{x})$. The expected value exceeds $\{e(\cdot)\}$.

C7: $e(\cdot)$ is a sequence of independent random variables.

C8: $\sum_{t=1}^{\infty} \gamma(t) = \infty$.

C9: $\sum_{t=1}^{\infty} \gamma(t) < \infty$ for some p.

C10: $\gamma(\cdot)$ is a decreasing sequence.

C11: $\lim_{t \to \infty} sup \left[1/\gamma(t) - 1/\gamma(t-1) \right] < \infty$.

Due to the dynamics of a single robot, the control of the proxy position is as follows:

$$q_i(t+1) = q_i(t) + \gamma(t)p_i(t) \tag{18}$$

where $p_i(t)$ is a control approach.

The algorithm is constrained by the regularity conditions C1–C11. Then, Z are denoted as follow:

$$Z := \left\{ q \in \mathcal{M}^{N_s} \middle| \sum_{j \in \{i\} \cup \mathcal{N}(i,q)} \phi(q_j)\phi^T(q_j) \neq 0, \forall i \in \ell \right\}. \tag{19}$$

Moreover, $f(x)$ 4 is defined as follows

$$f(x) = \left[\begin{array}{c} p \\ -\nabla U(q) - \left(\hat{L}(q) + K_d\right)p - \nabla C(q) \end{array} \right] \tag{20}$$

here $C(q) \in R \geq 0$ represents the cost function as follows:

$$C(q) := k_4 \sum_{i \in \ell} \left[\mu_{\max} - \mu(q_i) \right] \tag{21}$$

here $k_4 \in \mathbb{R}_{>0}$ represents a gain factor. $\mu_{max} \in \mathbb{R}_{>0}$ represents maximum of the field μ. Finally, the robot minimizes the global performance cost, as shown below:

$$V(q(\tau), p(\tau)) := C(q(\tau)) + U(q(\tau)) + \frac{p^T(\tau)p(\tau)}{2} \tag{22}$$

where $C(q)$ is the cost function of q. $U(q)$ is the cost function of q. The last item on the right side of the formula is the kinetic energy of the swarm robot system.

4 Simulation Results

4.1 Standard Environment

In the initial environment, we simulated swarm robots to search for task targets. Figure 1 shows that the number of robots n = 50, the balance distance D = 0.6, the noise level W = 1, and the color contour line represents the estimated error field when the iteration time t = 500. Figure 1 shows the maximum point where 50 robots can successfully search the task target. Figure 2 shows that after 500 iterations, all robots reduce the mean square error to the minimum.

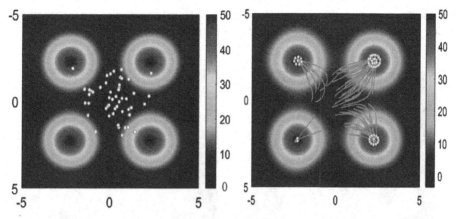

Fig. 1. Robots search for the task object in initial environment

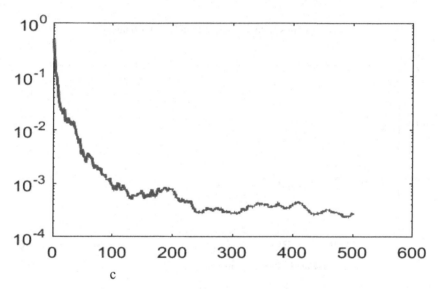

Fig. 2. Iterations in initial environment

4.2 Noise Environment

Noise is always an important problem in flocking control. Because noise usually leads to interruption of interaction or loss of communication. As shown in Fig. 4, when the noise level $W = 10$, each robot has strong noise noise, its stability and robustness are difficult to guarantee, as shown in Fig. 4. In the process of information data interaction between robots, if all robots are disturbed by strong noise, it will be very difficult to search the task target. Figure 3 shows that the increase of noise in the initial parameters makes it difficult for the swarm robot system to explore the task target. In addition, when each robot acts in a noisy environment, not only the sensor and information are distorted, but also there are a lot of communication delays and local communication interruptions, which are the common characteristics of swarm robot systems.

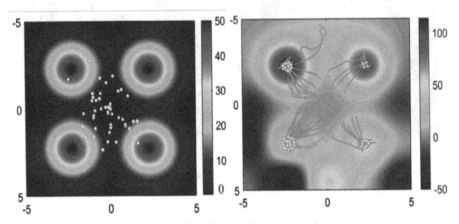

Fig. 3. Robot search task target in noisy environment

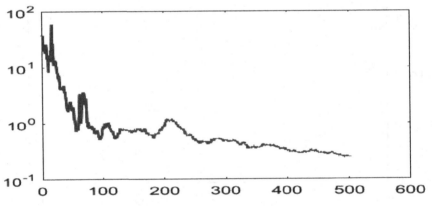

Fig. 4. Iterations in noisy environment

5   Conclusion

This paper presents a complex system for flocking control and distributed learning. Swarm robot systems can self-organize to find targets while avoiding collisions between robots in the swarm. The new mechanism, which combines attractive and repulsive forces, can drive robots to quickly avoid complex obstacles. Experiments show that the proposed distributed learning method can make swarm robots find effective targets faster. The simulation and experiment results show the validity of the system.

Acknowledgments. The research reported herein was supported by the NSFC of China under Grants Nos. 71571091 and 71771112.

References

1. Bak-Coleman, J.B., et al.: Stewardship of global collective behavior. Proc. Natl. Acad. Sci. **118**(27), e2025764118 (2021)
2. Berlinger, F., Gauci, M., Nagpal, R.: Implicit coordination for 3D underwater collective behaviors in a fish-inspired robot swarm. Sci. Robot. **6**(50), eabd8668 (2021)
3. O'Keeffe, K., Ceron, S., Petersen, K.: Collective behavior of swarmalators on a ring. Phys. Rev. E **105**, 014211 (2022)
4. Sha, J., Ebadi, A.G., Mavaluru, D., Alshehri, M., Alfarraj, O., Rajabion, L.: A method for virtual machine migration in cloud computing using a collective behavior-based metaheuristics algorithm. Concurr. Comput.: Pract. Exp. **32**(2), e5441 (2020)
5. Duarte, M., et al.: Evolution of collective behaviors for a real swarm of aquatic surface robots. PLoS ONE **11**(3), e0151834 (2016)
6. Ahmad, A., Licea, D.B., Silano, G., Báča, T., Saska, M.: PACNav: a collective navigation approach for UAV swarms deprived of communication and external localization. Bioinspir. Biomimet. **17**(6), 066019 (2022)
7. Bogue, R.: Disaster relief, and search and rescue robots: the way forward. Ind. Robot: Int. J. Robot. Res. Appl. **46**(2), 181–187 (2019)
8. Schranz, M., Umlauft, M., Sende, M., Elmenreich, W.: Swarm robotic behaviors and current applications. Front. Robot. AI **7**, 36 (2020)
9. Long, N.K., Sammut, K., Sgarioto, D., Garratt, M., Abbass, H.A.: A comprehensive review of shepherding as a bio-inspired swarm-robotics guidance approach. IEEE Trans. Emerg. Top. Comput. Intell. **4**(4), 523–537 (2020)
10. Gjeldum, N., Aljinovic, A., Crnjac Zizic, M., Mladineo, M.: Collaborative robot task allocation on an assembly line using the decision support system. Int. J. Comput. Integr. Manuf. **35**(4–5), 510–526 (2022)
11. Olfati-Saber, R.: Flocking for multi-agent dynamic systems: Algorithm and theory. IEEE Trans. Autom. Control **51**, 401–420 (2006)
12. Choi, J., Oh, S., Horowitz, R.: Distributed learning and cooperative control for multi-agent systems. Automatica **45**(12), 2802–2814 (2009)
13. Choi, J., Oh, S., Horowitz, R.: Cooperatively learning mobile agents for gradient climbing. In: 2007 46th IEEE Conference on Decision and Control, pp. 3139–3144. IEEE (2007)
14. Ljung, L.: Analysis of recursive stochastic algorithms. IEEE Trans. Autom. Control **22**(4), 551–575 (1977)

Exploration of Underwater Environments with a Swarm of Heterogeneous Surface Robots

Yifeng He[1]([✉])[iD], Kanzhong Yao[1][iD], Barry Lennox[1][iD], and Farshad Arvin[2][iD]

[1] Department of Electrical and Electronic Engineering, The University of Manchester, Manchester, UK
yifeng.he@manchester.ac.uk
[2] Swarm and Computational Intelligence Lab (SwaCIL), Department of Computer Science, Durham University, Durham, UK

Abstract. The main goal of swarm robotics is to control a large number of robots that interact together without a central controller. Swarm systems have a broad range of application areas, including the exploration and monitoring of extreme environments. This paper proposes a new approach to the coordination of a leader-follower system for use by a swarm of surface robots that are focused on the exploration of an unknown environment. The leader robot is controlled semi-autonomously by receiving a trajectory from a user and the swarm followers are then coordinated by a bio-inspired collective motion adapted from the state-of-the-art model, Active Elastic Sheet. To implement the proposed swarm controller, an open-source simulation platform, Gazebo, was used where the leader and follower robots, as well as a pond containing water, were simulated. The results demonstrate the feasibility of using the proposed swarm system in exploration applications.

Keywords: Collective Motion · Swarm Robotic · Autonomous System · Exploration

1 Introduction

In extreme environments, such as those characterised by nuclear radiation, high temperatures, or underwater conditions, measuring and monitoring environmental conditions can be crucial. While using humans for such missions is not always feasible, employing robots can offer a more reliable and safer option. For instance, the spent fuel storage pond in nuclear power stations is an example of an extreme environment that requires regular inspection to maintain safety. The International Atomic Energy Agency (IAEA) has traditionally used cameras that are moved around manually, such as the IAEA DCM-14 camera [1], to detect the condition of the storage. However, this approach is both time-consuming and costly, with over £25 million spent annually [2]. In contrast, underwater monitoring robots are a more efficient and cost-effective solution for inspecting large environments. While a single robot may have limitations in terms of battery power, reliability, and efficiency, multi-agent systems can provide improved performance.

Y. Tan et al. (Eds.): ICSI 2023, LNCS 13969, pp. 26–37, 2023.
https://doi.org/10.1007/978-3-031-36625-3_3

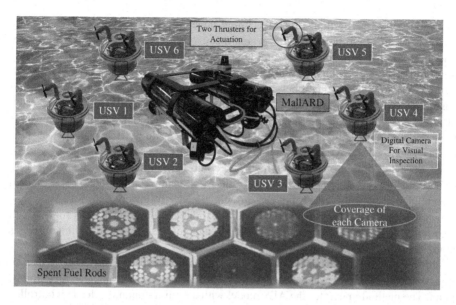

Fig. 1. A concept illustration of the exploration system for an underwater storage facility, where a leader is an autonomous aquatic surface vehicle that is surrounded by a swarm of surface robots.

There are many studies that have been conducted related to the exploration of swarm and multi-agent systems [3,4]. As an example, a multi-robot coverage problem was introduced by Green et al. [5] in a barrier-laden environment using a Pioneer P3-DX robot platform and a simulated barrier coverage robot. By having multiple robots follow a circular path, the team was able to establish a low-hardware, multi-agent coverage system that effectively addressed the problem at hand. In a similar topic, Hu et al. [6] studied collaborative coverage control, which involved utilising five vacuum cleaner robots to cover a specific area without any obstacles. To optimise the efficiency of the system, the researchers developed a fully-coordinated approach that allocated specific areas near each robot's starting point for exploration. To achieve this, a central controller was utilised to divide the area and generate paths for all the robots to follow.

Multi-agent systems can be implemented in different forms, with various implementations designed to solve the problems. One example is bio-inspired swarms, which draw on principles from social animals to enable self-organised flocking in mobile robots. Turgut et al. [7] proposed a swarm of real-world Kobot robots that were able to avoid collisions with walls while maintaining flocking behaviour [8]. Another work [9] proposed a bio-inspired exploration algorithm by combining the aggregation algorithm used by honeybees with pheromone tracking [10] to detect the source of a chemical leakage. The algorithm was implemented in the Webot simulation environment using a swarm robotic platform called Mona [11]. The experiments showed that the proposed exploration was effective in finding the source of the chemical leak.

Among bio-inspired swarm robotics, a collective motion mechanism called the Active Elastic Sheet (AES) model, which was developed by Ferrante et al. [12], has been proposed. This algorithm places simulated springs between neighbouring robots to enable collective motion. Since the initial development of the AES model, many researchers have built upon this theory and made various improvements. For instance, Raoufi et al. [13] used the Tabu Continuous Ant Colony System (TCACS) [14] optimisation technique to enhance the AES model. They applied the optimised collective motion to a simulated Mona robot, using two simulated forces to control the linear and angular motion of the swarm. The linear and angular movement experiments were tested on 100 simulated Mona robots arranged in a 10×10 square configuration, demonstrating improved behaviour of the AES model. In another study, Bahaidarah et al. [15] further improved the AES model by applying the particle swarm optimisation (PSO) method. In addition to bio-inspired methods, combining these approaches with classical control systems can yield successful results. For example, Ban et al. [16] presented a combination of a low-cost robot swarm with a leader-follower controller. Their system consisted of seven robots, with a leader positioned in the front and six followers in the rear. This system was capable of moving to a target position and avoiding obstacles along the path.

In this paper, we introduce a novel exploration mechanism for use in the inspection and exploration of water based storage facilities, such as a nuclear fuel storage pond. The method integrates the AES model with a semi-autonomous leader controller. Hence, a leader robot is positioned in the centre of the swarm and follows the user-defined trajectory, while the surrounding followers collaboratively inspect the facility and are coordinated using the bio-inspired collective motion, which considers the leader as an external force (Fig. 1).

2   Leader-Follower Collective Motion

2.1   Swarm Collective Motion of Followers

The collective motion algorithm used in this work is based on the AES model [12]. The algorithm was applied to the follower robots that are moving on the surface of the water. The angular $\dot{\theta}_i$ and linear $\dot{\vec{x}}_i$ velocities of the robot i are represented as:

$$\dot{\vec{x}}_i = v_0 \hat{n}_i + \alpha[(\vec{F}_i + D_r \hat{\xi}_r) \cdot \hat{n}_i]\hat{n}_i, \tag{1}$$

$$\dot{\theta}_i = \beta[(\vec{F}_i + D_r \hat{\xi}_r) \cdot \hat{n}_i^{\perp}] + D_\theta \xi_\theta, \tag{2}$$

where v_0 is the forward speed of the robots, α and β are the parameters related to linear and rotational movement, and $\hat{n}_i$ is the heading of the robot. The total elastic force, $\vec{F}$, applied to an individual robot relies on the spring constant k/l_{ij} and the initial distance l_{ij} presented in Eq. (3). D_r is the noise magnitude coefficient. $\hat{\xi}_r$ is a random orientated unit vector. The noise of robot's actuation is $D_\theta \xi_\theta$. D_θ is the coefficient of noise magnitude. ξ_θ is a random coefficient with a standard normal distribution between -1 and 1. In the ideal environment, where there is no noise, D_r and D_θ were set to 0.

$$\vec{F}_i = \sum_{j \in S_i} -\frac{k}{l_{ij}}(|\vec{r}_{ij}| - l_{ij})\frac{\vec{r}_{ij}}{|\vec{r}_{ij}|}, \tag{3}$$

$$\vec{r}_{ij} = \vec{x}_j - \vec{x}_i, \tag{4}$$

where l_{ij} is the equilibrium distances and r_{ij} is the length between robots i and j.

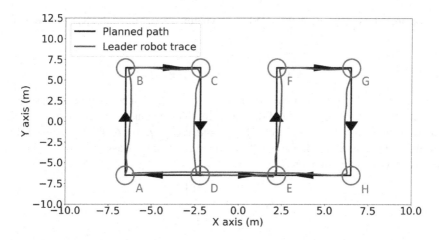

Fig. 2. Trajectory of the leader robot following a boustrophedon path started from point A and back to point A. Blue circles indicate the turning points of the robot. (Color figure online)

2.2 Controller of the Leader

In this study eight reference target points were set on the Y-axis boundaries, with the separation between points determined by the coverage area of the swarm. This resulted in the leader robot executing a boustrophedon path within a rectangular environment, as shown in Fig. 2. The leader robot used in this study, MallARD [17], has four thrusters in both perpendicular and horizontal directions, enabling it to move in a 2D plane without rotation. However, given that the direction of robots is crucial for collective motion and follower robots lack the ability to move laterally, the leader robot needs to rotate at each corner of the boustrophedon path. After turning to face the next corner, it proceeded to move in a straight line until it reached the next destination. Once the entire area was covered, the leader robot returned to the starting point.

3   Realisation of the Exploration

3.1   Robotic Platforms

The Leader Robot: MallARD, as depicted in Fig. 3, is an aquatic surface vehicle designed specifically for inspection and monitoring of extreme environments that can move with a velocity ranging from 0 to 1 m/s. The mechanical design of MallARD comprises two chassis plates, two pontoons, and four thrusters, with an area of 44 × 50 cm. The thrusters are positioned at the four corners of the vehicle, facing 45°C in relation to the forward direction. This particular layout was selected due to its inherent symmetry in both thrust and drag and its capacityto avoid obstruction of the central payload area. Based on the theory [18], the dynamic model of MallARD can be given by,

$$\dot{\vec{\eta}} = \vec{J}(\vec{\eta})\,\vec{v},$$
$$\vec{M}\dot{\vec{v}} + \vec{C}(\vec{v})\vec{v} + \vec{D}(\vec{v})\vec{v} + \vec{g}(\vec{\eta}) = \vec{\tau}. \tag{5}$$

where $\vec{J}(\vec{\eta})$ is the rotation matrix from the body frame to the world frame and $\vec{\eta} = [x,y,z,\phi,\theta,\psi]^{\top}$ defines the pose of the robot. MallARD's linear and angular velocities in robot body-fixed frames are denoted by $\vec{v} = [u,v,w,p,q,r]^{\top}$. As the manoeuvrability of MallARD is subject to 3 DoF on a 2D surface, $\vec{g}(\vec{\eta})$ can be assumed to be 0, hence, the mass matrix can be written as

$$\vec{M} = \begin{bmatrix} m_x & 0 & 0 \\ 0 & m_y & 0 \\ 0 & 0 & I_z \end{bmatrix}, \tag{6}$$

where I_z shows the rotation inertia along Z_b axis, let the mass of the vehicle be denoted by m, assume that both m_x and m_y are equal to m, and damping $\vec{C}$ is defined by

$$\vec{C}(\vec{v}) = \begin{bmatrix} 0 & 0 & -mv \\ 0 & 0 & mu \\ 0 & 0 & 0 \end{bmatrix}, \tag{7}$$

and the value of Coriolis rigid body effects $\vec{D}$ can be determined experimentally [19]. Control input $\vec{\tau}$ represents the force vector generated by actuators, which can be decomposed into corresponding force F_i and moment M_i generated by i^{th} thruster reading

$$\vec{\tau} = \sum_{i=1}^{n} \begin{bmatrix} F_i \\ M_i + r_i \times F_i \end{bmatrix} = [u_x, u_y, u_z, u_\phi, u_\theta, u_\psi]^{\top}, \tag{8}$$

where $\vec{r}_i$ is the vector from i^{th} thruster to the robot's centre of gravity, $u_{1:6}$ are inputs to controller affecting the $[x,y,z,\phi,\theta,\psi]$ rates. It is worth noting that for MallARD, $\vec{\tau}$ yields to $[u_x, u_y, 0, 0, 0, u_\psi]^{\top}$, therefore

$$\vec{\tau} = \begin{bmatrix} \tau_u \\ \tau_v \\ \tau_r \end{bmatrix} = \begin{bmatrix} X_F F_u \\ Y_F F_v \\ N_F F_r \end{bmatrix}, \tag{9}$$

where F_u, F_v and F_r are controlled by either the joypad or the trajectory tracking controller. The linear force coefficients are X_F, Y_F and N_F, thruster allocation algorithm used to generate F_1-F_4 from F_u, F_v and F_r can be found in [19].

Followers, Unmanned Surface Vehicles: The follower robots used in this study are unmanned-surface vehicles (USVs), as shown in Fig. 4. The USVs are designed to move on the water surface and capture photos using a modular camera that is faced downward to monitor the underwater facility. The robot platform features two propellers located on the top of the robot, which move the robot forward in the water, reducing the likelihood of contamination, which would be much greater if the propellers were in the water,

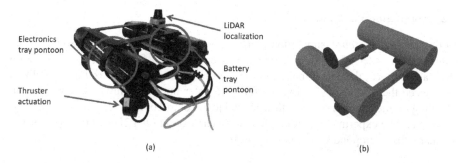

Fig. 3. (a) MallARD, an aquatic surface vehicle that is developed for inspection in known environments, and (b) simulated model in Gazebo.

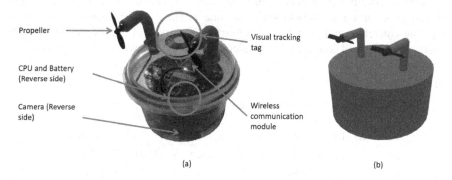

Fig. 4. (a) A low-cost unmanned surface vehicle designed for visual inspection of underwater facilities, and (b) 3D model of the robot in Gazebo.

which might contain mobile radioactive materials. The robot's enclosure is made of a 3D printed transparent plastic material. The dynamic model of the USV was developed by combining the AES model, drag force, and shear stress of water [20].

$$
\begin{cases}
\begin{bmatrix} V_i \\ \omega_i \end{bmatrix} = \begin{bmatrix} \overrightarrow{x}_i \\ \dot{\theta}_i \end{bmatrix} - \begin{bmatrix} V_i' \\ \omega_i' \end{bmatrix} \\
\begin{bmatrix} \dot{V_i'} \\ \dot{\omega_i'} \end{bmatrix} = \begin{bmatrix} \frac{1}{m} F_D \\ \frac{1}{m(\frac{l}{2})^2} \tau \end{bmatrix}
\end{cases}
, \tag{10}
$$

where V_i' is the induced linear velocity, ω_i is the induced angular velocity, V_i and ω_i' are the linear and angular velocities of the robot i, $\overrightarrow{x}_i$ and $\dot{\theta}_i$ is the linear and angular velocity output generated from the collective motion, F_D is the drag force given by the water, dt is the sampling frequency period of the system, m is the mass of a single robot, τ is the shear stress of water, and l is the distance between the propellers.

3.2 Simulation Platform

In this project, *Gazebo*[1] simulation platform was utilised as a 3D open-source robot simulator. In order to create a more realistic environment for the robots working on the water surface, the Free-Floating plugin was implemented into Gazebo. This plugin introduced the viscous and buoyancy force of water into the simulation environment, allowing for the simulation of the thruster and joint control of both surface and underwater robots. The experimental simulations were carried out on a 2D surface, specifically a square 20×20 m pond with bounding walls.

3.3 Experiments

The values of parameters and constants used in the experiments were obtained through trial and error within the simulation environment. The ratio between the parameters governing collective motion came from Ferrante's work [12] and hence the findings from this study should hold for future implementation in real-world robotic applications (Table 1).

Table 1. Values of parameters and constants in experiments

Parameters	Description	Value (units)
v_0	forward speed in collective motion of followers	2 (m/s)
Δt	sampling frequency	0.1 (s)
m_U	mass of USV	1 (kg)
m_M	mass of MallARD	10.5 (kg)
l	distance between the two propellers of USV	0.2 (m)
α	translational damping coefficient	0.1
β	rotational damping coefficient	1.2
k	spring constant	100 (N/m)
l_{ij}	set distance between robots	2 (m)

During the experiments, the robots were arranged in a hexagonal formation, with the leader robot at the centre and six followers surrounding it. To prevent collisions, all robots were kept 2 m away from their neighbours and 1.5 m from the surrounding walls. All experiments were repeated 10 times with random initial orientations (Fig. 5).

The Position Error of Leader: During the experiment, the leader robot, MallARD, followed a boustrophedon path 10 times. The trajectory of the robot was compared against the planned path, and the position error was calculated as the difference between these two trajectories. This analysis was used to evaluate the performance of the leader robot.

---

[1] https://gazebosim.org/.

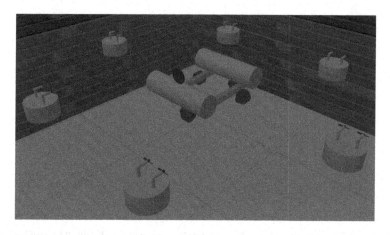

Fig. 5. An example of the experimental setup (a wet storage facility) to be explored by the proposed heterogeneous swarm system.

Force and Alignment of Swarm: The performance of the swarm in both a boustrophedon path and a straight line was investigated, and the positions of the follower robots were analysed. The force on each robot generated by the collective motion was recorded for analysis. In order to assess alignment, the differences between the orientation of the leader robot and each follower robot were also analysed. The leader travelled directly from point A to point H (see Fig. 2) without any rotation.

The Position Error of Follower: The position error of the swarm is defined as the level to which the followers maintained a hexagonal formation around the leader robot during the boustrophedon movement. This analysis was based on two aspects. Firstly, the distance between the leader and follower robots was tested. The difference between their actual distance and the set distance (2 m) was analysed. Secondly, the orientation difference between the leader and two neighbouring followers was measured.

4   Results and Discussion

4.1   Position Error of the Leader

Figure 6 shows that the position error of the leader MallARD was relatively small in comparison to the coverage size of the leader robot, with most errors falling under 0.4 m. The shaded area representing the position error is narrow due to the negligible difference observed between the ten repeated experiments. In each experiment, the leader began from the same position and followed the same path within the same environment, resulting in similar robot traces. When the leader robot passed target points along the path, a small error appeared on the curve, which was marked on the graph with the target points shown in Fig. 2. This result was caused by the trajectory of the leader robot, which required the robot to approach target points closely before advancing to the next point.

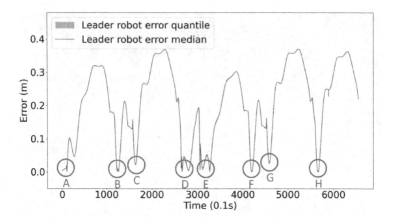

Fig. 6. The position error of leader robot when following a planned path. The target points in Fig. 2 had been marked.

4.2 Performance of Swarm in Simple Trajectory

Figure 7 shows the AES force and alignment of the USV swarm when moving in a straight line. The system took approximately 100 s to move at a distance of 13 m. Generally, the results converge after approximately $t = 20$ s, at which point both force and alignment tend towards a fixed value. During the first 20 s of movement, the system was aligning the followers, resulting in extreme values for both force and alignment. The force could reach up to 35, while the alignment error could increase up to 2.3 rad. In steady states, the AES force remained constant at around 10, and the alignment error tended towards 0 rad.

4.3 Performance of Swarm in Exploration

Figure 8 illustrates the AES force and alignment of the USV swarm when following a boustrophedon path. Both force and alignment showed significant fluctuation when the system passed the target points because the force within the swarm increases when the leader robot was rotating rather than moving in a straight line. However, when the leader robot resumed long-distance straight movement from rotation, the system became more stable again. The steady-state value of force was approximately 10, and the alignment error tended towards 0 rad, which is similar to the behaviour observed during straight-line movement.

4.4 The Position Error of Followers

Figure 9 shows the position errors, including distance and angular errors of the USV swarm when following a boustrophedon path. Compared to force and alignment, the distribution area of the position error is much larger, indicating that most of the time, the six USVs were not forming a tightly shaped hexagon around the leader robot. Even so, the distance error between the robots was acceptable, and the risk of collision between

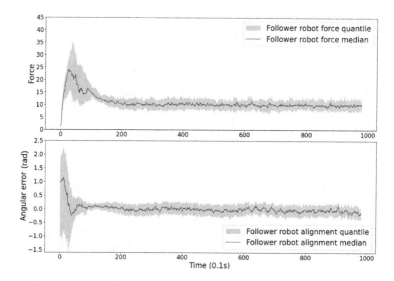

Fig. 7. (Top) Force of the follower robots, and (bottom) alignment of the follower robots in a straight movement.

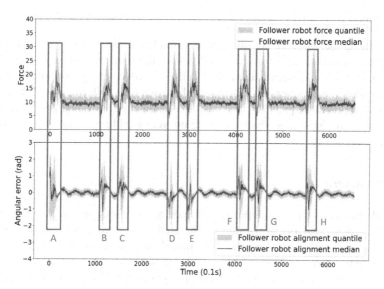

Fig. 8. (Top) Force of the follower robots, and (bottom) alignment of the follower robots in exploration scenario.

them was low. However, the system had difficulty dealing with rotations, and passing the target points caused the distance and angular errors to double in size. The swarm tended to become steady during long-distance straight movement, and both the distance and angular errors would stay close to zero until the next target point was reached.

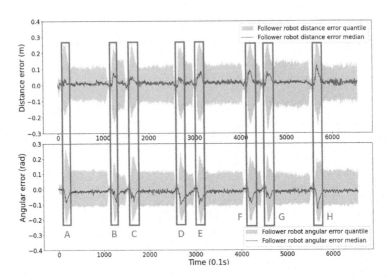

Fig. 9. (Top) The distance error of followers and (bottom) the angular error of followers.

4.5 Summary

By considering the observed results from the simulation experiments, the new exploration mechanism showed sufficient capability of the monitoring mission for use in underwater facilities. Moreover, the risk of collision between robots or bounding walls and the system becoming unstable is low. In the case of the follower robots' trajectory error on a long-distance path, the error in the rotation at each turning point was much larger. This could be a great challenge in future real-world applications if we require high-precision motion.

5 Conclusion

This work was focused on developing a new approach for using swarm of water-surface robots to explore underwater facilities. The study proposed a new leader-follower swarm scenario that is based on the AES model. We implemented a heterogeneous swarm including two different types of robots as the leader and followers. We used Gazebo to investigate the feasibility of the proposed method. In future work, the proposed exploration will be applied to real-world robots.

References

1. Doyle, J.: Nuclear Safeguards, Security and Nonproliferation: Achieving Security with Technology and Policy. Elsevier (2011)
2. Pepper, S., Farnitano, M., Carelli, J., Hazeltine, J., Bailey, D.: Lessons Learned in Testing of Safeguards Equipment. Brookhaven National Lab., Upton, NY (US), Tech. Rep. (2001)

3. Huang, X., Arvin, F., West, C., Watson, S., Lennox, B.: Exploration in extreme environments with swarm robotic system. In: IEEE international conference on Mechatronics (ICM), vol. 1, pp. 193–198. IEEE (2019)
4. Hu, J., Niu, H., Carrasco, J., Lennox, B., Arvin, F.: Voronoi-based multi-robot autonomous exploration in unknown environments via deep reinforcement learning. IEEE Trans. Vehicul. Technol. **69**(12), 14413–14423 (2020)
5. Green, T., et al.: A minimalist solution to the multi-robot barrier coverage problem. In: Fox, C., et al. (eds.) TAROS 2021. LNCS (LNAI), vol. 13054, pp. 349–353. Springer, Cham (2021). https://doi.org/10.1007/978-3-030-89177-0_35
6. Hu, J., Lennox, B., Arvin, F.: Collaborative coverage for a network of vacuum cleaner robots. In: Annual Conference Towards Autonomous Robotic Systems, pp. 112–115 (2021)
7. Turgut, A.E., Çelikkanat, H., Gökçe, F., Şahin, E.: Self-organized flocking in mobile robot swarms. Swarm Intell. **2**, 97–120 (2008)
8. Turgut, A.E., Gokce, F., Celikkanat, H., Bayindir, L., Sahin, E.: Kobot: A mobile robot designed specifically for swarm robotics research. Middle East Technical University, Ankara, Turkey, METU-CENG-TR Tech. Rep., vol. 5, no. 2007 (2007)
9. Amjadi, A., Raoufi, M., Turgut, A.E., Broughton, G., Krajnik, T., Arvin, F.: Cooperative pollution source exploration and cleanup with a bio-inspired swarm robot aggregation. In: International Conference on Collaborative Computing, pp. 469–481 (2020)
10. Na, S., et al.: Bio-inspired artificial pheromone system for swarm robotics applications. Adapt. Behav. 1059712320918936 (2020)
11. Arvin, F., Espinosa, J., Bird, B., West, A., Watson, S., Lennox, B.: Mona: An affordable open-source mobile robot for education and research. J. Intell. Robot. Syst. **94**(3–4), 761–775 (2019)
12. Ferrante, E., Turgut, A.E., Dorigo, M., Huepe, C.: Collective motion dynamics of active solids and active crystals. New J. Phys. **15**(9), 095011 (2013)
13. Raoufi, M., Turgut, A.E., Arvin, F.: Self-organized collective motion with a simulated real robot swarm. In: Towards Autonomous Robotic Systems, pp. 263–274 (2019)
14. Karimi, A., Nobahari, H., Siarry, P.: Continuous ant colony system and tabu search algorithms hybridized for global minimization of continuous multi-minima functions. Comput. Optimiz. Appl. **45**, 639–661 (2010)
15. Bahaidarah, M., Rekabi-Bana, F., Turgut, A.E., Marjanovic, O., Arvin, F.: Optimization of a self-organized collective motion in a robotic swarm. In: Swarm Intelligence: 13th International Conference, ANTS 2022, pp. 341–349 (2022)
16. Ban, Z., Hu, J., Lennox, B., Arvin, F.: Self-organised collision-free flocking mechanism in heterogeneous robot swarms. Mob. Netw. Appl. 1–11 (2021)
17. Groves, K., West, A., Gornicki, K., Watson, S., Carrasco, J., Lennox, B.: Mallard: An autonomous aquatic surface vehicle for inspection and monitoring of wet nuclear storage facilities. Robotics **8**(2), 47 (2019)
18. Fossen, T.I., Fjellstad, O.-E.: Nonlinear modelling of marine vehicles in 6 degrees of freedom. Math. Model. Syst. **1**(1), 17–27 (1995)
19. Groves, K., Dimitrov, M., Peel, H., Marjanovic, O., Lennox, B.: Model identification of a small omnidirectional aquatic surface vehicle: A practical implementation. In: International Conference on Intelligent Robots and Systems (IROS), pp. 1813–1818 (2020)
20. He, Y., Lennox, B., Arvin, F.: Exploration of underwater storage facilities with swarm of micro-surface robots. In: Towards Autonomous Robotic Systems, pp. 92–104 (2022)

Integrating Reinforcement Learning and Optimization Task: Evaluating an Agent to Dynamically Select PSO Communication Topology

Rodrigo Cesar Lira[1(✉)] , Mariana Macedo[2] , Hugo Valadares Siqueira[3] ,
and Carmelo Bastos-Filho[1]

[1] University of Pernambuco, Pernambuco, Brazil
{rcls,carmelofilho}@ecomp.poli.br
[2] University of Toulouse, Occitania, France
mmacedo@biocomplexlab.org
[3] Federal University of Technology, Paraná, Brazil
hugosiqueira@utfpr.edu.br

Abstract. A recent study catalogued hundreds of meta-heuristics proposed over the past three decades in Swarm Intelligence (SI) literature. This scenario makes it difficult for the practitioner to choose the most suitable meta-heuristic (RL) for a specific problem. This paper shows that Reinforcement Learning could be a powerful tool for SI. First, we describe a Reinforcement Learning environment to solve an optimization problem. Then, we investigate the usage of Proximal Policy Optimization to dynamically set the Particle Swarm Optimization topology accordingly to the simulation states. Our RL proposal reached competitive fitness values, even when evaluated in non-trained scenarios. In addition, we show the actions' distribution by simulation in the Rastrigin. The paper demonstrates how RL could be integrated to improve meta-heuristics capabilities, opening new research paths where RL will be used to improve meta-heuristics or select them accordingly to their strengths.

Keywords: Particle Swarm Optimization · Proximal Policy Optimization · Reinforcement Learning

1 Introduction

Swarm Intelligence (SI) is a branch of Computational Intelligence (CI) that comprises approaches inspired by the intelligent behaviour that arises in the interaction among living beings to solve complex problems [6]. These agents behave mostly without supervision, and their actions have a stochastic component according to the perception of their surroundings [5,23].

After three decades of the emergence of the research area, the number of meta-heuristics based on swarms increased from a few to hundreds [3]. Covid-19, African Buffalo, Buses and the FIFA World Cup are recent inspirations used in the proposal of new algorithms [3]. Knowing that even the same meta-heuristics

Y. Tan et al. (Eds.): ICSI 2023, LNCS 13969, pp. 38–48, 2023.
https://doi.org/10.1007/978-3-031-36625-3_4

can achieve weak results if we choose non-optimal hyperparameters. The current scenario makes complex the task of finding the most suitable meta-heuristic for each problem [10].

In order to overcome the manual and laborious task of choosing a meta-heuristic for a specific problem, a new approach based on meta-learning has emerged in the literature called Learning to Optimize (L2O) [13]. Recently, Yue et al. [4] created an L2O proposal by extending a point-based optimizer into a population-based optimizer. Gomes et al. [8] proposed the Learning-to-Optimize Partially Observable Markov Decision Process (LTO-POMDP), a framework based on reinforcement meta-learning to solve black-box problems.

On the other hand, we have Reinforcement Learning (RL) becoming an emergent machine learning tool to improve swarm intelligence capabilities. RL is a family of machine learning methods that can evolve from simple feedback. Seyyedabbasi [19,20] used a classical reinforcement learning technique, called Q-Learning, to guide the behaviour of several swarm intelligence algorithms, creating RL_{I-GWO}, RL_{Ex-GWO}, RL_{WOA} and RLSCSO. Wu et al. [24] employed reinforcement learning to improve PSO convergence, controlling the social and cognitive components. Almonacid integrated reinforcement learning and optimization problem, creating a method that automatically creates evolutionary meta-heuristics using reinforcement learning, named AutoMH [1]. Nevertheless, to the best of our knowledge, there is no approach to dynamically change the use of swarm communication topology using RL in the literature.

In this paper, we investigate the usage of Deep Reinforcement Learning (DRL) to dynamically set the Particle Swarm Optimization (PSO) [11] topology. We choose PSO because it is a well-known meta-heuristic widely used to solve several optimization problems [14,21]. In our approach, we first create a Reinforcement Learning environment where a PSO is integrated to solve an optimization task. Then, we train the RL agent to select the best-suited PSO topology. Finally, we evaluate the results in different scenarios.

This paper is divided as follows: Sect. 2 briefly describes Particle Swarm Optimization. Section 3 describes the methodology and parameterization for the experiments. Section 4 presents our findings and results. Lastly, Sect. 5 finishes the paper with some discussions and conclusions.

2 Particle Swarm Optimization

In 1995, Kennedy and Eberhart proposed Particle Swarm Optimization (PSO) [11], a novel algorithm to solve continuous optimization problems. PSO is composed of a set of particles (i.e., the agents) that move around the search space, influenced by the best solution that they individually have found (the cognitive component) and the best solution that any particle in their neighbourhood has found (social component).

In their movement, the particles update their velocity and position using Eq. 1 and Eq. 2, respectively [2].

$$\vec{v}_i(t+1) = \chi\{\vec{v}_i(t) + c_1 r_1[\vec{p}_i(t) - \vec{x}_i(t)] + \vec{c}_2 r_2[\vec{n}_i(t) - \vec{x}_i(t)]\} \qquad (1)$$

$$\vec{x}_i(t+1) = \vec{v}_i(t+1) + \vec{x}_i(t), \tag{2}$$

where $\vec{x}_i(t)$ and $\vec{v}_i(t)$ are the position and velocity in the iteration t of the particle i, respectively; c_1 is the cognitive acceleration coefficient, and c_2 is the social acceleration coefficient; r_1 and r_2 are random uniform numbers; $\vec{p}_i(t)$, and $\vec{n}_i(t)$ are, respectively, the best position found by the particle, and the best position found by any neighbour until the current iteration. Finally, χ is the constrictor factor, a mechanism to ensure convergence [21] defined in Eq. 3.

$$\chi = \frac{2}{|2 - \varphi - \sqrt{\varphi^2 - 4\varphi}|}, \varphi = c_1 + c_2. \tag{3}$$

PSO pseudocode is shown in Algorithm 1. While the stop criterion is not reached, each particle moves around the search space using Eq. 1 and 2. After each movement, the particle updates its best position, and the best position found by the neighbours is also updated.

Algorithm 1. PSO Pseudocode

1: Initialize the population randomly
2: **while** stop criterion is not reached **do**
3:      **for** each *particle* **do**
4:          Update particle velocity
5:          Update particle position
6:          Evaluate particle position
7:          Update $\vec{p}_i$ and $\vec{n}_i$
8:      **end for**
9: **end while**
10: Return the best solution

In the PSO, the communication topology is an essential hyperparameter because it defines the set of particles that will share information (neighbourhood). In a fully connected topology, the agents will quickly share information. A topology with limited sharing delays information propagation. Thus, communication topology directly affects the flow of information within the swarm and, consequently, the algorithm's performance.

Global (*gbest*) and Local (*lbest*) are two well-known topologies in the literature [21]. For the Global topology, each particle broadcasts information to the whole swarm. For the Local topology, each particle has a limited number of neighbours with whom they can communicate [16]. Local and Global topologies are static, meaning the neighbours do not change over the iterations.

Oliveira et al. proposed a new topology based in a dynamic neighbourhood [15]. The topology uses a mechanism that creates new connections between particles when the swarm stagnates. Every iteration, the particle that does not improve its $\vec{p}_i(t)$, increments a counter (p_k-$failure$) by one. When a threshold (p_k-$failure^T$) is reached, the particle modifies its neighbourhood by connecting to a new particle selected by the roulette wheel. p_k-$failure$ is set to zero in two situations: (i) every iteration where the particle improves its solution or (ii) after a new connection. On average, the dynamic topology showed better results than Local and Global.

3   Methodology

We developed a reinforcement learning environment using OpenAI Gym that encapsulates a PSO simulation to solve the optimization problem. The environment is one of the main components of a Reinforcement Learning task. In a trial-and-error process, the RL agent interacts with it, selecting an action and receiving feedback.

The RL agent is integrated into the PSO simulation by setting the PSO communication topology in our experiments. The agent acts in the environment by choosing Local, Global, or Dynamic topology and evaluating the improvement in the simulation. An action is taken every *step_ decision* iterations, an environment hyperparameter. We used *step_ decision=5*, meaning an action after five PSO iterations. There are 200 time steps where the RL agent interacts with the environment since in the PSO simulation, we used *max _iteration* = 1000 as the stop criterion. In our preliminaries simulations, *step_ decision=5* performed better and faster than *step_ decision=1*. Figure 1 depicts our simulation using the RL environment.

While the stop criterion is not reached, i.e. the episode is not over, the agent acts in the environment selecting the topology. Next, the agent receives a set of states representing the current simulation scenario and the reward that measures the action's effectiveness after a run of 5 iterations. The reward is calculated using the function proposed by Schuchardt et al. [17], showed in Eq. 4. It considers that multiple minor improvements should have the same effect as one considerable improvement to find the best solution.

$$reward(t) = \alpha_r \log_{10} \frac{F_{\max}(t)}{F_{\max}(t-1)} \tag{4}$$

where α_r is a scale factor that is set to 1 as indicated by the authors for solving continuous optimization, and $F_{\max}(t)$ is the fitness value of the fittest particle in time step t. The proposed reward function maximizes the maximum fitness gain between actions. However, as we deal with the problem as a minimization task, we transform the fitness using Eq. 5 [17].

$$F(x_i) = \frac{1}{\max(f(x_i), 10^{-20})} \tag{5}$$

where $f(x)$ is the value of our fitness function for the particle x_i. max() is used to avoid math errors in the division when the fitness is zero. Therefore, the reward function works in minimization and maximization problems.

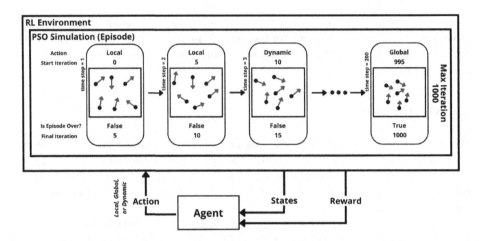

Fig. 1. Reinforcement Learning environment created for solving an optimization problem through Particle Swarm Optimization. Every time step, the RL agent selects a topology and evaluates its effectiveness.

We used a set of features retrieved from the simulation to inform the agent about the state. We based on a set of variables described in previous works [12,22], e.g. normalized euclidean distance among agents, remaining simulation budget, and the number of agents improved in the previous iterations. Additionally, we included a new feature that maps the topology chosen in the current action.

We choose Proximal Policy Optimization (PPO), a Reinforcement Learning agent famous for its stability and reliability [18]. We used PPO implementation from RLLib with default hyperparameters, being $\gamma = 0.99$ (discount factor), $timesteps\_per\_iteration = 20,000$, $train\_batch\_size = 4,000$, $num\_sgd\_iter = 30$, $clip\_param = 0.3$, $KL\_init = 0.2$, $KL\_target = 0.01$, $GAE\_parameter = 1.0$, $vf\_loss\_coeff = 1.0$, $entropy\_coeff = 0$, $learning_rate = $ 5e-05. During our preliminaries simulations, we also trained the Rainbow agent [9], but we removed it from our setup due to the lack of competitive results, supposedly a tuning problem.

We trained PPO during 500 iterations (epochs) to solve the Rastrigin function in two scenarios, 10 and 30 dimensions. In each training iteration, we run 600 PSO simulations (episodes). In the PSO simulation we used 20 particles, $c_1 = c_2 = 2.05$, $p_k\text{-}failure^T = 1$. Local topology allows communication with two neighbours (i.e., Ring topology). We used standard particle swarm optimization, but our proposal is not PSO version-dependent. We can switch to a different PSO version by simply changing the PSO code.

After training, we tested the agent in Rastrigin, Shifted Rastrigin and Shifted Rotated Rastrigin benchmark functions [7]. For each scenario, we run 50 simulations, collecting the best fitness and sequence of actions the RL agent takes.

4   Results

We present our results organized into two parts. First, we show the performance of PSO using RL to choose the topology dynamically compared to PSO using Local, Global or Dynamic topologies. Next, we evaluate the actions chosen by the approach using RL agent in Rastrigin.

Figure 2 shows the results found in Rastrigin with 10 (A) and 30 (B) dimensions, respectively. The box plots indicate that our RL approach found competitive results in each function, indicating that PPO learned to choose the suited PSO topology according to the simulation state.

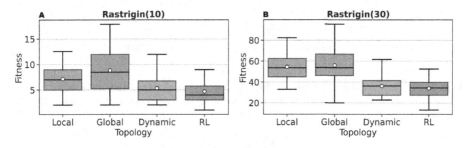

Fig. 2. Box plot of fitness values reached by our proposal (RL), and PSO using Local, Global and Dynamic topology in Rastrigin with 10 dimensions.

Even training only in Rastrigin, we tested the proposal in two variations to evaluate the model generalization when used in a non-simulated scenario. Figures 3, and 4 show the results found in the scenarios using Shifted Rastrigin and Shifted Rotated Rastrigin with 10 and 30 dimensions. Even though the RL agent was not trained with rotation and shifting, it could select good sequences of topologies that took it to reach low fitness values in these scenarios, even when we increase the complexity (Figs. 3 B and 4 B), we see our RL approach reaching good fitness results when compared with PSO using Local, Global or Dynamic topologies.

Figure 5 shows the convergence in the most complex trained scenario (30 dimensions). We see the same behaviour in Rastrigin with 10 dimensions. Our proposal and Dynamic topology in both scenarios reach the best solution around the final iteration. Differently, all approaches converge quickly to the final solution in Shifted Rastrigin and Shifted Rotated Rastrigin (non-trained scenarios).

We applied the Wilcoxon test to compare the efficiency across topologies using a confidence rate of 95%, as shown in Table 1. In the table, '–' indicates

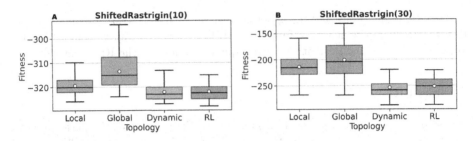

Fig. 3. Box plot of fitness values reached by our proposal (RL), and PSO using Local, Global and Dynamic topology in Shifted Rastrigin with 10 (A) and 30 (B) dimensions.

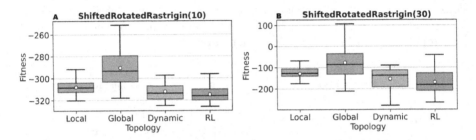

Fig. 4. Box plot of fitness values reached by our proposal (RL), and PSO using Local, Global and Dynamic topology in Shifted Rotated Rastrigin with 10 (A) and 30 (B) dimensions.

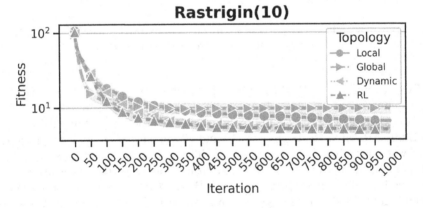

Fig. 5. Fitness per iteration for RL, and PSO using Local, Global and Dynamic topology in Rastrigin with 30 dimensions.

that there is no statistical difference between the results found by the topologies, '▲' indicates the proposal achieved better results than the other topology, and '▼' represents that our proposal reached worse results than the topology compared. We can conclude from this analysis that the RL approach performs better than Local and Global topologies for every simulated scenario. Our proposal overcomes the Dynamic topology only in Shifted Rotated Rastrigin with 30 dimensions. While in other scenarios, they are equivalent, with no statistical differences.

Table 1. Results of fitness values and Wilcoxon test with a confidence level of 95% comparing the RL with the other algorithms using 10 and 30 dimensions in Rastrigin, Shifted Rastrigin and Shifted Rotated Rastrigin.

Function		10 dimensions RL	Local	Global	Dynamic
Rastrigin	Mean Fitness	4.68	7.14	8.82	5.31
	STD	2.30	3.01	4.17	2.70
	Wilcoxon		▲	▲	–
Shifted Rastrigin	Mean Fitness	−322.00	−319.32	−313.48	−322.08
	STD	4.59	3.81	7.65	3.86
	Wilcoxon		▲	▲	–
Shifted Rotated Rastrigin	Mean Fitness	−314.88	−308.07	−290.47	−311.99
	STD	6.93	6.48	16.11	10.00
	Wilcoxon		▲	▲	–
Function		30 dimensions RL	Local	Global	Dynamic
Rastrigin	Mean Fitness	33.94	54.56	56.01	36.28
	STD	8.39	11.94	15.59	8.84
	Wilcoxon		▲	▲	–
Shifted Rastrigin	Mean Fitness	−251.67	−213.52	−200.68	−253.84
	STD	18.57	23.77	35.50	23.76
	Wilcoxon		▲	▲	–
Shifted Rotated Rastrigin	Mean Fitness	−169.35	−128.81	−77.30	−153.95
	STD	56.12	32.31	72.35	51.61
	Wilcoxon		▲	▲	▲

In addition, we evaluated the set of actions taken in each simulation and created Fig. 6. Figure 6 A shows that Dynamic topology was predominantly chosen in Rastrigin with ten dimensions. Nevertheless, in Rastrigin with 30 dimensions, Local and Dynamic topologies were selected almost similarly during the 200-time steps (Fig. 6 B). As the dimensions increase, the complexity grows, forcing the agent to behave differently.

Figure 7 shows the actions' distribution by simulation in Rastrigin. In Fig. 7 A, Dynamic (blue bar) is the most selected topology in each simulation. In turn, Local (green bar) and Dynamic are selected in a similar quantity in Fig. 7 B. Global topology (orange bar) is the less frequent choice in both scenarios.

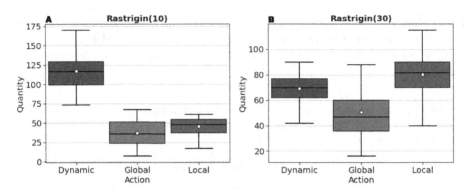

Fig. 6. Box plot of actions taken by RL agent in Rastrigin with 10 (A) and 30 (B) dimensions.

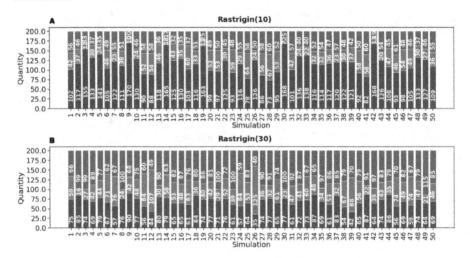

Fig. 7. Actions' distribution by simulation in Rastrigin with 10 (A) and 30 (B) dimensions.

5   Conclusions

We used Proximal Policy Optimization to set the Particle Swarm Optimization topology dynamically. Firstly, we developed an RL environment with a PSO simulation where the RL agent selects topology every five iterations. Then, we trained the agent using the Rastrigin function with 10 and 30 dimensions.

The proposal reached competitive results, overcoming PSO using Local and Global topology. Both scenarios' results were statistically equivalent compared to PSO using Dynamic topology. Even when evaluated in a non-simulated scenario, such as Shifted Rastrigin and Shifted Rotated Rastrigin, the RL approach reached good results, showing generalization capability. We also showed the choices' distribution by simulation using the Rastrigin benchmark function.

The results lead us to believe that the level of the problem's complexity forces the agent to act differently.

Despite our limited scenarios, the results found in this paper are essential for creating the foundation of new research paths where RL will be used to improve meta-heuristics or select them accordingly to their strengths. We utilized the standard PSO algorithm in our simulations, but our method could be applied to any PSO version. We aim to add new problem features in the environment state for future work to improve the agent's generalization. In addition, we plan to investigate in-depth the impact of simulation hyperparameters in different scenarios.

Acknowledgements. The authors thank the Brazilian National Council for Scientific and Technological Development (CNPq), processes number 40558/2018-5, 315298/2020-0, and Araucaria Foundation, process number 51497, for their financial support. This study was financed in part by the Coordenação de Aperfeiçoamento de Pessoal de Nível Superior - Brasil (CAPES) - Finance Code 001.

References

1. Almonacid, B.: Automh: Automatically create evolutionary metaheuristic algorithms using reinforcement learning. Entropy **24**(7) (2022). https://doi.org/10.3390/e24070957

2. Belotti, J.T., et al.: Air pollution epidemiology: A simplified generalized linear model approach optimized by bio-inspired metaheuristics. Environ. Res. **191**, 110106 (2020). https://doi.org/10.1016/j.envres.2020.110106

3. Campelo, F., Aranha, C.: EC Bestiary: A bestiary of evolutionary, swarm and other metaphor-based algorithms (2018). https://doi.org/10.5281/zenodo.1293352

4. Cao, Y., Chen, T., Wang, Z., Shen, Y.: Learning to optimize in swarms. In: Proceedings of the 33rd International Conference on Neural Information Processing Systems. Curran Associates Inc., Red Hook (2019)

5. Chandrasekaran, K., Simon, S.P., Padhy, N.P.: Binary real coded firefly algorithm for solving unit commitment problem. Inf. Sci. **249**, 67–84 (2013). https://doi.org/10.1016/j.ins.2013.06.022

6. Cruz, D.P.F., Maia, R.D., De Castro, L.N.: A critical discussion into the core of swarm intelligence algorithms. Evolution. Intell. **12**(2), 189–200 (2019). https://doi.org/10.1007/s12065-019-00209-6

7. Ding, K., Tan, Y.: A cuda-based real parameter optimization benchmark (2014). https://doi.org/10.48550/ARXIV.1407.7737

8. Gomes, H.S., Léger, B., Gagné, C.: Meta learning black-box population-based optimizers (2021). https://doi.org/10.48550/ARXIV.2103.03526

9. Hessel, M., et al.: Rainbow : Combining Improvements in Deep Reinforcement Learning (2013)

10. Houssein, E.H., Gad, A.G., Hussain, K., Suganthan, P.N.: Major advances in particle swarm optimization: Theory, analysis, and application. Swarm Evolution. Comput. **63**, 100868 (2021). https://doi.org/10.1016/j.swevo.2021.100868

11. Kennedy, J., Eberhart, R.: Particle swarm optimization. In: Proceedings of ICNN'95 - International Conference on Neural Networks. vol. 4, pp. 1942–1948 (1995). https://doi.org/10.1109/ICNN.1995.488968

12. de Lacerda, M.G.P., de Lima Neto, F.B., Ludermir, T.B., Kuchen, H.: Out-of-the-box parameter control for evolutionary and swarm-based algorithms with distributed reinforcement learning. Swarm Intell. (2023). https://doi.org/10.1007/s11721-022-00222-z

13. Li, K., Malik, J.: Learning to optimize (2016). https://doi.org/10.48550/ARXIV.1606.01885

14. Macedo, M., et al.: Overview on binary optimization using swarm-inspired algorithms. IEEE Access **9**, 149814–149858 (2021). https://doi.org/10.1109/ACCESS.2021.3124710

15. Oliveira, M., Bastos-Filho, C.J.A., Menezes, R.: Using network science to define a dynamic communication topology for particle swarm optimizers. In: Complex Networks, pp. 39–47. Springer, Heidelberg (2013). https://doi.org/10.1007/978-3-642-30287-9_5

16. Santos, P., et al.: Application of PSO-based clustering algorithms on educational databases. In: 2017 IEEE Latin American Conference on Computational Intelligence (LA-CCI), pp. 1–6. IEEE (2017). https://doi.org/10.1109/LA-CCI.2017.8285690

17. Schuchardt, J., Golkov, V., Cremers, D.: Learning to evolve (2019). https://doi.org/10.48550/ARXIV.1905.03389

18. Schulman, J., Wolski, F., Dhariwal, P., Radford, A., Klimov, O.: Proximal policy optimization algorithms (2017). https://doi.org/10.48550/ARXIV.1707.06347

19. Seyyedabbasi, A.: A reinforcement learning-based metaheuristic algorithm for solving global optimization problems. Adv. Eng. Softw. **178**, 103411 (2023). https://doi.org/10.1016/j.advengsoft.2023.103411

20. Seyyedabbasi, A., Aliyev, R., Kiani, F., Gulle, M.U., Basyildiz, H., Shah, M.A.: Hybrid algorithms based on combining reinforcement learning and metaheuristic methods to solve global optimization problems. Knowl. Based Syst. **223**, 107044 (2021). https://doi.org/10.1016/j.knosys.2021.107044

21. Shami, T.M., El-Saleh, A.A., Alswaitti, M., Al-Tashi, Q., Summakieh, M.A., Mirjalili, S.: Particle swarm optimization: A comprehensive survey. IEEE Access **10**, 10031–10061 (2022). https://doi.org/10.1109/ACCESS.2022.3142859

22. Sharma, M., Komninos, A., López-Ibáñez, M., Kazakov, D.: Deep reinforcement learning based parameter control in differential evolution. In: Proceedings of the Genetic and Evolutionary Computation Conference (GECCO '19), pp. 709–717. Association for Computing Machinery, New York (2019). https://doi.org/10.1145/3321707.3321813

23. Shunmugapriya, P., Kanmani, S.: A hybrid algorithm using ant and bee colony optimization for feature selection and classification (ac-abc hybrid). Swarm Evolution. Comput. **36**, 27–36 (2017). https://doi.org/10.1016/j.swevo.2017.04.002

24. Wu, D., Wang, G.G.: Employing reinforcement learning to enhance particle swarm optimization methods. Eng. Optimiz. **54**(2), 329–348 (2022). https://doi.org/10.1080/0305215X.2020.1867120

Swarm Multi-agent Trapping Multi-target Control with Obstacle Avoidance

Chenyang Li, Guanjie Jiang, Yonghui Yang, and XueBo Chen[✉]

School of Electronic and Information Engineering, University of Science and Technology
Liaoning, Anshan 114051, China
xuebochen@126.com

Abstract. In this paper, we investigate that swarm multi-agent can trap multi-target and avoid obstacles simultaneously through cooperative control. First, the control method proposed in this paper allows the number of multi-agent to trap each target evenly, without all or more than half of the number of agents trapping one of the targets. Second, a uniform number of agents track the target based on information about the target and local interactions with other agents. Introducing a repulsive potential function between the agent and the target can enclose the target. In addition, the control method designed in this paper can trap the target faster. Finally, agents trapping the same target converge their velocity to achieve the capture of the target after forming an enclosing state. In achieving this process, agents can simultaneously avoid obstacles well. The simulation results show the feasibility.

Keywords: Swarm · multi-agent · trapping · multi-target · avoid obstacles

1 Introduction

Swarm intelligence is currently one of the most popular fields, especially in the study of multi-agent. The main idea comes from the swarming behavior of animals in nature, such as flocks of sheep, schools of fish, and bees, which maintain group cohesion at all times for information sharing and processing and collective environmental perception [1]. Similarly, some animals, such as wolves and orangutans, hunt or fight against targets more vigorously than they utilize groups to survive [2]. It inspires the trapping control of swarm multi-agent. The advantage of swarm multi-agent is that it can turn a single complex system into multiple agents with simple and flexible features, which achieve distributed control and collaborative control through self-organization and local interaction. Swarm multi-agent trapping control has essential real-world applications, such as searching for survivors after earthquakes [3], capturing hostile targets [4], and fieldwork and resource exploration [5].

In 1987, Reynolds [6] proposed the Boid model based on bird swarming behavior, which demonstrates three basic rules for achieving swarming: (1) Global position constraint (2) Consistent velocity (3) Collision avoidance. In 1995, Vicesk [7] proposed the Vicesk model, which mainly shows the emergence of swarms and likewise considers the velocity alignment. Jadbabaie *et al.* [8] gave detailed proof and provided a theoretical justification for the Vicesk model. In 2006 the model proposed by Olfati-Saber [9] combined Reynolds' three basic rules and potential functions to study the swarming behavior of agents in the presence of obstacles. Since then, the study of swarming with multi-agent has received much attention from scholars.

For swarm multi-agent trapping control, there are mainly task assignment methods, potential field methods, and reinforcement learning methods. For the task assignment method, Jiang [10] *et al.* used a fairness-based task assignment theory of equitable distribution to achieve the trapping of a single target. Stouch [11] *et al.* propose a task assignment method for aerial flight targets and potential competing objects that can update system requirements while ensuring optimal form. Lope [12] *et al.* proposed a heterogeneous robot task assignment algorithm combining threshold corresponding and self-learning. For the potential field method, Barnes [13] *et al.* combined the potential field function and the restriction function to control the swarm's formation and overall movement direction and to trap the target by changing different siege shapes. Rezaee [14] proposed polygonal roundup control for mobile robots by exploiting the repulsive forces between charges and used potential fields for obstacle avoidance control. For the reinforcement learning method, the Q-learning method [15] and the DQN algorithm [16] are mainly used for optimization to achieve efficient trapping of targets. Most studies on trapping targets only consider trapping a single target, which is inconsistent with practical applications. The study of multi-target trapping is critical and of practical use. Luo [17] *et al.* studied the multi-target problem and proposed a target selection based on distance. However, there would be a deadlock problem, i.e., how will an individual select a target when it is the same distance from multiple targets? Zhang [18] *et al.* implemented trapping multiple targets with undefined shapes in advance, but the number of robots assigned to each target was not uniform. Obstacle avoidance is a necessary consideration during the multi-agent trapping of multiple targets and currently includes the use of different algorithms to solve this problem [19–21]. It achieves obstacle avoidance through the "attractive force" from each other and the "repulsive force" of the obstacle [22].

Unlike the previous trapping control, the trapping control method in this paper can solve the deadlock problem and the problem of uneven distribution when the swarm multi-agent trapping multiple targets. Furthermore, it can also achieve tracking and enclosing dynamic targets effectively by target information and repulsive potential function, and finally, capture the target by converging the velocity of agents. In the process of swarm multi-agent trapping multi-target, obstacle avoidance is also a problem that must be considered. This paper uses the artificial potential field method to solve the obstacle avoidance problem.

The structure of this paper is as follows. Section 2 introduces the basics of graph theory and problem description. Section 3 gives the details of trapping control with obstacle avoidance in this paper. Section 4 gives the simulation results and discussion, and Sect. 5 is the conclusion of this paper.

2 Modeling

Consider the swarm multi-agent trapping multi-target control system has n agents and m targets when they move freely in two-dimensional space and ignore their dimension and mass. Then the second-order continuous motion equation of agent i and the second-order continuous motion equation of the kth target t_k are shown in Eq. (1).

$$\begin{cases} \dot{p}_i = v_i \\ \dot{v}_i = u_i \end{cases}, \; i = \{1, 2, \cdots, n\} \quad \begin{cases} \dot{p}_{t_k} = v_{t_k} \\ \dot{v}_{t_k} = u_{t_k} \end{cases}, \; k = (1, 2 \cdots, m) \tag{1}$$

where p_i, p_{t_k}, represent the position of agent i and the target t_k, respectively. v_i and v_{t_k}, represent the velocity of agent i and the target t_k, respectively. u_i and u_{t_k}, represent the control input of agent i and the target t_k, respectively. Each agent has its safe distance d and finite perception radius r, usually $r > d > 0$, so local interaction and information sharing can be carried out between agents. The interaction forces between agents are all determined by the potential function. When the distance between two agents is less than d, they are repulsive. When the distance is greater than d and less than r, they are attractive, and when the distance is greater than r, there is no force between them (as shown in Fig. 1).

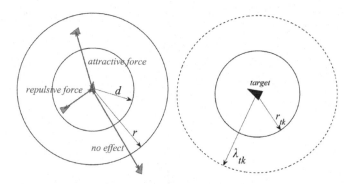

Fig. 1. Schematic diagram of the adjacent agent of agent i and target.

Other agents within the perception radius of agent i are called neighbors of agent i and can interact with each other for information, and the neighbor agents change with time. Then the set of neighbors of agent i is shown in Eq. (2).

$$N_i(t) = \{ j \in n, i \neq j | \; \|p_j - p_i\| < r \} \tag{2}$$

where $\| \; \|$ is the Euclidean norm. r_{t_k} in Fig. 1 is the radius of the circle in which agents enclose the target, and λ_{t_k} is the field of view of the target, which is a virtual range that changes with the number of agents around the target t_k. Its details will be mentioned below. Then the set of neighbors of target t_k is shown in Eq. (3).

$$N_{t_k}(t) = \{ k \in m, i \neq t | \; \|p_{t_k} - p_i\| < r_k \} \tag{3}$$

then, $L^t = D^t - A^t$, $A^t = [b_{it}] \in R^{n \times m}$ is the adjacency matrix between agents and the target, A^t is a symmetric matrix.

3 Details of Trapping Control with Obstacle Avoidance

Agents use simple rules and local interaction information to implement a complex swarm control. These rules can be represented as input vectors for each agent that control input u_i. Therefore, the control input of this paper is $u_i = u_i^\alpha + u_i^\beta$, where u_i^α is the swarm multi-agent trapping multi-target term and u_i^β is the obstacle avoidance term. The details are shown as follows.

$$u_i^\alpha = \underbrace{\sum_{i \in N_i(t)} f_\alpha(\|p_j - p_i\|_\sigma)e_{ij} + \sum_{i \in N_{t_k}(t)} f_\beta(\|p_t - p_i\|_\sigma)e_{it}}_{gradient\ term} + \underbrace{\sum_{i \in N_i(t)} a_{ij}(v_j - v_i) + \sum_{i \in N_{t_k}(t)} b_{it}(v_{t_k} - v_i)}_{velocity\ consensus\ item} + \underbrace{f_i^\gamma}_{feedback\ item}$$

(4)

u_i^α mainly consists of a gradient term, a velocity consensus term, and a feedback term, where f_α is the potential function acting between agents and f_β is the repulsive potential function acting between agents and targets, which aims to limit the target's movement and enclose the target. The specific expression is shown in Eq. (5). When $d = 5, r = 7$, $r_{t_k} = 7$, the image is shown in Fig. 2.

$$f_\alpha(x) = \begin{cases} -2k_1k_2e^{-k_2(x-d)}(e^{-k_2(x-d)}-1) & x \leq d \\ k_3(\frac{1}{d^2} - \frac{1}{x^2}) & d < x \leq r \\ 0 & oherwise \end{cases} \quad f_\beta(x) = \begin{cases} k_4(\ln(x) + \frac{r_{t_k}}{x}) & x \leq r_{t_k} \\ 0 & otherwise \end{cases}$$

(5)

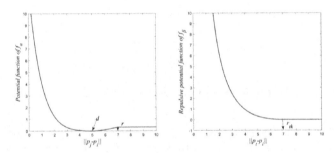

Fig. 2. Potential function of f_α and repulsive potential function of f_β.

where k_1, k_2 are attraction coefficients and k_3, k_4 are repulsion coefficients. The $\|x\|_\sigma$ in Eq. (4) is the σ norm, $\|x\|_\sigma = 1/\varepsilon * (\sqrt{1 + \varepsilon\|x\|^2} - 1)$, where $\varepsilon > 0$ is a constant. e_{ij} is the derivative of the corresponding σ norm, as shown in Eq. (6).

$$e_{ij} = \nabla\|p_j - p_i\|_\sigma = \frac{p_j - p_i}{\sqrt{1 + \varepsilon\|p_j - p_i\|^2}}, \quad e_{it} = \nabla\|p_t - p_i\|_\sigma = \frac{p_t - p_i}{\sqrt{1 + \varepsilon\|p_t - p_i\|^2}}$$

(6)

Remark 1: A potential function with smoothness allows the repulsive and attractive forces between agents to be quickly balanced to reach consensus. The attraction potential is proportional to its derivative attractive force $f_{attraction}(x) \propto d(f_{attraction}(x))/dt$. When the repulsive and attractive forces between agents are approximately equal $d(f_{repulsion}(x))/dt \approx d(f_{attraction}(x))/dt$ at safety distance d, agents will reach force balance quickly. If the rates of change of repulsion and attraction differ significantly at safety distance d, agents will repulse and attract repeatedly, delaying the formation of flocking.

$$\begin{cases} \text{when } \|p_j - p_i\| = d & \text{if } d(f_{repulsion}(x))/dt \approx d(f_{attraction}(x))/dt \quad \text{Force balance quickly} \\ \text{when } \|p_j - p_i\| = d \text{ if} \begin{cases} d(f_{repulsion}(x))/dt \gg d(f_{attraction}(x))/dt \\ \qquad\qquad or \qquad\qquad\qquad \text{Force balanced difficulty} \\ d(f_{repulsion}(x))/dt \ll d(f_{attraction}(x))/dt \end{cases} \end{cases}$$

This paper's main process of the control approach is (1) each target has a uniform number of agents that track the target based on its position and velocity information and can enclose the target according to the repulsive potential function f_β. (2) As the number of agents tracking the same target increases, the velocity of these agents starts to converge gradually to 0, to achieve the purpose of capturing the target. Therefore, the most critical factor of this control method is one is the distance between agents and the target, and the other is the number of agents tracking each target, so the feedback term f_i^γ is written as:

(1) If $\|p_{t_k} - p_i\| > \lambda_{t_k}$, agents select their target based on the shortest distance principle and start tracking and enclosing it. Then the details of the feedback term are shown below.

$$f_i^\gamma = -c_1(p_{t_k \min} - p_i) - c_2(v_{t_k \min} - v_i) \tag{7}$$

where $\lambda_{t_k} = \omega e^{-n_{t_k}/\omega}$ is the field of view of the target and is a monotonically decreasing function as n_{t_k} increases. $n_{t_k} > 0$ is the number of agents tracking target t_k. $p_{t_k \min}$ and $v_{t_k \min}$ are the position and velocity information of the closest target to agent i, respectively. ω, c_1 and c_2 are constants.

(2) If $\|p_{t_k} - p_i\| \leq \lambda_{t_k}$, agents tracking the same target start to converge velocity to capture the target, then the feedback item details are shown below.

$$f_i^\gamma = [-c_1(p_{t_k} - p_i) - c_2(\frac{v_{t_k}}{\tau} - v_i)]/n_{t_k} \tag{8}$$

where $\tau = \mu \tan(n_{t_k}/q)$, which is a monotonically increasing function with n_{t_k}. When the number of agents tracking the same target is increasing, τ will tend to infinity, then v_{t_k}/τ tends to 0, then the velocity of agents will tend to 0 to capture the target.

Adopting this feedback term allows an equal number of agents to track each target. It can be understood that each target has its initial equal field of view, and when the number of agents tracking target 1 is too much, then the field of view λ_{t_1} of target 1 will reduce as the number of tracking agents increases. Then the agent outside the field

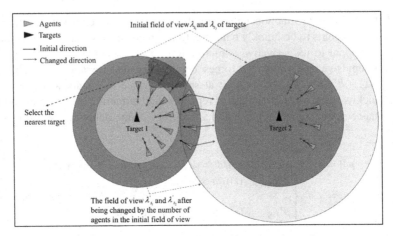

Fig. 3. Multi-agent trapping multi-target schematic

of view of target 1 will look for other targets to track. Conversely, when the number of agents tracking target 1 is few, then the field of view λ_{t_1} of target 1 will expand as the number of tracking agents decreases. A schematic diagram is shown in Fig. 3.

u_i^β is the obstacle avoidance term, and circular obstacles are considered in this paper, as shown in Fig. 4.

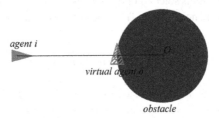

Fig. 4. Schematic diagram of the obstacle.

Agent δ is a virtual agent generated by the vertical projection connecting agent i to the center O_l of the obstacle on the obstacle's surface. We assign a repulsive potential to agent δ. When agent i approaches agent δ, it is subject to the repulsive potential, and agent i moves away from agent δ to avoid the obstacles. If there are multiple obstacles, agent i generates virtual agents δ on the surface of each obstacle. So that the swarm multi-agent can perfectly avoid obstacles while trapping multiple targets, we define the virtual agent δ to have the following properties.

(1) Create virtual agent δ on the edge of the circular obstacle surface, where the position needs to satisfy the following:

$$p_l = \arg\min_{p \in O_l}(\|p - p_i\|) \tag{9}$$

(2) Define the neighbor set of agent i and virtual agent δ as follows:

$$N_l(t) = \{l \in V : \|p_l - p_i\| < r'\} \tag{10}$$

where $l = 1, 2, \cdots, c$, r' is the interaction range of agent δ, and $r' < r$. c is the number of obstacles. $A^\beta = [c_{il}]$ is the adjacency matrix between agents, A^β is a symmetric matrix.

(3) Define the potential function F_δ of agent δ. The expression F_δ is shown as follows:

$$F_\delta(x) = \begin{cases} k_5(x(r'+1) - 1/2x^2 - r'\log(x)) & 0 < x < r' \\ 0 & otherwise \end{cases} \tag{11}$$

Therefore, the obstacle avoidance term is shown in Eq. (12).

$$u_i^\beta = \sum_{i \in N_l(t)} f_\delta(\|p_l - p_i\|_\sigma)e_{il} + \sum_{i \in N_l(t)} c_{il}(v_l - v_i) \tag{12}$$

4 Simulation Results and Discussion

4.1 Simulation Results

In this section, the effectiveness of the control method in this paper is verified by simulation. All parameters are set as shown in Table 1. To ensure that the perception radius of the agent varies together with the safety distance, we keep the two as follows $r = 1.4 \times d$.

Table 1. Parameter Setting

Parameters	Notation and Value
Safety distance and perception radius of agent	$d = 5, r = 7$
Enclosing radius of the target	$r_{t_k} = 7$
Parameters of the potential function f_α	$k_1 = 0.1, k_2 = 0.5, k_3 = 10$
Parameters of the potential function f_β	$k_4 = 10$
Parameters of the potential function f_δ	$k_5 = 5$
Target field of view parameter	$\omega = 25$
Limiting velocity variable parameter	$\mu = 20, q = 14$
Number of agents	$n = 70$
Number of targets	$m = 3$
Number of obstacles	$c = 4$
Obstacle coordinates and radius	$[15, 0;\ 25, 25;\ -10, -15;\ -30, 30]radius = [4;\ 5;\ 3;\ 6]$
Interaction range of virtual agent δ	$r' = 3$

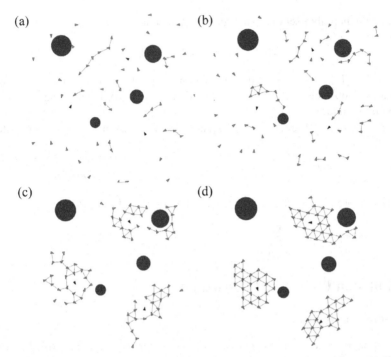

Fig. 5. Process of swarm multi-agent trapping multi-target with obstacle avoidance (a) Step $= 0$ (b) Step $= 100$ (c) Step $= 300$ (d) Step $= 500$.

The green triangles in Fig. 5 are agents, the black triangles are targets, and the blue circles are obstacles. The initial distribution of all agents obeys a Gaussian distribution with variance $[50, 50]^T$, and the velocity is chosen randomly between $[-1, 1]^T$. The initial positions of the three targets are $(-25, 6)$, $(20, -22)$, and $(10, 22)$, and the velocities are also chosen randomly between $[-1, 1]^T$. In Fig. 5(a) to (c), all agents track their targets and enclose them according to this paper's control method. In Fig. 5(d), a uniform number of agents reduce their velocity to capture the target. In this process, all agents can avoid the obstacles well. The position and velocity components of all agents are shown in Fig. 6.

The black curves in Fig. 6 are targets, and the other color curves are agents. Figure 6(a)(b) shows that agents have finished tracking and enclosing the target at 300 steps, and Fig. 6(c)(d) shows that the velocity of agents shrinks significantly at 300 steps. After 350 steps, the swarm multi-agent has finished capturing the target and tends to a stable state. Figure 6 also shows the movement paths of all agents, the black dots are the initial positions of agents and targets, and the blue dots are the end positions of agents and targets. The black curve in Fig. 6(e) shows the moving path of targets.

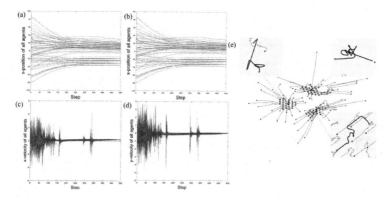

Fig. 6. Position and velocity components of the x and y axes and trajectories of all agents and targets during the multi-agent trapping multiple targets.

4.2 Discussion

We use the control method of [18] to conduct experiments to further verify the feasibility of the control method in this paper, as shown in Fig. 7. All parameters and the operating environment are kept constant, and the parameters are also shown in Table 1.

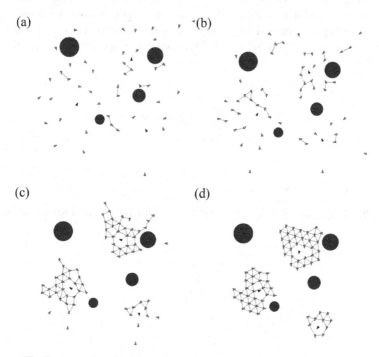

Fig. 7. Control method for multi-agent trapping multiple targets in [18]

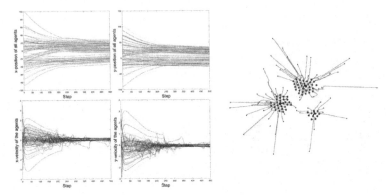

Fig. 8. Position and velocity components of the x and y axes and trajectories of all agents and targets during the multi-agent trapping multiple targets in [18].

The results in Fig. 5 and 8 show that the control method used in [18] causes the problem of uneven distribution of the number of multi-agents trapping each target. One of the targets might escape due to fewer multi-agents to trap, resulting in a failure to trap the target. The velocity profile in Fig. 8 also shows that the control method using the potential function in [18] converges more slowly after trapping a target than the control method using the potential function in this paper. To further compare the smooth potential function used in this paper with a faster convergence velocity, we used multi-agents trapping experiment on a single target without obstacles. The control method of combining potential functions in [18, 23, 24] is adopted, respectively, as shown below.

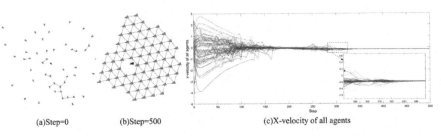

(a)Step=0   (b)Step=500   (c)X-velocity of all agents

Fig. 9. Multi-agent trapping target control method using potential functions in [23]

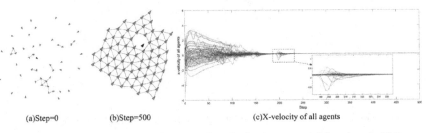

(a)Step=0   (b)Step=500   (c)X-velocity of all agents

Fig. 10. Multi-agent trapping target control method using potential functions in [24]

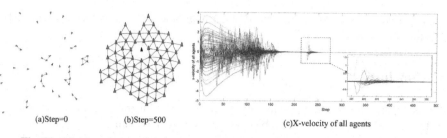

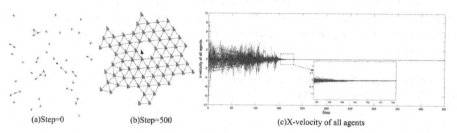

Fig. 11. Multi-agent trapping target control method using potential functions in [18]

Fig. 12. Multi-agent trapping target control method using potential functions in this paper

In the results in Fig. 9, 10, 11 and 12, the potential function designed in this paper converges faster when agents are trapped for one target. Then, agents converge equally fast in the case of trapping multiple targets, i.e., the time spent in the process of trapping multiple targets is short. We traversed the time taken to trap a target in the single target case from 10, 20, 30 to 100 agents and obtained Fig. 13. The time taken is the step size when the multi-agent velocity converges to zero (i.e., when the target is successfully captured) multiplied by the time taken for each step Δt.

Figure 13 shows the time of agents trapping targets for different control methods, and the horizontal coordinate is the number of agents. By comparison, we can see that as the number of agents increases, the control method combined potential function in

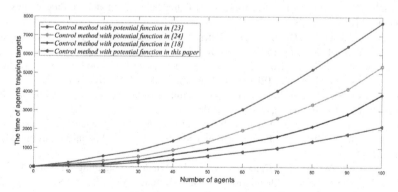

Fig. 13. The time of agents trapping targets for different control methods.

this paper can trap the target faster and better than other control methods [18, 23, 24], thus improving efficiency.

5 Conclusion

In this paper, we study the problem of swarm multi-agent trapping multi-target with obstacle avoidance, solving the problem that each target has a uniform number of agents to trap, which makes more practical sense. These uniform number of agents track the target based on its position and velocity information, enclose the target and restrict its movement through the repulsive potential function. Subsequently, these agents converge their velocities to capture the target. All agents can avoid obstacles in the process and spend less time trapping targets. Furthermore, simulation results verify the effectiveness of the method.

Acknowledgement. This research reported herein was supported by the NSFC of China under Grant No. 71571091 and 71771112.

References

1. Gómez-Nava, L., Bon, R., Peruani, F.: Intermittent collective motion in sheep results from alternating the role of leader and follower. Nat. Phys. **18**, 1494–1501 (2022)
2. Blum, C., Li, X.: Swarm intelligence in optimization. In: Blum, C., Merkle, D. (eds.) Swarm Intelligence. NCS, pp. 43–85. Springer, Heidelberg (2008). https://doi.org/10.1007/978-3-540-74089-6_2
3. Tian, Y., Liu, K.: Search and rescue under the forest canopy using multiple UAVs. Int. J Robot. Res. **39**, 1201–1221 (2020)
4. Wen, G., Duan, Z., Li, Z.: Flocking of multi-agent dynamical systems with intermittent nonlinear velocity measurements. Int. J. Robust Nonlinear Control **22**(16), 1790–1805 (2012)
5. Şahin, E.: Swarm robotics: from sources of inspiration to domains of application. In: Şahin, E., Spears, W.M. (eds.) SR 2004. LNCS, vol. 3342, pp. 10–20. Springer, Heidelberg (2005). https://doi.org/10.1007/978-3-540-30552-1_2
6. Reynolds, C.: Flocks, herds and schools: a distributed behavior model. In: Proceedings of the SIGGRAPH 1987, p. 25 (1987)
7. Vicsek, T., Cziroók, A., Ben-Jacob, E., Cohen, I., Shochet, O.: Novel type of phase transition in a system of self-driven particles. Phys. Rev. Lett. **75**(6), 1226–1229 (1995)
8. Jadbabaie, A., Jie, L., Morse, A.S.: Coordination of groups of mobile autonomous agents using nearest neighbor rules. IEEE Trans. Autom. Control **48**(6), 988–1001 (2003)
9. Olfati-Saber, R.: Flocking for multi-agent dynamic systems: algorithms and theory. IEEE Trans. Autom. Control **51**(3), 401–420 (2006)
10. Jiang, Z., Hu, W., Chen, L.: A specified-time multi-agent hunting scheme with fairness consideration. In: 2019 IEEE 28th International Symposium on Industrial Electronics (ISIE), pp. 1775–1780 (2009)
11. Stouch, D.W., Zeidman, E., Callahan, W., et al.: Dynamic replanning on demand of UAS constellations performing ISR missions. In: SPIE Defense, Security, and Sensing, pp. 238–243. International Society for Optics and Photonics (2011)

12. Lope, J.D., Maravall, D., Quiñonez, Y.: Response threshold models and stochastic learning automata for self-coordination of heterogeneous multi-task distribution in multi-robot systems. Robot. Auton. Systems **61**(7), 714–720 (2013)
13. Barnes, L.E., Fields, M.A., Valavanis, K.P.: Swarm formation control utilizing elliptical surfaces and limiting functions. IEEE Trans. Syst. Man Cybern. **39**(6), 1434–1445 (2009)
14. Rezaee, H., Abdollahi, F.: A decentralized cooperative control scheme with obstacle avoidance for a team of mobile robots. IEEE Trans. Ind. Electron. **61**(1), 347–354 (2014)
15. Xu, G., Zhao, Y., Liu, H.: Pursuit and evasion game between UVAs based on multi-agent reinforcement learning. In: 2019 Chinese Automation Congress (CAC), pp. 1261–1266. IEEE(2019)
16. Mnih, V., Kavukcuoglu, K., Silver, D.: Human-level control through deep reinforcement learning. Nature **518**(7540), 529 (2015)
17. Luo, X.Y., Li, S.B., Guan, X.P.: Flocking algorithm with multi-target tracking for multi-agent systems. Pattern Recogn. Lett. **31**(9), 800–805 (2010)
18. Zhang, S., Liu, M., Lei, X.: Multi-target trapping with swarm robots based on pattern formation. Robot. Auton. Syst. **106**, 1–13 (2018)
19. Karaman, S., Frazzoli, E.: Sampling-Based Algorithms for Optimal Motion Planning. Sage Publications, Inc. (2011)
20. Sathyaraj, B.M., Jain, L.C.: Multiple UAVs path planning algorithms: a comparative study. Fuzzy Optim. Decis. Mak. **7**(3), 257–267 (2008)
21. Koenig, S., Likhachev, M.: Fast replanning for navigation in unknown terrain. IEEE Trans. Robot. **21**(3), 354–363 (2005)
22. Zhu, B., Xie, L., Han, D., Meng, X., Teo, R.: A survey on recent progress in control of swarm systems. Sci. China Inf. Sci. **60**(7), 1–24 (2017). https://doi.org/10.1007/s11432-016-9088-2
23. Pei, H., Chen, S., Lai, Q.: A local flocking algorithm of multi-agent dynamic systems. Int. J. Control **88**(11), 1–8 (2015)
24. Shen, Y., Kong, Z., Ding, L.: Flocking of multi-agent system with nonlinear dynamics via distributed event-triggered control. Appl. Sci. **9**(7), 1336 (2019)

A Novel Data Association Method
for Multi-target Tracking Based on IACA

Yi Di, Guoyuan Zhou, Ziyi Tan, Ruiheng Li, and Zheng Wang[✉]

School of Information Engineering, Hubei University of Economics, Wuhan 430205, China
wz_hust_2012@126.com

Abstract. Aiming at the data association problem of multi-target tracking, a data association method for multi-target tracking is given based on an improved Ant Colony algorithm (IACA) in this paper. Firstly, according to the characteristics of multi-target data association problem, it turns the issue to combinatorial optimization problems. Secondly, to address the shortcomings of the ant colony algorithm, which tends to fall into local optimum, the elite strategies and sorting strategies is introduced into the traditional Ant Colony algorithm. In addition, for the problem of ant colony algorithm combined parameters selection, a model of optimum combined parameters selection based on game theory is introduced into the proposed IACA algorithm. Finally, the improved ant colony algorithm is applied to multi-target tracking data association problem. Simulation results show that IACA has better performance in solving multi-target data association problem than ACA and the ant colony system incorporating the game parameter estimation model is more practical in multi-target tracking.

Keywords: Multi-target tracking · Data association · Improved Ant Colony algorithm

1 Introduction

Data association is one of the key problems in the multi-sensor multi-objective tracking problem [1–3], as well as the most challenging step in a multi-objective tracking system. With the requirement of real-time and accuracy issues of target tracking, data association algorithms have been further explored by domestic and international researchers. The earliest proposed data association algorithm is the nearest neighbor (NN) analysis [4–6], in which a hard judgement method is used to pair the input data with an optimal track. This algorithm is simple and easy to implement. Nevertheless, when there are multiple closely spaced data points, the performance of the NN algorithm decreases dramatically, causing more false correlations. Then the Multiple Hypothesis Tracking (MHT) [7–9] was proposed, which considers each candidate measurement as a hypothesis and correlates it with each existing track until the hypothesis with the highest probability of association is obtained. Although the MHT method has low error rate, while the number of tracking objects become huge, the computational complexity will increase dramatically, therefore it is not effective for real-time tracking. Joint Probabilistic Data

© The Author(s), under exclusive license to Springer Nature Switzerland AG 2023
Y. Tan et al. (Eds.): ICSI 2023, LNCS 13969, pp. 62–73, 2023.
https://doi.org/10.1007/978-3-031-36625-3_6

Association (JPDA) [10, 11] uses each candidate observation set with individual weight for the trails update, all the individual weights represent the correct association probabilities of each candidate match. Joint Probabilistic Data Association is a Bayesian method that does not require any prior probabilities and is more mature in the field of multi-objective tracking, but also, the computational complexity will increase dramatically as the number of targets increases [11]. All the above traditional data association methods have some drawbacks, such as low tracking accuracy or long execution time.

In recent years, with the development of many bionic algorithms, genetic algorithms and neural networks have achieved better results in the field of multi-objective tracking [12–14].

Inspired by the collective behavior of ant colonies in nature, Italian scholar M. Dorigo first systematically proposed a new bionic algorithm based on ant populations, the Ant Colony Optimization (ACO) and solved a series of ACO solves a series of combinatorial optimization problems [15], such as the TSP problem, the quadratic allocation problem, and the scheduling problem. Since its success in the above problems, the ACO has penetrated target tracking problem areas and practical engineering applications [16, 17].

In this paper, a novel data association algorithm is introduced to address the drawbacks of the traditional data association algorithms which has insufficient tracking accuracy and long execution time. Firstly, the multi-objective tracking data association problem is re-modeled and transformed into a combinatorial optimization problem, then the elite strategy and optimal ranking strategy are introduced in the ant colony algorithm. In addition, for the problem of ant colony algorithm combined parameters selection, a model of optimum combined parameters selection based on game theory [18], is introduced into the proposed IACA algorithm. Finally, the improved ant colony algorithm is applied to the data association problem, which then form a multi-objective data association method based on the improved ant colony algorithm (IACDA).

2 Combinatorial Optimization Model of Data Association

Assume that the target number is M, $i = 0, 1, 2, ..., M$ where i is the target index. Firstly, set the target state vector to be estimated at time t as $X_t = \{x_{1,t}, x_{2,t}, ... , x_{i,t}, ... , x_{M,t}\}$, the observation vector is $Y_t = \{y_{1,t}, y_{2,t}, ... , y_{j,t}, ... , y_{mt,t}\}$, where m_t is the number of observations obtained at time t, $j = 1, 2,... , m_t$ is the observation number. The association vector $K_t = \{k_t^1, k_t^2, ... , k_t^j, ... , k_t^{mt}\}$ is introduced to describe the association between the observation and the track (target) at moment t, where the component $k_t^j \in [0,M]$ is an integer random variable, and when $k_t^j = i$, it means that the observation $y_{j,t}$ is associated with target i, when $i = 0$, then $k_t^j = 0$, it means that the observation is from clutter.

The objective function for data association can be expressed as:

$$z = \min \sum_{i=1}^{M} \sum_{j=1}^{m_t} u_{i,j} a_{i,j} \tag{1}$$

Binding conditions are defined as follows:

$$\sum_{i=1}^{M} a_{i,j} = 1, i = 1, ..., M \tag{2}$$

$$\sum_{j=1}^{m_t} a_{i,j} = 1, j = 1, ..., m_t \tag{3}$$

In Eq. (1)–(3), $a_{i,j}$ is a binary variable, from which the association matrix is formed. $a_{i,j}$ is 1 when the observation $y_{j,t}$ is associated with target i, otherwise $a_{i,j}$ is 0. Equation (2) and (3) are the binding conditions of data association, which means that at most one value of 1 in each row or column of the association matrix. This is consistent with the practical situation that each observation can be associated with at most one target. The cost value $u_{i,j}$, which means the association cost between the observation $y_{j,t}$ and target i, is defined as follows [5]:

$$u_{i,j} = -\log \left| \frac{P_D p(y_{j,t}|X_t, k_t^j = i)}{p(y_{j,t}|k_t^j = i)} \right| = -\log\{P_D p V \times p(y_{j,t}|X_t, k_t^j = i)\} \tag{4}$$

In Eq. (4), P_D is the sensor detection probability, V is the observation space, $k_t^j = i$ means that the observation $y_{j,t}$ is associated with target i, $p(y_{j,t}|X_t, k_{tj} = i)$ means the probability that the observation y is associated with target i at time t, i is only the real target, $p(y_{j,t}|k_{tj} = i)$ means the probability that the observation y is associated with target i at time t, when i is the real target or clutter, let the clutter obey uniform distribution in the observation space.

And in data association, the best association model is represented as the following equation:

$$z_{best} = \arg \max_{K_t} \{p(K_t|X_t, Y_t)\} \tag{5}$$

In Eq. (5), z_{best} represents the best association for the case where the target state vector Xt and the target observation vector Yt are known.

3   Improved Ant Colony Algorithm and Its Application in Data Association

3.1   Research on the Traditional Ant Colony Algorithm

Entomologists have discovered that ants can find the shortest path from their nest to a food source without any visible cues. While ants are searching for a food source, they can release the pheromone along the path they take, allowing other ants within a certain range to detect it and thus influence their subsequent path selection behavior. When more ants pass on a path, the intensity of the pheromone grows and the probability of the path being selected rises, then the intensity of the pheromone will increase even more, it is a positive feedback mechanism.

In the traditional ant system model, the total number of ants is m and the number of nodes is n. When artificial ants construct a feasible path, it is necessary to order these n nodes during path selection from 1 to n. Then probability selection formula is applied to determine the selection probability of each node, where nodes with transfer values are selected based on certain rules. The ant k in node i can calculate the probability of visiting node j according to Eq. (7):

$$p_{ij}^k(t) = \begin{cases} \dfrac{\tau_{ij}^\alpha \cdot \eta_{ij}^\beta}{\sum_{r \in allowed_k} \tau_{jr}^\alpha \cdot \eta_{rj}^\beta}, & j \in allowed_k \\ 0, & j \in allowed_k \end{cases} \tag{6}$$

where τ_{ij} is the concentration of pheromones on path (i, j) which consists of nodes i, j. α and β indicate the importance of pheromone concentration and the importance of heuristic information, respectively. $allowed_k$ is the node set available to ants currently. The selected nodes are stored in corresponding taboo table $tabu_k$. The records in table $tabu_k$ change as the ants select different paths.

After the ant colony completes the construction of the path, the pheromones are accumulated but not to an excess under which heuristic information is ignored. Here, ρ $(0 < \rho \le 1)$ represents the volatilization rate of pheromones and is updated as follows:

$$\tau_{ij}(t + T) = (1 - \rho)\tau_{ij}(t) + \sum_{k=1}^{m} \Delta \tau_{ij}^k(t, t + T) \tag{7}$$

where $\Delta \tau_{ij}^k$ represents the released pheromone of the ant k on the path (i, j). In this paper, we use the Ant-Cycle model for path selection:

$$\Delta \tau_{ij}^k = \begin{cases} \dfrac{Q}{L_k}, & (i, j) \in T \\ 0, & (i, j) \notin T \end{cases} \tag{8}$$

where Q is a constant, T is a path constructed by ant k, L_k is the path length. Ant-Cycle model updates pheromones for the global path, making it highly efficient and effective.

3.2 Our Improvement Strategies in Ant Colony Algorithm

The setting of important parameters in the traditional ant colony algorithm still lacks sufficient theoretical basis, and the convergence and practicality of the ant colony algorithm still need to be discussed. Therefore, we introduce three improvement strategies in the traditional ant colony algorithm, such as elitist strategy, optimal ranking strategy, and game strategy.

A. elitist strategy.

Ant system with elitist strategy (AS-elite) is the oldest improved ant system, which employs a similar elite strategy in genetic algorithms. In the AS-elite system, to make the optimal solution searched in this iteration more attractive to ants in the next iteration, a pheromone is added to the optimal solution after each iteration. This optimal solution is called a global optimal solution for one iteration, and the ants which searched this solution are defined as elitist ants (Elitist Ants). Therefore, the update rule for pheromone is re-established as:

$$\tau_{ij}(t+1) = (1-\rho)\tau_{ij}(t) + \Delta\tau_{ij} + \Delta\tau_{ij}^* \tag{9}$$

where $\Delta\tau_{ij} = \sum_{k=1}^{m}\Delta\tau_{ij}^k$ is the update of the amount of pheromone on the path (i, j) passed by the σ-1 excellent ants on that path based on the ranking order of the ant path length, ρ is the pheromone volatility factor, then $(1-\rho)$ denotes the information residual factor, and to prevent unrestricted pheromone increase, let $\ominus$.

$$\Delta\tau_{ij}^k = \begin{cases} \frac{Q}{L_k} & \text{If the kth ant pass the side (i,j)} \\ 0 & \text{Others} \end{cases} \tag{10}$$

$$\Delta\tau_{ij}^* = \begin{cases} \sigma\frac{Q}{L^*} & \text{If edge (i,j) is part of the optimal path} \\ 0 & \text{Others} \end{cases} \tag{11}$$

where $\Delta\tau_{ij}^*$ represents the pheromone increase caused by the elitist ants on path (i, j); σ represents the number of elitist ants; L^* is the length of the current optimal path.

The analysis shows that the ant system with the introduction of the elitist strategy can search for the optimal solution with fewer iterations and improve the solution results. However, the problem of how many elitist ants need to be set in an ant system needs to be discussed based on practical applications. A minimum number of elitist ants will decrease the performance of the algorithm. An excessive number will cause the search to rapidly concentrate around the local optimum, converge prematurely, and eventually the algorithm falls into a local optimum.

B. An elite ant system based on the idea of optimal ranking strategy.

In the ant system based on the elitist strategy, there is a shortcoming that an increase in total solution quality P leads indirectly to a narrower difference in the selection probability. This shortcoming makes the search occur to converge prematurely, thus affecting the search for the optimal solution, and this phenomenon is known as maintenance selection pressure in genetic algorithm. In genetic algorithms, this shortcoming is solved by using a selection mechanism such as ranking.

Extending this ranking idea to ant systems to develop a rank-based version of elitist ant system (EAS-rank). In the EAS-rank, After ants create paths, each ant is ranked according to its path length $(L_1 \leq L_2 \leq \cdots \leq L_m)$, and the ants are updated by pheromones on their respective paths and weighted according to the ranking μ of that ant in the sequence; In addition, only the top σ excellent ants are considered, which can effectively avoid the phenomenon of colony gathering caused by some local optimal over-attracting ants. The top $\sigma - 1$ ants in the ranking receive one additional enhancement of pheromone on the respective paths they travel. These pheromone updates are proportional to the ranking order of that ant, i.e., the higher the ranking ant, the more pheromone updates it has. Finally, when this iteration ends, all elite ants that searched for the optimal path undergo another increase in the pheromones on the path they passed through. Therefore, the ant system that introduces the idea of elite strategy and optimal ranking has its pheromone trace updated also according to Eq. (10), but $\Delta\tau_{ij} = \sum_{\mu=1}^{\sigma-1}\Delta\tau_{ij}^{\mu}$

denotes the update of the amount of pheromone trajectories by the σ-1 excellent ants on the side (i, j) according to their respective rankings, and,

$$\Delta \tau_{ij}^{\mu} = \begin{cases} (\sigma - \mu)\frac{Q}{L^{\mu}} & \text{If the } \mu\text{th ant associates the ith track with the jth observation} \\ 0 & \text{Others} \end{cases}$$

(12)

where $\mu \in [1,\sigma]$ is the ranking order of the excellent ants, L^{μ} is the length of the path searched by the μth best ant in this cycle, $\Delta \tau_{ij}^{\mu}$ is the pheromone left by the μth best ant on the associated pair (i, j), $\Delta \tau_{ij}^{*}$ is the pheromone left by the best ant that found the shortest path in the current colony on its path (i, j) after the path search process, L^{*} is the length of the optimal path found in the current colony path search process, σ denotes the number of elite ants in the colony participating in the path search, Q is a constant indicating the intensity of the pheromone on the current path.

3.3   IACDA Method

In our improved ant colony data association method. We assume that:

(1) A target produces at most one observation at one time.
(2) An observation can be associated with at most one target.
(3) Suppose there be three tracks in the observation space as A, B and C, and three other observations as a, b and c. If an ant performs a path search a possible path formed by (A, c) - (B, a) - (C, b). And the path length is defined as:

$$d = d_{Ac} + d_{Ba} + d_{Cb}$$

(13)

where d_{ij} is the distance between track i and observation j.

In addition, The characteristics of each ant in the colony are specified as follows:

(1) Ant k departs from each track and selects observations with a certain probability p.
(2) Ants leave volatile pheromones on each track-observation association as they choose their paths.
(3) Only the observation points that have not yet been selected in the observation set can be selected for constructing paths.
(4) Each trail can be associated with at most one observation.

Each ant selects a target-trail association pair by repeatedly applying the state transfer rule until a path search is completed. The path length is related to the degree of association of the observation-trail pair, which means the shorter the path length, the higher the degree of association of the observation-trail pair. In the data association problem, the ant selects the observation-track association pair based on the principle of minimizing the cost function and probabilistically assigns a reasonable observation to each target, while Eq. (4) shows that the cost value function is related to the likelihood estimate of the observation classification, so the selection of the appropriate association is actually based on the principle of maximizing the likelihood estimate of the observation classification.

In addition, the probability of the ant choosing a path and the amount of pheromone left on the observation-track association pair are both related to the likelihood function.

Thus, the optimal path searched by the ants is the set of observed divisions that maximizes the likelihood function.

The whole process of IACA-based data association starts with the entry of the target into the observation area, and the colony searches the shortest path from the first scan cycle until the target leaves. The specific steps are as follows:

(1) When the sensor acquires the current observation, k ants ($k = 1, \ldots, K$) are randomly placed on M trails and immediately start their respective trail-observation association pairs.

(2) Any ant r selects an observation with the transfer probability p, a table $tabu_k$ is set up to record the track-observation association pairs which are already selected by ant k and a temporary pool $temppool$ (k) is given to store the probability values of the track selected by ant k. The transfer probability p is expressed in Eq. (14)

$$p_{ij}^k(t) = \begin{cases} \frac{\alpha \cdot \tau_{ij} + \beta \cdot \eta_{ij}}{\sum_{r \notin tabu_k} \alpha \cdot \tau_{rj} + \beta \cdot \eta_{rj}}, & j \notin tabu_k \\ 0, & j \in tabu_k \end{cases} \tag{14}$$

Which α is the information heuristic factor, which indicates the relative importance of the pheromone τ_{ij}, and β is the expectation heuristic factor that indicates the importance of the visibility parameter η_{ij}, $\eta_{ij} = 1/d_{ij}$, where d_{ij} denotes the distance between a given association pair (i, j) in the path.

(3) When ants select paths, they leave volatile pheromones τ_{ij} on each association pair (i, j), and in order not to let the amount of pheromones on a particular path increase so dramatically that the algorithm falls into a local optimum solution, set the pheromone value $\tau_{ij} \in (\tau_{min}, \tau_{max})$, and the initial amount of pheromones is set to τ_{max}.

(4) Each ant will apply a local update rule to update the amount of pheromone on the identified observation-trail pair when it selects its own path. When all ants have completed their path selection, the colony will apply the global update rule to update the amount of pheromone on the path as shown in Eq. (9)–(11).

(5) Once the correlation matrix has been determined, all values of j corresponding to a_{ij} = 1 and $y_{j,t}$ are recorded to finally obtain the correct observation vector required.

4   Simulation Experiment and Analysis

The proposed algorithm is validated by Matlab simulation test of double target uniform motion under the linear Gaussian condition, and the experimental environment is: Dell i5-4210H CPU @ 2.3 GHz, Windows 10, Matlab 2015a.

The target state equation is:

$$x_{i,t} = A_{i,t-1} x_{i,t-1} + \Gamma w_t \tag{15}$$

where $x_{i,t-1}$ represents the state of i at moment $t-1$ and $A_{i,t-1}$ is the state transfer matrix, denoted as:

$$A_{i,t-1} = \begin{bmatrix} 1 & T & 0 & 0 \\ 0 & 1 & 0 & 0 \\ 0 & 0 & 1 & T \\ 0 & 0 & 0 & 1 \end{bmatrix} \tag{16}$$

Γ is the process noise matrix, denoted as:

$$\Gamma = \begin{bmatrix} T^2/T^2 & 0 & \\ 0 & 0 & \\ 0 & & T^2/T^2 \end{bmatrix} \tag{17}$$

The target observation equation is:

$$y_{i,t} = H_{i,t}x_{i,t} + v_t \tag{18}$$

where $x_{i,t} = (x_t, \dot{x}_t, y_t, \dot{y}_t)^T$ represents the state of target i at time t, x_t and y_t are the positions of the target on the x and y axes of the Cartesian coordinate system respectively, and $\dot{x}_t$, $\dot{y}_t$ are the current velocities of the target on the x and y axes respectively.

The measurement matrix is:

$$H_{i,t} = \begin{bmatrix} 1 & 0 & 0 & 0 \\ 0 & 0 & 1 & 0 \end{bmatrix} \tag{19}$$

In Eq. (15) and (18), w_t and v_t are system noise and observation noise respectively, and are both Gaussian white noise with zero mean, T is the sampling interval and is set to 1.The initial values of the two objectives are set as follows: $x_{1,0} = (1500, 300, 500, 400)$, $x_{2,0} = (500, 400, 1500, 300)$, initial variance $P_0 = (100^2, 100^2, 50^2, 50^2)$, sensor detection probability is $P_D = 0.9$, according to ant colony algorithm parameters optimization based on game theory strategy [18], set $\alpha = 3.2, \beta = 2.6, \rho = 0.2, K = 33$.

According to the principle that the number of ants involved in the problem search is slightly smaller than the number of problem size can get faster convergence. According to the principle of faster convergence speed, the number of ants $K = 30$ is set for the ant colony size. If the value of K is too large will cause the ant colony system to converge slower and increase the search time, if it is too small will not help the ant colony to converge to the optimal path, which will eventually cause the obtained path is not the optimal path. The two target motion tracks were sampled 50 times. Figure 1 shows the real track of the two targets set in the simulation experiment, Fig. 2 shows the observed and estimated trajectories. Figure 3 shows the mean squared difference between the measured and estimated values of the two targets. Figures 4 and 5 show the mean squared difference and maximum error of targets a and b respectively. From the analysis in Figs. 2, 3, 4 and 5, IACDA correctly achieves data association and high tracking accuracy for two targets in the case of crossed uniform motion.

In addition, the traditional Nearest Neighbor Data Association Algorithm (NNDA) was the first algorithms that uses a hard judgement method to match the input data to an optimal track. On the other hand, Joint Probabilistic Data Association (JPDA) is a Bayesian approach that uses each candidate observation set for the track update, except that the weights of each match are different. These weights are the correct association probabilities of each candidate match, and this approach is a more mature approach in the field of multi-objective tracking. The traditional ant colony data association algorithm (ACDA) converts the data association problem into a combinatorial optimization problem, then uses the traditional simple ant colony algorithm to search for the optimal solution for data association.

In this paper, we compare the Improved Ant Colony Algorithm data association method (IACDA) with the three typical data association algorithms (NNDA, JPDA) with respect to two indicators, such as the correct rate of association and the execution time. As shown in Table 1, the Improved Ant Colony Algorithm data association method has a more obvious improvement in accuracy than the traditional Ant Colony Algorithm data association method, but the execution time is slightly slower. Overall, the improved ant colony algorithm showed the expected timeliness in the data association problem.

Table 1. Performance comparison of data association algorithms

	NNDA	JPDA	ACDA $K = 33$	IACDA $K = 33$
Correct association rate	43.3%	91.6%	86.7%	92.1%
Execution time (s)	0.73	17.98	1.07	0.73

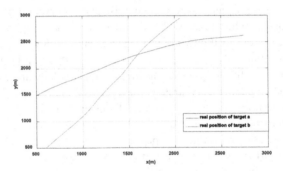

Fig. 1. The real track of target a and b

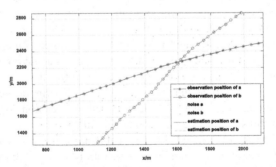

Fig. 2. Target tracking observation track and estimated trajectory

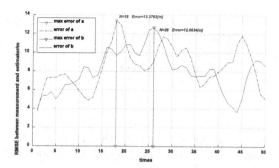

Fig. 3. Target tracking error and maximum error

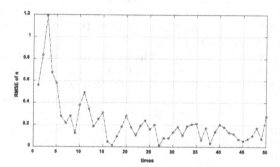

Fig. 4. Root mean square error of the target a

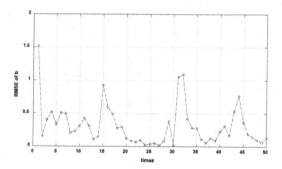

Fig. 5. Root mean square error of target b

5   Conclusion

For the multi-objective data association problem, a multi-objective data association method based on the improved ant colony algorithm is proposed. The method utilizes the ant colony algorithm with the elite strategy and optimal ranking strategy to achieve faster convergence speed and higher accuracy of global optimal solution in solving multi-target tracking data association problems. In addition, the improved ant colony algorithm uses the game strategy to set the key parameters more reasonably, which shows better performance than traditional ant colony algorithms.

In the future work, the IACDA method with more reasonable filter estimation algorithm will be using for maneuvering target tracking.

References

1. Chen, T., Yang, P., Peng, H., Qian, Z.: Multi-target tracking algorithm based on PHD filter against multi-range-false-target jamming. J. Syst. Eng. Electron. **31**(5), 859–870 (2020). https://doi.org/10.23919/JSEE.2020.000066
2. Wang, Z., Sun, J., Li, Q., Ding, G.: A new multiple hypothesis tracker integrated with detection processing. Sensors **19**, 5278 (2019). https://doi.org/10.3390/s19235278
3. Tian, Y.C., Dehghan, A., Shah, M.: On detection, data association and segmentation for multi-target tracking. IEEE Trans. Pattern Anal. Mach. Intell. **41**(9), 2146–2160 (2019). https://doi.org/10.1109/TPAMI.2018.2849374
4. Zhao, S., Huang, Y., Wang, K., Chen, T.: Multi-source data fusion method based on nearest neighbor plot and track data association. In: 2021 IEEE Sensors, pp. 1–4. IEEE (2021)
5. Mbelwa, J.T., Zhao, Q., Lu, Y., Liu, H., Wang, F., Mbise, M.: Objectness-based smoothing stochastic sampling and coherence approximate nearest neighbor for visual tracking. Vis. Comput. **35**(3), 371–384 (2018). https://doi.org/10.1007/s00371-018-1470-5
6. Singh, K., Karar, V., Poddar, S.: Radius nearest neighbour based feature classification for occlusion handling. Pattern Recogn. Image Anal. **30**, 416–427 (2020). https://doi.org/10.1134/S1054661820030268
7. Zheng, L., Wang, X.: Improved multiple hypothesis tracker for joint multiple target tracking and feature extraction. IEEE Trans. Aerosp. Electron. Syst. **55**, 3080–3089 (2019). https://doi.org/10.1109/TAES.2019.2897035
8. Sheng, H., Chen, J., Zhang, Y., et al.: Iterative multiple hypothesis tracking with tracklet-level association. IEEE Trans. Circuits Syst. Video Technol. **29**, 3660–3672 (2019). https://doi.org/10.1109/TCSVT.2018.2881123
9. Yamada, T., Gocho, M., Akama, K., et al.: Multiple hypothesis tracking with merged bounding box measurements considering occlusion. IEICE Trans. Inf. Syst. **E105.D** (2022). https://doi.org/10.1587/transinf.2021EDP7197
10. Memon, S.A., Kim, M., Shin, M., et al.: Extended smoothing joint data association for multi-target tracking in cluttered environments. IET Radar Sonar Navig. **14**, 564–571 (2020). https://doi.org/10.1049/iet-rsn.2019.0075
11. Li, Q., Song, L., Zhang, Y.: Multiple extended target tracking by truncated JPDA in a clutter environment. IET Signal Proc. **15**, 207–219 (2021). https://doi.org/10.1049/sil2.12024
12. Ding, C., Chen, L., Zhong, B.: Exploration of intelligent computing based on improved hybrid genetic algorithm. Clust. Comput. **22**(4), 9037–9045 (2018). https://doi.org/10.1007/s10586-018-2049-7
13. Fadel, I.A., Alsanabani, H., Oz, C., et al.: Hybrid fuzzy-genetic algorithm to automated discovery of prediction rules. J. Intell. Fuzzy Syst. **40**, 43–52 (2021). https://doi.org/10.3233/JIFS-182729
14. Liu, Z., et al.: Target detection and tracking algorithm based on improved mask RCNN and LMB. In: 2021 International Conference on Control, Automation and Information Sciences (ICCAIS), Xi'an, China, pp. 1037–1041 (2021). https://doi.org/10.1109/ICCAIS52680.2021.9624519
15. Dorigo, M., Maria, L.: Ant colony system: a cooperative learning approach to the traveling salesman problem. IEEE Trans. Evol. Comput. **1**(1), 53–66 (1997)

16. Kang, L., Xie, W.X., Huang, J.X.: ACA based data association method for multi-target tracking. Acta Electron. Sin. **36**(3), 586–589 (2008). (In Chinese)
17. Zhang, K.Q., Huang, J.X., Kang, L.: Data association method based on ant colony algorithm. J. Syst. Simul. **21**(22), 7127–7129 (2009). (In Chinese)
18. Gao, Q., Sun, S., Han, Q., et al.: Combination of ant colony algorithm parameters optimization based on game theory. Comput. Eng. Appl. **49**(21), 51–55 (2013). (In Chinese)

MACT: Multi-agent Collision Avoidance with Continuous Transition Reinforcement Learning via Mixup

Pu Feng[1], Xin Yu[1], Junkang Liang[1], Wenjun Wu[2(✉)], and Yongkai Tian[1]

[1] School of Computer Science and Engineering, Beihang University,
Beijing 100083, China
[2] Institute of Artificial Intelligence, Beihang University, Beijing 100083, China
wwj09315@buaa.edu.cn

Abstract. Autonomous collision avoidance is a critical component in various multi-robot applications. While deep reinforcement learning (RL) has demonstrated success in some robotic control tasks, it remains challenging to apply it to real-world multi-agent collision tasks due to poor sample efficiency. The limited amount of transition data available and its strong correlation with multi-agent task characteristics have restricted recent research in this area. In this paper, we propose Multi-agent Collision Avoidance with Continuous Transition Reinforcement Learning via Mixup (MACT) to address these challenges. MACT generates new continuous transitions for training by linearly interpolating consecutive transitions. To ensure the authenticity of constructed transitions, we develop a discriminator that automatically adjusts the mixup parameters. Our proposed approach is evaluated through simulations and real-world swarm robots consisting of E-pucks, demonstrating its practical application. Our learned policies outperform existing collision avoidance methods in terms of safety and efficiency.

Keywords: Multi-agent Collision Avoidance · Reinforcement Learning · Continuous Transition. · Data Augmentation

1 Introduction

With the increasing use of unmanned vehicles, mobile robots are becoming more common for a variety of tasks, including rescue operations, final-mile delivery, and clustering tasks. However, navigating swarm robots safely and efficiently in complex and ever-changing real-world environments presents significant coordination challenges. Unlike single robots, swarm robots must take into account the movements of other robots in the cluster. To address these challenges, researchers have extensively studied distributed control, which can be categorized into two main approaches. The first approach is the sensor-based approach, which includes methods such as Artificial Potential Field (APF) [7]

Supported by National Key R&D Program of China 2022ZD0116401.

Y. Tan et al. (Eds.): ICSI 2023, LNCS 13969, pp. 74–85, 2023.
https://doi.org/10.1007/978-3-031-36625-3_7

and Optimal Reciprocal Collision Avoidance (ORCA) [1]. These methods generate policies based on local observations, but they may fall into local optimum points and have low efficiency. The second approach is the distributed control method based on end-to-end reinforcement learning. This approach has shown promise in dealing with the deadlock problem and achieving better task completion efficiency. Various researchers have employed reinforcement learning for collision avoidance [2,3], with these methods typically taking the robot's own position, velocity information, and information from other agents in the surrounding domain as input to the neural network. However, one of the main challenges with reinforcement learning is its low sample efficiency.

Efforts have been made to improve sample efficiency in reinforcement learning (RL) for multi-agent collision avoidance. One method is to use inverse reinforcement learning to model the environment, which more fully utilizes environmental information [14]. Another classic approach for improving sample efficiency is off-policy RL, which includes methods such as Deep Deterministic Policy Gradient (DDPG) [8], Twin Delayed Deep Deterministic Policy Gradient (TD3) [4], and Soft Actor-Critic (SAC) [5]. These methods collect samples during the training process and store them in a replay buffer. During policy updates, batch-sized data (s, a, r, s', d) are randomly selected from the buffer. Compared to on-policy methods, off-policy methods reuse collected data and have a higher sample utilization rate. Furthermore, selecting effective data from the replay buffer is also an important research point to further improve sample efficiency [6]. Prioritized sampling methods, which select samples with greater differences for policy learning, have achieved some improvement. However, these methods analyze the discrete transitions collected in training for continuous action tasks. Dividing a segment into several discrete transition pairs still has limited sample efficiency improvement, as the number of transitions is often small. Transforming the trajectories of a group of agents in obstacle avoidance navigation into continuous transitions may lead to better performance.

In recent years, data augmentation techniques have achieved tremendous success in augmenting data and improving sample efficiency. Various data augmentation methods have been extensively researched and demonstrated to be effective in computer vision applications. Among them, mixup has gained increasing attention in recent years. Mixup was first introduced as a way to generate new training examples by linearly interpolating pairs of randomly selected samples and their labels [15]. The resulting mixed samples and labels are then used to train a deep neural network. Since its introduction, mixup has been shown to improve the performance of various neural network architectures across a range of tasks, such as image classification, object detection, and natural language processing. However, most of these strategies have been developed for image inputs. To the best of our knowledge, only a few papers have investigated mixup for single-agent reinforcement learning training [9], and no research has been conducted on mixup for multi-agent collision avoidance tasks. Therefore, there is a need for further research to explore the potential of mixup for multi-agent collision avoidance tasks, which may improve sample efficiency and enhance the performance of the trained models.

For multi-agent collision avoidance tasks, as the number of agents increases, the probability of mutual influence among agents in the field increases. In the process of heading towards the target point, agents not only are affected by fixed obstacles but also experience greater interference from other moving agents. This is similar to the increased driving difficulty on congested roads. In this paper, we introduce mixup to multi-agent reinforcement learning, where we use the state-action-reward tuples of the Markov decision process as the object for data augmentation. We construct continuous transitions between discrete steps using linear interpolation, where the similarity between continuous states is high, and action states and rewards are close. Through mixup, we create new transitions that are likely to exist in the actual transition manifold or are close to it. To make the transition more consistent with the true manifold, we use interpolation ratios to measure the relative weights between two consecutive transitions. However, when the interpolation ratio approaches 0 or 1, the state will degenerate into the original sample collected in the experiment. The interpolation ratio serves as a coefficient for adjusting the proportion of different parts of the data. By changing the value of the ratio, we can make the continuous transitions as close as possible to the authentic transition manifold. To achieve this, we constructed an energy discriminator to automatically adjust the value of the ratio.

The contributions of this paper are summarized as follows:

- We propose a novel method named MACT by constructing new continuous transitions for multi-agent collision avoidance for optimizing the sampling efficiency and asymptotic performance.
- We construct a multi-agent collision avoidance scenario in the Webots simulation environment, which is suitable for training. We apply planning algorithms such as APF and reinforcement learning algorithms such as SAC and validate that the performance of MACT is superior to other algorithms in Webots.
- We establish a platform using real robots and validate the proposed algorithm in real-world E-puck2 swarms, thus verify the applicability of MACT in the real-world.

2   Preliminaries

2.1   Problem Formulation

In this paper, the multi-agent collision avoidance problem is formulated in the context of a nonholonomic differential drive robot moving on the Euclidean plane and avoiding collision with obstacles and other neighbor robots.

To tackle this problem, we assume all robots in this task are homogeneous. To be specific, all of N agents are modelled as the agents with the same radius R and same max velocity V_{max}. Each robot i has access to its observation that only provides partial information of the environment and then according to its policy, computes a collision-free action a_i to drive itself towards the target g_i.

$$\mathbf{a}^t \sim \pi_\theta \left(\mathbf{a}^t \mid \mathbf{o}^t \right), \tag{1}$$

In our experiment, θ denotes the policy parameters, and the computed action $\mathbf{a}^t$ is a vector representing the left and right wheel speed of the E-puck2 mobile car. Each robot sequentially makes its decision until it reaches its target, and robots cannot access the states and targets of other robots, which is in line with the real-world scenario. In the initial task setting, robots are placed at the edge of the environment and assigned a goal on the opposite side.

2.2   Mixup

MixUp [15] is a data-augmentation strategy originally developed for image classification. In essence, MixUp improves the training of deep learning models by creating synthetic samples through linear interpolation of pairs of training samples and their corresponding labels. Specifically, given a pair of training samples x^i and x^j, which consist of input data and their corresponding labels, MixUp generates synthetic samples by linearly interpolating between the two samples based on an interpolation ratio ϵ

$$\mathbb{M}_\epsilon\left(x^i, x^j\right) = \epsilon x^i + (1 - \epsilon)x^j \tag{2}$$

where the interpolation ratio ϵ denotes the weights of two samples. The resulting mixed data is a linear combination of the two original data, while the mixed label is a linear combination of the two original labels. This process generates a new training data point that lies on the line connecting the two original data points

3   Methodologies

In the process of training a reinforcement learning algorithm, simulators such as Webots [11] and MPE (Multi-Agent Particle Environment) [10] are commonly used to simulate state transitions at regular intervals known as time-steps. As the duration of these intervals is often quite short, there is typically a strong continuity between the states of adjacent time-steps. In the multi-agent obstacle avoidance task, we implement parameter sharing for each agent. Agents explore the environment to collect new states, and we use the Multi-Agent Continuous Transition (MACT) method to construct previously unknown continuous transitions.

3.1   Reinforcement Learning Setup

In our research, we present a decentralized partially observable Markov decision process (Dec-POMDP) as a common framework for multi-agent reinforcement learning. In this context, Dec-POMDPs are represented as a tuple, namely, $(I, S, A, P, R, r, O, Z, \gamma)$. The index set $(I = 1, \ldots, N)$ corresponds to the collection of agents. It is crucial to note that due to the partial observability of the environment state s_t, each agent relies on their individual observation o_i in accordance with the local observation function $o_{i,t} = \mathcal{Z}i(s_t) : \mathcal{S} \rightarrow \mathcal{O}$. Every agent

i selects its action based on its policy $ai, t \sim \pi_i (\cdot \mid O_t)$. The joint action space A encompasses the union of all agents' action spaces $\bigcup_{i=1}^{N} Ai$. We introduce a deterministic transition function $T : S \times A \rightarrow S$ and a discount factor $\gamma \in (0, 1)$. Assuming the agent policy $ri, t = \mathcal{R}_i (s_t, \mathbf{a}_t)$ is parameterized by θ_i, the individual goal of each agent is established as the discounted sum of individual rewards:

$$R_i(\tau) = \sum_{t=t_i^s}^{t_i^e} \gamma^{t-t_i^s} r_{i,t}$$

Reward Function. In the proposed algorithm, we aim to avoid collisions during navigation and minimize the mean arrival time of all robots. Each agent obtains its own reward at each time step, which is defined as follows:

$$r_i^t = (g_r)_i^t + \sum (c_r)_i^t \tag{3}$$

The reward r_i^t received by the robot i at time-step t consist of two terms, including g_r and c_r. In particular, the robot is awarded by $(g_\gamma)_i^t$ for getting closer to its target:

$$(g_r)_i^t = w_g \left(\left\| p_i^{t-1} - g_i \right\| - \left\| p_i^t - g_i \right\| \right) + r_{\text{arrival}} \tag{4}$$

$$\begin{cases} r_{\text{arrival}} = 0.05 & \text{if } \left\| p_i^t - g_i \right\| < 0.03\,m \\ r_{\text{arrival}} = 0 & \text{otherwise.} \end{cases} \tag{5}$$

where p_i^t denotes the position of the robot i at time t, g_i denotes the target position of agent i, and w_g represents the distance reward weight. When the robot collides with other robots or obstacles in the surrounding environment, it is penalized by $(c_r)_i^t$:

$$(c_r)_i^t = \begin{cases} c_{\text{collision}} & \text{if } \left\| p_i^t - c_i \right\| < R_{\text{agent}} + R_{\text{col}} \\ & \text{or } \left\| p_i^t - p_j^t \right\| < 2R_{\text{agent}} \\ 0 & \text{otherwise.} \end{cases} \tag{6}$$

where c_i denotes the position of collision around agent i. We set the reward for arriving the target point $r_{\text{arrival}} = 0.05$, the weight factor $w_g=100$, the penalty for collision $c_{\text{collision}}=-0.35$, the agent's influence radius $R_{\text{agent}}=0.04$, the collision's influence radius $R_{\text{col}}=0.11$ during the training process.

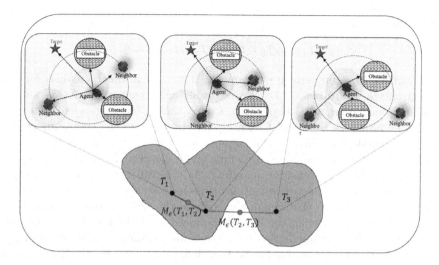

Fig. 1. The sketch of the multi-agent collision avoidance transition manifold.

3.2 Multi-agent Continuous Transition

Reinforcement learning (RL) has emerged as a prevalent approach for training agents to tackle complex tasks. Nevertheless, during the training process, a pervasive challenge is low training efficiency, primarily due to the performance constraints of various simulators employed for training, such as Webots [11] and AirSim [13]. These simulators operate on discrete time intervals for simulation, yielding discrete transitions that form a complete trajectory. For tasks with continuous action states, the volume of data accessible for training is considerably diminished. This issue is further intensified in multi-agent tasks, where state changes are influenced not only by the agent's actions but also by the actions of other agents, leading to an expanded state space that demands broader exploration and heightened sample efficiency. To tackle this challenge, we suggest employing off-policy methods, such as SAC and DDPG, which store the gathered data in a replay buffer and repurpose samples to enhance sample efficiency. In this paper, we present the mixup method for creating a Multi-agent Continuous Transition, which supplements authentic transitions with constructed transitions to bridge gaps between discrete transitions. Our approach offers a solution for boosting sample efficiency and generating a more comprehensive set of training data.

To achieve a smooth and uninterrupted progression, we employed mixup linear interpolation to bridge the gap between two discrete transitions and create new continuous transitions. The new transitions build upon the foundation of the original transitions, ensuring that it changes seamlessly along the actual trajectory. Specifically, in the multi-agent obstacle avoidance task, where parameter sharing is utilized, we constructed continuous transitions T for each agent's sample using mixup interpolation. The new transition $M_\epsilon(T_1, T_2)$ can be obtained as following equation:

$$\mathbb{M}_\epsilon\left(T_1, T_2\right) = \epsilon T_1 + (1 - \epsilon)T_2, where \epsilon \sim \mathbb{B}(\beta, \beta) \tag{7}$$

We utilize the temperature parameter β to adjust the beta distribution. In continuous control tasks, successive transitions frequently contain similar dynamic information, such as position. Consequently, for multi-agent tasks, we store the continuous samples of each agent separately in the environment and create independent continuous transitions. However, as the similarity between samples from different agents is often low, mixup-generated transitions may not reside within the true transition manifold. As illustrated in Fig. 1, the interpolation of two closely situated transitions is more likely to yield a new transition (blue dot in the figure) that genuinely exists in the task manifold, whereas the interpolation (yellow dot in the figure) produced by T_2 and T_3 is situated outside the manifold. To provide two straightforward examples: consider a scenario where a robot maintains stable forward motion, resulting in relatively consistent transitions. In this case, the constructed transition is likely to remain within the task manifold. Conversely, if the robot experiences a collision during its movement, the state changes between transitions may be more substantial, causing the newly constructed transition to appear outside the manifold.

In the MACT algorithm, the most crucial hyperparameter is the temperature β of the beta distribution. When β approaches 0, the beta distribution resembles a 2-point Bernoulli distribution, and as β approaches 1, it converges towards the uniform [0, 1] distribution. Following the insights from [9], we observe a positive correlation between the expected distance of continuous transitions and authentic transitions manifold with the value of β. To maintain the distance between two transitions below the tolerance threshold M, we can adjust the β accordingly.

An energy-based discriminator is employed to estimate the distance. We utilize a Multilayer Perceptron (MLP) f_ϕ, parameterized by ϕ, to encapsulate the information of state s_t and action a_t. We denote (s_t, a_t) as x_t and (s_{t+1}, r_t, d_t) as y_t. Consequently, the distance estimator $D(T_t)$ can be expressed as follows:

$$D\left(T_t\right) = \|E\left(x_t, y_t\right)\| = \|f_\phi\left(s_t, a_t\right) - y_t\|_2^2 \tag{8}$$

The discriminator is enhanced by optimizing ϕ to minimize Eq. 8. The new transition can be represented as x_t' and y_t'. By incorporating the constructed new transition $\mathbb{M}\epsilon\left(T_t, Tt + 1\right)$ into the estimator, we can compute the loss between the current D and the actual discrete transitions $E(x_t, y_t)$ and $E(x_{t+1}, y_{t+1})$. By constraining the loss (calculated using Eq. 9) within the tolerance range M, we transform the problem into a constrained optimization problem.

$$Loss\left(\mathbb{M}_\epsilon\left(T_t, T_{t+1}\right)\right) = \|E(x_t', y_t') - \mathbb{M}_\epsilon\left(E(x_t, y_t), E(x_{t+1}, y_{t+1})\right)\|_2^2 \tag{9}$$

Formally, we can solve the constrained optimization problem to optimize the β, as shown in Equ. 10.

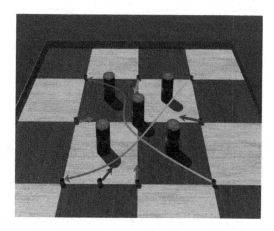

Fig. 2. Multi-robots collision avoidance scenario: Each agent is placed around the field, and the target point is set on the opposite side of the field. The agent aims to go to the target point in the shortest time while avoiding static obstacles and other agents.

$$\max \beta, \ \text{s.t.} \ \mathbb{E}\left[Loss\left(\mathbb{M}_\epsilon\left(T_t, T_{t+1}\right)\right)\right] \leq M, 0 < \beta \leq 1$$
$$\text{where } \epsilon \sim \mathbb{B}(\beta, \beta) \tag{10}$$

4 Experiments

In this study, our primary focus is on multi-agent navigation within confined spaces containing swarm agents and static obstacles, a scenario that is anticipated to become increasingly prevalent in the practical applications of UAVs and unmanned vehicles. Each agent is positioned around the field, with the target point set on the opposite side of the field. The agents strive to reach the target point in the shortest possible time while evading static obstacles and other agents, as depicted in Fig. 2. In this section, we conduct simulations and experiments to showcase the effectiveness of the proposed MACT algorithm. Through these simulations, we verify the benefits of MACT by comparing it with APF [7], ORCA [1], SAC [5], and PPO [12]. ORCA and APF represent two state-of-the-art traditional planning methods.

4.1 Simulation

We train our methods using Webots, a three-dimensional simulation platform. We configure E-puck2 robots, obstacles, and destinations as depicted in Fig. 2. In the simulation environment, we establish a 2m x 2m area, mirroring the real-world environment dimensions, and place five cylinders as static obstacles with a radius of 0.1m. Corresponding to the real vehicle, we employ the E-puck2 model as the agent in Webots. The E-puck2 robot has a diameter of 0.08m and a maximum speed of 0.14m/s. The action space of MACT is continuous and

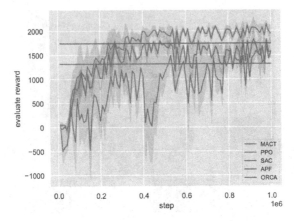

Fig. 3. The evaluate reward of 10 agents task

represents the speed of both wheels. Each episode runs for a maximum of 1000 steps. We consider a collision to occur when agents come within 0.08m of other agents or within 0.15m of obstacles.   We conducted experiments in a dense environment with 5 and 10 agents. In the task involving five agents, we observed that both traditional methods and reinforcement learning methods were effective in completing the task and achieving satisfactory policy performance due to the relatively low task difficulty.

As shown in Fig. 3, the task becomes increasingly challenging as the number of agents grows. Under this reward function design, MACT exhibits superior performance compared to planning methods. MACT is capable of learning a more effective policy when facing difficult task scenarios. Additionally, MACT demonstrates faster sample efficiency and higher reward returns compared to reinforcement methods such as SAC and PPO.

As shown in Table 1, all methods perform well in the task involving 5 agents, which has a lower difficulty level. However, in the more challenging task with 10 agents in a limited area, reinforcement learning methods SAC and PPO are more likely to encounter dangerous states. The APF and ORCA algorithms exhibit excellent safety and rarely collide, but the task requires a larger number of steps and a longer distance to be completed due to frequent stalemates. As seen in Fig. 4, MACT outperforms both reinforcement learning and planning methods in terms of travelled distance and success rate. It achieves better performance in the challenging task of 10 agents, demonstrating its effectiveness in improving multi-agent collision avoidance.

4.2   Real World

To validate our approach in a real-world setting, we utilized a motion capture system to obtain the coordinates of each E-puck robot in physical space and convert them into positions within the virtual space of Webots. Position and

Table 1. Performance analysis of APF, ORCA, SAC, PPO, MACT in collision avoidance task

Agents Number	5					10				
Methods	APF	ORCA	SAC	PPO	MACT	APF	ORCA	SAC	PPO	MACT
Extra Travelled Distance(m)	2.801	2.724	2.655	2.865	**2.615**	2.881	2.946	2.912	3.021	**2.739**
Extra Travelled Steps	985	873	735	980	**733**	1850	1240	1190	1505	**964**
Danger State Rate(%)	**0.0**	**0.0**	1.1	2.1	1.0	**0.0**	**0.0**	2.0	3.3	0.9
Success Rate in 1000 Steps(%)	100	100	100	100	100	20	60	90	60	**100**

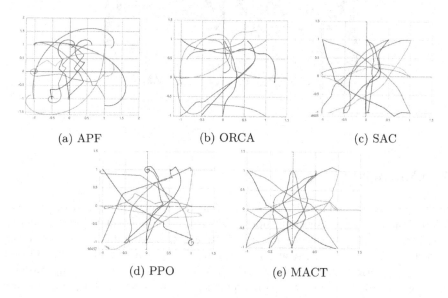

(a) APF   (b) ORCA   (c) SAC

(d) PPO   (e) MACT

Fig. 4. Trajectory of five methods

velocity values from the physical space were transmitted to the E-puck2 model in Webots via ROS communication. We set the perception range for each E-puck2 agent to 0.4m, which means that location coordinates are broadcasted to neighboring robots and obstacles within a 0.4m radius from the agent.

Figure 5 displays the demonstration of our real-world experiment. In Fig. 5(a), eight agents are positioned at the starting state around the environment's edge. To reach the target point quickly, the agents move towards the center while avoiding static obstacles (safety barrels). As shown in Fig. 5(c), when all eight agents approach the center, they encounter significant mutual interference and dynamic changes, leading to actions such as braking to prevent collisions and successfully avoiding the deadlock problem that often occurs in this region. Ultimately, as depicted in Fig. 5(d), the agents successfully complete obstacle avoidance and approach the target point. Our algorithm's effectiveness is thus validated in a real-world environment.

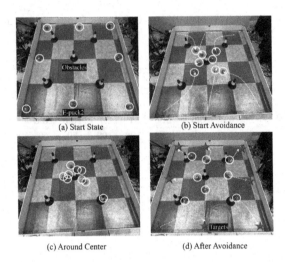

(a) Start State   (b) Start Avoidance

(c) Around Center   (d) After Avoidance

Fig. 5. Demonstration of real experiment

5   Conclusion

In this paper, we employed reinforcement learning to tackle a multi-agent collision avoidance task. For continuous tasks like this, we constructed continuous transitions using linear interpolation, effectively leveraging information from the trajectory. To generate transitions closer to the true manifold, we introduced an energy discriminator to adjust the temperature coefficient of mixup, providing more accurate samples for reinforcement learning training. Our experiments demonstrated that, compared to reinforcement learning methods such as SAC, our approach improved both sampling efficiency and training performance in the multi-agent obstacle avoidance task. Furthermore, we conducted experiments on real E-puck robots to validate our method beyond simulations. Our current approach is based on parameter sharing, and in future work, we plan to incorporate mixup data augmentation in other multi-agent algorithms, such as MAPPO.

References

1. Alonso-Mora, J., Breitenmoser, A., Rufli, M., Beardsley, P., Siegwart, R.: Optimal reciprocal collision avoidance for multiple non-holonomic robots. In: Distributed Autonomous Robotic Systems, pp. 203–216. Springer, Heidelberg (2013). https://doi.org/10.1007/978-3-642-32723-0_15
2. Chen, Y.F., Everett, M., Liu, M., How, J.P.: Socially aware motion planning with deep reinforcement learning. In: 2017 IEEE/RSJ International Conference on Intelligent Robots and Systems (IROS), pp. 1343–1350. IEEE (2017)
3. Everett, M., Chen, Y.F., How, J.P.: Motion planning among dynamic, decision-making agents with deep reinforcement learning. In: 2018 IEEE/RSJ International Conference on Intelligent Robots and Systems (IROS), pp. 3052–3059. IEEE (2018)

4. Fujimoto, S., Hoof, H., Meger, D.: Addressing function approximation error in actor-critic methods. In: International Conference on Machine Learning, pp. 1587–1596. PMLR (2018)

5. Haarnoja, T., Zhou, A., Abbeel, P., Levine, S.: Soft actor-critic: Off-policy maximum entropy deep reinforcement learning with a stochastic actor. In: International Conference on Machine Learning, pp. 1861–1870. PMLR (2018)

6. Hessel, M., et al.: Rainbow: Combining improvements in deep reinforcement learning. In: Proceedings of the AAAI Conference on Artificial Intelligence, vol. 32 (2018)

7. Li, G., Yamashita, A., Asama, H., Tamura, Y.: An efficient improved artificial potential field based regression search method for robot path planning. In: 2012 IEEE International Conference on Mechatronics and Automation, pp. 1227–1232. IEEE (2012)

8. Lillicrap, T.P., et al.: Continuous control with deep reinforcement learning. arXiv preprint arXiv:1509.02971 (2015)

9. Lin, J., Huang, Z., Wang, K., Liang, X., Chen, W., Lin, L.: Continuous transition: Improving sample efficiency for continuous control problems via mixup. In: 2021 IEEE International Conference on Robotics and Automation (ICRA), pp. 9490–9497. IEEE (2021)

10. Lowe, R., Wu, Y.I., Tamar, A., Harb, J., Pieter Abbeel, O., Mordatch, I.: Multi-agent actor-critic for mixed cooperative-competitive environments. Adv. Neural Inf. Process. Syst. **30** (2017)

11. Michel, O.: Cyberbotics ltd. webots^{TM}: Professional mobile robot simulation. Int. J. Adv. Robot. Syst. **1**(1), 5 (2004)

12. Schulman, J., Wolski, F., Dhariwal, P., Radford, A., Klimov, O.: Proximal policy optimization algorithms. arXiv preprint arXiv:1707.06347 (2017)

13. Shah, S., Dey, D., Lovett, C., Kapoor, A.: AirSim: High-fidelity visual and physical simulation for autonomous vehicles. In: Hutter, M., Siegwart, R. (eds.) Field and Service Robotics. SPAR, vol. 5, pp. 621–635. Springer, Cham (2018). https://doi.org/10.1007/978-3-319-67361-5_40

14. Yu, X., Wu, W., Feng, P., Tian, Y.: Swarm inverse reinforcement learning for biological systems. In: 2021 IEEE International Conference on Bioinformatics and Biomedicine (BIBM), pp. 274–279. IEEE (2021)

15. Zhang, H., Cisse, M., Dauphin, Y.N., Lopez-Paz, D.: mixup: Beyond empirical risk minimization. arXiv preprint arXiv:1710.09412 (2017)

Research on UAV Dynamic Target Tracking with Multi-sensor Position Feedback

Peng Liu[1(✉)], Junchao Wei[1], Hao Wang[2], Chunjun He[2], Jianxia Zhang[2], Haiyue Li[2], Ziyuan Liu[2], and Baorui Jiang[1]

[1] Chengdu University of Information Technology, 610225 Chengdu, China
liupeng@cuit.edu.cn
[2] China Construction Third Engineering Bureau Installation Engineering CO., LTD., 430000 Wuhan, China

Abstract. The path planning and control of unmanned aerial vehicles (UAVs) is an important part of autonomous flight. High-precision positioning and navigation is one of the key technologies to achieve adaptive flight, efficient reconnaissance and precise attack. A control method of adaptive target tracking and obstacle avoidance for rotorcraft UAVs is proposed in the paper to solve the problem. According to the characteristics of the nonlinear UAVs' system, the artificial potential field method and backstepping method are used for path planning and flight control respectively. The unscented Kalman filter combined with UWB and INS is used to obtain accurate position and feedback to the artificial potential field to ensure the stability of the system and the accuracy of target tracking. This paper describes the dynamic target simulation of multiple obstacles. It is shown that the satisfactory tracking performance with the new algorithm, and the stability of obstacle avoidance is also improved.

Keywords: Artificial potential field method · Backstepping · Unscented Kalman filter · UWB positioning

1 Instruction

With the development of UAV technology and the demand of logistics, television, military and other aspects, the ability of UAVs to avoid obstacles independently becomes a major influence condition to ensure flight safety and solve technical barriers. Obstacle avoidance of UAV refers to the method of making UAVs safely reach the target point from the starting point under the specified flight environment [1]. During the process from the starting point to the target point, it may encounter multiple constraints such as path boundaries, obstacles, passing points and so on. The common obstacle avoidance methods for robots are divided into global obstacle avoidance planning method and local obstacle avoidance planning method [2]. The former includes self-space method, grid method, posture

This research was supported by the Opening Project of Unmanned System Intelligent Perception Control Technology Engineering Laboratory of Sichuan Province(WRXT2021-004).

space method, ant colony algorithm, etc. The latter includes artificial potential field method [3,4], probabilistic road map method [9], genetic algorithm [6],
ANN method, etc. Generally, the global optimal solution can be obtained, but
the calculation efficiency is low. Compared with the global obstacle avoidance
algorithm, the local planning obstacle avoidance algorithm abandons the com-
pleteness, but can be effectively applied to real-time obstacle avoidance, and is
mostly used to deal with the environment with dynamic obstacles. In practical
application, the combination of the two methods can also be used as required
[7].

The UAV real-time target tracking, obstacle avoidance and path planning
algorithms are various, and can match the appropriate algorithm to achieve
the desired planning effect in most situations. However, in the actual operating
environment, it is often difficult to obtain the accurate position information of
the controlled UAV, which affects the subsequent path planning effect and even
makes the UAV in a dangerous position. At the very least, it can deviate from
the target trajectory and affect the subsequent planning accuracy, In serious
cases, it may even collide with obstacles. Xiao Ronghao, Ma Xu and others proposed the AB line path planning algorithm with position correction to correct
the coordinates of agricultural machinery and provide real-time reference information for the auxiliary driving system [8]; Chen Xi designed a UAV obstacle
avoidance path control system based on laser sensors, which monitors the environment and scene changes through the UAV environment perception module,
calculates the flight path in real time, and realizes the calculation and control
of the UAV obstacle avoidance path, with high accuracy [9]; Afandi of Glasgow
University tried to use A * and backstepping method to establish path planning
and control model for four-rotor aircraft model, and achieved proud tracking
performance in static scenarios [10].

In this paper, the artificial potential field(APF) method is used to plan the
path of the UAV, the backstepping is used to control the UAV to fly according to
the planned path, and the inertial navigation system (INS) and ultra-wideband
(UWB) location system [11] are used to locate the UAV during the flight process,
and the real-time position of the UAV is obtained through filtering. Finally, the
real-time position is used as feedback to calibrate the actual position.

2   Path Planning Algorithm Design

The APF method takes the target point of the UAV and its surrounding obstacles as the attractive and repulsive points. According to the nature of attractive
and repulsive force, an APF is constructed. In this potential field, the resultant
force of attractive and repulsive force forces the object to generate acceleration,
velocity and displacement, thus realizing independent path planning [12]. The
APF designed by Gu Yujin can effectively guide the UAV to complete path planning and obstacle avoidance control, and has good adjustment and adaptability
in case of unexpected interference [13].

The attractive potential field and the repulsive potential field are constructed from the universal attractive field formula(U_{at}) and the repulsion field formula(U_{re}) [14], and the force is obtained by calculating the negative gradient of the potential field:

$$
\begin{cases}
\begin{aligned}
\boldsymbol{F}_{\text{at}}(\boldsymbol{X}_p) &= -\nabla\left(U_{\text{at}}(\boldsymbol{X}_p)\right) \\
&= -k_{\text{at}}\,\text{diss}\left(\boldsymbol{X}_p, \boldsymbol{X}_{\text{goal}}\right)\frac{\partial\,\text{diss}\left(\boldsymbol{X}_p, \boldsymbol{X}_{\text{goal}}\right)}{\partial \boldsymbol{X}_p}
\end{aligned} \\[2ex]
\begin{aligned}
\boldsymbol{F}_{\text{rei}}(\boldsymbol{X}_p) &= -\nabla\left(U_{\text{rei}}(\boldsymbol{X}_p)\right) \\
&= \begin{cases}
k_{\text{re}}\left(\frac{1}{\text{diss}(\boldsymbol{X}_p,\boldsymbol{X}_{\text{obi}})} - \frac{1}{\rho_0}\right)\frac{1}{\text{diss}^2(\boldsymbol{X}_p,\boldsymbol{X}_{\text{obi}})}\frac{\partial\,\text{diss}(\boldsymbol{X}_p,\boldsymbol{X}_{\text{obi}})}{\partial \boldsymbol{X}_p} \\
\quad , \text{diss}\left(\boldsymbol{X}_p, \boldsymbol{X}_{\text{obi}}\right) \le \rho_0 \\
0 \\
\quad , \text{diss}\left(\boldsymbol{X}_p, \boldsymbol{X}_{\text{obi}}\right) > \rho_0
\end{cases}
\end{aligned}
\end{cases}
\tag{1}
$$

where, the attractive force of the UAV is F_{at}, the repulsive force of each obstacle on the UAV is F_{re}, $\boldsymbol{X}_p = (x, y, z)^T$ is the x/y/z axis coordinate of the UAV, $X_{obi} = (x_{obi}, y_{obi}, z_{obi})^T$ is the coordinate of the i_{th} obstacle, $X_{goal} = (x_{goal}, y_{goal}, z_{goal})^T$ is the coordinate of the target, k_{at} and k_{re} are the field coefficients of attractive and repulsion respectively, u_{at} and U_{re} is the calculated attractive potential field and repulsive potential field, $diss(X_p, X_{obi})$ is the distance between UAV(X_p) and the i_{th} target point(X_{obi}), and ρ_0 is the repulsive range of the obstacle.

The resultant force of UAV is the vector sum of attractive force and repulsive force:

$$
\boldsymbol{F}(\boldsymbol{X}_p) = \boldsymbol{F}_{at}(\boldsymbol{X}_p) + \sum_{i=1}^{n}\boldsymbol{F}_{rei}(\boldsymbol{X}_p)
\tag{2}
$$

where n is the total number of obstacles, and the action diagram of each force is shown in Fig. 1.

At each moment, the resultant force of the UAV directly affects the acceleration of the aircraft, thus affecting the speed and position.

3   Design of Control Law Based on Backstepping Method

In this paper, a four-rotor UAV is selected, making several assumptions:

(1) The ambient wind speed is zero, or the wind speed can be ignored;
(2) The flight altitude and yaw angle remain unchanged;
(3) With the object thrown as the background, when the ground projection point of the UAV coincides with the target point, it means that it reaches the target point.

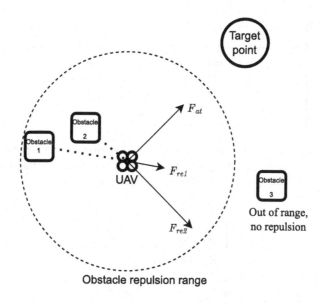

Fig. 1. Diagram of APF

First of all, the kinematics model is established. The solution formula of acceleration$(\ddot{x}, \ddot{y}, \ddot{z})$ and angular acceleration$((\ddot{\phi}, \ddot{\theta}, \ddot{\psi}))$ of UAV is as follows [15]:

$$
\begin{cases}
\ddot{x} = (\cos\phi\sin\theta\cos\psi + \sin\psi\sin\phi)\frac{1}{m}U_1 - \frac{k_1\dot{x}}{m} \\
\ddot{y} = (\cos\psi\sin\theta\cos\phi - \cos\psi\sin\phi)\frac{1}{m}U_1 - \frac{k_2\dot{y}}{m} \\
\ddot{z} = -g + (\cos\theta\cos\phi)\frac{1}{m}U_1 \\
\ddot{\phi} = \dot{\theta}\dot{\psi}\left(\frac{I_y - I_z}{I_x}\right) + \frac{l}{I_x}U_2 - \frac{J_r}{I_x}\dot{\theta}\omega_r \\
\ddot{\theta} = \dot{\phi}\dot{\psi}\left(\frac{I_z - I_x}{I_y}\right) + \frac{l}{I_y}U_3 + \frac{J_r}{I_y}\dot{\phi}\omega_r \\
\ddot{\psi} = \dot{\phi}\dot{\theta}\left(\frac{I_x - I_y}{I_z}\right) + \frac{1}{I_z}U_4
\end{cases}
\tag{3}
$$

where, (ϕ, θ, ψ) is the roll angle, pitch angle and yaw angle, (I_x, I_y, I_z) is the moment of inertia of the x/y/z axis in the body coordinate system, J_r is the moment of inertia of the motor, l is the distance from the motor to the center of attractive force of the UAV, $(k_1 = k_2 = 0.1)$ is the damping coefficient, where 0.1 is taken, and $\omega_r = \omega_2 + \omega_4 - \omega_1 - \omega_3$ is the interference of the motor.

Assuming that the expected value of the attitude angles are $(\phi_d, \theta_d, \psi_d)$. Because the yaw angle is constant, ψ_d may be assumed as 0. Because the height is constant, it is the total lift, the formula is listed as:

$$
U_1 = m\sqrt{\left(\frac{F_x}{m}\right)^2 + \left(\frac{F_y}{m}\right)^2 + g^2}
\tag{4}
$$

From the dynamic model, we can get:

$$
\begin{cases}
F_x = (\cos\phi_d \sin\theta_d \cos\varphi_d + \sin\varphi_d \sin\phi_d)\,U_1 - k_1\dot{x} \\
F_y = (\cos\varphi_d \sin\theta_d \cos\phi_d - \cos\varphi_d \sin\theta_d)\,U_1 - k_2\dot{y} \\
F_z = -g + (\cos\theta_d \cos\phi_d)\,U_1 = 0 \\
\phi_d = \sin^{-1}\left(\frac{-F_y}{U_1}\right) \\
\theta_d = \tan^{-1}\left(\frac{F_x}{mg}\right)
\end{cases}
\tag{5}
$$

In the obstacle avoidance control of UAV described in this paper, the controlled object is clear and the external force is known, so the backstepping method is selected for control. The tracking error is:

$$
\begin{cases}
Z_1 = \phi_d - \phi \\
Z_2 = \dot{\phi} - \dot{\phi}_d - \alpha_1 Z_1 \\
Z_3 = \theta_d - \theta \\
Z_4 = \dot{\theta} - \dot{\theta}_d - \alpha_3 Z_3 \\
Z_5 = \psi_d - \psi \\
Z_6 = \dot{\psi} - \dot{\psi}_d - \alpha_5 Z_5
\end{cases}
\tag{6}
$$

4   UWB and INS Fusion Location Algorithm Based on UKF

The real-time position estimation algorithm of UAV is divided into three parts: UWB positioning, INS positioning and unscented Kalman filtering. UWB positioning uses the receiving time or time difference of the pulse signal to calculate the distance between the object to be located and the pre-arranged base station, and uses the weighted least squares method to calculate the distance value of the UWB base station to calculate the coordinates to be measured. Suppose a total of **n** positioning base stations are used, the coordinates of the ith base station are defined as $(x_i, y_i, z_i)^T$, and the distance from the object to be measured is r_i. The coordinates of the object to be measured can be calculated by the following formula:

$$
\mathbf{P} = (\mathbf{H}^T\mathbf{H})^{-1}\mathbf{H}^T\beta
\tag{7}
$$

where, $H = \begin{bmatrix} x_2 - x_1 & y_2 - y_1 & z_2 - z_1 \\ x_3 - x_1 & y_3 - y_1 & z_3 - z_1 \\ \vdots & \vdots & \vdots \\ x_n - x_1 & y_n - y_1 & z_n - z_1 \end{bmatrix}$, $\beta = \frac{1}{2}\begin{bmatrix} R_2^2 + R_1^2 + r_1^2 - r_2^2 \\ R_3^2 + R_1^2 + r_1^2 - r_3^2 \\ \vdots \\ R_n^2 + R_n^2 + r_n^2 - r_n^2 \end{bmatrix}$, $R_i = \sqrt{x_i^2 + y_i^2 + z_i^2}$

UWB signals have limited anti-interference capability, especially in the case of obstacles, radio signals can only reach the receiving end through reflection or refraction, resulting in inevitable positioning error [17]. INS uses inertial measurement elements carried by UAV to measure the instantaneous three-axis acceleration and three-axis angular acceleration. Using certain filtering algorithm is a common means to improve the positioning accuracy of UWB and INS.

Because the positioning systems of UWB and INS are both nonlinear systems, extended Kalman filter (EKF), unscented Kalman filter (UKF) and other methods can be used for data fusion. Especially, UKF can perform higher-order approximation of the model with higher accuracy. With UKF-based data fusion, the state vector and input vector are generated from INS data, and the predicted state is corrected by UWB position information.

Establish a 3-dimensional coordinate system (x, y, z), where the east direction the positive direction of the axis x, the north the positive direction of the axis y, and the right hand rule determine that the vertical ground is the positive direction of the axis z. In this coordinate system, order the three-axis position, speed, acceleration and attitude angle (roll angle, pitch angle, yaw angle) of the aircraft as the state vector, that is: $\mathbf{X} = (P_x, P_y, P_z, V_x, V_y, V_z, A_x, A_y, A_z, \phi, \theta, \psi)^T$, building the state equation of the filtering algorithm:

$$\mathbf{X}(k+1) = F(\mathbf{X}(k)) + \mathbf{u}(k)T + \mathbf{w} \tag{8}$$

where, T is the sampling time, F is the state transfer equation:

$$F\left(\mathbf{X}(k+1)\right) = \begin{pmatrix} \mathbf{X}_1(k) + \mathbf{X}_4(k) \cdot T \\ \mathbf{X}_1(k) + \mathbf{X}_5(k) \cdot T \\ \mathbf{X}_1(k) + \mathbf{X}_6(k) \cdot T \\ \mathbf{X}_4(k) + \mathbf{C}_{B1}^N(\mathbf{X}_7(k), \mathbf{X}_8(k), \mathbf{X}_9(k))^{\mathrm{T}} \cdot T \\ \mathbf{X}_5(k) + \mathbf{C}_{B2}^N(\mathbf{X}_7(k), \mathbf{X}_8(k), \mathbf{X}_9(k))^{\mathrm{T}} \cdot T \\ \mathbf{X}_6(k) + \mathbf{C}_{B3}^N(\mathbf{X}_7(k), \mathbf{X}_8(k), \mathbf{X}_9(k))^{\mathrm{T}} \cdot T \\ 0 \\ 0 \\ 0 \\ \mathbf{X}_{10} \\ \mathbf{X}_{11} \\ \mathbf{X}_{12} \end{pmatrix} \tag{9}$$

where: $\mathbf{X}_1(k)$ is the first item of $\mathbf{X}$ at time$=k$, and $\mathbf{X}_2(k)$ is the second item at time $\mathbf{X}$, and so on. $\mathbf{u}(k)$ is the input vector at time k, ω is the process noise, and its corresponding covariance matrix is Q, $\mathbf{X}(k)$ and $\mathbf{X}(k+1)$ are the state vectors at time k and $k+1$, respectively. The specific form is as follows:

$$\mathbf{u(k)} = [0, 0, 0, 0, 0, 0, \mathbf{A_x}, \mathbf{A_y}, \mathbf{A_z}, \phi, \theta, \psi]^{\mathrm{T}} \tag{10}$$

where, $A_{x/y/z}$ is the three-axis acceleration, which is directly measured by IMU, so it is the value in the body coordinate system.

$(C_{B1}^N, C_{B2}^N, C_{B3}^N)$ refers to the first, second and third rows of the rotation matrix of the machine-ground coordinate system, and its value is:

$$\begin{cases} C_{B1}^N = [\cos\psi\cos\theta, \sin\psi\cos\theta, -\sin\theta] \\ C_{B2}^N = [-\sin\psi\cos\phi + \cos\psi\sin\theta\sin\phi, \cos\psi\cos\phi + \sin\psi\sin\theta\sin\phi, \cos\theta\sin\phi] \\ C_{B3}^N = [\sin\psi\sin\varphi + \cos\psi\sin\theta\cos\phi, -\cos\psi\sin\varphi + \sin\psi\sin\theta\cos\varphi, \cos\theta\cos\varphi] \end{cases} \tag{11}$$

The observation equation is constructed as follows:

$$\mathbf{Z(k) = HX(k) + v} \tag{12}$$

where: $Z(k)$is the observation vector at time k, H is the observation matrix, $v$is the observation noise, and its corresponding covariance matrix is R.

The specific values of the observation matrix(H) is as follows:

$$H = \begin{bmatrix} 1\,0\,0\,0\,0\,0\,0\,0\,0\,0\,0\,0 \\ 0\,1\,0\,0\,0\,0\,0\,0\,0\,0\,0\,0 \\ 0\,0\,1\,0\,0\,0\,0\,0\,0\,0\,0\,0 \end{bmatrix} \tag{13}$$

The system state equation and measurement equation determined above are used for UKF, and the initial coordinates and filtering initial values of the target carrier are selected as follows:

$$\begin{cases} X(0){=}[x_p(1), y_p(1), z_p(1), 0, 0, 0, A_x, A_y, A_z, \phi, \theta, \psi]^T \\ P(0){=}\ 0.1{*}diag[10, 10, 10, 1, 1, 1, 1, 1, 1, 1, 1, 1] \\ Q{=}0.1 * diag[1, 1, 1, 1, 1, 1, 1, 1, 1, 1, 1, 1] \\ R{=}0.1 * diag[1, 1, 1] \end{cases} \tag{14}$$

After UKF calculation, the obtained X_k is the output value of state esti-mation at time=k. K_k and P_k are used as the gain matrix and state variance matrix. The state mean $X^-_{k+1|k}$ and state variance matrix $P^-_{k+1|k}$ can be further predicted. After the measured value Z_{k+1} is obtained, the output valueX_{k+1} at time $k + 1$ can be calculated. The first three items$(X(1), X(2), X(3))$ of X at each moment are the position coordinate$(\hat{P}_x, \hat{P}_y, \hat{P}_z)$ of the filtered UAV, which can be brought into the artificial potential field to obtain the attractive and repulsive force at the current moment. After calculating the resultant force, it is controlled by the backstepping method again, and then circulated in turn until the UAV reaches the target point.

5   Simulation and Analysis

In order to evaluate the effectiveness of the algorithm in this paper, the simu-lation experiment of path planning and control is carried out, and the aircraft is set to fly autonomously in a flat 1 m from the ground, and reach the tar-get point directly above the target without colliding with obstacles along the way. Set the starting point of the UAV as (3,2)m, and the target is on a circle with a center of (12,10)m and a radius of 1 m. The flat of the circle is parallel to the XoY flat, and the position changes with time. The specific coordinate $(x_{goal}(t), y_{goal}(t), z_{goal}(t))$ m parameter is as follows:

$$\begin{cases} x_{goal}(t) = 12 + \cos(t) \\ y_{goal}(t) = 10 + \sin(t) \\ z_{gosl}(t) = 1 \end{cases} \tag{15}$$

where: t(s) is the time.

Because the UAV flies in the same plane, the environment is simplified to a two-dimensional plane for display, and the map used for the test is drawn as shown in Fig. 2.

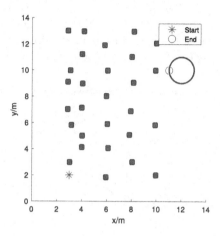

Fig. 2. Map

The Fig. 2 shows the starting point(∗) and the ending point(○), and the red rectangle shows the obstacle. The basic parameter settings of UAV are shown in Table 1.

Table 1. Basic parameters of the UAV

Items	Value	Unit	Items	Value	Unit
m	0.65	Kg	J_r	9×10^{-6}	Ns^2/rad
I_x, I_y	0.00397	Ns^2/rad	d	7.5×10^{-7}	Ns^2/rad
I_z	0.0006	Ns^2/rad	b	0.0000313	Ns^2
g	9.8	m/s^2	K_1, K_2	0.1	Ns/m

The noise parameters are set as follows: the standard deviation of acceleration noise measured by INS is $0.1 \, m/s^2$, and the standard deviation of UWB ranging is 0.001m [18].

Using the above parameters, run the simulation program and compare the UKF fusion algorithm described in this paper with INS and UWB positioning methods. The actual flight path of the UAV is shown in Fig. 3. The position error curve of each method is shown in the Fig. 4. The flight accuracy with UWB positioning is the worst, with the maximum error of 1.647 (m) and the average absolute error of 0.769 (m). The accuracy of INS is in the middle, with the maximum absolute error of 1.390 (m) and the average absolute error of 0.433 (m). The maximum error of the UKF fusion algorithm described in this paper is 0.406 (m), and the average value of the absolute value of the error is 0.140 (m).

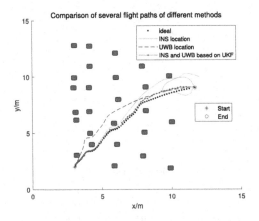

Fig. 3. Comparison of flight paths of different methods

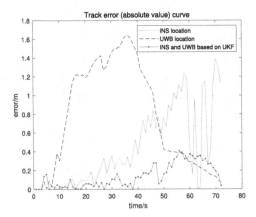

Fig. 4. Comparison of trajectory error of different methods

When INS or UWB diverges, UKF fusion feedback can still maintain convergence and achieve higher accuracy. Figure 5 is the trajectory of INS diverges, and Fig. 6 is the corresponding distance error curve. Because INS diverges, the error is meaningless, and it is replaced by -1.

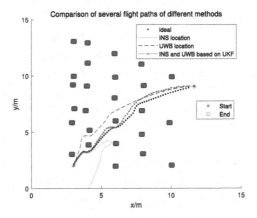

Fig. 5. Comparison of flight paths of different methods (INS inconvergent)

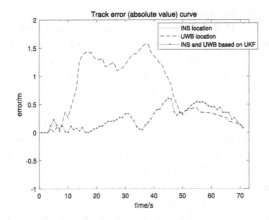

Fig. 6. Comparison of trajectory error of different methods (INS inconvergent)

It is defined as 20 experiments consisted a group. During 3 groups of experiments, the maximum error, divergence times and collision times of obstacles under different positioning methods are recorded, as shown in Table 2.

Table 2. Statistical table of the results of multiple trials

Items	Maximum error (m)				Mean error (m)				Number of divergent			
	INS	UWB	EKF	UKF	INS	UWB	EKF	UKF	INS	UWB	EKF	UKF
1	2.34	3.30	0.62	0.59	0.55	0.79	0.45	0.31	2	2	1	0
2	2.60	2.03	1.51	1.42	0.42	0.99	0.42	0.39	4	3	0	0
3	2.24	1.64	0.54	0.23	1.04	0.53	0.25	0.28	2	4	0	0
4	3.04	5.32	1.21	0.93	0.86	0.15	0.74	0.40	0	1	1	0
5	0.45	0.96	0.15	0.13	0.18	0.39	0.05	0.06	4	2	1	0
6	0.59	0.73	0.08	0.09	0.19	0.12	0.07	0.05	4	2	1	0
7	1.24	1.64	0.12	0.09	0.22	0.07	0.28	0.28	1	2	0	0
8	1.57	1.27	1.72	2.25	0.52	4.76	1.56	1.12	2	1	0	0
9	0.55	1.77	0.21	0.11	0.22	0.71	0.08	0.06	2	2	1	0
10	1.22	2.63	1.10	0.70	0.04	0.07	0.07	0.03	2	1	1	0
Average	1.58	2.13	0.73	0.65	0.42	0.86	0.40	0.30	2	2	1	0

6   Conclusion

In this paper, the problem of UAV tracking dynamic target and autonomous obstacle avoidance is discussed. After the path planning of UAV using artificial potential field method, the backstepping method is used to control the UAV, and the UKF algorithm is used to estimate the real-time position of UAV, which is used as feedback to plan the next path. Accurate position feedback enhances the accuracy of obstacle avoidance and flight safety of UAV. In order to verify the feasibility of the algorithm, a simulation experiment was carried out in MAT-LAB. The experimental results show that the algorithm can control the UAV to track the dynamic target stably under the condition of low sensor accuracy. Compared with the traditional single sensor position feedback, the maximum error in the flight process is reduced by about 59% compared with INS positioning, 69% compared with UWB positioning, and the ability to avoid obstacles has also been significantly improved.Compared to traditional EKF filtering, the positioning error of UKF is reduced by 10%.

The method proposed in this paper will feedback the position of the controlled object back to the path planning algorithm in real time. The feedback position is obtained by fusing the data of multiple sensors, which plays a positive role in the path planning at the subsequent time. At the same time, it provides a reference for the further study of the application of Kalman filter in path planning.

References

1. Radmanesh, M., Kumar, M., Guentert, P.H., et al.: Overview of path-planning and obstacle avoidance algorithms for UAVs: A comparative study. Unman. Syst. **6**(2): 95–118 (2018)
2. Li, Y.: Research on Obstacle Avoidance of Mobile Robots Based on Artificial Potential Field Method. Hefei University of Technology, Hefei (2013)
3. Haiyun, C., Huazhou, C., Qiang, L.: Multi-UAV 3D formation path planning based on improved artificial potential field method. J. Syst. Simulat. **32**(03), 414–420 (2020)
4. Khatib, O.: Real-time obstacle avoidance for manipulators and mobile robots. In: Autonomous Robot Vehicles, pp. 396–404. Springer, New York (1986). https://doi.org/10.1007/978-1-4613-8997-2_29
5. Chen, G., et al.: Path planning for manipulators based on an improved probabilistic roadmap method. Robot. Comput. Integrat. Manuf. **72**, 102196 (2021)
6. Sonmez, A., Kocyigit, E., Kugu, E.: Optimal path planning for UAVs using genetic algorithm. In: 2015 International Conference on Unmanned Aircraft Systems (ICUAS), pp. 50–55. IEEE (2015)
7. Jianfa, W., Honglun, W., Yiheng, L., Peng, Y.: A review of research on UAV obstacle avoidance route planning methods. Unman. Syst. Technol. **3**(01), 1–10 (2020)
8. Xiao, R., et al.: Design and experiment of UWB-based positioning aided navigation system for agricultural machinery. J. South China Agricult. Univ. **43**(3), 116–123 (2022)
9. Chen, X.: Design of laser sensor-based obstacle avoidance path control system for agricultural UAVs. Mach. Manuf. Automat. **50**(06), 170–173 (2021)
10. Afandi, M.N.R.B.M., et al.: Comparison of backstepping, fuzzy-PID, and PID control techniques using X8 model in relation to A* path planning. In: 2017 2nd IEEE International Conference on Intelligent Transportation Engineering (ICITE), pp. 340–345. IEEE (2017)
11. Wang, C., Shen, C., Zhang, K.: TDOA/AOA fusion algorithm combining steepest descent algorithm for indoor NLOS scenarios. J. Hainan Univ. (Nat. Sci. Edn.) **37**(04), 299–305 (2019)
12. Ma, Y., Liu, G., Hu, Y., et al.: Research on UAV pre-defined route obstacle avoidance based on improved artificial potential field. High Technol. Commun. **30**(1), 64–70 (2020)
13. Yuzin, G., Xiaocheng, S., Xiaopei, L., et al.: UAV obstacle avoidance control based on Laplace artificial potential field. J. Univ. Chin. Acad. Sci. **37**(05), 681–687 (2020)
14. Tu, K., Hou, H., Su, W.: Improved artificial potential field method for UAV obstacle avoidance path planning. J. Xi'an Univ. Technol. **42**(2), 170–177 (2022)
15. Chunxiao, G.: Research on quadrotor target tracking technology based on ultra-wideband positioning. Comput. Digit. Eng. **50**(02), 294–299 (2022)
16. Wan, W., Liu, Y.: Design of attitude control system for four rotor UAV. IOP Conf. Ser. Mater. Sci. Eng. IOP Publ. **452**(4), 042165 (2018)
17. Zhang, T.: Research on Ultra-Wideband Localization Technology in Non-visual Range Environment. Xi'an University of Science and Technology, Xi'an (2018)
18. Jiang, S., Qiu, Q., Yuan, S., et al.: Stochastic resonance for a bistable system with multiplicative noise and additive recycled noise. Indian J. Phys. **2022**, 1–7 (2022)

Multiple Unmanned Aerial Vehicles Path Planning Based on Collaborative Differential Evolution

Yao Lu[1,2] and Xiangyin Zhang[1,2(✉)]

[1] Faculty of Information Technology, Beijing University of Technology, Beijing 100124, China
xy_zhang@bjut.edu.cn
[2] Engineering Research Center of Digital Community, Ministry of Education, Beijing 100124, China

Abstract. In this paper, we propose a collaborative differential evolution algorithm (CODE) to solve the problem of the multiple unmanned aerial vehicles (multi-UAVs) path planning in the three-dimensional (3D) environment. Because the centralized differential evolution algorithm (DE) solves the multi-UAVs path planning problem, the dimension is too high and the amount of computation is too large, CODE divides the entire population of DE into several groups equally, and each group searches in parallel, calculates the individual cost of each UAVs and establishes an information exchange mechanism to calculate the cooperation cost between UAVs, and outputs the path corresponding to a UAV. Experimental results show that the proposed algorithm is significantly better than other comparative algorithms.

Keywords: Multiple unmanned air vehicles (multi-UAVs) · Differential evolution (DE) · Path planning · Collaborative evolution

1 Introduction

With the advancement of technology and the extension of application fields, aerospace field has developed rapidly in recent years. The advantages of UAVs are that they do not require human piloting, are flexible and easy to operate, and can replace humans to perform dangerous tasks. However, in practical applications, due to the high energy consumption, many operating equipment and other factors, a single UAV can not be able to complete all tasks. So multi-UAVs are more suitable for cooperative operations [1–3].

The purpose of path planning is to plan the shortest and lowest cost path from the start point to the end point in a defined amount of time. When planning the flight path, it is necessary to consider the performance of the UAV and avoid threat zones and no-fly zones in the flight environment. In addition, the path planning of multi-UAVs also needs to consider the time and space coordination (such as the collisions between UAVs) and the allocation between multi-UAVs and targets. Artificial potential field algorithm, fireworks algorithm (FWA) [4] and random tree algorithm (RRT) [5] can solve multi-UAVs path planning problems, but they are prone to fall into local optimum. In addition, particle swarm optimization (PSO) [6], differential evolution (DE) [7], genetic algorithm (GA) [8], ant colony optimization (ACO) [9] can solve multi-UAVs path planning problems with multiple constraints. PSO has been improved in [10] to solve the collaborative path planning problem in multi-UAVs four-dimensional environment. ACO has improved in [11] to solve the problem of multi-UAVs collaborative path planning using the idea of clustering. DE has been improved in [12], and the improved adaptive DE improves the convergence speed of shared information between particles, which can solve the problem of multi-UAVs collaborative path planning.

This paper proposes the CODE algorithm to solve the problem that the standard DE cannot plan suitable paths within the limited time due to the high complexity of multi-UAVs path planning problem. The main idea of CODE is to divide the entire population of DE evenly into several groups, one group represents a UAV, and each group evaluates and outputs its own individual costs in parallel, including terrain constraints, threat zones constraints, etc. The groups then share information and assess the cost of cooperation between UAVs to ensure that collisions do not arise between UAVs. Based on this algorithm, we can plan the paths of multi-UAVs. Experimental results show that the proposed algorithm is superior to other algorithms in solving the problem of multi-UAVs cooperative path planning.

The rest of this article is organized as follows. The second section describes the multi-UAVs path planning problem and how to model it. The third section gives a detailed process of how to plan multi-UAVs path with the CODE algorithm. The fourth section design experiment verifies the feasibility of the algorithm, and compares it with the existing algorithm to verify the superiority of the CODE algorithm. The fifth section briefly summarizes the article.

2 Mathematical Model of Multi-UAVs Path Planning

2.1 Path Representation

The purpose of multi-UAVs cooperative path planning is to plan multiple paths from the start point to the target point with high safety and low flight costs according to the performance of the UAV itself and the space and time coordination between the UAVs. The multi-UAVs cooperative path planning described in this paper belongs to global path planning, which is characterized by the known location of all threat zones and no-fly zones and the constant flight environment.

Flight cost and flight safety are the main performance indicators to evaluate the flight path of UAV. The flight cost is related to the length and altitude of the flight. Assume that in 3D flight space, the path of UAV is described as $Path = \{S, P_1, P_2, \ldots, P_D, T\}$ in the order of flight start point S, D waypoints and target point T, which can be represented by D control points and predefined trajectory smoothing strategy, as shown in Fig. 1. In order to ensure the flight ability and smoothness of UAV path, this paper uses cubic B-spline curve to describe its paths.

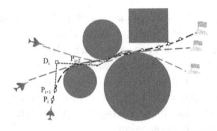

Fig. 1. Schematic diagram of the multi-UAVs paths

2.2 Cost and Constraint

The problem of multi-UAVs cooperative path optimization should consider flight cost and flight safety, and plan multiple paths with short flight length, flight height compliance, and lowest threat costs. During the flight, in order to avoid the impact of various threats, it is often necessary to process the obtained flight environment information and threat information. Threats such as terrain threats, no-fly zones, climbing and sliding angles, and the risk of collision with other UAVs are mainly considered.

Therefore, the flight costs of UAV path is calculated by

(1) Length cost

If the flight speed of the UAV is constant, the shorter the flight distance, the less time it takes to complete the flight mission. The cost function J_L represents the length of the path, which is calculated by

$$J_L = \sum_{t=0}^{D} l_t$$

(1)

with

$$l_t = \sqrt{(x_{t+1} - x_t)^2 + (y_{t+1} - y_t)^2 + (z_{t+1} - z_t)^2}$$

where (x_t, y_t, z_t) is the coordinate of the waypoint P_t.

(2) Altitude cost

Low-altitude flight is beneficial for UAV to avoid enemy attack by taking advantage of terrain and other environmental advantages. Therefore, to make UAV better search for low-altitude path, which is calculated by

$$J_H = \int_{path} H_p dl$$

(2)

with

$$H_p = \begin{cases} 0 & \text{if } z_t < 0 \\ z_p & \text{otherwise} \end{cases}$$

where is the altitude of the UAV path at the waypoint P.

The constraints of UAV is calculated as follows:

(1) Threat constraint

The article considers three types of threats during the flight of UAV: radars, missiles and anti-aircraft guns. The threat zones are represented by cylinders, and the threat probability for any point on the path is calculated by

$$PR_{,t} = \frac{R_{max}^4}{d^4 + R_{max}^4} \quad d \leq R_{max} \tag{3}$$

where $PR_{,t}$ can represent the threat value of radars, missiles, anti-aircraft guns, etc. d represents the distance of the UAV relative to the respective threat center. R_{max} indicates the scope of the threat.

The threat constraint function C_1 for all detection probabilities and kill probabilities is calculated by

$$C_1 = \sum_{t=1}^{D} PR_{,t} \tag{4}$$

(2) Forbidden flying area constraint

During the flight of the UAV, it should avoid entering some user-defined unknown zones, harsh climate zones and other no-fly zones, and its constraint function C_2 is calculated by

$$C_2 = \sum_{t=1}^{N_z} (L_{in,t}) \tag{5}$$

where $L_{in,t}$ is the length of the UAV path in the no-fly zone, and N_z is the number of no-fly zones.

(3) Terrain constraint

During the flight of the UAV, the set safe flight altitude should be less than the minimum altitude of the path above the ground. The terrain constraint function C_3 is calculated by

$$C_3 = \sum_{t=1}^{N} q_t \tag{6}$$

with

$$q_t = \begin{cases} 1 \text{ if } H_{safe} < z_t - H_{surf}(x_t, y_t) \\ 0 \text{ otherwise} \end{cases}$$

where $H_{surf}(x_t, y_t)$ is the height at (x_t, y_t) relative to the ground. H_{safe} is the minimum safe flight altitude of the UAV.

(4) Turning angle constraint

For a smoother route, the maximum turning angle is applied to each waypoint P_t. The turning angle constraint function C_4 is calculated by

$$C_4 = \sum_{t=1}^{N} q_t \tag{7}$$

with

$$q_t = \begin{cases} 1 & \text{if } \varphi_t < \varphi_t^{max} \\ 0 & \text{otherwise} \end{cases}$$

where φ_t and φ_t^{max} represents the turning angle and maximum turning angle of the UAV at P_t, respectively, calculated by

$$\varphi_t = \pi - arccos\frac{l_{t-1,t}^2 + l_{t,t+1}^2 + l_{t+1,t-1}^2}{2 \times l_{t-1,t} \times l_{t,t+1}}$$

with   (8)

$$l_{i,j} = \sqrt{(x_i - x_j)^2 + (y_i - y_j)^2}$$

$$\varphi_t^{max} = \frac{n_{max} * g}{V} \cdot \sqrt{(x_{t+1} - x_t)^2 + (y_{t+1} - y_t)^2}$$

where n_{max} is the maximum lateral overload, V is the UAV flight speed, and g is the acceleration due to gravity.

(5) Climbing/sliding angle constraints

The UAV slope S_t is related to the minimum glide slope β_t and the maximum climb slope α_t, and the size of these is related to the altitude, and the constraint function C_5 is calculated by

$$C_5 = \sum_{t=1}^{N} q_t$$

with   (9)

$$q_t = \begin{cases} 1 & \text{if } S_t \geq \alpha_t || S_t \leq \beta_t \\ 0 & \text{otherwise} \end{cases}$$

where

$$\alpha_t = -1.5377 \times 10^{-10} z_t^2 - 2.6997 \times 10^{-5} z_t + 0.4211$$
$$\beta_t = 2.5063 \times 10^{-9} z_t^2 - 6.3014 \times 10^{-6} z_t - 0.3257$$ (10)
$$S_t = \frac{z_{t+1} - z_t}{\sqrt{(x_{t+1} - x_t)^2 + (y_{t+1} - y_t)^2}}$$

(6) Collision constraint

In multi-UAVs path planning, UAVs need to coordinate in time and space, and maintain a safe distance between UAVs to avoid collisions. The coordinates of the waypoint p_i of the UAV are (x_i, y_i, z_i), and the coordinates of the other UAV waypoint p_j are (x_j, y_j, z_j). Therefore, the distance between these two UAV waypoints is calculated by

$$L_{ij} = \sqrt{(x_i - x_j)^2 + (y_i - y_j)^2 + (z_i - z_j)^2}$$ (11)

The collision constraint C_6 is calculated by

$$C_6 = \sum_{i=1}^{N} \sum_{j=1}^{N} q_{ij}$$

with   (12)

$$q_{ij} = \begin{cases} 1 & \text{if } L_{ij} < L_{safe} \wedge |t_i - t_j| < t_{safe} \\ 0 & \text{otherwise} \end{cases}$$

where t_i and t_j are the time that takes for the UAVs to fly to the waypoint p_i and p_j, respectively, L_{safe} represents the safe distance between UAVs, and t_{safe} represents the safe time to avoid collisions between UAVs.

Taking the total constraint as the objective function, the mathematical model of the multi-UAVs path planning problem can be represented by

$$min\ f = J + C$$

with

$$J = \sum_{n=1}^{N\_uav} (J_L(n) + J_H(n))$$ (13)

$$C = \sum_{n=1}^{N\_uav} \sum_{t=1}^{6} C_t(n)$$

where N_uav is the number of UAVs, C is the sum of the security costs and J is the sum of the flight costs. When the value of the objective function f is finally less than 1000, multiple feasible UAV paths can be generated.

3 Path Planning Based on CODE

3.1 Standard Differential Evolution

DE is an evolutionary algorithm encoded with real numbers proposed by Storn and Price [13], which evolves as mutation, crossover, and selection, and its principles are very similar to genetic algorithms. Algorithm 1 describes the general framework of a standard DE.

Algorithm 1 Standard differential evolution
/*Input*/
Dimension: D, generation: NC_{max}, population number: M_{pop}
/*Initialization*/
Generate individuals x_i of the initial population randomly based on a uniform distribution in the search space
While $NC < NC_{max}$ **do**
for $i=1$:M_{pop} **do**
/*Mutation*/
Generate a mutant vector v_i by using different individuals
/*Crossover*/
Generate trail vectors u_i by a random crossover scheme
/*Selection*/
Evaluate u_i and make greedy choices
End for
$NC=NC+1$
End while

3.2 Collaborative Differential Evolution

The standard DE uses the entire population to plan the paths of multi-UAVs in a centralized way, but this is too high in dimension and computationally intensive. Therefore, this paper proposes a CODE to divide the centralized large-scale optimization problems into several sub-problems, which can reduce the dimension and the amount of computation when solving the problem of multi-UAVs path planning [14].

CODE algorithm uses a collaborative search framework. The cost is evaluated in two parts: (1) the personal cost of a single UAV and (2) the cost of cooperation between UAVs (considering issues such as the collisions between UAVs). When planning the path, the population is evenly divided into several groups (one group represents a UAV), each UAV evaluates its own individual cost in parallel, and then establishes an information sharing mechanism with other UAVs to evaluate the cooperation cost between UAVs. This algorithm is divided into the following five steps.

Step 1. Initialization: The entire population X is divided into N groups X^n, and the individual in the n-th group is $x_i^n = \left\{ x_{i,1}^n,\ x_{i,2}^n,\ ...,\ x_{i,j}^n,\ ...,\ x_{i,3*D}^n \right\}, i = 1,2,...,M_{pop}/N$. When $NC = 0$, a random number distribution is used for initialization by

$$x_{i,j}^n(0) = x_{j,min}^n + rand \cdot (x_{j,max}^n - x_{j,min}^n) \tag{14}$$

where *rand* is a uniformly random number, D is the number of control points, $3*D$ is the number of control points in a 3D environment, n is the number of UAV and the number of small populations.

Step 2. Mutation: For each target vector x_i^n of each group, the mutation operation is performed by

$$v_i^n = x_{r_1^n} + F \cdot (x_{r_2^n} - x_{r_3^n}) \tag{15}$$

where v_i^n is mutant vector. $r_1^n, r_2^n, r_3^n (r_1^n \neq r_2^n \neq r_3^n \neq i)$ are the three distinct random integers selected in n groups. Parameter F is the positive control parameter of the amplification factor.

Step 3. Crossover: After the mutation, a binomial cross operator is performed between x_i^n and v_i^n to generate the trail vector u_i^n by

$$u_{i,j}^n = \begin{cases} v_{i,j}{}^n & \text{if} \quad rand < CR \text{ or } j = randr \\ x_{i,j}{}^n & \text{otherwise} \end{cases} \tag{16}$$

where the crossover rate CR is a cross operator between [0,1], $randr = \text{ceil}(3*D*rand)$.

Step 4. Evaluation:

(1) Each UAV calculates its own individual cost $f(x_i^n)$ by Eq. (14) to determine the optimal solution and sends it to other UAVs.
(2) Based on the information sent by each UAV, the cost of cooperation between UAVs $f(u_i^n)$ is calculated by Eqs. (12) (13).
(3) Consider the individual cost and cooperation cost of UAVs (total cost $f(u_i^n)$) to evaluate particle performance in population.

Step 5. Selection: In order to bring vectors with low fitness values into the next generation, the fitness values of vectors u_i^n and x_i^n are compared by

$$x_i^n(NC + 1) = \begin{cases} u_{i^n}(NC) & \text{if} \quad f(u_{i^n}) \leq f(x_{i^n}) \\ x_{i^n}(NC) & \text{otherwise} \end{cases} \tag{17}$$

3.3 Path Planning for Multi-UAVs

The pseudo-code of multi-UAVs path planning in a 3D complex environment based on CODE algorithm is shown in Algorithm 2.

Algorithm 2 CODE based path planning

/*Input*/
Set the maximum number of iterations NC_{max}, number of control points D, and algorithm parameters $F_{max}, F_{min}, CR, M_{pop}, N_{uav}$.
/*Initialization*/
Set the generation number $NC=1$.
Initialize the position X_i randomly, $i=1, 2, ..., M_{pop}$.
/*Evaluation*/
Generate the smooth trajectory $Path=\{p_0, p_1, p_2, ..., p_{D+1}\}$.
Calculate the individual cost $f(x_i^n)$ of each UAV by Eq. (13).

/*Iteration computation*/
While $NC<NC_{max}$ **do**
  For $n=1:N_{uav}$ **do**
    For $i=1:M_{pop}$ **do**
      Calculate the trial vectors u_i^n by Eq. (16).

      Evaluate individual costs $f(u_i^n)$ by Eq. (13).

      For $m=1:N_{uav}$ **do**
        If $m \neq n$
          Evaluate total cost $f(u_i^n)$ according to Step 4.

        End if
      End for
      If $f(u_i^n) < f(x_i^n)$
$$f(x_i^n) = f(u_i^n); \; x_i^n = u_i^n$$
      Else
$$f(x_i^n) = f(x_i^n); \; x_i^n = x_i^n$$
      End if
    End for
  End for
  $NC=NC+1$

End while
/*Output*/
Output the best paths for all UAVs.

---

4 Experiment Evaluation and Comparison

To evaluate the performance of the CODE algorithm in solving multi-UAVs path planning problems, two scenarios are designed to compare the performance of the CODE algorithm with other algorithms, including PSO and DE. The algorithm is implemented using Matlab-R2022 a without using any commercial algorithm tools.

The main parameters of CODE and DE are set as follows: $NC_{max} = 400$, $M_{pop} = 40$, $F_{max} = 0.9$, $F_{min} = 0.1$, $CR = 0.85$. The main parameters of PSO are set as follows: $w_{max} = 0.9$, $w_{min} = 0.1$, $c_1 = c_2 = 1.4$.

The optimal planar path generated in cases I and II is shown in Fig. 2 and Fig. 6, and the optimal path generated in the 3D environment is shown in Fig. 3 and Fig. 7. Table 1 show the statistical results of CODE, DE and PSO during 50 independent runs of Case I and Case II, respectively, recording the best, median, mean, worst, standard deviation cost values and average run time (AT).

By observing Table 1, it can be seen that under case I, CODE outperforms other algorithms in planning the optimal path of multi-UAVs. Although the algorithm takes longer to run once, it is significantly better than other algorithms in terms of the optimal cost of the best path and the average cost of all generated paths. From the Fig. 4, it can be seen that the average cost of CODE always remains optimal during iteration. In general, CODE outperforms the other three algorithms in multi-UAVs global path planning.

Table 1. Performance comparison of various algorithms on Case I and Case II

	Algorithm	Best	Mean	Median	Std	Worst	AT
Case I	CODE	501.59	770.72	536.01	672.31	3656.59	278.32
	DE	593.43	3444.13	2854.77	2272.15	10273.01	194.72
	PSO	3710.78	15845.66	15948.50	8333.44	53678.55	198.68
Case II	CODE	539.64	1321.78	792.46	811.07	2890.46	172.57
	DE	813.54	1290.78	968.74	662.43	3239.32	131.01
	PSO	885.26	57631.34	32252.27	52059.09	181578.65	127.88

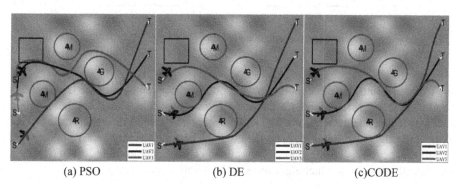

| (a) PSO | (b) DE | (c)CODE |

Fig. 2. Plane view of the best flight paths on Case I

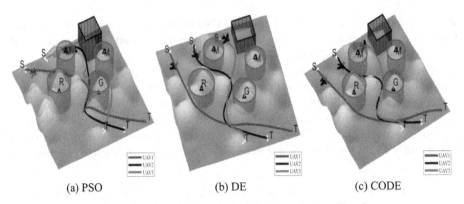

(a) PSO   (b) DE   (c) CODE

Fig. 3. The best flight paths in 3D environment on Case I

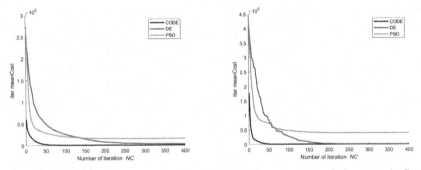

Fig. 4. Average cost evolution curve in Case I **Fig. 5.** Average cost evolution curve in Case II

The start points, the target points and environmental parameters of the UAV in Case I and Case II are different, but CODE and DE have found the global optimal solution after fifty independent operations. The second column of Table 1 represents the best cost of the three algorithms, and it can be seen that the best path of multi-UAVs planned by CODE has the lowest cost. Figure 5 shows that in Case II, the average cost of CODE is significantly superior to other algorithms in the iterative process.

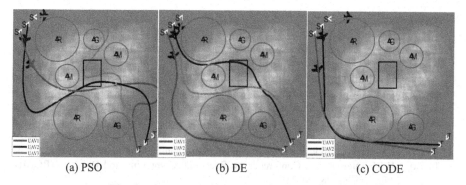

(a) PSO   (b) DE   (c) CODE

Fig. 6. Plane view of the best flight paths on Case II

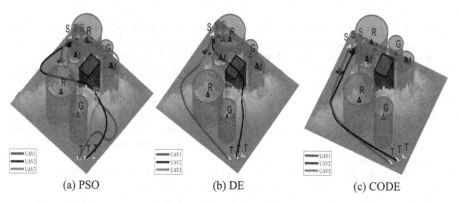

(a) PSO   (b) DE   (c) CODE

Fig. 7. The best flight paths in 3D environment on Case II

5   Conclusion

To summarize, in view of the complexity of multi-UAVs path planning problem, a CODE algorithm is proposed to divide the DE population into multiple groups, each group represents a UAV and evaluates the cost in parallel, generates the individual cost corresponding to the UAV, and establishes an information exchange mechanism between populations to avoid the interlacing of time and space between UAVs. The algorithm improves DE and is able to meet the path planning tasks of the UAVs better.

The CODE algorithm was tested in two different scenarios, and evaluated simultaneously with other algorithms through comparative testing. By observing the experimental data, it is found that the paths of the CODE scheme are the closest to the optimal solution and do not violate any constraints. This shows that the algorithm can effectively solve the problem of multi-UAVs cooperative path planning. Next, we will adjust the parameters of the improved algorithm. In addition, we will also study high-dimensional multi-UAVs path planning.

References

1. Zhao, Y., Zheng, Z., Liu, Y.: Survey on computational-intelligence-based UAV path planning. Knowl.-Based Syst. **158**, 54–64 (2018)
2. Alotaibi, E.T., Alqefari, S.S., Koubaa, A.: LSAR: Multi-UAV collaboration for search and rescue missions. IEEE Access **7**, 55817–55832 (2019)
3. Yu, H., Meier, K., Argyle, M., Beard, R.W.: Cooperative path planning for target tracking in urban environments using unmanned air and ground vehicles. IEEE/ASME Trans. Mechatron. **20**(2), 541–552 (2015)
4. Qi, Y., Liu, J., Yu, J.: A fireworks algorithm based path planning method for amphibious robot. In: 2021 IEEE International Conference on Real-time Computing and Robotics (RCAR), Xining, China, pp. 33–38 (2021)
5. Li, Z., Hu, C., Ding, C., Liu, G., He, B.: Stochastic gradient particle swarm optimization based entry trajectory rapid planning for hypersonic glide vehicles. Aerosp. Sci. Technol. **76**, 176–186 (2018)
6. Fu, Y.G., Ding, M.Y., Zhou, C.P., Hu, H.P.: Route planning for unmanned aerial vehicle (UAV) on the sea using hybrid differential evolution and quantum-behaved particle swarm optimization. IEEE Trans. Syst. Man Cybern.-Syst. **43**(6), 1451–1465 (2013)
7. Zhang, X.Y., Duan, H.B.: An improved constrained differential evolution algorithm for unmanned aerial vehicle global route planning. Appl. Soft Comput. **26**, 270–284 (2015)
8. Pehlivanoglu, Y.V., Pehlivanoglu, P.: An enhanced genetic algorithm for path planning of autonomous UAV in target coverage problems. Appl. Soft Comput. **112**, 107796 (2021)
9. Sun, J., Yu, Y., Xin, L.: Research on path planning of AGV based on improved ant colony optimization algorithm. In: 2021 33rd Chinese Control and Decision Conference (CCDC), Kunming, China, pp. 7567–7572 (2021)
10. Yang, L., Zhang, X.J., Zhang, Y.: Collision free 4D path planning for multiple UAVs based on spatial refined voting mechanism and PSO approach. Chin. J. Aeronaut. **6**, 1504–1519 (2019)
11. Wu, Y., Low, K.H., Pang, B., Tan, Q.: Swarm-based 4D path planning for drone operations in urban environments. IEEE Trans. Veh. Technol. **8**, 7464–7479 (2021)
12. Lu, L., Dai, J., Ying, J.: Distributed multi-UAV cooperation for path planning by an NTVPSO-ADE algorithm. In: 2022 41st Chinese Control Conference (CCC), Hefei, China, pp. 5973–5978 (2022)
13. Storn, R., Price, K.: Differential evolution: a simple and efficient adaptive scheme for global optimization over continuous spaces. J. Glob. Optim. **23**(1) (1995). TR-95-012
14. Gong, Y.J., Chen, W.N., Zhan, Z.H., et al.: Distributed evolutionary algorithms and their models: a survey of the state-of-the-art. Appl. Soft Comput. **34**, 286–300 (2015)

Design and Analysis of VLC-OCC-CDMA Rake System

Qiu Yang[1(✉)], Si Yujuan[2(✉)], Yu Xiaoyu[3(✉)], Yang Dong[4(✉)], Chen Yuexin[2], and Yang Wenke[2]

[1] Zhuhai College of Science and Technology, Zhuhai 519041, China
qiuy@zcst.edu.cn
[2] Jilin University, Changchun 130012, China
siyj@jlu.edu.cn, {cyx21,yangwk21}@mails.jlu.edu.cn
[3] Sun Yat-Sen University, Zhuhai 519082, China
yuxy69@mail.sysu.edu.cn
[4] Zhuhai Micro Creative Technology Limited Liability Company, Zhuhai 519082, China
yd@microcreative.org

Abstract. Visible light communication (VLC) is one of green communication technologies, which realizes dual functions of lighting and communication on the basis of indoor lighting facilities. Code Division Multiple Access (CDMA) has a wealth of signature codes, which can effectively resist narrowband interference and multipath fading by using signature codes with ideal correlation. In this paper, optical complementary code (OCC) is used as the signature code of VLC-CDMA system to meet nonnegative characteristics of visible light communication. This paper mainly studies VLC-OCC-CDMA system in multi light sources scenario. Channel model expression that only considers channel impulse response is proposed. The VLC-OCC-CDMA Rake receiver structure is designed and system performance of different combination criteria is analyzed. The results show that OCCs used in the VLC-CDMA system have good correlation characteristics, which can superpose separated multipath information effectively.

Keywords: VLC · CDMA · Channel model · Multiple sources · Rake

1 Introduction

VLC adopts indoor basic lighting facilities, which has dual functions of lighting and communication, and has become one of the key technologies of 6G [1–3]. In

This research was funded in part by the 2022 Featured Innovation Projects of General Colleges and Universities in Guangdong Province, the 2022 Enhancement of Key Construction Discipline Research Ability Project of Guangdong Province, Natural Science Foundation of Guangdong Province, Project of Major Health Industry Related Disciplines at Zhuhai College of Science and Technology (Nos. 2022KTSCX189, 2022ZDJS140, 2023A1515011302, 2023DJKCY010) and Doctoral Promotion Program of Zhuhai College of Science and Technology.

the multi-user scenario, a suitable multiple access method is needed to meet the requirements of multi-user communication. CDMA technology has the function of satisfying the simultaneous access of multiple users to network [4–7]. OCC is a kind of unipolar complementary code, which can meet the physical characteristics of VLC. Meanwhile, OCC has ideal autocorrelation characteristics and minimum cross-correlation sidelobe interference, which can ensure that VLC-CDMA systems can completely eliminate multi-user interference (MUI) in line of sight (LOS) [5].

Intensity modulation/direct detection (IM/DD) is usually used in VLC systems. Compared with the system communication rate, the movement rate of transmitters, receivers and other objects is very slow, and there is no multi-path fading [8–11]. Therefore, the impulse response can be used to represent the channel model [10,12–14]. In VLC-OCC-CDMA system, the transmitting end of VLC-OCC-CDMA system usually adopts LED array composed of multiple LED chips [4,15]. The multiple light source system in this paper adopts a four light source structure. Each light source contains 38 groups of RGB LED arrays, namely 152 LED chips. It can meet the international lighting standards, the illumination is more moderate, and will not cause waste of resources.

The main principle of Rake receiver is to collect the energy of multi-path transmission. Then, the received signal is enhanced by superimposing the multi-path energy using the good correlation characteristics of signature codes [16–18]. Partial Rake receivers collect the optical signals of L paths. This method can directly collect part of the multipath energy, and directly combine the previous L paths without the process of selection [19,20].

The above discussions motivate us to work out a better way to implement VLC-OCC-CDMA Rake systems. The major contributions of this work can be summarized as follows. In this work, we introduce VLC-OCC-CDMA using Rake receiver, which is named as VLC-OCC-CDMA Rake system; We give the transmitter structure of VLC-OCC-CDMA system and analyze channel model of VLC system. We propose the channel model expression that only considers the channel impulse response; We design the VLC-OCC-CDMA Rake receiver. We conduct a theoretical analysis on the system performance with different combing methods, and show that the signature codes used in the VLC-OCC-CDMA system have good correlation characteristics, which can superpose the separated multipath information effectively.

2   System Model

2.1   Transmitter

The multiple sub-codes are used in VLC-OCC-CDMA system, and different sub-codes are transmitted by LEDs with different wavelengths at the transmitter. The system transmitter model is shown in Fig. 1.

Transmitters are white LEDs and use OOK modulation. $b^{(k)}$ is signal of the kth user after modulation. $b_m^{(k)}$ is the mth data stream of the kth user, where $m \in \{1, 2, \cdots, M\}$, $k \in \{1, 2, \cdots, K\}$. A set of complementary codes $\mathcal{C}(K, M, N)$ are used as signature codes for K users in the system. Assume that $\mathbf{C}^{(k)} = \{\mathbf{c}_m^{(k)}\}_{m=1}^{M}$ is signature code for user k [5].

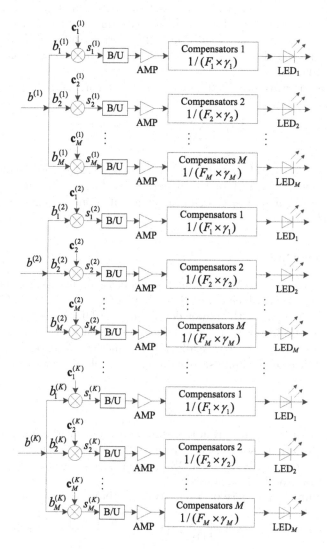

Fig. 1. Transmitters structure of the VLC-CC-CDMA system.

The number of active users in the system is K, each user is assigned a OCC, the length of each sub-code is N. B/U module is used to convert bipolar signals into unipolar signals. AMP is used to amplify signals. The optical filter gain at the receiving end is F_m, whose center wavelength is the same as the peak wavelength of LEDs approximately. The PD response sensitivity is γ_m. We add compensator circuits for LEDs with the gains of $1/(F_m \times \gamma_m)$.

Each data stream of a user is assigned a separate LED to ensure that users can work in linear regions of LEDs [21]. Each sub-code of a user is transmitted by a LED with a different peak wavelength to avoid interference among sub-codes of the same user. Therefore, M LEDs with different peak wavelengths are required to send the data of M sub-codes. The corresponding sub-codes of each user are transmitted through indoor VLC channels, and the specific process is explained as follows.

The spreading waveform of the mth sub-code for user k is,

$$C_m^{(k)}(t) = \sum_{n=1}^{N} c_{m,n}^{(k)} q(t - nT_c + T_c), \tag{1}$$

where $m \in \{1, 2, \cdots, M\}$, $n \in \{1, 2, \cdots, N\}$, $k \in \{1, 2, \cdots, K\}$, which are the same definitions in the rest of the paper. T_c is duration of one chip, $q(t)$ is chip pulse waveform with $q(t)$ being a rectangular pulse. When $0 \leq t < T_c$, $q(t) = 1/\sqrt{MNT_c}$.

2.2   Channel Model

Lee and others established VLC channel model based on infrared communication channel model in 2011, which is widely used in VLC systems [22,23]. Considering LOS and reflection paths, the power delay profile (PDP) is used to characterize channel model of VLC system. In practical analysis, if we want to get the pure channel impulse response, we usually use the method of removing the power term. The LOS channel impulse response $h^{(0)}(t)$ obtained by this method is shown as,

$$h^{(0)}(t) = \frac{A_{PD}(m+1)}{2\pi d_0^2} \cos^m \varphi_0 \cos \theta_0, \tag{2}$$

where d_0 represents distance from transmitter to receiver in LOS, φ_0 represents radiation angle of light source, and θ_0 represents incidence angle of receiver. It is the same as the expression of LOS channel impulse response in [8], which indicate that our method is correct.

Channel impulse response of p reflections is obtained shown as,

$$h^{(p)}(t) = \int_S \left[\mathcal{L}_1 \mathcal{L}_2 \cdots \mathcal{L}_{p+1} \mathrm{rect}\left(\frac{\theta_{p+1}}{\mathrm{FOV}}\right) \delta\left(t - \frac{d_1 + d_2 + \cdots d_{p+1}}{c}\right) \right] dA_{\mathrm{ref}}, \quad p \geq 1. \tag{3}$$

The channel impulse response can be obtained as,

$$h(t) = \sum_{q=1}^{\mathcal{Q}_{LED}} \sum_{p=0}^{\infty} h^{(p)}(t), \tag{4}$$

where $\mathcal{Q}_{LED}$ is the number of LEDs.

Based on the analysis of channel model proposed by Lee et al., the channel impulse response expression (4) using in this paper is obtained [22]. The correctness of this representation is illustrated by comparing existing literature. The model expression will be used as channel model in the system design and simulation analysis in this paper.

3   Design and Analysis for VLC-OCC-CDMA Rake Receiver

3.1   Design of Rake Receiver for VLC-OCC-CDMA System

The main principle of Rake receiver is to collect energy of multi-path transmission, and use good correlation characteristic of signature codes to superimpose the multi-path energy to enhance received signal. Figure 2 shows block diagram of VLC-OCC-CDMA Rake receiver.

The number of paths L is the tap coefficient of Rake receiver. The difference between the previous receiver and Rake receiver is that the received signal $r(t)$ delayed by multipath firstly and aligned each path. The method is to delay the received signal $\tau^L - \tau^{l'}$, $l' = 1, 2, \cdots L$ is the total number of paths. After signal delay and alignment, the desired receiving signal of the l'th path is expressed as $r^{l'}(t)$, that is, $r^{l'}(t) = r\left[t - (\tau^L - \tau^{l'})\right]$.

The next step is despreading process of the desired signal on each path. $c_m^{(g)}$ indicates signature code of the mth sub-code for the gth user. $\widehat{b}_m^{l'(g)}$ is result of equal gain combination of sub-codes in the l'th path for the gth user. $\widehat{b}_c^{l'(g)}$ representes combination coefficient of the signal from the l'th path. The final output signal is obtained by setting an appropriate decision threshold. The decision threshold is usually set to half of detection peak, i.e. $w_M/2$.

3.2   Analysis of Rake Receiver for VLC-OCC-CDMA System

The transmitted signal of the kth user is shown as,

$$s^{(k)}(t) = \sum_{m=1}^{M} s_m^{(k)}(t) S_m(\lambda) \frac{1}{F_m \gamma_m}, \tag{5}$$

where λ is wavelength, $S_m(\lambda)$ is spectrum function of LED at transmitter.

The received signal is shown as,

$$r(t) = \sum_{k=1}^{K} \sum_{l=1}^{L} h^{l(k)}(t) s^{l(k)}(t) + n(t), \tag{6}$$

where, $h^{l(k)}(t)$ is channel impulse response of the kth user on the lth path. Therefore, the channel impulse response of the kth user can be expressed as $h^{(k)}(t) = \sum_{l=1}^{L} h^{l(k)}(t)$. $s^{l(k)}(t) = s^{(k)}(t - \tau^l)$ represents the signal of the kth user after delay on the lh path, and τ^l is delay of the lth path. $n(t)$ is noise.

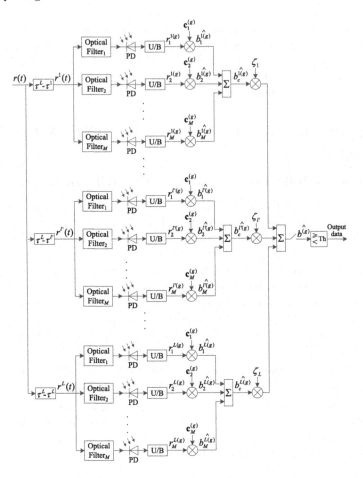

Fig. 2. Diagram of VLC-CC-CDMA Rake system.

$r^{l'}(t) = r[t-(\tau^L - \tau^{l'})]$ is the received signal after delay alignment, indicating the signal of the l'th path among received signals. It is required to multiply gains of optical filters and PDs. Taking despreading a single user as an example, the received signal of the gth user on the l'th path is shown as,

$$r_m^{l'(g)}(t) = \sum_{k=1}^{K} \sum_{l=1}^{L} h^{l(k)}(t) s_m^{l(k)} \left[t - \tau_k - (\tau^L - \tau^{l'}) \right] + n_m(t), \tag{7}$$

where, the channel impulse response of the kth user on the lth path should be expressed as $h_m^{l(k)}(t)$. Assuming that the impulse response of each sub-code is the same, which is $h^{l(k)}(t)$. $n_m(t)$ is noise of the mth data stream.

$\mathbf{c}_m^{(g)}$ indicates the mth sub-code of the desired user g, where the same signature code is used in each path. $\widehat{b}_m^{l'(g)}(t)$ is despreaded result of the mth sub-code for the jth data on the l'th path, shown as,

$$\widehat{b}_m^{l'(g)}(j) = \int_0^{NT_c} r_m^{l'(g)}(t + jT_b + \tau_g) C_m^{(g)}(t) dt = \sqrt{P_t} h^{l'(g)} b^{l'(g)}(j) + I_m^{l'(g)} + \mathfrak{I}_m^{l'(g)} + v_m^{l'}.$$
(8)

The multi-path transmission of VLC channel models only has energy gain. When the system adopts selective combining, the weighting coefficient on the selected path $l' = 1$, $\zeta_1 = 1$. When the system adopts maximum ratio combining, the weighting coefficient $\zeta_{l'} = \frac{E^{l'}}{\sum_{l'=1}^L E^{l'}}$. Since the path coefficient is normalized, the combination coefficient $\zeta_{l'} = h^{l'}$. When the system adopts equal gain combining, the combining coefficients $\zeta_1 = \cdots = \zeta_{l'} = \cdots = \zeta_L = \frac{1}{L}$. We can get the jth data of the gth user $\widehat{b}_{\mathrm{SC}}^{(g)}(j)$, $\widehat{b}_{\mathrm{MRC}}^{(g)}(j)$, and $\widehat{b}_{\mathrm{EGC}}^{(g)}(j)$ above three combination modes are shown as,

$$\begin{cases} \widehat{b}_{\mathrm{SC}}^{(g)}(j) = \zeta_1 \widehat{b}_c^{1(g)}(j) = \widehat{b}_c^{1(g)}(j) = \sqrt{P_t} \sum_{m=1}^M h^{1(g)} b_m^{1(g)}(j) + I^{1(g)} + \mathfrak{I}^{1(g)} + V^1, \\ \widehat{b}_{\mathrm{MRC}}^{(g)}(j) = \sqrt{P_t} \sum_{l'=1}^L \sum_{m=1}^M \left[h^{l'(g)} \right]^2 b_m^{l'(g)}(j) + \sum_{l'=1}^L h^{l'(g)} I^{l'(g)} \\ + \sum_{l'=1}^L h^{l'(g)} \mathfrak{I}^{l'(g)} + \sum_{l'=1}^L h^{l'(g)} V^{l'}, \\ \widehat{b}_{\mathrm{EGC}}^{(g)}(j) = \frac{\sqrt{P_t}}{L} \sum_{l'=1}^L \sum_{m=1}^M h^{l'(g)} b_m^{l'(g)}(j) + \frac{1}{L} \sum_{l'=1}^L I^{l'(g)} + \frac{1}{L} \sum_{l'=1}^L \mathfrak{I}^{l'(g)} \\ + \frac{1}{L} \sum_{l'=1}^L V^{l'}. \end{cases}$$
(9)

On the basis of the assumption that transmission power is normalized, the desired decision variable, namely $w_M/2$, can be obtained. Thus, the appropriate decision variable cannot be set by interference problem can be avoided.

3.3 Theoretical Analysis on Elimination of Interference and BER

The first item in the decision variable is user data to be recovered, and the second item is interference from other users, where, $I^{l'(g)} = \sum_{m=1}^M I_m^{l'(g)}$, $I_m^{l'(g)}$ is interference of the gth user from other users in the mth sub-code on the l'th path. The third term is multipath interference, where $\mathfrak{I}^{l'(g)} = \sum_{m=1}^M \mathfrak{I}_m^{l'(g)}$, $\mathfrak{I}_m^{l'(g)}$ is interference of the gth user from other paths in the mth sub-code on the l'th path, and the multi-user interference $I_m^{l'(g)}$ be expressed as:

$$I_m^{l'(g)} = \sum_{k=1, k \neq g}^K \sum_{l=1}^L \sqrt{P_t} h^{l(k)} \left\{ \alpha_m^{l(k)} b_m^{l(k)}(j) + \beta_m^{l(k)} b_m^{l(k)} \left[j + \mathrm{sgn}(\delta_k^l) \right] \right\}, \quad (10)$$

where, $\mathrm{sgn}(x)$ is equal to 1 when $x \geq 0$ and -1 when $x < 0$, $\delta_k^l = (\tau_g - \tau_k + \tau^{l'} - \tau^l)/T_c$. The values of $\alpha_m^{l(k)}$ and $\beta_m^{l(k)}$ are determined by relative delay of data stream of the lth path of the kth user and the l'th path of the lth user. Since L paths need to be considered, $\alpha_m^{l(k)}$ and $\beta_m^{l(k)}$ of the lth path are shown as,

$$
\begin{cases}
\delta_k^l > 0 : \alpha_m^{l(k)} = \phi(\mathbf{c}_m^{l(g)}, \mathbf{c}_m^{l(k)}; \delta_k^l), & \beta_m^{l(k)} = \phi(\mathbf{c}_m^{l(k)}, \mathbf{c}_m^{l(g)}; N - \delta_k^l) \\
\delta_k^l < 0 : \alpha_m^{l(k)} = \phi(\mathbf{c}_m^{l(k)}, \mathbf{c}_m^{l(g)}; -\delta_k^l), & \beta_m^{l(k)} = \phi(\mathbf{c}_m^{l(g)}, \mathbf{c}_m^{l(k)}; N + \delta_k^l) \\
\delta_k^l = 0 : \alpha_m^{l(k)} = \phi(\mathbf{c}_m^{l(g)}, \mathbf{c}_m^{l(k)}; 0), & \beta_m^{l(k)} = 0
\end{cases}
\tag{11}
$$

where $\phi(\mathbf{a}, \mathbf{b}; \delta)$ is aperiodic correlation function of $\mathbf{a}$ and $\mathbf{b}$.

Multipath interference can be expressed as:

$$
\mathfrak{I}_m^{l'(g)} = \sum_{l=1, l \neq l'}^{L} \sqrt{P_t} h^{l(g)} b^{l(g)}(j) \phi(\mathbf{c}_m^{l(g)}, \mathbf{c}_m^{l(g)}; \delta^l)
\tag{12}
$$

$$
+ \sum_{l=1, l \neq l'}^{L} \sqrt{P_t} h^{l(g)} b^{l(g)}[j + \mathrm{sgn}(\delta^l)] \phi(\mathbf{c}_m^{l(g)}, \mathbf{c}_m^{l(g)}; N - \delta^l),
$$

where, $\delta^l = (\tau^{l'} - \tau^l)/T_c$. When path variable $l' = 1$, the result is interference expression of selective combining.

Assuming $\Delta^{l(g)}$ and $\mathfrak{d}^{l(g)}$ are expressed as,

$$
\begin{cases}
\Delta^{l(g)} = \sum_{m=1}^{M} \sum_{k=1, k \neq g}^{K} \sum_{l=1}^{L} \sqrt{P_t} h^{l(k)} \left\{ \alpha_m^{l(k)} b_m^{l(k)}(j) + \beta_m^{l(k)} b_m^{l(k)}[j + \mathrm{sgn}(\delta_k^l)] \right\}, \\
\mathfrak{d}^{l(g)} = \sum_{m=1}^{M} \left\{ \sum_{l=1, l \neq l'}^{L} \sqrt{P_t} h^{l(g)} b^{l(g)}(j) \phi(\mathbf{c}_m^{l(g)}, \mathbf{c}_m^{l(g)}; \delta^l) \right. \\
\left. + \sum_{l=1, l \neq l'}^{L} \sqrt{P_t} h^{l(g)} b^{l(g)}[j + \mathrm{sgn}(\delta^l)] \phi(\mathbf{c}_m^{l(g)}, \mathbf{c}_m^{l(g)}; N - \delta^l) \right\}.
\end{cases}
\tag{13}
$$

Interference of the three combination modes can be shown as,

$$
\begin{cases}
I_{\mathrm{SC}}^{(g)} = \frac{1}{h^{l(g)}} \Delta^{l(g)}, \; \mathfrak{I}_{\mathrm{SC}}^{(g)} = \frac{1}{h^{1(g)}} \mathfrak{d}^{l(g)} \\
I_{\mathrm{MRC}}^{(g)} = \frac{\sum_{l'=1}^{L} h^{l'(g)}}{\sum_{l'=1}^{L} [h^{l'(g)}]^2} \Delta^{l(g)}, \; \mathfrak{I}_{\mathrm{MRC}}^{(g)} = \frac{\sum_{l'=1}^{L} h^{l'(g)}}{\sum_{l'=1}^{L} [h^{l'(g)}]^2} \mathfrak{d}^{l(g)} \\
I_{\mathrm{EGC}}^{(g)} = \frac{\sum_{l'=1}^{L}}{\sum_{l'=1}^{L} h^{l'(g)}} \Delta^{l(g)}, \; \mathfrak{I}_{\mathrm{EGC}}^{(g)} = \frac{\sum_{l'=1}^{L}}{\sum_{l'=1}^{L} h^{l'(g)}} \mathfrak{d}^{l(g)}.
\end{cases}
\tag{14}
$$

According to the correlation characteristics of OCC, the maximum value of cross-correlation of sub-codes is 1. When there is only one LOS, VLC-OCC-CDMA system can eliminate interference by designing different positions of 1 in sub-codes. However, VLC-OCC-CDMA system is still affected by interference in case of multipath due to poor cross-correlation and side lobe.

The last item of the decision variable is noise, where $V^{l'} = \sum_{m=1}^{M} v_m^{l'}$. It is still a Gaussian process and its distribution is the same as $v_m^{l'}$. $v_m^{l'}$ is noise of the mth sub-code. The mean value is zero, and the variance is shown as,

$$
\sigma_{\mathrm{thermal}}^2 = \frac{8\pi k_B \mathcal{T}_\mathcal{K} \eta A_{\mathrm{PD}} I_2 B^2}{\mathcal{G}} + \frac{16\pi^2 k_B \mathcal{T}_\mathcal{K} \varepsilon \eta^2 A_{\mathrm{PD}}^2 I_3 B^3}{\mathfrak{g}},
\tag{15}
$$

where k_B is Boltzmann constant, $\mathcal{T}_\mathcal{K}$ is absolute temperature, η is fixed capacitance per unit area of the PD, A_{PD} is the area of the PD, and I_2 is noise bandwidth factor. B is noise bandwidth equal to the value of data rate (R_b). $\mathcal{G}$ is openloop voltage gain. ε is the channel noise factor of field effect transistor (FET), I_3 is noise bandwidth factor, and $\mathfrak{g}$ is the transconductance of the FET,

$$\sigma_{\text{shot}}^2 = 2qP_rB + 2qI_{bg}I_2B, \tag{16}$$

where q is the electron charge, P_r is received optical power, and I_{bg} is background current.

The variances of the noise term ($\text{var}[V_{\text{SC}}]$, $\text{var}[V_{\text{MRC}}]$ and $\text{var}[V_{\text{EGC}}]$) in the three combination modes can be updated as,

$$\begin{cases} \text{var}[V_{\text{SC}}] = \text{var}\left[\dfrac{V^1}{h^1(g)}\right] = \dfrac{1}{[h^1(g)]^2}(\sigma_{\text{shot}}^2 + \sigma_{\text{thermal}}^2), \\[2ex] \text{var}[V_{\text{MRC}}] = \text{var}\left[\dfrac{\sum_{l'=1}^{L} h^{l'}(g)V^{l'}}{\sum_{l'=1}^{L}\left[h^{l'}(g)\right]^2}\right] = \left\{\dfrac{\sum_{l'=1}^{L} h^{l'}(g)}{\sum_{l'=1}^{L}\left[h^{l'}(g)\right]^2}\right\}^2(\sigma_{\text{shot}}^2 + \sigma_{\text{thermal}}^2), \\[2ex] \text{var}[V_{\text{EGC}}] = \text{var}\left[\dfrac{\sum_{l'=1}^{L} V^{l'}}{\sum_{l'=1}^{L} h^{l'}(g)}\right] = \sum_{l'=1}^{L}\left\{\dfrac{1}{\sum_{l'=1}^{L} h^{l'}(g)}\right\}^2(\sigma_{\text{shot}}^2 + \sigma_{\text{thermal}}^2). \end{cases} \tag{17}$$

Assume E_b is bit energy. The energy is $E_{b1} = P_t PG T_c$ when sending "1". Where, P_t is chip transmission power, $PG = MN$ is processing gain, T_c is chip time. There is only one "1" in each sub-code of OCC. The processing gain does not need to consider the length of sub-code (N). Assuming that "0" and "1" are transmitted as equal probabilities, the average bit energy is $E_b = (E_{b1} + E_{b0})/2$. Signal to interference and noise ratio (SINR) can be expressed as:

$$\begin{cases} \text{SINR}_{\text{SC}} = \dfrac{E_b}{\text{var}\left[I_{\text{SC}}^{(g)}\right] + \text{var}\left[\mathfrak{I}_{\text{SC}}^{(g)}\right] + \text{var}\left[V_{\text{SC}}\right]}, \\[2ex] \text{SINR}_{\text{MRC}} = \dfrac{E_b}{\text{var}\left[I_{\text{MRC}}^{(g)}\right] + \text{var}\left[\mathfrak{I}_{\text{MRC}}^{(g)}\right] + \text{var}\left[V_{\text{MRC}}\right]}, \\[2ex] \text{SINR}_{\text{EGC}} = \dfrac{E_b}{\text{var}\left[I_{\text{EGC}}^{(g)}\right] + \text{var}\left[\mathfrak{I}_{\text{EGC}}^{(g)}\right] + \text{var}\left[V_{\text{EGC}}\right]}. \end{cases} \tag{18}$$

OOK modulation is applied in the system, BER can be expressed as: $T_{\text{BER}_{\text{OOK}}} = Q\left(\sqrt{\dfrac{E_b/N_0}{2}}\right)$, where Q function is $Q(x) = \dfrac{1}{2\pi}\int_x^\infty e^{\frac{-t^2}{2}} dt$.

The decision variable after sub-code combination is multiplied by corresponding coefficient when the VLC-OCC-CDMA Rake system uses different combination modes. The theoretical BER ($T_{\text{BER}_{\text{SC}}}$, $T_{\text{BER}_{\text{MRC}}}$ and $T_{\text{BER}_{\text{EGC}}}$) can be expressed as:

$$\begin{cases} T_{\text{BER}_{\text{SC}}} = Q\left(\sqrt{\dfrac{E_b/N_0}{2}(h^1)^2}\right), \\[2ex] T_{\text{BER}_{\text{MRC}}} = Q\left(\sqrt{\dfrac{E_b/N_0}{2}\sum_{l'=1}^{L}(h^{l'})^2}\right), \\[2ex] T_{\text{BER}_{\text{EGC}}} = Q\left(\sqrt{\dfrac{E_b/N_0}{2}\dfrac{(\sum_{l'=1}^{L} h^1)^2}{L}}\right), \end{cases} \tag{19}$$

where, $T_{BER_{SC}}$, $T_{BER_{MRC}}$ and $T_{BER_{EGC}}$ are single user theoretical BER expressions when the system adopts selectivity, maximum ratio and equal gain combining, respectively.

4   Simulation Details and Results

4.1   Simulation Setup

The simulation parameters are shown in Table 1. Assumed that data rate $R_b = 125$ Mbps and data length is 10^6. The positions of 1s for OCC (16, 1, 4, 0, 1) are: [1 1 1 1], [2 3 4 5], [3 2 6 8], [4 6 2 10], [5 8 10 13], [6 4 3 2], [7 11 8 4], [8 5 12 15] [9 15 5 11], [13 9 7 12] and [16 10 15 9], respectively. The length of sub-code $N = 16$, the number of sub-code $M = 4$, and the detection peak value is also 4. The autocorrelation sidelobe is 0, the maximum value of the cross-correlation interference is 1, and the maximum number of cross-correlation interference is 4. The chip rate $R_c = 2$ Gcps and chip time $T_c = 0.5$ ns. The channel impulse response calculation method in Sect. 2.2. OOC (64, 4, 1) is selected as the system performance comparison, the positions of 1s for the five codes are: [8 15 27 37], [1 2 6 26], [4 7 25 34], [9 17 23 40] and [3 5 16 31], respectively.

Table 1. Simulation parameters of channel impulse response for visible light communication.

Parameters	Values
Room size [8]	$5 \times 5 \times 3 \, \text{m}^3$
Locations of sources [8]	(1.25, 1.25, 3) (3.75, 1.25, 3)
	(1.25, 3.75, 3) (3.75, 3.75, 3)
Height of receiver (h) [8]	0.85 m
Reflection index of walls [22]	0.8
half-power angle of LED ($\varphi_{1/2}$) [24]	60°
Receiving area of PD (A_{PD}) [24]	1 cm^2
Field of view of PD (FOV) [24]	70°
Reflection index of optical concentrator (n) [24]	1.5
transmitted power of LED [24]	249.6 mW
transmitted power of RGBY LED [24]	4.308 W
luminous intensity of RGBY LED [24]	76.2 cd

4.2   Simulation Results

The designed OCCs and OOCs are used as signature codes of VLC-CDMA Rake system for comparative analysis. The system performance of selective, maximum ratio and equal gain combining are given as Figs. 3, 4 and 5. It can be seen that BER performances get better with increase of SNR and worse with increase of K, which is caused by multi-user interferences. When the number of users $K = 1$, the simulation results using the two signature codes coincide with theoretical curve. This is because there are no multi-user interferences when there is only one user in the system, which also shows the accuracy of our theoretical BER analysis. With the increase of the number of users, the performance of using OCCs is obviously better than that of using OOCs. This is because the OCCs has ideal autocorrelation characteristics and minimum cross-correlation interference.

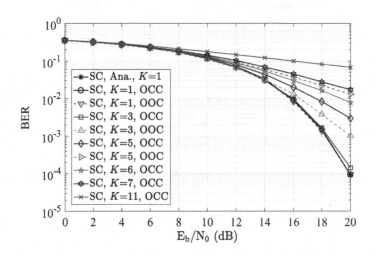

Fig. 3. SC results of VLC-CC-CDMA Rake system with OCCs and OOCs.

The figures also show performance of using OCCs when $K = 6$ and $K = 7$. Since OCC supported users of OCCs and OOCs are 11 and 5, respectively, the performance using OCCs ($K = 6$) is better than using OOCs ($K = 5$). However, when $K = 7$, the performance using OCCs starts to be worse than that of OOCs. The main purpose of showing this trend is to conclude that VLC-OCC-CDMA system has more advantages than VLC-OOC-CDMA system in terms of number of users. In addition, the performance of the maximum number of users ($K = 11$) using OCCs is also given in order to observe the performance changes with different users. The chip rates of OCCs and OOCs are assumed as the same in VLC-CDMA system simulations. The length of OCC and OCC are 16 and 64, respectively, the communication rate of the system using OCCs is 4 times than that of OOCs.

From Figs. 3, 4 and 5, BER performance is getting worse with the increases of number of users. BER performance with a large SNR is generally better than with small SNR. The result of equal gain combining is the worst. When the number of users is small, BER performance with the maximum ratio combining is the best. However, with the increase of users, the best BER performance becomes the system with selective combining. The main reason is that when the number of users is relatively large, the multi-user interference has a significant impact on the system performance.

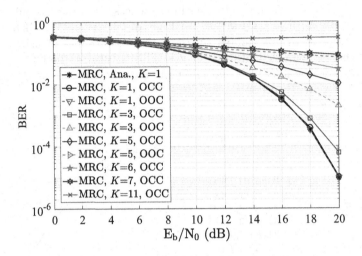

Fig. 4. MRC results of VLC-CC-CDMA Rake system with OCCs and OOCs.

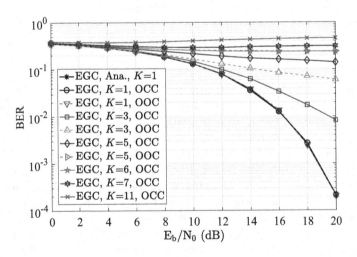

Fig. 5. EGC results of Rake-VLC-CC-CDMA system with OCCs and OOCs.

In the above analysis, corresponding coefficients have been multiplied to maintain the judgment variable as $w_M/2$. The original multi-user and multi-path interferences also change, and the interferences with selectivity, maximum ratio and equal gain combining shown as (14). The multipath energy gain in VLC only comes from each path loss and no gain generated by multipath fading.

5   Conclusions

This paper mainly studies the design and analysis of Rake receiver in VLC-OCC-CDMA system with multiple light sources. In this paper, the optical complementary code (OCC) is used as signature codes of VLC-CDMA system to meet the nonnegative characteristics of visible light communication. In practical application environment, the illumination and receiving power of multiple light sources are more uniform than that of single light source. Therefore, this paper mainly studies VLC-OCC-CDMA system in multi light sources scenario. Rake receiving technology is used to make full use of multi-path energy, which is caused by multi-path transmission in the multi-light sources. The VLC-OCC-CDMA Rake receiver structure is designed and the system performance of different combination criteria is analyzed. The results show that the signature codes used in the VLC-OCC-CDMA system have good correlation characteristics, which can superpose the separated multipath information effectively.

References

1. Naser, S., et al.: Toward federated-learning-enabled visible light communication in 6G systems. IEEE Wirel. Commun. **29**(1), 48–56 (2022)
2. Abumarshoud, H., Mohjazi, L., Dobre, O.A., Di Renzo, M., Imran, M.A., Haas, H.: LiFi through reconfigurable intelligent surfaces: A new frontier for 6G? IEEE Vehicul. Technol. Magaz. **17**(1), 37–46 (2022)
3. Stefano, C., Lorenzo, M., Muhammad, A.U., Marco, M., Marco, S., Jacopo, C.: The role of bidirectional VLC systems in low-latency 6G vehicular networks and comparison with IEEE802.11p and LTE/5G C-V2X. Sensors **22**(22), 8618 (2022)
4. Krohn, A., Harlakin, A., Arms, S., Pachnicke, S., Hoeher, P.A.: Impact of liquid crystal based interference mitigation and precoding on the multiuser performance of VLC massive MIMO arrays. IEEE Photon. J. **14**(5), 1–12 (2022)
5. Qiu, Y., Chen, H.H., Li, J.Q., Meng, W.X.: VLC-CDMA systems based on optical complementary codes. IEEE Wirel. Commun. **27**(1), 147–153 (2020)
6. Wang, X., Chen, H.H., Liu, X.Q., Guo, Q.: Complementary coded CDMA with multi-Layer quadrature modulation. IEEE Trans. Vehicul. Technol. **71**(3), 2991–3007 (2022)
7. Qiu, Y., Chen, S.Y., Chen, H.H., Meng, W.X.: Visible light communications based on CDMA technology. IEEE Wirel. Commun. **25**(2), 178–185 (2018)
8. Komine, T., Nakagawa, M.: Fundamental analysis for visible-light communication system using LED lights. IEEE Trans. Consum. Electron. **50**(1), 100–107 (2004)
9. Kahn, J.M., Krause, W.J., Carruthers, J.B.: Experimental characterization of non-directed indoor infrared channels. IEEE Trans. Commun. **43**(2/3/4), 1613–1623 (1995)
10. Higgins, M.D., Green, R.J., Leeson M.S.: A. Genetic algorithm method for optical wireless channel control. J. Lightwave Technol. **27**(6), 760–772 (2009)

11. Yang, S.H., Jung, E.M., Han, S.K.: Indoor location estimation based on LED visible light communication using multiple optical receivers. IEEE Commun. Lett. **17**(9), 1834–1837 (2013)

12. Barry, J.R., Kahn, J.M., Krause, W.J., Lee, E.A., Messerschmitt, D.G.: Simulation of multipath impulse response for indoor wireless optical channels. IEEE J. Select. Areas Commun. **11**(3), 367–379 (1993)

13. Higgins, M.D., Green, R.J., Leeson, M.S.: Optical wireless for intravehicle communications: A channel viability analysis. IEEE Trans. Vehicul. Commun. **61**(1), 123–129 (2012)

14. Turan, B., Narmanlioglu, O., Koc, O.N., Kar, E., Coleri, S., Uysal, M.: Measurement based non-line-of-sight vehicular visible light communication channel characterization. IEEE Trans. Vehicul. Technol. **71**(9), 10110–10114 (2022)

15. Niu, W.Q., et al.: Phosphor-free golden light LED array for 5.4-Gbps visible light communication using MIMO tomlinson-harashima precoding. J. Lightwave Technol. **40**(15), 5031–5040 (2022)

16. Ali, W., Manousiadis, P.P., O'Brien, D.C., Turnbull, G.A., Samuel, I.D.W., Collins, S.: A gigabit VLC receiver that incorporates a fluorescent antenna and a SiPM. J. Lightwave Technol. **40**(16), 5369–5375 (2022)

17. Demeslay, C., Rostaing, P., Gautier, R.: Simple and efficient LoRa receiver scheme for multipath channel. IEEE Internet Things J. **9**(17), 15771–15785 (2022)

18. Weisman, R., Shlomo, T., Tourbabin, V., Calamia, P., Rafaely, B.: Robustness of acoustic Rake filters in minimum variance beamforming. IEEE/ACM Trans. Audio Speech Lang. Process. **29**, 3668–3678 (2021)

19. Ershadh, M.: A computationally lightest and robust neural network receiver for ultra wideband time hopping communication systems. IEEE Trans. Vehicul. Technol. **70**(5), 4657–4668 (2021)

20. Hamza, A., AlShammary, H., Hill, C., Buckwalter, J.F.: A full-duplex Rake receiver using RF code-domain signal processing for multipath environments. IEEE J. Solid-State Circuits **56**(10), 3094–3108 (2021)

21. Qian, H., Dai, S.C., Zhao, S., Cai, S.Z., Zhang, H.: A robust CDMA VLC system against front-end nonlinearity. IEEE Photon. J. **7**(5), 7801809 (2015)

22. Lee, K., Park, H., Barry, J.: Indoor channel characteristics for visible light communications. IEEE Commun. Lett. **15**(2), 217–219 (2011)

23. Lee, K., Park, H.: Channel model and modulation schemes for visible light communications. In: 54th International Midwest Symposium on Circuits and Systems, pp. 1–4. IEEE, Seoul (2011)

24. Komine, T., Lee, J.H., Haruyama, S., Haruyama, S., Nakagawa, M.: Adaptive equalization system for visible light wireless communication utilizing multiple white LED lighting equipment. IEEE Trans. Wirel. Commun. **8**(6), 2892–2900 (2009)

Machine Learning

Noise-Tolerant Hardware-Aware Pruning for Deep Neural Networks

Shun Lu[1,2], Cheng Chen[2], Kunlong Zhang[1], Yang Zheng[3], Zheng Hu[3], Wenjing Hong[1], Guiying Li[1,2(✉)], and Xin Yao[1,2]

[1] Guangdong Provincial Key Laboratory of Brain-Inspired Intelligent Computation, Department of Computer Science and Engineering, Southern University of Science and Technology, Shenzhen 518055, China
ligy@sustech.edu.cn
[2] Research Institute of Trustworthy Autonomous Systems, Southern University of Science and Technology, Shenzhen 518055, China
[3] TTE Lab, Huawei Technologies Co., Ltd., Shenzhen, China

Abstract. Existing hardware-aware pruning methods for deep neural networks do not take the uncertain execution environment of low-end hardware into consideration. That makes those methods unreliable, since the hardware environments they used for evaluating the pruned models contain uncertainty and thus the performance values contain noise. To deal with this problem, this paper proposes noise-tolerant hardware-aware pruning, i.e., NT-HP. It uses a population-based idea to iteratively generate pruned models. Each pruned model is sent to realistic low-end hardware for performance evaluations. For the noisy values of performance indicators collected from hardware, a threshold for comparison is set, where only the pruned models with significantly better performances are kept in the next generation. Our experimental results show that with the noise-tolerant technique involved, NT-HP can get better pruned models in the uncertain execution environment of low-end hardware.

Keywords: Deep neural networks · Pruning · Hardware-aware · Noise-tolerant neural network pruning

1 Introduction

Deep Neural Networks (DNN) have shown impressive performance in many realistic tasks, like object detection [1], video tracking [2], etc. These significant

This work was supported the National Natural Science Foundation of China (Grant 62106098, Grant 62272210), the Guangdong Provincial Key Laboratory (Grant 2020B121201001), the Program for Guangdong Introducing Innovative and Entrepreneurial Teams (Grant 2017ZT07X386), the Shenzhen Peacock Plan (Grant KQTD2016112514355531), and the Stable Support Plan Program of Shenzhen Natural Science Fund (Grant 20200925154942002).

Y. Tan et al. (Eds.): ICSI 2023, LNCS 13969, pp. 127–138, 2023.
https://doi.org/10.1007/978-3-031-36625-3_11

appearances attract more attention of deploying DNN on mobile devices like smartphones or electric vehicles. Due to the over-parameterized nature, DNN always consume lots of hardware resources, i.e., CPU cores, memory allocations. This makes DNN difficult to be deployed on low-end hardware with limited computational resources directly. A commonly used solution is DNN compression, that means lowering down the computing overhead while maintaining the accuracy of DNN. Based on the DNN compression, the practical pipeline of deployment is training a DNN on GPU servers for high accuracy first, and then compressing the trained DNN to fit the requirement of target devices.

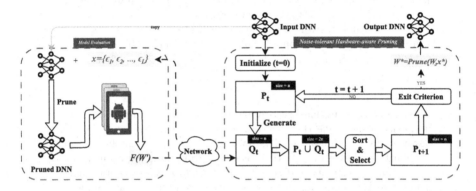

Fig. 1. The framework of NT-HP.

Pruning is a famous type of DNN compression methods. This kind of method simplifies DNN by removing redundant sub-structures. These sub-structures could be connections, neurons, filters or even a whole hidden layer inside DNN. Lots of studies on DNN pruning focus on designing algorithms to find these redundant structures [3], and the key to identifying the redundancy lies on the performance indicators (i.e., latency, memory, CPUs, energy). The performance values could be affected if the selected sub-structure is pruned. Obviously, the more affected the performance values are, the more important the sub-structure is, which means the sub-structure should not be easily pruned, and vice versa.

Hardware-aware pruning uses the feedback of DNN models on the hardware as the performance indicators. Unlike hardware-aware pruning, many existing DNN pruning methods [3] adopt surrogate indicators. Because the real hardware feedback is hard to collect and contains noise. These surrogate indicators, such as compression ratio or Multiply-ACcumulate operations (MACs), are designed by experts and highly relating to the DNN model's complexity. However, the DNN models pruned by the surrogate indicators may suffer from the poor performances on hardware environment, since the indicators cannot represent the true running process on devices. Note that the model's performance is affected by many factors, such as computing framework, hardware environment, etc. The expert designed indicators only reveal parts of these factors, which can not actually represent the real performance of the model on the target platform. Many studies [4] have shown that such shortage of surrogate indicators does exist, for example, models requiring smaller MACs lead to larger inference latency [5]. Hardware-aware pruning uses indicators such as memory allocation and inference latency to guide the model compression algorithm directly. These indicator values are collected by running pruned models on the target platform. Compared with the compression algorithm using only the surrogate indicators, hardware-aware pruning can achieve better performance on target devices.

However, the indicator values collected are often unreliable, which is a critical challenge for hardware-aware pruning. That is, the running performance of a specific DNN model on a low-end device, e.g., memory allocation during the running, could be different values among multiple runs. This phenomenon could be caused by multiple factors, like operating system scheduling, cache management, and memory garbage collection. Therefore, the indicator values are affected by noise, and the noise is complex, multifaceted, and difficult to model in math. Due to the existence of the noise, the pruning algorithm may get inaccurate evaluation values of the model performance, thus making the final pruned model unreliable on the target devices. Some of the existing hardware-aware pruning methods notice this phenomenon, but they simply use some statistical approaches to preprocess the feedback values before applying their algorithms. For example, NetAdapt [6] runs each pruned model several times in the device and uses the median or average as the performance indicator value. But these approaches do not take the noise seriously and not have mechanism in algorithm to counter the noise.

In this paper, we propose a pruning algorithm which can resist the noise in memory evaluation while generating pruned models. The method, called **N**oise-**T**olerant **H**ardware-aware **P**runing for DNN (NT-HP), organizes the pruning process as an optimization problem and solve it by a newly designed evolutionary algorithm [7–9]. Specifically, each pruned model would run on a target device during the fitness evaluation, and the noise-tolerant technique is used in the individual selection, which is the key design of NT-HP. In the experiments, NT-HP achieved the best memory usage of 334.02MB under the given accuracy constraint.

The main contributions of this paper are as follows:

1. For the first time, the noise is considered in designing hardware-aware pruning algorithm, and the memory compression in hardware-aware pruning is specifically considered.
2. A new noise-tolerant mechanism is introduced in hardware-aware pruning, which enables the pruning process to filter the noise in performance values collected from the hardware environment, to prune the DNN models more effectively.
3. We have verified the proposed algorithm on edge computing platform Jetson NX, and it has outperformed the comparison hardware-aware pruning methods.

The rest of the paper is organized as follows. Section 2 presents the related work. Section 3 introduces the proposed NT-HP algorithm. Section 4 presents the empirical studies. Section 5 finally concludes the paper.

2   Related Work

There are many ways to utilize the hardware feedback in hardware-aware pruning. Some methods use the feedback to build a agent for the target device, so that the pruned model can be estimated by the agent. In this way, the pruned model does not need to be truly deployed on devices, and the noise evaluation problem can be avoided. CompactNet [10] automatically optimizes a pre-trained DNN model on a specific resource-limited platform given a specific target of inference speedup, guided by a simulator of the target platform. ChamNet [11] proposed a novel algorithm to search for optimal architectures aided by efficient accuracy and resource (latency and/or energy) predictors. The accuracy and energy predictors incorporate Gaussian process regressors augmented with Bayesian optimization and imbalanced quasi Monte-Carlo sampling. It also includes an operator latency look-up table (LUT) in the latency predictor for fast, yet accurate, latency estimation. It is a common idea of building LUT based on hardware feedbacks and using it in pruning. HFP [12] builds up its LUT for fast estimating the latency of target network about filter configuration layer by layer, and then it prunes the network by leveraging information gain to globally evaluate the filters contribution to network output distribution. NetAdapt [6] also incorporates LUT into its adaptation algorithm. The LUT are built using empirical measurements for fast resource consumption estimation, so that the pruning process can be sped up.

Some methods use the direct feedback from devices while pruning. AMC [13] leverages reinforcement learning (RL) to efficiently sample the design space and can improve the model compression quality. AMC uses latency as the resource constraint. Auto-prune [14] is an automated DNN pruning framework. It also uses RL to automatically determine the pruning policy considering the constraint of accuracy loss. The reward function of RL agents is designed using hardware's direct feedback.

Existing hardware-aware methods [6,10,11] hardly notice the noise in the direct feedback from devices, or only averages the results for noise filtering with no specific design in their algorithm. In this paper, NT-HP firstly proposes noise-tolerant mechanism in hardware-aware pruning, and gets more reliable performances.

3   Methodology

3.1   Pruning as Optimization

DNN pruning can be regarded as an optimization problem. In this paper, we focus on the problem of pruning DNN for smaller memory consumption. A DNN

W with $L + 1$ layers can be viewed as:

$$W = \{W_l^t | l \in [1...(L+1)], t \in [1...D_l]\}. \tag{1}$$

For the l-th convolutional layer, D_l represents the number of filters in the l-th layer, and W_l^t represents the t-th filter parameter matrix. For each $W_l^t \in \mathbb{R}^{K_l \times H_l \times C_l}$, its spatial structure can be represented as a three-dimensional matrix of size $K_l \times H_l \times C_l$ where $K_l \times H_l$ represents the size of kernel in the filter, and C_l represents the number of input channels in the l-th layer, that is, the number of kernels in the filter. For the l-th fully-connected layers, D_l represents the number of neurons in the layer, and W_l^t represents corresponding parameters of the t-th neuron. Noted that the $(L + 1)$-th layer is the output layer.

We prune W separately in each layer, where the unimportant filters or neurons in a layer are pruned according to a certain proportion ϵ where $\epsilon \in [0, 1)$. For convolutional layers, all the filters are sorted by their $\ell 2$-norm in ascending order, and the top ϵ of filters are pruned. For fully-connected layers, all neurons are sorted by the $\ell 2$-norm of parameters in ascending order, and the top ϵ of neurons are pruned. Let $\mathrm{Rank}(\|W_l^t\|_2)$ denote the ranking proportion of the t-th filter/neuron, the pruning can be described as follows

$$\mathrm{Prune}(W, \epsilon) = \{W_l^t | \mathrm{Rank}(\|W_l^t\|_2) > \epsilon_l, l \in [1...L], t \in [1...D_l]\}, \tag{2}$$

where $\epsilon = \{\epsilon_l | \epsilon_l, l \in [1...L]\}$, ϵ_l is the portion to be pruned in the l-th layer. $\|W_l^t\|_2$ indicates the $\ell 2$-norm.

Therefore, the optimization problem can be formulated as:

$$W^* = \underset{W'=\mathrm{Prune}(W,\epsilon),\epsilon \in [0,1)^L}{argmin\, f(W')} \quad s.t.\ \mathrm{ACC}(W') > \theta, \tag{3}$$

where W^* indicates the best pruned model, W' indicates a candidate pruned model. Function $f(W')$ indicates the memory usage of W' on the target device, function $\mathrm{ACC}(W')$ indicates the accuracy of W' on the validation data-set, θ is the accuracy constraint that the pruned model must hold. It is easy to see that the smaller value of $f(W')$, the better. And the larger value of $\mathrm{ACC}(W')$, the better. Function $\mathrm{Prune}(W, \epsilon)$ indicates pruning the model W by thresholds ϵ. By solving the Eq. (3), we can get the best pruned model W^*, which has the lowest memory usage and fits the accuracy constraint.

However, there are two major problems in solving the Eq. (3). Firstly, the value $f(W')$ of which we can collect, contains noise. The next subsection will discuss this problem. Secondly, we need an algorithm to solve the Eq. (3), which takes the noise into consideration. The final subsection will give the algorithm.

3.2   Noise in Hardware-aware Pruning

Observations Affected by Noise. In a hardware-aware scenario, collecting the value of performance indicator often suffer from noise. Taking memory as an example, the evaluation value of the memory usage within a single process is uncertain. It can be easily affected by frequent memory exchange, OS process scheduling or software frameworks algorithms depend on.

In this paper, we assume the memory evaluations in an edge device will receive multiplicative noise. Because it is intuitive that the more memory allocated, the more temporary resources are required. And these temporary resources (like cache, swap, etc.) may cause the instability of memory occupation. Let the noise environment be a multiplicative noise model, which follows the definition in [15]:

$$f(t) \cdot (1 - \varepsilon) \leq f'(t) \leq f(t) \cdot (1 + \varepsilon), \tag{4}$$

where $f(t)$ represents the real value of the indicator, $f'(t)$ represents the value collected from the edge device with noise, which fluctuates within a certain range. $\varepsilon \in [0, 1]$ denotes the multiplicative factor, which is used to describe the degree of this noise. Obviously, when there is noise in the evaluation environment, we can hardly guarantee that the model really occupies less memory.

Suppose there are two models x and y to be compared, we have

$$\begin{cases} f(x) \cdot (1 - \varepsilon) \leq f'(x) \leq f(x) \cdot (1 + \varepsilon) \\ f(y) \cdot (1 - \varepsilon) \leq f'(y) \leq f(y) \cdot (1 + \varepsilon). \end{cases} \tag{5}$$

When selecting a model with better memory consumption, we need to get a confident comparison between $f'(x)$ and $f'(y)$. That means we need to conduct the relationship between $f(x) < f(y)$ and $f'(x) < f'(y)$. Because we only observe the values of $f'(x)$ and $f'(y)$, and what we want are the values of $f(x)$ and $f(y)$.

According to Eq. (5), it is easy to conclude that if

$$f'(x)/f'(y) < (1 - 2\varepsilon) \tag{6}$$

holds, then $f(x) < f(y)$ holds.

Similarly, we can conclude that if

$$f'(x)/f'(y) > \frac{1}{1 - 2\varepsilon} \tag{7}$$

holds, then $f(x) > f(y)$ holds.

Equations (6) and (7) indicate that, if we need to select a better model from x and y with their memory observations $f'(x)$ and $f'(y)$, then the x is a better model if Eq. (6) meets, or the y is a better model if Eq. (7) meets. Otherwise, x and y are not comparable, since their difference is not stronger enough to conquer the noise.

Objective Functions for Memory Evaluation. Since the collected value of performance indicator will fluctuate around the true value in the noise scene, simply comparing the value of a certain run is not reliable. Therefore, some statistics can be carried out by rerunning the model multiple times. With the sampled values of $f'_i(W')$, NT-HP adopts new performance indicators for memory evaluation of model W', i.e., the minimum value, the average value and the

---

Algorithm 1: Compare Models under Noise

Input: model x and model y; The noise parameter ε.
Output: The smaller model.

Let smaller_model = None;
Calculate $F(x)$ and $F(y)$ by Eq. (8);

if $\dfrac{\min_i f_i'(x)}{\min_i f_i'(y)} < (1 - 2\varepsilon)$ *and* $\dfrac{avg_i f_i'(x)}{avg_i f_i'(y)} < (1 - 2\varepsilon)$ *and* $\dfrac{\max_i f_i'(x)}{\max_i f_i'(y)} < (1 - 2\varepsilon)$ **then**

$\quad \lfloor$ smaller_model = x;

else if $\dfrac{\min_i f_i'(x)}{\min_i f_i'(y)} > \dfrac{1}{1-2\varepsilon}$ *and* $\dfrac{avg_i f_i'(x)}{avg_i f_i'(y)} > \dfrac{1}{1-2\varepsilon}$ *and* $\dfrac{\max_i f_i'(x)}{\max_i f_i'(y)} > \dfrac{1}{1-2\varepsilon}$ **then**

$\quad \lfloor$ smaller_model = y;

`// Otherwise x and y are not comparable.`
return smaller_model.

---

maximum value of all the sampled $f_i'(W')$. Therefore, for a given model W', the objectiveness of memory for W' is defined as a tuple:

$$F(W') = (\min_i f_i'(W'), avg_i f_i'(W'), \max_i f_i'(W')), i \in [1, ..., N], \qquad (8)$$

where $f_i'(W')$ is a sampled value of $f'(W')$, N is the number of times that rerun the W' on target device. It should be noted that, Eq. (8) is not a definition for multi-objective optimization because the three indicators are highly related to memory allocations. By using Eq. (8), NT-HP collects not only a specific value of memory but a range of the memory's change.

Comparing Models Under Noise. As discussed above, Eq. (6) and Eq. (7) give the way to comparing the observed values under noise, Eq. (8) gives the observations we need to collect for comparing the memory of models. For two given pruned model x and y, NT-HP defines a process in Algorithm 1 to determine whose memory consumption is better with their observations.

Firstly, NT-HP runs the x and y on the hardware platform for N times to calculate their objectiveness values in Eq. (8). Secondly, NT-HP checks whether x has the better memory consumption. It is done by comparing each value of indicators following Eq. (6). If all the memory indicator values of x are better, x is better. Thirdly, NT-HP checks whether y has the better memory consumption by Eq. (7). If there is no result after 3 steps then no model is better, that means x and y cannot be compared.

3.3   Optimization Algorithm

Based on Eq. (8), now the optimization problem Eq. (3) becomes

$$W^* = \underset{W'=Prune(W,\epsilon),\epsilon \in [0,1)^L, i \in [1,..,N]}{argmin} (\min_i f_i'(W'), avg_i f_i'(W'), \max_i f_i'(W'))$$

$$s.t.\ ACC(W') > \theta, \qquad (9)$$

where different models of W' are compared by Algorithm 1. Now, an optimization algorithm is needed to solve Eq. (9) for the best pruned model W^*.

For optimization problems with multiple objectives, Multi-Objective Evolutionary Algorithms (MOEAs) [16] can effectively solve them. Compared with traditional optimization algorithms, MOEAs have two advantages. First, MOEAs do not require special assumptions, such as differentiable or continuity, thus black-box optimization problems can be solved. The second is the ability to find multiple Pareto optimal solutions in a single run, which enables models with advantages in different indicators to be found together in the model compression scenario, which allows the algorithm to retain more flexibility. Although the three indicators in Eq. (9) are all related to memory, we can still use the idea of Pareto optimal to design the algorithm for Eq. (9). To solve Eq. (9), an evolutionary optimization algorithm is proposed in Algorithm 2.

Given a trained DNN W, it can be encoded as a vector $x^0 = 0^L$ based on Eq. (2) where x^0 is the ϵ with 0 values. This means that

$$\text{Prune}(W, x^0) = W, \tag{10}$$

so that each pruned model W' can be encoded as a vector:

$$x = \epsilon = \{\epsilon_l | \epsilon_l \in [0, 1), l \in [1 \ldots L]\}.$$

Now a population-based evolution is employed to search the best x^*, which is identical to the best pruned model, i.e., $W^* = \text{Prune}(W, x^*)$. For the generation t in evolution, let $P_t = \{x^1, x^2, ..., x^k, ..., x^n\}_t$ denote the parent, where x^k is the k-th individual, n is the population size. Q_t of size n is the offspring generated by P_t. The optimization starts by accepting a pre-trained model W as input, and obtains an initial individual x^0. Then the initial population P_0 with population size n needs to be constructed, where x^0 is included and the other $n-1$ individuals are random initialized.

Then the iteration of optimization starts. In the t-th iteration, crossover and mutation methods are used to generate Q_t based on P_t. Then, the new generation P_{t+1} is generated as follows:

- **Step1:** all the individuals are compared to each other by Algorithm 1 with three objectives ($\min_i f_i'(W')$, $\text{avg}_i f_i'(W')$, $\max_i f_i'(W')$)and the individuals in Pareto front are selected into P_{t+1}. If the size of P_{t+1} equals to n, then P_{t+1} is ready; if the size is larger than n, randomly discard individuals until the size is n; if the size is smaller than n, turn to **step 2**;
- **Step2:** rest individuals are then compared to each other following the same process in **step 1**, with new objectives($\text{avg}_i f_i'(W')$, $\text{ACC}(W')$). If the size of P_{t+1} is smaller than n, turn to **step 3**;
- **Step3:** sort the rest individuals by $\text{ACC}(W')$ in descending, and fill P_{t+1} with top ranking individuals until the size of P_{t+1} equals to n.

After P_{t+1} is ready, new iteration starts. And the optimization ends for the given number of iterations. Finally, all the individuals in the final population

---

Algorithm 2: Noise-tolerant Hardware-aware Pruning

Input: The pre-trained network W; the number of iterations T; the population
 size n.
Output: The best pruned model.

% Let $x^0 = 0^L$ denote the base network W following Eq. (10);
Let $P_0 = \{x^0, x^1, x^2, ..., x^{n-1}\}$, where $x^1, ..., x^{n-1}$ are randomly initialized;
Let $t = 0$, $x^* = x^0$;
// Generate better pruned model within T iterations.
while $t < T$ **do**
    Generate Offspring population Q_t using crossover and mutation operators
    from P_t;
    Set $R_t = P_t \cup Q_t$;
    // Evaluate each generated model.
    foreach x in R_t **do**
        calculate accuracy on x;
        calculate the corresponding $F(\text{Prune}(W, x))$ by Eq. (8);
    Generates P_{t+1} following the 3 steps in Sect. 3.3;
    Set $t = t + 1$;
// Select the best model in population P_T.
foreach x in P_T **do**
    Rerun x for hundreds of times to get the new value of $avgf'_i(\text{Prune}(W, x))$;
    if $avgf'_i(\text{Prune}(W, x)) < avgf'_i(\text{Prune}(W, x^*))$ **then**
        Set $x^* = x$
Let $W^* = \text{Prune}(W, x^*)$
return W^*.

---

are send to run on the target devices for hundreds of times, and the model with
the smallest average memory usage is the best found one. The whole pipeline is
illustrated in Fig. 1.

4 Experiments

4.1 Experimental Setup

In this section, we apply NT-HP to AlexNet, which consists of five convolutional
and three fully connected layers totaling 61 million parameters. The performance
of our algorithm will be examined on a well-known image classification data-set
CIFAR-10 [17], which consists of 50,000 training photos and 10,000 testing photos. Note that a validation set with 1,000 photos from training set is generated.
The experiment environment is a cluster consists of two kinds of devices. Master
node is a GPU server with 256 GB memory and a RTX 3080Ti GPU for running
optimization process. The rest are 10 NVIDIA Jetson Xavier NX, on which the
pruned DNN models to be deployed and evaluated. Devices are connected by
network and interact with each other by HTTP. The whole pruning process is

implemented by PyTorch. We use NetAdapt [6] and HAMP [16] as the comparison algorithms. They are the latest hardware-aware pruning methods that can adapt the memory evaluation scenario where other hardware-aware pruning methods can't. It should be noted that the official NetAdapt source code only implemented for inference latency, while our NT-HP mainly focuses on memory usage. Therefore, we modify the optimization objective of NetAdapt to memory consumption, which is the same as NT-HP. The memory usage is recorded by jetson-stats, a package for monitoring and controlling NVIDIA Jetson.

4.2   Effectiveness of Noise-Tolerant Hardware-Aware Pruning

In this section, we show the effectiveness of noise-tolerant technique used in NT-HP. For the same model, the collected memory values are variant during model execution. The proposed NT-HP uses ε to estimate the noise in devices, i.e., Eq. (4), where the larger value, the more noise. As illustrated in Fig. 2(a), using the noise assumption and introducing the noise-tolerant technique can significantly outperform the one without such setting.

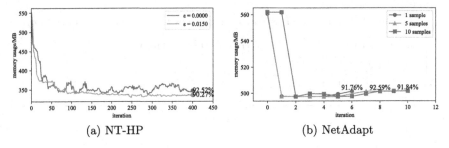

(a) NT-HP   (b) NetAdapt

Fig. 2. (a)Performances of NT-HP with (i.e., $\varepsilon > 0$) and without (i.e., $\varepsilon = 0$) the noise-tolerant technique. (b)Performances of NetAdapt with (i.e., average the memory for 10 runs) and without (i.e., use the memory of a single run) the statistical measurements for memory estimation. Obviously, NT-HP with the noise-tolerant technique performances the best. The horizontal axis is the number of iterations of the algorithm and the vertical axis is the average memory usage.

When $\varepsilon > 0$, NT-HP adapts the noise-tolerant comparison technique in Algorithm 1 and when $\varepsilon = 0$, it doesn't. Figure 2(a) compares the situations of using the noise-tolerant approach, i.e., $\varepsilon > 0$ or not, i.e., $\varepsilon = 0$. Obviously, when there is no treatment for noise, it is hard for optimization process to converge, and it is difficult to get a good solution.

We also show the method of filtering the noise used in NetAdapt. By incorporating LUT into its adaptation algorithm, NetAdapt is potentially taking the noise into account and trying to filter it. In our experiments, we also built a LUT of memory usage that sampled several times of feedback values to estimate the real memory usage. As illustrated in Fig. 2(b), the sampling times vary from 1 to 10, which directly determine the reliability of the memory usage estimation in the uncertain execution environment. Empirically, the more we sample, the more reliable the estimation, and the more likely the algorithm is to get better results. However, as shown in Fig. 2, with the 10% accuracy constraint on validation set, NT-HP can find more pruned models while NetAdapt failed quickly in several iterations. Furthermore, with the similar accuracy, the pruned model generated by NT-HP has smaller memory consumption.

4.3   Comparison of Pruning Performances

We compare our performance with other two hardware-aware pruning methods. AlexNet used in the experiment is trained on CIFAR-10 by PyTorch Example project on GitHub. Its accuracy on the test set is 86.42%, and its memory usage is 561.32MB. For NT-HP, the run times N is set to 10, which means it would rerun the model on the device for 10 times. Considering the influence of noise, when the final iteration of the algorithm is achieved, the individuals are tested 100 times for selecting the best individual. In Table 1, NT-HP gets the best pruned model with a memory usage of 334.02MB with the lowest standard deviation. All the result fit the constraint of 10% accuracy loss on the validation set.

Table 1. Comparison result. The best values among pruning algorithms are in **bold** value.

Network	Top-1 Acc (%)	Mean of Memory (MB)
AlexNet	86.42	561.32 ± 2.14
NetAdapt (2018) [6]	74.58	502.88 ± 1.71
HAMP (2022) [16]	75.32	342.26 ±1.05
NT-HP	75.71	**334.02 ± 0.90**

5   Conclusion

In this paper, we have discussed the noise in hardware-aware pruning and proposed a noise-tolerant pruning algorithm NT-HP. In order to filter the noise in direct hardware feedback, NT-HP proposes new noise-tolerant technique which compares pruned models under multiplicative noise assumption, so that only the models with confident performances are selected. To find the optimal pruned model, NT-HP employs an MOEA framework with newly proposed memory indicators. In empirical studies, the proposed NT-HP gets the best pruned model with 334.02MB in memory compared with other state-of-the-art methods under the same accuracy constraint. The result indicates that with the noise-tolerant technique involved, NT-HP is more suitable for DNN pruning in the uncertain execution environment on low-end platforms.

References

1. Zou, Z., Chen, K., Shi, Z., Guo, Y., Ye, J.: Object detection in 20 years: A survey. In: Proceedings of the IEEE, pp. 1–20 (2023)
2. Yao, R., Lin, G., Xia, S., Zhao, J., Zhou, Y.: Video object segmentation and tracking: a survey. ACM Trans. Intell. Syst. Technol., 11(4), 36:1–36:47 (2020)
3. He, Y., Liu, P., Wang, Z., Hu, Z., Yang, Y.: Filter pruning via geometric median for deep convolutional neural networks acceleration. In: CVPR 2019, Long Beach, CA, pp. 4340–4349 (2019)
4. Marculescu, D., Stamoulis, D., Cai, E.: Hardware-aware machine learning: modeling and optimization. In: ICCAD 2018, San Diego, CA, pp. 1–8 (2018)
5. Yu, J., Lukefahr, A., Palframan, D.J., Dasika, G.S., Das, R., Mahlke, S.A.: Scalpel: Customizing DNN pruning to the underlying hardware parallelism. In: ISCA 2017, Toronto, ON, Canada, pp. 548–560 (2017)
6. Yang, T., et al.: Netadapt: Platform-aware neural network adaptation for mobile applications. In: 15th European Conference, Munich, Germany, pp. 289–304 (2018)
7. Yao, X.: Evolving artificial neural networks. Proc. IEEE 87(9), 1423–1447 (1999)
8. Yang, P., Yang, Q., Tang, K., Yao, X.: Parallel exploration via negatively correlated search. Front. Comp. Sci. 15(5), 1–13 (2021). https://doi.org/10.1007/s11704-020-0431-0
9. Yang, P., Zhang, H., Yu, Y., Li, M., Tang, K.: Evolutionary reinforcement learning via cooperative coevolutionary negatively correlated search. In: SWEC 2022, vol. 68, p. 100974 (2022)
10. Li, W., Wang, R., Qian, D.: Compactnet: Platform-aware automatic optimization for convolutional neural networks. In: PMAM@PPoPP 2021, Virtual Event, Republic of Korea, pp. 11–20 (2021)
11. Dai, X., et al.: Chamnet: Towards efficient network design through platform-aware model adaptation. In: CVPR 2019, Long Beach, CA, pp. 11398–11407 (2019)
12. Yu, F., Han, C., Wang, P., Huang, R., Huang, X., Cui, L.: HFP: hardware-aware filter pruning for deep convolutional neural networks acceleration. In: ICPR 2020, Virtual Event, Italy, pp. 255–262 (2020)
13. He, Y., Lin, J., Liu, Z., Wang, H., Li, L., Han, S.: AMC: automl for model compression and acceleration on mobile devices. In: 15th European Conference, Munich, Germany, pp. 815–832 (2018)
14. Yang, S., Chen, W., Zhang, X., He, S., Yin, Y., Sun, X.: AUTO-PRUNE: automated DNN pruning and mapping for reram-based accelerator. In: ICS 2021, pp. 304–315. Virtual Event (2021)
15. Qian, C.: Distributed pareto optimization for large-scale noisy subset selection. IEEE Trans. Evol. Comput. 24(4), 694–707 (2020)
16. Hong, W., Li, G., Liu, S., Yang, P., Tang, K.: Multi-objective evolutionary optimization for hardware-aware neural network pruning. In: Fundamental Research 2022 (2022) (in press)
17. Krizhevsky, A., Hinton, G.E.: Learning multiple layers of features from tiny images (2009)

Pyrorank: A Novel Nature-Inspired Algorithm to Promote Diversity in Recommender Systems

Doruk Kilitcioglu, Nicholas Greenquist, and Anasse Bari$^{(\boxtimes)}$

Computer Science Department, Courant Institute of Mathematical Sciences, New York University, New York, NY 10012, USA
{doruk.kilitcioglu,abari}@nyu.edu

Abstract. Recommender systems are essential to many of the largest internet companies' core products. Today's online users expect sites offering a vast assortment of products to provide personalized recommendations. Although traditional recommender systems optimize for prediction accuracy, such as RMSE, they often fail to address other important aspects of recommendation quality. In this paper, we explore the crucial issue of diversity in the recommendations generated by recommender systems. We explain why diversity is essential in recommender systems and review related work on diversifying recommendations. We quantify and classify various diversity metrics into logical categories. Then, we introduce Pyrorank, a novel bio-inspired re-ranking algorithm designed to improve recommendation diversity. Pyrorank is inspired by the positive effects of pyrodiversity in nature and is optimized to increase user-based diversity and mitigate the systemic bias that traditional recommender system models learn from the data. Our experimental results on multiple large datasets indicate that Pyrorank can achieve better user-based diversity metrics than state-of-the-art re-ranking methods, with little decrease in prediction accuracy.

Keywords: Recommender System · Artificial Intelligence · Predictive Analytics · Recommendation Diversity

1 Introduction

Recommender systems provide users with recommendations for new content that they might be interested in, such as music, movies, books, and more. Companies that rely on recommendation systems to drive much of their sales can see significant increases in sales or customer retention. Studies have shown that 35% of Amazon.com's revenue is generated by its recommendation engine [1].

A common pitfall of these recommender systems is that they are susceptible to push the user into a filter bubble [2], a phenomenon where a user always gets recommended content extremely similar to the content they have previously consumed. It is also common for a set of recommendations to be extremely similar to each other, such as a set of songs that are from the same artist or genre, or videos that are all from one channel or all about the same topic. In terms of getting positive feedback actions from a user (such as

Y. Tan et al. (Eds.): ICSI 2023, LNCS 13969, pp. 139–155, 2023.
https://doi.org/10.1007/978-3-031-36625-3_12

increased click-thru-rate, purchases, or RMSE), this may be desirable, leading to lower error metrics, but this experience ignores serving a diverse range of content.

Traditional metrics such as click-thru-rates can fail to identify the long-term implications and ethics of cornering users into a filter bubble. There are instances where recommender systems have recommended self-harm videos to susceptible individuals [3]. While YouTube has been optimizing their recommendation engine for watch time [4], they have recently had to manually tweak their algorithm to fight issues such as conspiracy theories [5], anti-vax movement [6], and trolling of celebrities [7]. Successful recommender systems need to be able to recognize the biases of a user, leading to more effective recommendations and long-term health of the platform.

Diversification is an important step in combating the negative effects of recommender systems. Intuitively, diverse recommendations should be able to show users a wide range of content. However, since recommendation systems are often optimized for a small, fine-tuned set of business-critical metrics, modifying the recommendations often leads to slightly worse performance in these metrics. This has caused diversity in recommender systems to not be a popular area of research. Despite this, it is more paramount today than ever for recommender systems to be more ethical and less prone to creating filter bubbles for its users. Most importantly, it has also been shown that homogeneity and filter bubbles decrease the long-term utility of recommender systems [8] and that diversification decreases acceptance rate but increases user satisfaction [9].

Lastly, diversity is also a critical part of natural, real-world ecosystems. Inspired by how biodiversity can be affected by non-natural, manmade actions, we developed our novel re-ranking algorithm with these ideas in mind. It is worth noting that mild and controlled disturbances in ecosystems can often lead to an increase in biodiversity levels across various ecosystems [10].

In this study, we outline recent work on diversity in recommender systems, introduce our novel nature-inspired Pyrorank algorithm, and evaluate its performance by analyzing the effects of changing various hyperparameters and relevant metrics.

2 Related Work

The term recommender system stems from Information Retrieval theory, and it was first used for retrieving relevant documents to match text queries [11]. It should be no surprise that the earliest mentions of diversity in terms of recommendations were also through the context of information retrieval [12–14]. The task in those cases was in trying to maximize the similarity of the returned items to a given query while trying to increase diversity in the list of items returned. These two tasks were often at odds with each other. Of particular note is Smyth and McClave [12], which introduced the intra-list dissimilarity (ILD) metric (explained in detail later) and the algorithm to greedily maximize the combination of diversity and similarity.

Since then, recommender systems have been modified to handle not only relevant items, but also more desirable items. Most implementations of recommender systems rely on a predicted (sometimes binary) rating to gauge how good a recommended item is to a user. In that case, the diversity goes from being an opposing force to a complementary force, meaning there doesn't have to be a clear decrease in the fitness of the returned results when trying to increase recommendation diversity.

2.1 Related Work: Recommendation Diversity Metrics

Although there have been many recommendation diversity metrics proposed in past research, there does not appear to be any consensus regarding which metrics are more useful for real world systems [15]. We divide these metrics into two categories:

- List-context metrics take the list of recommended items as their source and aim to improve the internal diversity of the recommended items. We call this diversity level 1.
- User-context metrics take the items the user has rated as their source, and aim to diversify the recommended items with respect to user's previously consumed content. We call this diversity level 2.

1) List-context metrics: The most popular metric among researchers has remained intra-list dissimilarity, which was introduced in Smyth and McClave [12]. Researchers have also used the Gini coefficient, which is an indicator of inequality, as a means of measuring diversity[16]. Clarke et al. [17] model each item as a collection of topics and use a Normalized Discounted Cumulative Gain (nDCG) metric where the gain is the number of topics, optimizing which leads to more topics being recommended. Zhou et al. [18] model diversity as the probability of recommending a long-tail item and calculate it by the mean inverse user frequency of an item.

  It is also possible to look at the temporal aspect of diversity. Lathia et al. [19] model diversity as the change of the recommended items between two consecutive recommendations. Nguyen et al. [20] show that the intralist dissimilarity goes down over time even without any intervention from a recommender system and show that taking recommendations slows down this decrease in diversity.

2) User-context metrics: The user-based approaches to measuring diversity is not as common as the list-based approaches. Adamopoulos and Tuzhilin [21] define diversity as the average"unexpectedness" of an item, wherein each item may be expected if it comes from a primitive (i.e.. Popularity based) algorithm. They measure the unexpectedness of a set of items first as the set difference between the expected items and the recommended items, and then improve it to be the average pairwise dissimilarity between the two sets. They also introduce the centroid dissimilarity, which is the dissimilarity of each item with the cluster centroid of the expected items.

  Hurley and Zhang [22] build up on these ideas by introducing the concept of user novelty, which substitutes the expected set in [21] with the set of items user has consumed. They define the novelty of an item as the average dissimilarity of the item to the user's items, and diversity as the average novelty. A similar metric was also proposed by L'Huillier et al. [23], where the diversity of a user was denoted by the intralist diversity of the last k items the user has rated.

In this study, we will focus more on user-context metrics, as they explicitly model how biased a user's recommendations are towards the content they have already consumed in the past or in the same session.

2.2 Related Work: Diversification Algorithms

As mentioned in the previous section, the first algorithm designed to increase diversity was the greedy optimization of both diversity and relevancy by Smyth and McClave [12]. These are generally known as re-ranking algorithms, as they aim to re-rank the generated list of recommendations to promote diversity while maintaining a high aggregate score.

Since then, Adomavicius and YoungOk Kwon [24] has introduced a variety of baseline re-ranking algorithms, such as reverse predicted rating, item popularity with rating threshold, item rating variance, random re-ranking, and linear combinations of these methods. All of these methods are shown to improve diversity similarly.

Aytekin and Karakaya [25] introduced a method in which items are placed in pre-built clusters and only the best performing items from each cluster are used in recommendations (based on the user's preference for said clusters). This work is often used as one of the standard comparisons for different metrics and algorithms.

2.3 Biological Inspired Approach: Pyrodiversity

Our approach to promoting diversity in recommendations is inspired by natural Pyrodiversity in real world ecosystems. Ponisio et al. [26] define pyrodiversity as the diversity of fires in a region, and it has been hypothesized to increase biodiversity. Ponisio et al. [26] discovered that increases in pyrodiversity (where diversity is based on fire characteristics such as extent, severity, and frequency) was positively related to richness in the diversity of plant life. Ponisio et al. [26] explain how pyrodiversity is an example of Disturbance Diversity, which is thought to 'promote biodiversity because shifting environmental conditions discourage dominance'. This idea is at the center of the Intermediate Disturbance Hypothesis, first described by Connell [27], which claims that biodiversity is maximized at intermediate levels of disturbance in the environment (Fig. 1). Biswas and Mallik [10] 'found that, like species richness and diversity, functional richness and diversity reached peaks at moderate disturbance intensity; functional diversity followed the predictions of the IDH'.

We hypothesize that modelling recommended items by their features as a 'virtual forest' and introducing a 'burning mechanism' could lead to an increase in diversity measurements such as diversity level 1 and diversity level 2. Introducing this burning mechanism is akin to 'Controlled Burning', which Webster and Halpern [28] showed promotes biodiversity. Webster and Halpern [28] found that after ten years, burned plots supported more than twice as many native plant species as non-burned plots, and the once-burned plots showed a nearly threefold increase by year twenty.

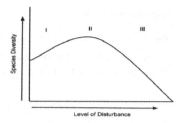

Fig. 1. Properties of the Intermediate Disturbance Theory

3 Pyrorank: Algorithm Design and Implementation

3.1 Pyrorank: High Level Design

Pyrorank is inspired by both the positive effects of pyrodiversity in the natural world, and the need for user-based recommendation diversity in real world digital recommendation systems. Pyrodiversity includes controlled burning techniques that burn off certain sections of a forest in order to protect the areas from future wildfires. However, a secondary benefit of this burning is increased long-term biodiversity, which in our algorithm gets translated into increased user-based diversity.

1) Nature Inspired Model: Here, we define Pyrorank in terms of a virtual forest made up of item 'trees'.

- Every possibly recommended item is modelled as a 'tree' in a virtual forest

  - A tree is defined by its position in an n-dimensional space created by the item features.

- The biological disturbance in this virtual forest is a 'fire'

  - A fire is a sphere in the n-dimensional feature space.
  - Fires spread with an evenly growing radius.
  - We model time as a series of discrete iterations, where every fire in the forest 'grows' by the same amount in each iteration.
  - Every iteration, any 'tree' that is within the radius of any fire is 'burned' and will not be included in the final recommendations.

- Pyrorank starts a fire at every item a user has interacted with

  - User items are also represented as 'trees' in the same virtual forest the candidate list trees exist in.

- How many 'trees' are burned off is controlled by a parameter passed into the algorithm, making it similar to 'controlled burning' in actual forests
- After the desired amount of trees are burned off, the virtual forest is left with a more diverse selection of trees.
- From the surviving trees, the algorithm then selects the ones with the highest predicted rating.

3.2   Pyrorank Algorithm: Implementation

Typical recommendation pipelines simply train a prediction model, which inputs N items and outputs N predictions (one for each item in the input corpus). Then, these N items are sorted by the predicted rating and a final subset of K items is presented to the user as shown Fig. 2.

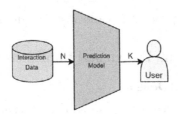

Fig. 2. Standard Prediction Model Pipeline

Pyrorank requires a candidate list, of length C possible items, in order to re-rank them and output a more diverse set. First, we train a standard recommendation algorithm to produce a candidate list of C possible recommended items instead of K. When producing the intermediate candidate list, typically N >> C > K. YouTube, for example, uses a re-ranking step on a candidate list to improve recommendations, and for YouTube's process N = millions, C = hundreds, and K = dozens[4].

In addition to a candidate list, Pyrorank also requires the item feature matrix, where every row is the feature vector for every item. This feature matrix should be reduced to a more manageable number of dimensions either through dimensionality reductions techniques such as SVD or PCA, or more sophisticated techniques such as Autoencoders. If item features are not available, you can also simply pass in the learned representations of each item (the Q matrix from matrix factorization, or learned hidden layers from a Neural Network recommender system approach), as outlined in Kunaver et al. [29]. Together with the candidate list and the item features, Pyrorank can filter down the candidate list to K items and present those to the user (Fig. 3).

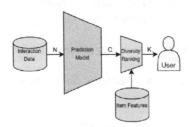

Fig. 3. Revised Prediction Model Pipeline

3.3 Pyrorank: Pseudo-code

Pyrorank is analogous to starting a 'controlled burn' from every item a user has already interacted with. Because the user has already interacted with these items, unseen items extremely close to them should not be shown to the user again and again, and should be 'burned off' from the final recommended set of items.

Pyrorank starts by computing the pair-wise distance from every item the user has interacted with to every item in the input set. We then repeat the 'burn' step until we reach the desired number of remaining items. The burn step loops over every item the user has interacted with and then removes any items in the set that within a minimum distance to that item. We then slightly increase the 'burn' distance for the next iteration, so more items are considered as too similar. A detailed pseudo-code implementation is seen below.

Algorithm 1 Pyrorank

```
 1: function PYRORANK(predictions, k, item_features,
        user_ratings, ratio_keep, start_dist, increment)
 2:     num_items = NUMROWS(item_features)
 3:     num_keep = num_items * ratio_keep
 4:     nonzeros = GETNONZERORATINGS(user_ratings)
 5:     nonzero_features = item_features[nonzeros]
 6:     D = PAIRWISEDISTS(item_features, nonzero_features)
 7:     curr_dist = start_dist
 8:     possible_items = [0 ... num_items]
 9:     while NUMROWS(possible_items) > num_keep
        do
10:         dead_indices = []
11:         for i = 0 to num_items - 1 do
12:             min_dist = MIN(D[i])
13:             if min_dist < curr_dist then
14:                 APPEND(dead_indices, i)
15:             end if
16:         end for
17:         possible_items = possible_items - dead_indices
18:         curr_dist+ = increment
19:     end while
20:     return GETTOPITEMS(predictions[possible_items], k)
21: end function
```

Once Pyrorank has the candidate list, it decides how many items to keep using a ratio keep parameter. If ratio keep = 0.5, we would keep half the items with the lowest distances to a user's existing items. In this way, Pyrorank explicitly optimizes for diversity level 2.

We provide two versions of our algorithm. The original Pyrorank algorithm (algorithm 1) follows the bio-inspired setup exactly. The optimal Pyrorank algorithm (algorithm 2) skips the intermediate spread of the fire and directly aims to find the list of possible items. Both are provided here for the sake of completion. All evaluations below assume the optimal algorithm.

Algorithm 2 Pyrorank-OPTIMAL

1: **function** PYRORANK(*predictions, k, item features, user_ratings, ratio_keep*)
2:    *num_items* = NUMROWS(*item features*)
3:    *num_keep = num_items * ratio_keep*
4:    *nonzeros* = GETNONZERORATINGS(*user_ratings*)
5:    *nonzero_features = item_features[nonzeros]*
6:    *D* = PAIRWISEDISTS(*item features, nonzero features*)
7:    *min_dists* = []
8:    **for** *i* = 0 **to** *num_items* − 1 **do**
9:        APPEND(*min_dists, MIN(D[i])*)
10:    **end for**
11:    // Using a partitioning algorithm for fastest selection
12:    *possible_items* = GETTOPK(*min_dists, num_keep*)
13:    **return** GETTOPITEMS(*predictions[possible items], k*)
14: **end function**

3.4 Pyrorank: Runtime Analysis

Assuming the following parameters:

- N = num_items to rank from recommendation model output
- K = final number of items to keep/show to user
- U = number of items user has already seen/interacted with
- F = size of feature vector for each item

The runtime of Algorithm 2 Pyrorank is the runtime of computing the pairwise distances between every input item and every item the user has interacted with (which is $O(N*U*F)$) added to the runtime of computing the final top-k selection (which is $O(K*log(N))$).

4 Experiments and Results

4.1 Recommendation Model

For our experiments, we chose to use the popular Matrix Factorization approach for producing recommendations. Matrix Factorization (MF) decomposes a rating matrix R of shape m × n into two feature matrices, P and Q. P (m × f) and Q (n × f), where f is the number of factors, are the learned feature matrices. The goal of MF is to train the best P and Q matrices such that R = P × Q. SGD is the most popular form of optimizer algorithm for this recommender system specific MF, and was popularized by Koren [30].

4.2 Datasets

To test the performance of Pyrorank, we use the following datasets.

1) MovieLens-20M: MovieLens-20M [31] is one of the goto datasets for testing recommender systems. This dataset contains 20 million ratings and 465,000 tag applications applied to 27,000 movies by 138,000 users. The ratings are explicit and between in the range of 1–5.

2) Goodbooks-10k: Goodbooks-10k [32] is a book ratings dataset collected from the book cataloging website Goodreads. This dataset contains 6 million ratings for 10,000 books. The data also includes book features such as author, year, etc., and user created tags, shelves, and genres.

3) Goodreads-105M: Goodreads-105M [33] is a newer and larger book ratings dataset again collected from the book cataloging website Goodreads. This dataset contains 225,394,930 interactions, 104,713,520 ratings, 1,561,465 books, and 808,749 users. This dataset is extremely large and closer in size to what might be run in production for large companies. We wanted to train and run Pyrorank on this dataset to show that our algorithm is not affected by the size of the dataset and instead can be attached at the end of any recommendation pipeline.

The Goodreads-105M dataset is too large to fit into the main memory of a single machine, so we trained it using a GPU accelerated Matrix Factorization library.

4.3   Feature Engineering

To measure diversity, we need feature vectors for each item. Each dataset contains different features for each item, so each dataset required a different way to prepare the feature vectors.

For MovieLens movies, we used the movie tags. Each tag is a binary feature i.e. is this movie comedy or not.

For Goodbooks-10k, we used the TF-IDF of the book tags and shelves. Users tag books with genres or other useful markers ('quick-read', 'funny') and also put books on shelves ('abandoned', 'must-read') for organizational purposes.

For Goodreads-105M, only user shelves were used as features. We again used the TF-IDF of the shelves attributed to each book.

Once we get the feature matrices for each dataset, we perform Singular Value Decomposition (SVD) on them to reduce them to more manageable dimensions.

4.4   Diversity Measures

For measuring diversity, we identified two primary metrics. For measuring diversity level 1, we use the Mean Intralist Dissimilarity (MID) [12] of a candidate list C, which is the average feature vector distance of all the items in the list C.

$$MID(C) = \frac{\sum_{i=1}^{|C|} \sum_{j=i+1}^{|C|} dist(C_i, C_j)}{\frac{|C|^2 - |C|}{2}}. \tag{1}$$

For measuring diversity level 2, we use user unexpectedness, which is defined as the Mean User Dissimilarity (MUD) [22] between candidate list C and user history U, which is the average feature vector distance of every item in list C to every item in user history U.

$$MUD(U, C) = \frac{\sum_{i=1}^{|U|} \sum_{j=1}^{|C|} dist(U_i, C_J)}{|U||C|}. \tag{2}$$

We use two common pairwise metrics, Cosine distance and Euclidean distance. Using the same pairwise metrics for both diversity types allows us to compare the difference between the two metrics.

1) *Euclidean Distance:*

$$dist(p, q) = \sqrt{\sum_{i=1}^{n}(q_i - p_i)^2}. \tag{3}$$

2) *Cosine Distance:*

$$\cos(p, q) = \frac{1}{2}\left(1 - \frac{\sum_{i=1}^{n} p_i q_i}{\sqrt{\sum_{i=1}^{n}(p_i)^2}\sqrt{\sum_{i=1}^{n}(q_i)^2}}\right) \tag{4}$$

We also calculate the following diversity metrics in addition to the two primary ones above.

3) *Mean Centroid Dissimilarity:*

$$centroid(C) = \frac{1}{|C|}\sum_{i=1}^{|C|} C_i. \tag{5}$$

$$MCD(C) = \frac{\sum_{i=1}^{|C|} dist(C_i, centroid(C))}{|C|}. \tag{6}$$

4) *Mean User Centroid Dissimilarity:*

$$centroid(U) = \frac{1}{|U|}\sum_{i=1}^{|U|} U_i. \tag{7}$$

$$MCUD(U, C) = \frac{\sum_{i=1}^{|C|} dist(C_i, centroid(U))}{|C|}. \tag{8}$$

5) *Mean Inverse User Frequency:* $freq(i)$ represents the number of users who have rated item i out of the total n users.

$$MIUF(C) = \frac{1}{|C|}\sum_{i=1}^{|C|} -\log(\frac{freq(C_i)}{n}). \tag{9}$$

To measure diversity, we need feature vectors for each item. Each dataset contains different features for each item, so each dataset required a different way to prepare the feature vectors.

4.5 Algorithms

We compare our algorithm to the following four algorithms:

- Best: The traditional method of taking the top k items with the highest ratings.
- Random: Just taking items randomly from the candidate set.
- KMeansPre: The KMeansPre method is a special case of ClusDiv [25] where the threshold is equal to 1. That means that with k clusters, only a single item is selected per cluster. This is done to ensure that we compare with the maximum level of diversity reachable by the ClusDiv method. The pre suffix denotes that the clusters are precomputed.
- KMeans: This algorithm is a variation of KMeansPre where it does not precompute the clusters, but does it at re-ranking time. This allows the algorithm to fit to any candidate list, while KMeansPre is locked to considering all items.

4.6 Comparisons

To measure diversity, we need feature vectors for each item. Each dataset contains different features for each item, so each dataset required a different way to prepare the feature vectors.

For metrics we use ILD [12], user unexpectedness [22] (shown in the table as user dissimilarity), user centroid unexpectedness [21], and mean inverse user frequency [18]. We also introduce list centroid unexpectedness, which is a blend of ILD and user centroid unexpectedness.

Due to space constraints, we have abbreviated the metrics as below:

- Mean Predicted Rating: MPR
- Mean Euclidean Centroid Dissimilarity: MECD
- Mean Cosine Centroid Dissimilarity: MCCD
- Mean Euclidean Intralist Dissimilarity: MEID
- Mean Cosine Intralist Dissimilarity: MCID
- Mean Euclidean User Dissimilarity: MEUD
- Mean Euclidean User Centroid Dissimilarity: MEUCD
- Mean Inverse User Frequency: MIUF
- Elapsed time in seconds: TIME

The Tables 1, 2, 3 show the comparison of all metrics for the Goodbooks-10k, MovieLens-20M, and Goodreads-105M datasets respectively. The Pyrorank results are calculated with a keep ratio of 0.3, 0.3, and 0.25 for the three datasets. The errors and the resulting confidence intervals are calculated for 95% confidence from a random selection of 300 users.

For comparison with the pre-computed KMeans method, we also pre-computed the pairwise distances necessary for Pyrorank. This leads to a fair comparison of running times. Due to problem of mixing a variable candidate size and precomputation of clusters in KMeans, we opted to not restrict candidate size, meaning the candidate size is equal to the total number of items for each dataset. It should be noted that precomputation on the Goodreads-105M dataset was not feasible, and was therefore ignored in this comparison. The results are available in Table 4.

Table 1. Goodbooks-10k.

METHOD METRIC	BEST	RANDOM	KMEANS	PYRORANK
MPR	6.134± 0.062	5.003± 0.048	5.989± 0.059	5.899± 0.058
MECD	6.557± 0.019	6.576± 0.017	6.728± 0.013	6.725± 0.015
MCCD	0.606± 0.004	0.617± 0.003	0.650± 0.002	0.645± 0.003
MEID	9.490± 0.027	9.519± 0.025	9.763± 0.019	9.738± 0.022
MCID	0.888± 0.003	0.897± 0.003	0.923± 0.002	0.919± 0.002
MEUD	9.472± 0.015	9.483± 0.014	9.550± 0.011	9.695± 0.013
MEUCD	6.763± 0.019	6.783± 0.017	6.872± 0.014	7.030± 0.018
MIUF	4.958± 0.027	5.001± 0.029	4.987± 0.027	5.051± 0.026
TIME	0.001± 0.000	0.000± 0.000	0.362± 0.002	0.019± 0.001

Table 2. MovieLens-20M.

METHOD METRIC	BEST	RANDOM	KMEANS	PYRORANK
MPR	5.181± 0.069	4.234± 0.047	5.020± 0.065	4.975± 0.067
MECD	1.193± 0.017	1.197± 0.014	1.453± 0.006	1.395± 0.014
MCCD	0.413± 0.007	0.439± 0.006	0.497± 0.003	0.477± 0.006
MEID	1.711± 0.025	1.723± 0.021	2.108± 0.009	1.988± 0.023
MCID	0.680± 0.010	0.713± 0.008	0.783± 0.003	0.752± 0.007
MEUD	1.890± 0.018	1.895± 0.017	2.029± 0.011	2.085± 0.012
MEUCD	1.343± 0.016	1.344± 0.015	1.516± 0.009	1.572± 0.013
MIUF	4.502± 0.148	5.485± 0.100	4.276± 0.107	4.826± 0.138
TIME	0.000± 0.000	0.000± 0.000	0.118± 0.001	0.004± 0.001

Table 3. Goodreads-105M

METHOD METRIC	BEST	RANDOM	KMEANS	PYRORANK
MPR	6.684± 0.104	5.670± 0.073	6.417± 0.095	6.437± 0.096
MECD	0.481± 0.008	0.468± 0.008	0.705± 0.004	0.714± 0.007
MCCD	0.196± 0.006	0.186± 0.006	0.399± 0.006	0.408± 0.007
MEID	0.701± 0.011	0.680± 0.011	1.030± 0.005	1.036± 0.010
MCID	0.368± 0.010	0.350± 0.010	0.669± 0.007	0.679± 0.009
MEUD	0.725± 0.007	0.708± 0.008	0.887± 0.005	0.928± 0.006
MEUCD	0.567± 0.007	0.547± 0.007	0.742± 0.005	0.787± 0.008
MIUF	11.829± 0.037	11.267± 0.082	11.852± 0.045	11.947± 0.053
TIME	0.002± 0.000	0.001± 0.000	0.364± 0.007	0.025± 0.006

In order to explore how the variation of keep ratio affects the diversity, we looked at how diversity level 1 and 2 change as we slide the keep ratio from 0.1 to 1 (Fig. 4a). Do note that when keep ratio is equal to 1, the algorithm is equivalent to getting the best possible candidates in terms of ratings. The candidate list was kept at 1000 items.

In addition, we also looked at the prediction accuracy and diversity tradeoff when changing the keep ratio (Fig. 4b). Intuitively, we would expect keep ratio to be inversely correlated with diversity and positively correlated with prediction accuracy (RMSE).

We evaluated the effect of changing the size of the candidate list when doing recommendations on the Goodbooks-10k dataset. We looked at both the change in diversity (Fig. 4c) and the change in speed (Fig. 4d) as the size of the candidate list is increased from 100 to 10000 (which is maximal of Goodbooks-10k).

Table 4. Pre-computed Comparison

METRIC \ METHOD	GOODBOOKS-10K		MOVIELENS-20M	
	KMEANSPRE	PYRORANKPRE	KMEANSPRE	PYRORANKPRE
MPR	5.957± 0.059	5.871± 0.057	4.877± 0.059	4.859± 0.062
MECD	6.796± 0.013	6.759± 0.015	1.483± 0.005	1.421± 0.016
MCCD	0.668± 0.002	0.653± 0.003	0.534± 0.003	0.482± 0.006
MEID	9.863± 0.018	9.786± 0.021	2.155± 0.007	2.020± 0.025
MCID	0.936± 0.001	0.925± 0.002	0.821± 0.003	0.756± 0.007
MEUD	9.618± 0.011	9.731± 0.012	2.073± 0.010	2.126± 0.011
MEUCD	6.960± 0.014	7.077± 0.017	1.554± 0.008	1.609± 0.012
MIUF	5.005± 0.026	5.071± 0.026	4.842± 0.116	5.038± 0.143
TIME	0.017± 0.023	0.040± 0.062	0.005± 0.006	0.020± 0.004

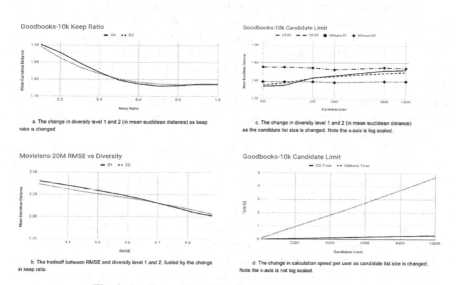

a. The change in diversity level 1 and 2 (in mean euclidean distance) as keep ratio is changed

c. The change in diversity level 1 and 2 (in mean euclidean distance) as the candidate list size is changed. Note the x-axis is log scaled.

b. The tradeoff between RMSE and diversity level 1 and 2, fueled by the change in keep ratio

d. The change in calculation speed per user as candidate list size is changed. Note the x-axis is not log scaled.

Fig. 4. Prediction Model Pipeline with Diversity Ranking

Although it is now shown here, reducing the dimensionality of item features also results in a major speedup, as the calculation of the pairwise distances is a bottleneck for both the KMeans and the Pyrorank approach.

5   Discussion

When changing the keep ratio, we found that using smaller keep ratios led to higher level 1 and level 2 diversity (Fig. 4). This was in line with our expectations, as burning more of the items starting from the user's items by definition has a non-decreasing effect on level 2 diversity. It was also within expectations that there is no meaningful difference between having 0.7 to 1, as there is a higher chance of keeping the traditionally highest ranking items in the candidate set. Although our method does not explicitly optimize for diversity level 1, it also followed a similar curve to diversity level 2.

We could also see the direct tradeoff between prediction accuracy (RMSE) and diversity in both diversity metrics (Fig. 5). This tradeoff was caused again by a change of keep ratios, and it was again in line with our expectations: RMSE correlated positively with keep ratio and correlated negatively with diversity, which led to a clear inverse correlation between RMSE and diversity metrics.

When comparing the effects on changing the size of the candidate list, it is obvious that increasing the size leads to better diversity metrics for both diversity level 1 and 2 (Fig. 6). The diversity of the KMeans method (without pre-computation) does not show a significant difference with different candidate list sizes. When the two methods are compared, Pyrorank quickly surpasses it in terms of diversity level 2, as expected by the aim of our algorithm. It also comes close to matching KMeans in diversity level 1 but falls slightly short.

In terms of speed, we can see that both methods scale linearly with the size of the candidate set. When the clusters are not pre-computed, our method has a significant advantage in terms of time. This becomes useful if the upfront cost of training an extremely large KMeans model with millions of items become infeasible, or if the candidate list is always changing. Both our method and KMeans have been shown to benefit from pre-computing. Note that the pre-computation in our case needs to store the pairwise distances of all items, meanwhile KMeans only needs to store the cluster assignments of items. Both methods can benefit from using a fast approximate nearest neighbors algorithm such as NMSLIB [34].

While KMeansPre is faster than our method, it is affected by the size of the dataset. Even if the initial clusters are precomputed, there will always be the need of finding the item with the highest rating per cluster, meaning the algorithm will scale linearly with the number of items in the dataset. Since Pyrorank does not scale with the size of the dataset (but simply with the candidate set), this creates an advantage with larger datasets.

It should also be noted that while we are at the maximum level diversity that can be offered by the ClusDiv approach, our ceiling for diversity is significantly higher. Our diversity level 2 metrics equal that of KMeans' around a keep ratio of 0.5 with a higher sum of predicted ratings, and surpasses the maximum level of diversity level 2 when the ratio is lowered even further, while giving similar sums of predicted ratings.

The Goodbooks-10k and the MovieLens-20M datasets show the tradeoff of using KMeans vs. Pyrorank, as Pyrorank only reaches better diversity metrics with a lower prediction accuracy. For the Goodreads-105M, Pyrorank performs better in both levels of diversity metrics and in prediction accuracy. We also note that in terms of recommending long-tail items, which is what the Mean Inverse User Frequency measures, Pyrorank performs better than the KMeans method for all datasets.

6   Conclusion

In this paper we have explored the motivation behind diversifying the recommendations of a recommender system, explored various metrics that quantify recommendations, and introduced a novel bio-inspired re-ranking algorithm called Pyrorank. Our algorithm is optimized for diversification in the context of a user, and it outperforms even the most diverse instantiation of the ClusDiv algorithm in terms of user diversification and recommending long-tail items. Moreover, it has a higher ceiling in terms of diversification compared to ClusDiv.

Pyrorank can easily be appended to the results of an existing recommender system that outputs a candidate list of recommendations. This makes it highly adaptable compared to replacing the whole recommendation pipeline to promote diversification, potentially saving many engineer-hours. As such, Pyrorank is a low-cost and easy to productionize solution for reducing systemic bias in recommender systems that need to be held accountable for their recommendations and for the way they influence users.

6.1 Future Work and Research

It has been shown that combining and ensembling different recommender systems can lead to better diversity even without explicitly optimizing for them [19]. We predict that ensembling re-ranking algorithms such as Pyrorank or ClusDiv may lead to even better diversity metrics. Ensembling algorithms that optimize for different levels or metrics of diversity may lead to a more robust system at the cost of more computation.

In some domains, completely removing items close to the user's existing items might not be what the user would respond best to. To mitigate this, one could modify Pyrorank to leave some 'trees' alive inside the burned off areas and control this amount with a new parameter. This way, a user would get multiple items with high predicted ratings recommended that are similar to what they typically interact with, but still much less than with traditional prediction models. This approach would lower diversity level 2 scores but could increase diversity level 1 as items inside the burned areas would be highly isolated and far from other surviving candidate trees. More testing would be necessary to measure the impact of these remaining items.

Acknowledgements. We would like to thank Mengting Wan and Julian McAuley for the new Goodreads dataset and responding promptly to our emails about the dataset.

References

1. MacKenzie, I., Meyer, C., Noble, S.: How retailers can keep up with consumers. McKinsey Comp. **18**(1), 1–10 (2013)
2. Pariser, E.: The Filter Bubble: What the Internet is Hiding from You. Penguin ,UK (2011)
3. Gerrard, Y., Gillespie, T.: When algorithms think you want to die. Wired Maga. **21** (2019)
4. Covington, P., Adams, J., Sargin, E.: Deep neural networks for YouTube recommendations. In: Proceedings of the 10th ACM Conference on Recommender Systems, pp. 191–198. ACM (2016)
5. Dwoskin, E.: Youtube is changing its algorithms to stop recommending conspiracies. Washington Post (2019)
6. Youtube takes ads off 'anti-vax' video channels. BBC UK (2019)
7. Alexander, J.: Youtube fought brie larson trolls by changing its search algorithm. Verge Maga. (2019)
8. Chaney, A.J.B., Stewart, B.M., Engelhardt, B.E.: How algorithmic confounding in recommendation systems increases homogeneity and decreases utility. In: Proceedings of the 12th ACM Conference on Recommender Systems, Series, RecSys 2018, pp. 224–232. ACM, New York (2018)

9. Castagnos, S., Brun, A., Boyer, A.: When diversity is needed... but not expected. In: International Conference on Advances in Information Mining and Management, pp. 44–50. IARIA XPS Press (2013)
10. Biswas, S.R., Mallik, A.U.: Disturbance effects on species diversity and functional diversity in riparian and upland plant communities. Ecology **91**(1), 28–35 (2010)
11. Salton, G., McGill, M.J.: Introduction to Modern Information Retrieval. McGraw-Hill, New York (1986)
12. Smyth, B., McClave, P.: Similarity vs. diversity. In: Aha, D.W., Watson, I. (eds.) ICCBR 2001. LNCS (LNAI), vol. 2080, pp. 347–361. Springer, Heidelberg (2001). https://doi.org/10.1007/3-540-44593-5_25
13. Bradley, K., Smyth, B.: Improving recommendation diversity (2001)
14. McSherry, D.: Diversity-conscious retrieval. In: Craw, S., Preece, A. (eds.) ECCBR 2002. LNCS (LNAI), vol. 2416, pp. 219–233. Springer, Heidelberg (2002). https://doi.org/10.1007/3-540-46119-1_17
15. Kunaver, M., Poz̆rl, T.: Diversity in recommender systems – a survey. Knowl.-Based Syst. **123**, 154–162 (2017)
16. Fleder, D.M., Hosanagar, K.: Recommender systems and their impact on sales diversity. In: Proceedings of the 8th ACM Conference on Electronic Commerce, Series, EC 2007, pp. 192–199. ACM, New York (2007)
17. Clarke, C.L., et al.: Novelty and diversity in information retrieval evaluation. In: Proceedings of the 31st Annual International ACM SIGIR Conference on RESEARCH and Development in Information Retrieval SIGIR 2008, p. 659. ACM, New York (2008)
18. Zhou, T., Kuscsik, Z., Liu, J.-G., Medo, M., Wakeling, J. R., Zhang, Y.-C.: Solving the apparent diversity-accuracy dilemma of recommender systems. In: Proceedings of the National Academy of Sciences, vol. 107, no. 10, pp. 4511–4515, (2010)
19. Lathia, N., Hailes, S., Capra, L., Amatriain, X.: Temporal diversity in recommender systems. In: Proceeding of the 33rd International ACM SIGIR Conference on Research and Development in Information Retrieval SIGIR 2010, p. 210. ACM, New York (2010)
20. Nguyen, T.T., Hui, P.-M., Harper, F.M., Terveen, L., Konstan, J.A.: Exploring the filter bubble. In: Proceedings of the 23rd International Conference on World Wide Web WWW 2014, pp. 677–686. ACM, New York (2014)
21. Adamopoulos, P., Tuzhilin, A.: On unexpectedness in recommender systems: or how to better expect the unexpected. ACM Trans. Intell. Syst. Technol, (TIST) **5**(4), 54 (2015)
22. Hurley, N., Zhang, M.: Novelty and DIVERSITY in top-N recommendation-analysis and EVALUATION. ACM Trans. Internet Technol. **10**(4), 1–30 (2011)
23. L'Huillier, A., Castagnos, S., Boyer, A.: Understanding usages by modeling diversity over time. In: Proceedings of the 22nd Conference on user modeling, Adaptation, and Personalization UMAP 2014, vol. 1181 (2014)
24. Adomavicius, G., Kwon, Y.: Improving aggregate recommendation diversity using ranking-based techniques. IEEE Trans. Knowl. Data Eng. **24**(5), 896–911 (2012)
25. Aytekin, T., Karakaya, M.O.: Clustering-based diversity improvement in top-N recommendation. J. Intell. Inf. Syst. **42**(1), 1–18 (2014)
26. Ponisio, L.C., et al.: Pyrodiversity begets plant–pollinator community diversity. Glob. Change Biol. **22**(5), 1794–1808 (2016)
27. Connel, J.H.: Diversity in tropical rain forests and coral reefs. Science **199**(4335), 1302–1310 (1978)
28. Webster, K.M., Halpern, C.B.: Long-term vegetation responses to reintroduction and repeated use of fire in mixed-conifer forests of the Sierra Nevada. Ecosphere 1(5), art9 (2010)
29. Kunaver, M., Dobravec, S., Kos̆ir, A.: Using latent features to measure the diversity of recommendation lists. In: 2015 38th International Convention on Information and Communication Technology, Electronics and Microelectronics (MIPRO), pp. 1230–1234. IEEE (2015)

30. Koren, Y.: The bellkor solution to the Netflix grand prize. Netflix Prize Doc. **81**, 1–10 (2009)
31. Harper, F. M., Konstan, J. A.: The movielens datasets: history and context. ACM Trans. Interact. Intell. Syst. **5**(4), 19:1–19:19 (2015)
32. Zajac, Z.: Goodbooks-10k: a new dataset for book recommendations. In: FastML (2017)
33. Wan, M., McAuley, J.: Item recommendation on monotonic behavior chains. In: Proceedings of the 12th ACM Conference on Recommender Systems, pp. 86–94. ACM (2018)
34. Boytsov, L., Naidan, B.: Engineering efficient and effective non-metric space library. In: Brisaboa, N., Pedreira, O., Zezula, P. (eds.) SISAP 2013. LNCS, vol. 8199, pp. 280–293. Springer, Heidelberg (2013). https://doi.org/10.1007/978-3-642-41062-8_28

Analysis of SIR Compartmental Model Results with Different Update Strategies

Mijat Kustudic[1], Maja Gulan[1], Junrui Lu[1,2(✉)], and Ben Niu[1,2]

[1] College of Management, Shenzhen University, Shenzhen 518060, China
luxunrui2021@email.szu.edu.cn
[2] Greater Bay Area International Institute for Innovation, Shenzhen University, Shenzhen 518060, China

Abstract. Mathematical modeling is crucial for analyzing how intensively a disease spreads across a population so that an adequate counterstrategy can be prepared. In this paper, we employ agent-based modeling (ABM) to analyze the SIR model (susceptible – infected – recovered). The computational implementation requires the use of rulings, one of them being the iteration procedure. The observed procedures are based on two principles of iteration: sequential and synchronized. We use four different iteration strategies: linear, random, parallel, and concurring with the first two adhering to the sequential and the second two to the synchronized principle. We implement the strategies by modeling an unobstructed area with three congestion scenarios. Finally, we analyze the outcomes and observe what tendencies the approaches have. Our results show that different iteration procedures do affect the outcomes of simulations. Synchronized strategies are more conservative when it comes to the infection spreading due to the agent's ability to check the environment before taking action resulting in fewer contacts with others. Different congestion scenarios show similar results. Knowing this is crucial to make a realistic prediction of the situation, underestimating it will lessen the countermeasures enabling the pathogen to spread quicker while overestimating may create public panic.

Keywords: SIR Model · Agent Based Modeling · Swarm Intelligence · Social Computing

1 Introduction

In epidemiology, a mathematical modeling approach is used to analyze how contagious diseases spread across a population. Models analyze and predict how diseases spread, their subsequent speed, and possible response strategies [1, 2]. To achieve this, modelers need to know mathematical frameworks, understand the pathogen and also have insights into human behavior.

The first equation-based models were developed and employed to analyze malaria parasites and their spread from mosquitoes to humans [3]. They have evolved into the well-established and often used SIR compartmental model [4] (letter abbreviations stand

© The Author(s), under exclusive license to Springer Nature Switzerland AG 2023
Y. Tan et al. (Eds.): ICSI 2023, LNCS 13969, pp. 156–167, 2023.
https://doi.org/10.1007/978-3-031-36625-3_13

for susceptible, infected, and recovered) in which the population is divided into three compartments. The primary model has been adapted to include more compartments to include additional types of individuals such as exposed, treated, latent, asymptotic, and others. Apart from the linear setup, models can utilize a circular notion so that individuals can be infected multiple times because no immunity is gained.

Simulation is a third way of doing science through thought experiments [5]. It starts with deducing assumptions and creates data based on those assumptions. Induction is then used for analyzing outcomes. One of the common approaches used is agent-based modeling (ABM). It is an adaptive modeling tool used decision making by practitioners and policymakers for assessing the relative impact of infection control strategies [6].

This paper delves deep into the essential notion of computational experimentation by checking the iteration procedure of each setup and experiment. It is based on the use of autonomous agents which resemble real-world entities in order to observe emergent behavior and draw conclusions. The primary advantage of this approach is that it can simulate limited rationality and lack of information. The secondary advantage is that the agents act locally and the ensuing results are seen globally. Both phenomena are found in the real world. ABM in this regard has been already used to model various diseases such as measles [7], Ebola [8], H1N1 [9], H5N1 [10], and HPV [11]; its advanced use has been similar to swarm intelligence algorithms (SI) since there is a necessity for observing emergent properties of systems. Recent advances in machine learning have allowed an additional layer into ABM thus creating intelligent agents [12, 13]. By using this symbiotic relationship, in conjunction with real-world data, better results and more accurate predictions can be obtained.

Since ABMs are programs and use computers for implementation, they need to follow programming rules and methods and are thus bound by them. A crucial element of their approach is the updating procedure of agents. This procedure comprises updating the individual's positions, situations, and positions of all agents like in the SI algorithms. Depending on how it is implemented, it can give an advantage to certain individuals. For example, the advantage can be based on movement since moving first can lead to better choices. On the other hand, moving last can have the advantage of having a complete overview of choices made by others. Based on the previous, it is crucial to take into account how the program implementation plays on the simulation results [14]. The results can change based on their dynamics, intensity, or timing and thus can result in a big difference when conclusions are made. Hence choosing the best implementation technique is crucial.

This paper focuses on two questions. The first question analyzes how different types of iterations affect simulation results. The second question asks whether different congestion (concentration) levels of agents play a role in disease spreading in conjunction with iteration strategies. The questions are resolved in several ways. Firstly, we implement different iteration procedures which follow global and local synchronization rules taking into account how agents perceive their surroundings and their sequence for making actions. Secondly, we implement these iteration strategies within different scenarios of the SIR model where agents are differently congested so that their interaction intensity follows congestion levels. This will also influence the dynamics of the pathogen spread demanding different countermeasures for combating the infection.

The rest of the paper is organized as follows. Section 2 will give a brief overview of compartmental models including SIR. Section 3 will define different approaches for making iterations within a simulation. Section 4 will present and analyze the graphical and numerical outcomes of the simulations. Section 5 is the conclusion.

2 Overview of the SIR Model

2.1 Universal Aspects of Models

Epidemiology boasts numerous models, as previously shown. Still, the basic and primary one is the SIR model from which all other compartmental models are derived. Due to their similarity, in the next part, the basics corresponding to all models will be explained after which the SIR specifics will be analyzed.

Sciences are based on collecting data after which the data is used for experimenting and testing hypotheses. Epidemiology is one science where the object of observation, the pathogen, cannot be spread through a population to test the hypotheses and check the corresponding outcomes. Data is collected by observing natural phenomena and their dynamics, this limits its total accuracy and appropriate dynamical observations. Observations of this kind are useful to estimate a range of values for some parameters [15], later these parameters are used for analysis and prediction.

One important element for mathematical epidemiologists is to determine disease dynamics to formulate a counterstrategy. It is important to define the current lifecycle step of the disease on a micro or macro level. The micro level focuses on an individual to prescribe adequate treatment methods. The macro level is equivalent to the social level intending to stop the infection before an outbreak or manage its consequences.

2.2 Specifics of the SIR Model

In this paper we will focus on the classical Kermack–McKendrick SIR model [4]. The agents within the model are assigned linearly to a single compartment: Susceptible (S), Infectious (I), and Removed or Recovered (R) with the last type achieving immunity from the infection. Throughout the process of disease transmission, individuals move between different compartments, as indicated by the following equations. The population stays constant and the only change is number-wise between the disjointed groups. Infective ones transmit the disease to others. Recovered are those who are removed from the susceptible-infective interaction by recovery with immunity, isolation, or death. The transmission rate, denoted by β, , determines the rate at which susceptible individuals become infectious, while the recovery rate (γ) determines how quickly infected individuals recover from the disease. The force of infection can be expressed as $\beta I = N$, where N is the total population. It is assumed that in this basic scenario, the disease transmission dynamics are considerably faster than demographic processes such as births, deaths, and migration. Therefore, these demographic factors can be disregarded. The following formulae are specific to the SIR model.

$$\frac{dS}{dt} = \frac{-\beta SI}{N} \tag{1}$$

$$\frac{dI}{dt} = \frac{\beta SI}{N} - \gamma I \tag{2}$$

$$\frac{dR}{dt} = \gamma I \tag{3}$$

In a system where the population is limited (such as ABM) the most likely outcome for a disease, in the long run, is its eradication from the population. An important determinant is the basic reproduction number which shows the speed of disease spread, denoted by R_0. . In a given population (N) where there are no infected individuals all are susceptible, which means $S_0 = N$. When an infected individual comes into the population then that person can infect βN other individuals at a given moment. Since every individual is susceptible and because the expected length of infection is given as $\frac{1}{\alpha}$ it means that the person can infect $\frac{\beta N}{\alpha}$ individuals so

$$R_0 = \frac{\beta N}{\alpha} \tag{4}$$

where R_0 tells on average how many individuals can an infected person infect. If $R_0 > 1$ then an epidemic will occur since the disease spreads rapidly. If $R_0 < 1$ the disease will eventually be eradicated from the population. If $R_0 = 1$ then the disease becomes endemic and will stay within the population with a relatively unchanged number of infected. Figure 1 shows complete population transfer options.

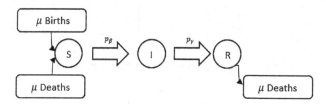

Fig. 1. Basics of the SIR compartmental model

3   Theoretical Setup

Simulations can be considered to be a way of doing thought experiments. The assumptions can be simple but the goal is to find complexity in the results, which is also the main point of SI. Large-Scale effects of locally interacting agents are called "emergent properties" of the system. Such properties are often surprising because it can be hard to anticipate the full consequences of even simple forms of interaction. These properties and consequences are not obtainable through analytical or linear analysis. ABM models are more flexible than linear ones since they can include multiple characteristics of a single agent. This will make the agent resemble a real individual enabling elements like their social mixing, density, and other entomological-related parameters to be observed [16]. These interactions generate multiple nonlinear relationships and interactions such as feedback, learning, and adaptation, integrating spatially and temporally variable information, as well as cross-scale and cross-discipline data and methods [17].

As a result, ABMs are a suitable add-on to the compartmental analysis because they allow them to surpass the limitations of equation modeling which relegated to the setup as a smooth continuum, without special singularities or irregularities. Additionally, one of the main advantages of simulation is that it allows the analysis of adaptive as well as rational agents [5]. They can be programmed to populate a social network, a city, or a multinational region. Their autonomous interactions can follow a daily itinerary, for example going to school or work; they also suffer from limited (often local) information, and limited cognitive capacity [18]. Figure 2 describes options that agents have, regarding actions, their position, and health states, this diagram will form the base of our experiment. The global options are the four iteration procedures and the probability of infection p_β, when interacting with other agents.

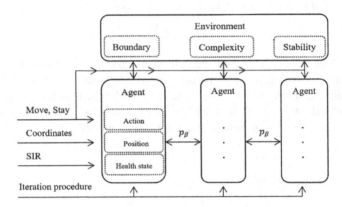

Fig. 2. Agent environment interaction potential

The agents are free to move randomly around their environment, with respect to the congestion scenario. Emergence in this case is perceived as a bottom-up pattern that agents form when interacting and transferring the disease across the population. This type of phenomenon is observed by using ABM because of its intrinsic capacity to incorporate individual-level (e.g., heterogeneous subsystems, autonomous entities) information, to allow for multiple nonlinear relationships and interactions such as feedback, learning, and adaptation, to account for spatially and temporally variable information, and to integrate cross-scale and cross-discipline data and methods [17].

3.1 Approaches to Iterating

It is crucial to understand that the way of defining a program has big effects on its performance and outcomes, let alone the program itself. There is research done specifically regarding the sequence of updates [19, 20]. Work done by [21] analyses how synchronous, random-asynchronous and incentive-asynchronous updating has great implications on the results of the simulations. There are numerous ways of defining the updating procedures, for example, based on different movement initiatives or message exchanges. Work done by [16] concludes that all procedures can be summed to form the sequential strategy, parallel strategy, and concurrent strategy. As a side note, due to their broader basis, I will consider sequential and synchronized strategies also as principles or primary strategies. In this paper, we use four different types of iteration strategies which are derived from two primary ones, as shown in Fig. 3.

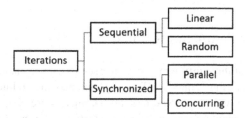

Fig. 3. Different types of iteration strategies

Sequential Strategy. This is the simplest and most used strategy and often the most distant one from the real world. The idea is that all messages and movements are implemented immediately after each agent's turn. They become instantly visible so each other individual can prepare a response strategy. Sequential order means an agent acts only after the previous one has finished an action. This has several drawbacks, the main being that the first acting agent has the most freedom and options to choose from. Based on the first acting agent, all others form their response strategy putting them in a lower-tier position. On the other hand, the last acting agent has "witnessed" all other strategies and has the most knowledge, but maybe the least options.

These drawbacks can be diminished if the order of movement switches randomly between iterations. This relatively balances out the lack of information and the benefits of the initiative. It gives each agent a certain chance to move first and/or to move last. For each iteration and before the agents act, their movement turn/order is randomly chosen for that iteration specifically. Based on the way of implementing the updating sequence we can divide the strategy into the linear one and the random one. Despite its drawbacks, in some cases, the results do not need perfect, such as for a real-world simulation, and are still useful despite the linearity.

Parallel Strategy. This strategy is based on simulating real-world parallel implementation and abstracts how simultaneous systems function. For each iteration, at its beginning, all agents are informed about the current environment, overall situation, and agent position. Each agent gets a turn to act, this may be counterintuitive from parallelism but is necessary due to computational reasons. Acting is perceived as reading the current environment situation and "declaring" the intended move. The intended action is memorized and the next agent acts by declaring his intended action. At the end of the iteration, all declarations are "resolved" by implementing them simultaneously into the environment forming a new environment.

Concurrent Strategy. This strategy can be viewed as a combination of previous strategies. Like the parallel strategy, it has global synchronization. But unlike it, agents can check for declarations before they are resolved in unison. The process is linear because each agent firstly checks what have the previous agents done and then declares the answer strategy. The process is also parallel because the environment is updated simultaneously. The main idea is that updates made by all agents are in parallel but all messages sent and changes to the environment are immediately visible [16].

4 Experimental Setup and Results

In this part, we present three different simulation scenarios and analyze obtained results depending on the iteration sequence. The implemented methodology is kept simple due to the focus on universality. Experimental parameters are chosen according to our experience regarding agent movement speed and terrain coverage while considering the notion of simplicity. The terrain used in the simulation is limited and consists of discreet fields. Each agent moves across it according to the Moore neighborhood logic only once per iteration. At the start of each simulation, agents will be randomly generated across the terrain, after which they will have the option to randomly move in all directions relatively keeping the congestion levels, as shown in the next figure, Fig. 4 the first congestion strategy intends to keep the agents within the central 50% of the terrain. The second strategy keeps them 75% and the third enables agents to move randomly across the whole terrain. Having the same number of agents differently concentrated will cause more interactions and thus more infections.

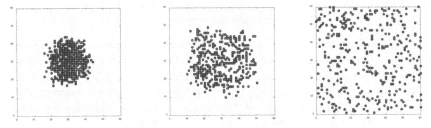

Fig. 4. Spatial congestion options: left to right are high, medium, and low congestion

The infection can be transmitted only by direct contact with the infected individual. The β probability of infection is set to 100% at each contact with another individual. The duration of the disease α is set to 10 iterations and γ is set to 90%. The number of simulations will be set to 100 with 10 runs with the result being the mean of all simulations and for each iteration strategy. The number of agents is set to 350 with each being allowed a single move during the iteration. Each run starts with 10 initially infected individuals randomly dispersed. The size of the experimental space is set to 60 x 60 with a single agent per Moore square space.

4.1 Dynamics of the Model

The following figure shows the dynamics of the model while using different scenario starting variables. It is important to note that the outcomes do not vary, compared to other runs of the same scenarios (Fig. 5).

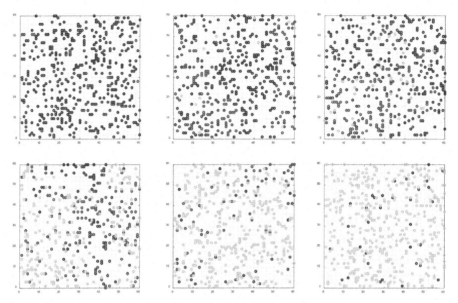

Fig. 5. Shows the dynamics of the SIR model, how the susceptible (blue) become infected (red) finally reaching the recovered state (yellow). (Color figure online)

In the final iteration notice two interesting epidemiological phenomena both related to residual susceptible agents. Firstly, there are some agents unaffected by the infection which leaves them without immunity: they are still susceptible to the next outbreak, if or when it happens. Considering this is important when it comes to infection incidence and reservoirs [8]. In the meantime, we see that they are protected by herd immunity, a term which is referring to the preexisting immunity of the majority.

In reality, indefinite measure maintenance is unlikely and it is important to consider the honeymoon and divorce effects which are referring to non-immunizing control measures used against endemic infections. The honeymoon refers to the effect of measures having a temporary impact on reducing the prevalence of the infection. However, they can result in a decrease in herd immunity, which ultimately leaves the population vulnerable to larger outbreaks once the control measures are lifted [22]. After the control is lifted a severe outbreak can occur which is referred to as the divorce effect.

4.2 Graphical Results

In the following figures, susceptible, infected, and recovered agents are presented in blue, red, and yellow respectively. As stated before, in a closed population after a sufficient number of iterations, the infection will be eliminated, as shown by the receding red line.

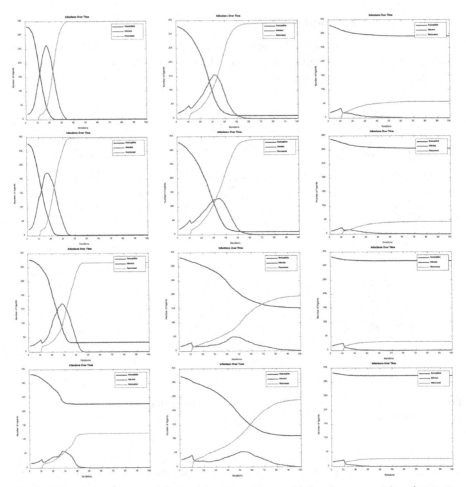

Fig. 6. For each row from top to bottom: linear, random, parallel, and concurrent iteration strategies. From left to right are high, medium, and low congestion scenarios, respectively. (Color figure online)

We observe the dynamics of the disease are greater within the linear strategy in all scenarios (Fig. 6). This is followed by a quicker decline and pathogen elimination from population. In high and medium congestion scenarios, the majority of all susceptible agents are infected and afterward, they recover. The parallel strategy performs differently from the concurrent one where a greater number of agents are being infected and recovered meaning that its dynamics are greater when it comes to resolvent of the disease. The concurrent strategy, in the highly congested scenario, shows the greatest difference with quick disease elimination together with the least infected agents, keeping it the most conservative when it comes to disease spreading. In the medium congested scenario, this strategy shows greater disease-spreading dynamics.

Comparing two primary strategies in the highly congested scenario we can observe similar results regarding both sequential and parallel strategies with the last one being the most conservative. A bigger difference is seen in the medium congested scenario where the disease is eliminated before spreading through the entire population. In synchronized strategies, it can be seen as long-lasting. Synchronized strategies do not perform significantly different when it comes to low-congested scenarios.

4.3 Numerical Results

Table 1 shows mean numerical results of S, I, and R agents according to their respective runs and scenarios. The linear iteration strategy has the most infected and recovered agents, and the least susceptible number of agents. Similar results are obtained by using the randomized iteration strategy, which has the second-largest number of infected and recovered agents. The parallel strategy has the second highest number of susceptible agents followed by the second least infected and recovered agents. The concurrent iteration strategy has the largest pool of susceptible agents which means the infected and recovered numbers are the least since their numbers are zero-sum.

Table 1. Comparison of mean numerical results based on SIR conditions of agents, according to different scenarios and approaches to iterating

	Linear	Random	Parallel	Concurrent
Susceptible	140.5493	146.2683	212.7833	256.3500
Infected	27.4973	26.8857	20.0013	14.2637
Recovered	181.9533	176.8460	117.2153	79.3863

From the numerical results, we see how the presented strategies can be observed as variations of two main principles with the first one following sequential and the second synchronized actions.

Table 2. Mean time needed for simulations according to different congestion scenarios

	Mean seconds
High congestion	228.51
Medium congestion	115.13
Low congestion	105.44

Table 2 shows how different scenarios have different computational complexities. They are measured in seconds needed to perform a simulation run, and the numbers are the mean results of all respected simulations. This happens because in low-density scenarios there is not a lot of need for calculating infection transfers, since contact between agents is not often. The opposite is seen in the highly congested scenario where agents often make contact and spread the infection which requires computation. Computational results show that it is possible to differentiate scenarios according to their complexity and to define which need to be simplified to make them efficient and effective.

The conclusion to be made from the numerical results is that sequential strategies can produce different results from synchronous strategies. More conservative results, regarding susceptibility, are obtained from the synchronous strategies while the sequential strategies show more dynamic infection-oriented strategies. One reason for this is the advantage or disadvantage the agents have while moving sequentially. There are certain benefits and drawbacks to taking action at any particular moment. This is an important fact since different national or regional strategies to combat the spread of infection can be formulated resulting in different outcomes of the disease dynamics.

5   Conclusion

Computer modeling is crucial for adequately understanding disease spreading and its effect on the population. ABM is an upgrade compared to the analytical methods because it allows for a more realistic approach even on the individual level (individual-based modeling). SI principles were used in this paper to observe how the disease can spread with increased interaction and to observe emergent properties. Both approaches require computers so iteration procedures must be used.

In this paper, we define primary and secondary approaches to iterating. The first example of primary ones is the sequential iteration procedure which follows each agent's turn to perform actions. Their results and dynamics are greater while the infection is eliminated sooner. One reason is the added initiative by certain individuals, bringing them closer to others and thus infecting them at a greater rate.

Synchronized strategies are the parallel and the concurring one, these strategies employ global synchronization and environment update at the end of each iteration. The goal is to enable synchronized movement by all individuals. These approaches are more conservative when it comes to infection spreading. Since the concurrent strategy allows for an environment check before the action, there are fewer contacts with other agents thus the disease is eliminated quicker resulting in low numbers of recovered agents.

The future perspective of this work will follow the exploration of how ABM simulation results vary if the programming, is implemented differently. This is important in order to adequately perceive the disease dynamics and characteristics.

References

1. McVernon, J., McCaw, C., Mathews, J.: Model answers or trivial pursuits? the role of mathematical models in influenza pandemic preparedness planning. Influenza Other Respir. Viruses **1**, 43–54 (2007)
2. De Jong, M.C.M., Hagenaars, T.J.: Modeling control of avian influenza in poultry: the link with data. Revue Scientifique et technique **28**, 371–377 (2009)
3. Ross, R.: The Prevention of Malaria, pp. 651–686. John Murray, London (1911)

4. Kermack, W.O., McKendrick, A.G.: Contributions to the mathematical theory of epidemics. III. Further studies of the problem of endemicity. Proc. Roy. Soc. A: Math. Phys. Eng. Sci. **141**, 94–122 (1933)
5. Axelrod, R.: Advancing the art of simulation in the social sciences. In: Conte R., Hegselmann R., Terna P. (eds) Simulating Social Phenomena. Lecture Notes in Economics and Mathematical Systems, vol 456. Springer, Heidelberg (1997)
6. Laskowski, M., Demianyk, B.C.P., Witt, J., Mukhi, S.N., Friesen, M.R., McLeod, R.D.: Agent-based modeling of the spread of influenza-like illness in an emergency department: a simulation study. IEEE Trans. Inf. Technol. Biomed. **15**(6), 877–889 (2011)
7. Perez, L., Dragicevic, S.: An agent-based approach for modeling dynamics of contagious disease spread. Int. J. Health Geogr. **8**(1), 1–17 (2009)
8. Kustudic, M., Niu, B., Liu, Q.: Agent-based analysis of contagion events according to sourcing locations. Sci. Rep. **11**, 16032 (2021)
9. Frias-Martinez, E., Williamson, G., Frias-Martinez, V.: An agent based model of epidemic spread using human mobility and social network information. In: IEEE Conference on Social Computing (2011)
10. Dibble, C., Wendel, S., Carle, K.: Simulating pandemic influenza risks of us cities. In: Proceedings of the 2007 Winter Simulation Conference, pp. 1548–1550 (2007)
11. Olsen, J., Jepsen, M.R.: Human papillomavirus transmission and cost-effectiveness of introducing quadrivalent HPV vaccination in Denmark. Int. J. Technol. Assess. Health Care **26**(2), 183–191 (2010)
12. Zhang, W., Valencia, A., Chang, N.B.: Synergistic integration between machine learning and agent-based modeling: a multidisciplinary review. IEEE Trans. Neural Netw. Learn. Syst. **34**, 1–21 (2021)
13. Yang, Z., von Briesen, E. (eds.): Proceedings of the 2020 Conference of The Computational Social Science Society of the Americas. Springer Proceedings in Complexity. Springer, Cham (2021). https://doi.org/10.1007/978-3-030-83418-0
14. Thaler, J., Siebers, P.O.: The art of iterating: update-strategies in agent-based simulation. Payne D. et al. (eds) Social Simulation for a Digital Society. SSC 2017. Springer Proceedings in Complexity. Springer, Cham (2019). https://doi.org/10.1007/978-3-030-30298-6_3
15. Levin, S., Hallam, T., Gross, L.: Applied Mathematical Ecology, vol. 121, p. 18. Springer, Heidelberg (1989). https://doi.org/10.1007/978-3-642-61317-3
16. Zhang, Q., Sun, K., Chinazzi, M., Pastore, Y., Piontti, A., Dean, N.E., et al.: Spread of Zika virus in the Americas. Proc. Natl. Acad. Sci. U.S.A. **114**, E4334–E4343 (2017)
17. Bersini, H., Detours, V.: Asynchrony induces stability in cellular automata based models. In Proceedings of Artificial Life I, pp. 382–387. MIT Press. (1994)
18. Parker, J., Epstein, J.M.: A distributed platform for global-scale agent-based models of disease transmission. ACM Trans. Model. Comput. Simul. **22**(1), 1–25 (2011)
19. An, L., Grimm, V., III B.: Editorial: meeting grand challenges in agent-based models. J. Artif. Soc. Soc. Simul. **23**(1), 13 (2020)
20. Railsback, S., Grimm, V.: Agent-Based and Individual-Based Modeling: A Practical Introduction. Princeton University Press, Princeton (2011)
21. Page, S.E.: On incentives and updating in agent based models. Comput. Econ. **10**(1), 67–87 (1997)
22. Hollingsworth, B., Okamoto, K.W., Lloyd, A.L.: After the honeymoon, the divorce: unexpected outcomes of disease control measures against endemic infections. PLoS Comput Biol **16**(10), e1008292 (2020)

Research on Location Selection of General Merchandise Store Based on Machine Learning

Boyu Lin[1], Feipeng Li[1,2(✉)], Jiqiang Feng[1,2], and Shengbing Xu[3]

[1] College of Mathematics and Statistics, Shenzhen University, Shenzhen 518000, China
mathfei@163.com
[2] Shenzhen Institute of Computing Science, Shenzhen 518000, China
[3] School of Computer and Information, Dongguan City College, Dongguan 523401, China
1900201003@email.szu.edu.cn

Abstract. In recent years, the continuous spread of the COVID-19 virus, the constant impact of Internet giants, and the influence of the external economy have put general merchandise store chains under unprecedented competitive pressure. Therefore, this paper attempts to help companies reduce corporate costs and improve economic efficiency through rational and scientific site selection decisions. Firstly, feature extraction is performed on the point of interest (POI) dataset and residential dataset, and Boruta is used for feature screening to find 49 important features out of 208 features. Second, several machine learning algorithms, such as logistic regression, random forest, LightGBM, and CatBoost, were used to train the dataset for the site selection model. Then, to combine the advantages of the four models, the four models are integrated using Stacking integration. The experimental results show that the models obtained by Stacking integration are more effective and stable. Finally, the application case of Hangzhou City shows that the model can judge the reasonableness of alternative addresses or find suitable areas to provide alternative addresses. This study provides a new way of selecting general merchandise store addresses, which is of great significance, and also provides a reference for other site selection problems.

Keywords: Location Selection · Feature screening · Machine Learning

1 Introduction

General merchandise stores (GMS), adopt the self-selection sales method, mainly selling valuable and popular goods. They combine the operating advantages of GMS and discount stores with a complete variety of products that meet customers' needs for one-stop shopping in retail. Comprehensive GMS are generally located in commercial centers, urban-rural junctions, residential areas, and transportation hubs.

The global economy is declining, and consumer demand is shrinking. In 2022, with the continuation of the COVID-19 pandemic and external economic impact, chain general merchandise store enterprises faced the most complex environment and challenges in history. In the first half of 2022, the situation between Russia and Ukraine continued

Y. Tan et al. (Eds.): ICSI 2023, LNCS 13969, pp. 168–180, 2023.
https://doi.org/10.1007/978-3-031-36625-3_14

to ferment. China also faced a COVID-19 outbreak that exceeded expectations, making it difficult for the Chinese economy to move forward. The growth of household income is slowing down, and consumer demand is shrinking. The impact of Internet giants has increased competition. In recent years, the rapid development of e-commerce platforms such as Tmall Supermarket and JD Supermarket has become an undeniable new force in the "online supermarket" industry. Alibaba's retail platform "Hema Fresh" and others have entered the market, making competition in the general merchandise store industry even more intense. In addition, community group-buying platforms such as Meituan Buy and Xingsheng Youxuan have greatly solved the problem of community residents' daily necessities supply but also brought some pressure to traditional GMS. With the deep integration of online and offline, Internet giants such as Alibaba, JD, and Meituan are increasing their investment in "instant retail," which is increasingly significant in traditional physical retail enterprises, affecting customer traffic and business performance. In this case, exploring a large-scale, comprehensive, scientific, and quantifiable location selection of general merchandise store evaluation method suitable for China's national conditions is of great practical significance.

In the current context of slowing market growth and declining general merchandise store numbers, it is necessary to establish a reasonable location selection model for general merchandise stores. Western countries have extensively researchedon location selection, proposing multiple theories, evaluation methods, and models. In contrast, research on location selection in China is comparatively scarce and dispersed due to a lack of theoretical guidance. Nelson (1958) [1] proposed the method of listing and put forward a relatively detailed indicator system, which includes eight major categories and 36 subdivisions, among which the changes in population and the size of the area of alternative sites are highly valued and concerned. Moutinho (1993) [2] used multiple regression analysis to identify seven key factors related to selection, including population size, purchasing power, transportation convenience, competition, and distance from competitors within the shopping area, as well as the average spending on a particular product. Kuo (2002) [3] discussed the location selection indicators in seven categories, including population characteristics, attractiveness, store features, competition, accessibility, convenience, and economic stability, among which competition and attractiveness are considered the most important factors. Karamshuk (2013) [4] used the support vector machine algorithm to characterize the correlation between regional foot traffic density and retail store sales results by mining the geographic coordinates left by consumers on Twitter near the stores. Sever and Ali (2016) [5] trained an end-to-end model of regional features-store sales performance using machine learning algorithms and obtained the feature importance of each characteristic for sales performance within the region. Baumbach (2019) [6] uses OSM's road network to calculate catchment areas and combine them with the location factors of a geographic data model, enabling dynamic customer, competitor and supplier analysis and its application to online food delivery scenarios. Xiao (2019) [7] used hierarchical analysis to calculate the weights of five geographic factors and then used GIS spatial analysis to assign location scores to make site recommendations for hypermarkets. Rincón (2020) [8] has developed a spatial model indicator for estimating household food expenditure demand that can be used to advise on the location of conventional GMS in low-income cities. Han (2022) [9] used a relevance-based feature subset selection algorithm and Best-First search method to perform feature screening. Then he used an improved gray evaluation method and kernel regression to obtain the potential of candidate locations. Lan (2022) [10] constructed a geospatial traffic map based on Singapore land features and traffic network dataset, constructed a traffic network connectivity map, and used a graph convolutional network (GCN) to integrate the traffic network to predict stores well as attraction sites in connected areas.

In this paper, we studied the literature related to the location selection model of GMS and identified the influencing factors for location selection. Firstly, we gridded the original dataset of first-tier cities in China, then determined positive and negative samples and conducted feature engineering. Finally, we used Boruta to select features and obtained the final dataset. We experimented with four machine learning algorithms on the dataset and integrated the four models using the stacking method. The experimental results showed that the stacking method effectively improves the models. Based on the location selection model constructed, we selected Hangzhou City as a case study for application and analysis, verifying the model's practicality.

2   Feature Engineering

2.1   Original Dataset

Residential Data. The data records the housing price data of residential neighborhoods in major cities nationwide, including 17 first-tier and quasi-first-tier cities such as Beijing, Shanghai, Guangzhou, Shenzhen, Chengdu, and Chongqing. The data records the number of residential neighborhoods in the city and the average price of residential areas in a 250 m × 250 m grid as a unit. Each data entry records: the grid center point longitude, grid center point latitude, grid minimum longitude, grid maximum longitude, grid minimum latitude, grid maximum latitude, number of residential neighborhoods, the average price of residential areas, highest price of residential neighborhoods, lowest price of residential neighborhoods, etc.

POI Data. This data records the point of interest (POI) data of the Gaode Map, which refers to the latitude and longitude information and the classification of any geographical data on the map. The facilities on the map, including scenic spots, companies, residences, bus stops, etc., are all POI data. In the acquired data, a total of 15 categories of POI data are included: restaurants and food, tourist attractions, public facilities, companies and enterprises, shopping and consumption, transportation facilities, financial institutions, hotels and accommodation, science, education and culture, automobile-related, business housing, living services, sports and leisure services, health care services, sports and fitness, where each piece of data records the type, name, major category, medium category, longitude, latitude, etc.

2.2   Feature Engineering

Population Factors. Population factors can be divided into two categories: the population's quantity and structure and the population's consumption level.. Among them, the number and structure of the population can reflect the concentration of the population, the age structure of the population, the education level of the population, and the nature of the work of the population. Therefore, in the dataset used in this paper, the number and distance of residential houses in the district in the region, the number of kindergartens in primary and secondary schools in the region and the distance, the number of universities and vocational schools in the region, and the number and distance of companies and enterprises in the region can be used to represent the number and structure of the population. The average house price in the region is used to indicate the spending power of the population in the region.

Competitive Factors. The competitive factors can be divided into the intensity and concentration of competitors. Competitors include mini-supermarkets, fresh supermarkets, convenience stores, markets, fish and bug markets, etc. In constructing the characteristics, this paper considers both based on the number of competitors and distance, where the number of competitors indicates the intensity of competitors and the distance of competitors suggests the concentration of competitors.

Transportation Factors. Transportation factors can be divided into the degree of transportation density and the degree of transportation convenience. In this paper, the number and distance of traffic factors in the region are calculated for the construction of transportation factor features as above. The number of traffic factors indicates the density of transportation, and the distance of transportation factors suggests the convenience of transportation.

Other Factors. Other factors can include shopping centers, performing arts centers, cultural plazas, convention centers, sports venues, government agencies, entertainment venues, etc., in the region. Thereforethe indicators are still constructed using the number and distance approach for other facilities that may affect the location of hypermarkets.

3 Location Selection of General Merchandise Store Based on Machine Learning

3.1 Dataset Construction

This article's research area focuses on the four first-tier cities in China: Beijing, Shanghai, Guangzhou, and Shenzhen. By using publicly available data and machine learning algorithms, a location selection model of GMS is trained based on derived features. Each grid area is set at 2000*2000 m, resulting in a total of 3589 grid areas across the four cities, which are assigned numbers ranging from 1 to 3589. A general merchandise store in each grid is used as a criterion for dividing the samples into positive and negative categories.

The features of the general merchandise store location model in this article are divided into four major categories: population, transportation, competition, and other factors. The datasets used to construct these features included POI data and residential property price data from the four first-tier cities. For each type of POI and residential property, features are derived and a total of 208 indicators are constructed. In addition, for zero values deemed unreasonable in terms of distance, a value of 2 km is used for filling in the gaps. For other unreasonable zero values and missing data, the K-Nearest Neighbor (KNN) interpolation method fills in the gaps. Specifically, this article selected the mean value of the non-empty corresponding positions of the ten nearest neighbor samples to fill in the missing values.

In the previous feature engineering step, 208 feature indicators are constructed. Due to a large number of features, it is necessary to perform feature selection. Therefore, we conducted a Boruta feature selection on these 208 features. In this study, the base estimator of Boruta is set to a random forest with a maximum depth of 5, and the maximum number of iterations is limited to 200. Finally, 49 important features are selected. Please refer to Appendix Table 7 for details.

3.2   The Single Models of Location Selection of General Merchandise Store

We divide our dataset into a training set and a testing set in a 3:1 ratio, where the training set is used to train the models. We then used Bayesian hyperparameter tuning to set the model parameters to their optimal values and obtain the best-performing model. The testing set is used to evaluate the accuracy of the trained model.

This paper selects four models, Logistic Regression, Random Forest, LightGBM, and CatBoost. The parameters of each model are adjusted using Bayesian optimization, and the final parameter settings are shown in Table 1.

Table 1. Four machine learning algorithm training parameters.

Algorithm	Parameter
CatBoost	random_state = 520, iterations = 150, cat_features = cat_de_ind, depth = 5, learning_rate = 0.03, rsm = 0.7, subsample = 0.7, bagging_temperature = 0.5
Random Forest	n_estimators = 200, min_samples_leaf = 5, max_depth = 15, max_features = 0.3, min_samples_split = 21, random_state = 1111
LightGBM	random_state = 520, learning_rate = 0.015, max_depth = 6, min_child_samples = 52, num_leaves = 400, subsample = 0.7, colsample_bytree = 0.9, n_estimators = 300
Logistic regression	solver = 'liblinear', class_weight = "balanced", C = 1.0, intercept_scaling = 1, max_iter = 100, multi_class = 'ovr', n_jobs = 1, penalty = 'l2', random_state = None, tol = 0.0001, verbose = 0, warm_start = False

After training and validation on the training and testing sets, the results are shown in Fig. 1, 2, 3 and 4 (Table 2).

AUC is the area under the ROC curve, which is a good performance indicator for measuring the quality of a learning algorithm. From the above model comparison results, the performance of the four models is CatBoost $\approx$ Random Forest > LightGBM > Logistic. From the results, overfitting occurs in Random Forest, LightGBM, and CatBoost,

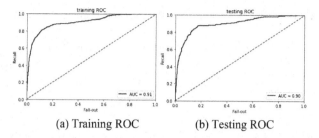

(a) Training ROC   (b) Testing ROC

Fig. 1. ROC curve of logistic regression model.

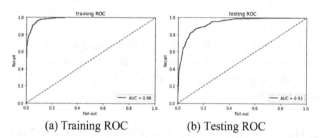

(a) Training ROC   (b) Testing ROC

Fig. 2. ROC curve of Random Forest model.

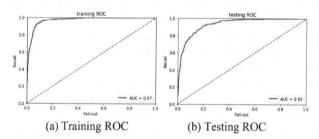

(a) Training ROC   (b) Testing ROC

Fig. 3. ROC curve of LightGBM model.

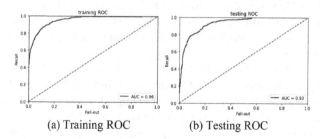

(a) Training ROC   (b) Testing ROC

Fig. 4. ROC curve of CatBoost model.

and the test set performs slightly worse than the training set, possibly due to the higher model complexity. Although the relatively simple Logistic regression model does not

Table 2. The AUC value of the four machine learning algorithms.

	Logistic	Random forest	LightGBM	CatBoost
Training AUC	0.905	0.984	0.978	0.965
Testing AUC	0.897	0.932	0.928	0.932

show overfitting, its performance is poorer due to its simplicity. Overall, the CatBoost and Random Forest models performed the best, followed by the LightGBM model, while the performance of the Logistic regression model is average. Therefore, the decision-maker could consider CatBoost or Random Forest models site recommendation in the real-world location selection.

Considering that the above models all have their advantages and disadvantages, in the following section, we will consider whether we can improve the model's performance, reduce the overfitting of good models, and improve the predictive performance of poor models by using ensemble learning to integrate several models.

3.3   Stacking Model of Location Selection of General Merchandise Store

We have constructed a more complex and more extensive GMS location model based on the four models in the previous section. In this article, we adopt the stacking ensemble method to combine multiple learning models to achieve better training results, reduce model overfitting, and improve prediction accuracy.

Stacking is modeled by modeling the stacking of the original data fit. It first learns the original data by base learners, then these base learners all output the original data, then the output of these models are stacked according to the column, which constitutes the new sample, then the new sample data is given to the second layer model for fitting.

Generally, in the Stacking algorithm, the first layer uses several models with higher fitting degrees, such as random forests and GBDT, to thoroughly learn from the training data. Because different models have differences in principle and on the training set, the first-layer model can be regarded as automatically extracting effective features from the original data. However, due to using complex nonlinear algorithms to extract features, Stacking is more prone to overfitting. Therefore, to reduce the risk of overfitting, the second-layer model in Stacking tends to use simpler models, such as the Logistic regression algorithm.

In this article, we select several models with better performance in the previous section as the first-layer models of the Stacking algorithm: random forest, CatBoost, and LightGBM. We also choose the relatively simple Logistic regression as the second-layer model. The integrated model is shown in the figure below (Fig. 5).

The experimental results of the Stacking model are obtained by feeding the training dataset into it, which are shown in Fig. 6.

Based on the results in Fig. 6, and the data in Table 3, it can be seen that the Stacking ensemble learning model achieved an AUC of nearly 0.94, which is an improvement over any of its base learners. Moreover, the performance on the training and testing sets

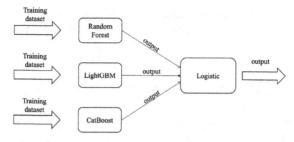

Fig. 5. Stacking model of Location Selection of General Merchandise Store.

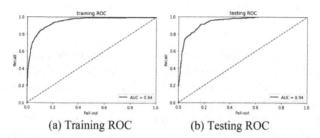

(a) Training ROC   (b) Testing ROC

Fig. 6. ROC curve of Stacking model.

Table 3. The Training and Testing AUC values of the Stacking model.

Stacking	AUC
Training AUC	0.944
Testing AUC	0.939

is similar, successfully reducing model overfitting and making the model's performance more stable.

4 The Application of the Location Selection Model

To further verify the usability and general applicability of the model, this study selected Hangzhou City as a case for analysis. Hangzhou city, ranked fifth in the comprehensive strength ranking of China in 2022, following only the four first-tier cities, indicates that the development of Hangzhou has approached that of the four first-tier cities. Therefore, selecting Hangzhou City as the validation of this model is appropriate. The given rectangular range divides the grid into 1095 grid objects based on a 2000 * 2000-m grid. Following the same data processing method mentioned before, the original dataset is processed to obtain the dataset of Hangzhou city. Various models are used to predict the dataset, and the predicted probabilities and results for each grid object are compared with the true values to validate the practicality of the model. Finally, the prediction performance of each model is presented in the Table 3, with the prediction results for each grid object.

Table 4. Experimental Results on Hangzhou City.

	Logistic	Random Forest	LightGBM	CatBoost	Stacking
precision	0.83	0.85	0.84	0.85	0.85
recall	0.84	0.85	0.85	0.85	0.86
F1 score	0.82	0.83	0.83	0.83	0.84

The model's results are similar to the analysis results in the previous section. The Stacking ensemble model has the best performance, followed by CatBoost and Random Forest, which are slightly better than the LightGBM model. In contrast, the Logistic Regression model has the worst performance.

Due to space limitations, this paper randomly selected six grid areas in Hangzhou City as examples to show the final data of each grid and the predicted probability and results of each model and compared them with the true values of the grid area, as shown in Table 5.

From the predicted results shown in Table 4, it can be seen that the Stacking ensemble model has the best performance. Therefore, we use the Stacking model as the standard and identify the top ten regions with the highest predicted probability results. These fifteen regions are considered to have the most suitable business environment and ecology for building GMS in Hangzhou, as shown in Table 6.

Based on the predicted results of the location model of the general merchandise store constructed, we can provide two suggestions to enterprises:

Firstly, based on the selected candidate areas provided by the enterprise and the data within the region, predict whether the area is suitable for constructing a general merchandise store and provide a reasonable suggestion to the enterprise, which can verify the rationality of the selected points.

Secondly, we can conduct a large-scale scan of the city or administrative area to find the center points of suitable regions for building GMS and provide a preliminary location selection suggestion to the enterprise. It allows them to conduct field research and determine whether building a general merchandise store in that area is suitable. This can reduce the workload of enterprise field research and help them find potential locations of general merchandise stores more accurately.

Table 5. Prediction data (partial).

Region	1	2	3	4	5	6
Longitude of the central point	120.096998	120.143856	120.143856	119.839279	120.237572	120.237572
Latitude of the central point	30.285012	30.303024	30.321036	30.248988	30.17694	30.375072
Number of KTVs	2	6	5	0	0	0
...	...	...	...	...	...	...
Minimum housing price	22135.00	15500.00	24812.00	14979	13730.00	16004.00
Maximum housing price	127580.00	79531.00	154613.00	15875	42645.00	29029.00
CatBoost predicted value	1	1	1	0	1	1
Random forest predicted value	1	1	1	0	0	0
LightGBM predicted value	1	1	1	0	0	0
Logistic regression predicted value	1	1	1	1	1	1
Stacking model predicted value	1	1	1	0	1	1
Actual value	1	1	1	0	1	1

Table 6. Top 10 predicted probabilities (stacking).

Stacking model predicted probability	Longitude of the central point	Latitude of the central point
0.990	120.096999	30.285012
0.989	120.143857	30.321036
0.989	120.143857	30.303024
0.989	120.120427	30.303024
0.987	120.214143	30.267000
0.987	120.096999	30.303024
0.987	120.167285	30.285012
0.987	120.073569	30.285012
0.986	120.214143	30.194952
0.986	120.120427	30.285012

5   Conclusion

In this paper, we use high-resolution Amap city POI data and residential data of urban neighborhoods to construct features and use the Boruta feature selection method to screen the original dataset. Then, we trained, predicted, and evaluated the data using Logistic Regression, Random Forest, and GBDT separately. Additionally, we used the stacking ensemble method to integrate the models mentioned above. The experimental results show that the performance and generalization ability of the Location Selection is better after stacking. After the model construction is completed, we select Hangzhou as the application case for the model, verifying and demonstrating two functions of the location selection model: determining the rationality of candidate sites according to the given candidate sites and scanning the entire city to provide preliminary suggestions for candidate sites. This research provides a new idea and method for the location selection of general merchandise stores, which is significant for improving GMS performance and competitiveness. It also provides a reference for other location selection issues.

Acknowledgements. This work is supported by Natural Science Foundation of Shenzhen under Grant 20200821143547001 and Social science and technology development project in Dongguan under Grant 2020507151806.

Appendix

Table 7. The Important Features of location selection model.

Features	Features	Features	Features
Number of KTVs	Number of shopping malls	Number of coffee and cake shops	Average distance to fast food restaurants
Number of specialized hospitals	Number of GMS	Number of metro stations	Average distance to sports and entertainment stores
Number of primary and secondary schools and kindergartens	Number of bars	Number of home building materials stores	Average distance to chess and internet cafes
Number of Chinese restaurants	Number of banks	Number of home appliance and digital stores	Average distance to amusement parks
Number of residential areas	Number of emergency medical clinics	Number of industrial parks	Average distance to flower, bird, fish, and insect markets
Number of parking lots	Average distance to KTV	Number of factories	Average distance to Western-style restaurants
Number of charging and gas stations	Average distance to specialized hospitals	Number of cinemas and theaters	Average distance to shopping malls
Number of bus stops	Average distance to animal hospitals	Number of stationery stores	Average distance to sports and fitness facilities
Number of public toilets	Average distance to coffee and cake shops	Number of hotels	Average distance to bars
Number of companies	Average distance to subway stations	Number of chess and internet cafes	Average distance to banks
Number of animal clinics	Average distance to cinemas and theaters	Number of amusement parks	Lowest housing price
Number of flower and bird fish insect markets	Number of Western-style restaurants	Number of vocational schools and universities	Highest housing price
Average housing price			

References

1. Nelson, R.L.: The Selection of Retail Locations. Dodge (McGraw-Hill), New York (1958)
2. Moutinho, L., Curry, B., Davies, F.: Comparative computer approaches to multi-outlet retail site location decisions. Serv. Ind. J. **13**(4), 201–220 (1993)
3. Kuo, R.J., Chi, S.C., Kao, S.S.: A decision support system for selecting convenience store location through integration of fuzzy AHP and artificial neural network. Comput. Ind. **47**(2), 199–214 (2002). https://doi.org/10.1016/S0166-3615(01)00147-6
4. Karamshuk, D., Noulas, A., Scellato, S., et al.: Geo-spotting: mining online location-based services for optimal retail store placement. ACM (2013)
5. Sever, A.: A new efficient machine learning algorithm to solve facility location selection problem of geoinformatics. Curr. J. Appl. Sci. Technol. **8**, 1–10 (2016)
6. Baumbach, S., Rubel, C., Ahmed, S., et al.: Geospatial customer, competitor and supplier analysis for site selection of supermarkets. In: Proceedings of the 2019 2nd International Conference on Geoinformatics and Data Analysis, pp. 110–114 (2019)
7. Xiao, D., Ye, W.: Combining GIS and the analytic hierarchy process to analyze location of hypermarket. In: IOP Conference Series: Earth and Environmental Science, vol. 2019, p. 032012. IOP Publishing (2019)
8. Rincón, E.T., Tiwari, C.: Demand metric for supermarket site selection: a case study. Papers Appl. Geogr. **6**(1), 19–34 (2020)
9. Han, S., Jia, X., Chen, X., et al.: Search well and be wise: a machine learning approach to search for a profitable location. J. Bus. Res. **144**, 416–427 (2022)
10. Lan, T., Cheng, H., Wang, Y., et al.: Site Selection via learning graph convolutional neural networks: a case study of Singapore. Remote Sens. **14**(15), 3579 (2022)

CF-PMSS: Collaborative Filtering Based on Preference Model and Sparrow Search

Wei Song[⊠] [iD] and Shuya Li

School of Information Science and Technology, North China University of Technology,
Beijing 100144, China
songwei@ncut.edu.cn

Abstract. Collaborative filtering (CF) is a widely used recommendation method that can be evaluated according to the rating prediction or top-N recommendations. In rating prediction-based CF, it is important to refine the bias of the original ratings and select users with a greater degree of similarity. To achieve this rating refinement, we transform the countable raw ratings into real numbers using a preference model. To identify similar users, we add item types to the similarity calculation, and use the sparrow search algorithm to enhance the effect of user clustering. The performance of the proposed CF based on a preference model and sparrow search is compared with that of related heuristic-based CFs, and the results demonstrate the superiority of the proposed approach in terms of accuracy and correlation.

Keywords: Recommender system · Collaborative filtering · Preference model · Sparrow search algorithm · Comprehensive similarity

1 Introduction

Recommender systems (RSs) are increasingly popular software techniques for providing suggestions regarding items of interest to a user. Although several RSs have been developed, such as historical-behavior-based methods [8] and matrix factorization techniques [4], collaborative filtering (CF) [5, 14] remains the most popular approach. Methods based on CF have been widely applied in the fields of road transport networks [3] and text classification [11], among others.

CF can be categorized into two main types: user-based CF and item-based CF. Grouping similar users first is an effective way of providing high-quality recommendations [8]. In this paper, we apply the K-means algorithm to cluster similar users.

Heuristic algorithms (HAs) are a problem-solving approach whereby a practical process is used to produce a feasible solution that is sufficient to quickly solve a problem and achieve the immediate goals, but do not necessarily reach the optimal solution. Because different users have different degrees of satisfaction with the recommendation results, the problem of providing suitable recommendations does not necessarily have a definite optimal solution. This is consistent with the essence of HAs providing feasible solutions. Therefore, applying HAs to improve the performance of RSs is a promising

© The Author(s), under exclusive license to Springer Nature Switzerland AG 2023
Y. Tan et al. (Eds.): ICSI 2023, LNCS 13969, pp. 181–192, 2023.
https://doi.org/10.1007/978-3-031-36625-3_15

direction of study, and HAs based on swarm intelligence [1] and simulated annealing (SA) [7] have been successfully applied in RSs. In this paper, we use the sparrow search algorithm (SSA) [13] to optimize the centers used by the K-means algorithm. This reduces the influence of random initial centers on the results of the K-means algorithm.

From the perspective of accuracy, CF can be evaluated in terms of rating prediction and top-N recommendations. We mainly focus on improving the accuracy of rating prediction, i.e., minimizing the prediction errors for all unobserved (or missing) user–item pairs.

This paper describes the improvement of the CF performance from three perspectives. First, we use a preference model to refine the bias of different users for the same rating. Second, we exploit the SSA to determine the initial centers of the K-means algorithm. Thus, the users participating in rating calculations can be accurately identified. Third, we incorporate the item types into the similarity calculation. Experimental results show that the proposed CF based on preference model and sparrow search (CF-PMSS) method improves the recommendation accuracy over existing approaches.

2   Problem Description

CF is typically described using a *rating matrix* (RM). Let $\mathcal{U} = \{u_1, u_2, \ldots, u_M\}$ be a set of users and $\mathcal{I} = \{i_1, i_2, \ldots, i_N\}$ be a set of items. The associated RM $\boldsymbol{R}$ is an $M \times N$ matrix. Each entry $r_{j,k}$ of $\boldsymbol{R}$ corresponds to the preference of user u_j $(1 \leq j \leq M)$ for item i_k $(1 \leq k \leq N)$. If $r_{j,k} \neq 0$, then u_j has rated i_k; otherwise, this user has not rated the item.

We study CF using the similarity between users, which predicts the target user's interest in a candidate item based on rating information from similar users. A greater similarity between users means that the ratings will make a greater contribution to predicting the test item's rating. Specifically, the set of neighboring users that contribute to the prediction of ratings for target user u_t is formed as

$$SR(u_t) = \{u_a | rank(simR(u_t, u_a)) \leq num\}, \tag{1}$$

where $rank()$ is a function that returns the rank of users according to their similarity to the target user u_t, num is an integer that limits the number of similar users, and $simR(u_t, u_a)$ is the similarity between u_t and u_a. We use the Pearson correlation coefficient (PCC) as the similarity measure:

$$simR(u_a, u_t) = \frac{\sum_{ij \in I}(r_{a,j} - \overline{r_a}) \times (r_{t,j} - \overline{r_t})}{\sqrt{\sum_{ij \in I}(r_{a,j} - \overline{r_a})^2} \times \sqrt{\sum_{ij \in I}(r_{t,j} - \overline{r_t})^2}}, \tag{2}$$

where I is the set of items rated by both u_a and u_t, $\overline{r_a}$ is the average rating of u_a, and $\overline{r_t}$ is the average rating of u_t. The value of $simR(u_a, u_t)$ in Eq. 2 lies in the range $[-1, 1]$, and is called the *similarity by rating* (SR). A higher value of $simR(u_a, u_t)$ indicates greater similarity between u_a and u_t.

The predicted rating of a candidate item i_s by target user u_t is computed as.

$$r_{t,s}^* = \overline{r_t} + \frac{\sum_{u_a \in SR(u_t)} simR(u_a, u_t) \times (r_{a,s} - \overline{r_a})}{\sum_{u_a \in SR(u_t)} simR(u_a, u_t)}. \tag{3}$$

3   Preference Model

Although user ratings are often predicted based on Eq. 3, problems are encountered when the user bias is not appropriately processed [6]. For example, some users may always give high ratings, whereas other users may prefer to give low ratings. For a dataset in which the ratings are integers in the range [1, 5], a rating of 3 may represent a dislike for the item by users who always give high ratings, whereas the same rating may represent a like for the item by users who always give low ratings. To overcome this problem, we first transform the original ratings using a preference model [2].

Let $\{C_1,\ldots, C_k\}$ be a set of rating categories in a specific range. The items in a high rating category are ranked higher than those in a low rating category. Suppose that r_{ui} belongs to category C, i.e., $r_{ui} \in C$. To compute the preference score of r_{ui}, we simply count the number of items that are in low rating category C', i.e., $C > C'$, and the number of items in the same category C. The preference score (PS) of user u for category C is computed as

$$pref\,(u,\, C) = \alpha \times \sum\nolimits_{C' \in \{C_1,\ldots,C_k\}} \frac{count(C > C')}{|\mathcal{R}_u|} + \beta \times \frac{count(C)}{|\mathcal{R}_u|}, \qquad (4)$$

where α and β are weight parameters, $count(C > C')$ is the number of items that belong to the lower rating category C', $count(C)$ is the number of items that belong to C, and $|\mathcal{R}_u|$ is the total number of items rated by u. Following a previous study [2], α and β are set to 1.0 and 0.5, respectively, in our method.

We can see from Eq. 4 that PS maps the original countable rating categories to real numbers in the range (0, 1). Thus, the preference model refines the raw ratings according to the users' rating preferences.

4   SSA-Enhanced K-Means Clustering

4.1   Basic Principle of SSA

The SSA [13] is inspired by the collective movement of sparrows and their typical social behaviors. Unlike most swarm intelligence algorithms, e.g., particle swarm optimization (PSO) [10], which only consider foraging behavior, SSA also considers anti-predation behavior.

Formally, the position of the individual sparrows can be represented by the following matrix:

$$X = \begin{bmatrix} x_{1,1} & x_{1,2} & \ldots & x_{1,d} \\ \vdots & \vdots & \cdots & \vdots \\ x_{S,1} & x_{S,2} & \cdots & x_{S,d} \end{bmatrix}, \qquad (5)$$

where S is the number of sparrows and d is the dimension of each sparrow. The position of the ith $(1 \leq i \leq S)$ sparrow can be represented by a vector $\mathbf{X}_i = <x_{i,1}, x_{i,2}, \ldots, x_{i,d}>$. The fitness value of this sparrow is f_i.

To realize SSA, three kinds of sparrows are used iteratively.

Producer. The producer sparrows are responsible for searching for food and guiding the movement of the entire population. During each iteration, the location of the producers is updated as

$$x_{i,j}^{t+1} = \begin{cases} x_{i,j}^t \times \exp(\frac{-i}{\gamma \times t_{max}}) & , \text{if } \delta < ST \\ x_{i,j}^t + \varepsilon, & \text{if } \delta \geq ST \end{cases}, \tag{6}$$

where t is the current iteration, $x_{i,j}^t$ is the value of the jth dimension of the ith sparrow at iteration t, t_{max} is a constant representing the maximum number of iterations, $\gamma \in (0, 1]$ is a random number, $\delta \in [0, 1]$ is a random number representing the alarm value, $ST \in [0.5, 1.0]$ represents the safety threshold, and ε is a random number that obeys a normal distribution.

When $\delta < ST$, there are no predators around. If $\delta \geq ST$, some sparrows have discovered the predator, and all sparrows need to fly to safe areas as quickly as possible.

Scrounger. Scrounger sparrows monitor the producers. Once a producer has found a good food source, they immediately leave their current position to compete for the food. The position update for the scroungers is

$$x_{i,j}^{t+1} = \begin{cases} \varepsilon \times \exp(\frac{XW_j^t - x_{i,j}^t}{i^2}) & , \text{if } i > S/2 \\ x_{p,j}^t + \frac{1}{d}\sum_{j=1}^{d} rand(-1, 1) \times |x_{i,j}^t - x_{p,j}^t|, & \text{otherwise} \end{cases}, \tag{7}$$

where $\mathbf{X}_p^t$ is the optimal position occupied by the producers at iteration t, $x_{p,j}^t$ is the jth element of $\mathbf{X}_p^t$, $\mathbf{XW}^t$ denotes the global worst location at iteration t, XW_j^t is the jth element of $\mathbf{XW}^t$, and $rand(-1, 1)$ randomly selects a value of either -1 or 1 at each time step. When $i > S/2$, the scrounger with the ith-worst fitness value is likely to be starving.

Scouter. Randomly selected sparrows within the total population are aware of danger. They make up 10%–20% of the population. The position update for the scouters is

$$x_{i,j}^{t+1} = \begin{cases} XB_j^t + \theta \times |x_{i,j}^t - XB_j^t| & , \text{if } f_i > f_{best} \\ x_{i,j}^t + \rho \times (\frac{|x_{i,j}^t - XW_j^t|}{(f_i - f_{worst}) + \tau}), & \text{if } f_i = f_{best} \end{cases}, \tag{8}$$

where $\mathbf{XB}^t$ is the global optimal location at iteration t, XB_j^t is the jth element of $\mathbf{XB}^t$, θ is a normal distribution of random numbers with a mean value of 0 and a variance of 1 that controls the step size, $\rho \in [-1, 1]$ is a random number representing the direction in which the sparrow moves, f_i is the fitness value of the present sparrow, f_{best} and f_{worst} are the current global best and worst fitness values, and τ is a small constant to prevent division by zero.

The case in which $f_i > f_{best}$ indicates that the sparrow is at the edge of the group. When $f_i = f_{best}$, the sparrows in the middle of the population are aware of the danger and need to move closer to the others.

4.2 Determining Cluster Centers with SSA

K-means is a classical partition-based clustering algorithm. The K-means algorithm starts with K randomly selected centers. At each iteration, every data object is assigned to its nearest cluster center, and each cluster center is updated to the average of the data points assigned to it. This process is repeated until either the cluster centers have stabilized or a predefined number of iterations has been performed.

The main limitation faced by K-means is that the algorithm is sensitive to the initial cluster centers: choosing different initial centers often leads to very different results. To solve this problem, we use SSA to optimize the K initial centers before clustering the users. The objective is to find K centers such that the sum of the distances between each user and each cluster center is minimized.

To perform K-means clustering, a row of RM is used to represent a user, that is, each user is represented as a vector of length N. However, in real-world applications, user ratings are very sparse. Therefore, using the original ratings to cluster users directly incurs a high computational cost. Furthermore, the similarity results are often poor because very few items are co-rated among users.

To overcome this problem, we transform the original user vectors to a lower dimension with dense values by incorporating the item types. Compared with the huge number of specific items, the item types are relatively limited, so clustering users based on item types is a feasible solution to the problem of sparse ratings.

Specifically, let $T = \{t_1, t_2, ..., t_L\}$ be a set of item types and represent each user by an L-dimensional vector $\mathbf{U}_i$ ($1 \le i \le M$), where each dimension u_{ij} ($1 \le j \le L$) of $\mathbf{U}_i$ is the mean rating of u_i for all items within the jth type. Correspondingly, each sparrow X is a $K \times L$ matrix, each row of X is a *center vector* $\mathbf{C}_j$ ($1 \le j \le K$). The optimization objective is to find a center vector such that

$$f = \genfrac{}{}{0pt}{}{argmin}{j} \sum_{i=1}^{M} \sum_{j=1}^{K} \|\mathbf{U}_i - \mathbf{C}_j\|_2. \tag{9}$$

The method for determining the optimal cluster centers by SSA is described in Algorithm 1.

Algorithm 1	Function SSA-Center( )
Input	Rating matrix R, maximal number of iterations t_{max}, population size S, number of producers S_p, number of scouters S_c, safety threshold ST.
Output	The optimal sparrow composed of K cluster centers.
1	Initialize S sparrows X_i $(1 \leq i \leq S)$ randomly;
2	Calculate the initial fitness values of each sparrow using Eq.9;
3	$t = 1$;
4	**while** $(t < t_{max})$ **do**
5	Calculate XB^t and XW^t using Eq. 9;
6	**for** $i = 1$ to S_p **do**
7	Update producers' positions using Eq.6;
8	**end for**
9	**for** $i = S_p+1$ to S **do**
10	Update scroungers' positions using Eq.7;
11	**end for**
12	Select S_c sparrows at random as scouters;
13	Update scouters' positions using Eq.8;
14	t ++;
15	**end while**
16	Return the current XB.

In Algorithm 1, S sparrows are initialized in Step 1. The value range of each element of each center vector is consistent with the value range of the user ratings for the items. In Step 2, the initial fitness values of the S sparrows are calculated, and then the iteration number is set as 1 in Step 3. The main loop (Steps 4–15) updates the population until the maximal number of iterations is reached. In Step 5, the current best and worst sparrows are determined. The next two loops (Steps 6–11) update the positions of the S_p producers and the remaining scroungers, respectively. Scouters are selected in Step 12, and their positions are updated in Step 13. In Step 14, the iteration counter is incremented by 1. Finally, the global best sparrow is output as the K optimal cluster centers in Step 16.

5   Synthesis Similarity Measure

5.1   Type-Based Similarity

As stated in Sect. 4.2, besides user representation, the item type is used to improve the accuracy of similarity calculation in CF-PMSS. We use the type-based similarity measure starting from the rating probabilities of item types [12]. Let $|\mathcal{R}_u(t_j)|$ be the number of items with type t_j $(1 \leq j \leq L)$ rated by u and $|\mathcal{R}_u|$ be the total number of items rated by u. The user rating type probability (URTP) of u to t_j is calculated as

$$\Pr(u, t_j) = |R_u(t_j)|/|R_u|. \tag{10}$$

We can see from Eq. 10 that URTP is the fraction of the number of ratings for one item type to the total number of ratings made by the target user.

Using URTP, the *similarity by type* (ST) is calculated as

$$sim_T(u_a, u_t) = \frac{\sum_{t_j \in T}(\Pr(u_a, t_j) - \overline{Pr_a}) \times (\Pr(u_t, t_j) - \overline{Pr_t})}{\sqrt{\sum_{t_j \in T}(\Pr(u_a, t_j) - \overline{Pr_a})^2} \times \sqrt{\sum_{t_j \in T}(\Pr(u_t, t_j) - \overline{Pr_t})^2}}, \quad (11)$$

where T is the set of item types rated by both u_a and u_t, $\overline{Pr_a}$ is the average URTP of u_a, and $\overline{Pr_t}$ is the average URTP of u_t. The value of $sim_T(u_a, u_t)$ in Eq. 11 is in the range $[-1, 1]$. A higher value of $sim_T(u_a, u_t)$ indicates greater similarity between u_a and u_t.

5.2 Combining Rating and Type

We calculate the similarity among users from two perspectives: based on ratings (Eq. 2) and based on item types (Eq. 11). For a user u_a within the same cluster as the target user u_t, their *comprehensive similarity* is calculated as

$$sim_C(u_a, u_t) = \lambda \times sim_R(u_a, u_t) + (1 - \lambda) \times sim_T(u_a, u_t), \quad (12)$$

where λ is a *balance factor* in the range $(0, 1)$.

Using the comprehensive similarity, Eq. 3 can be rewritten to approximate the predicted rating of i_s by u_t as

$$r_{t,s}^* = \overline{r_t} + \frac{\sum_{u_a \in S_C(u_t)} sim_C(u_a, u_t) \times \left(r_{a,s} - \overline{r_a}\right)}{\sum_{u_a S_C(u_t)} sim_C(u_a, u_t)}, \quad (13)$$

where $S_C(u_t)$ is the set of neighboring users that contribute to the prediction of the target user u_t using the comprehensive similarity.

6 Overall Algorithm

The general algorithm of the proposed CF-PMSS method is described in Algorithm 2.

Algorithm 2	CF-PMSS
Input	Rating matrix R.
Output	Predicted ratings of the target user u_t to unrated items.
1	Transform each rating in R using Eq.4;
2	$C =$ SSA-Center( );
3	Perform K-means clustering with K cluster centers in C;
4	**For each user u_a within the same cluster as u_t do**
5	Calculate $sim_C(u_a, u_t)$ using Eq.12;
6	**end for**
7	Predict ratings of u_t for unrated items using Eq.13.

In Algorithm 2, the original user ratings are transformed into PS in the range $(0, 1)$ in Step 1. Algorithm 1 is then called to determine the K centers in Step 2. Next, Step 3 clusters the users into K groups according to the optimized cluster centers. In the loop from Steps 4–6, the similarity between the target user and all other users within the same group is calculated. Finally, in Step 7, the predicted rating of each item that has not been rated by the target user is calculated.

7   Performance Evaluation

We compare our CF-PMSS algorithm with the conventional CF algorithm [5] and an SA-based CF algorithm, CF-SC [7]. To evaluate the influence of the SSA on the performance of K-means, we replace the SSA-based K-means with PSO-based clustering [10] in our method, denoted by CF-PSO. CF-SC, CF-PSO, and CF-PMSS all use stochastic optimization algorithms i.e., SA, PSO, and SSA. Thus, we take the average values over 10 runs as the initial cluster centers.

The algorithms are evaluated on three datasets, MovieLens 100K, MovieLens 1M, and FilmTrust. The two MovieLens datasets can be downloaded from https://groupl ens.org/datasets/movielens/, and the FilmTrust dataset can be downloaded from https://www.kaggle.com/datasets/abdelhakaissat/film-trust. Table 1 presents the characteristics of the datasets used in the experiments. The datasets are divided into two parts, with 80% used as the training set TR and 20% used as the test set TS.

Table 1. Characteristics of the datasets

Dataset	Users	Items	Ratings	Types	Rating scales
MovieLens 100K	943	1,682	100,000	19	1–5
MovieLens 1M	6,040	3,900	1,000,209	18	1–5
FilmTrust	1,508	2,071	35,497	19	1–5

7.1   Evaluation Metrics

We study the accuracy of the estimations provided by our model using the mean absolute error (MAE). This metric measures the average absolute error between the actual values and predicted values, and is calculated as

$$\text{MAE} = \frac{\sum_{(u_j, i_k) \in TS} |r_{j,k} - r_{j,k}^*|}{|TS|},\tag{14}$$

where $r_{j,k}$ and $r_{j,k}^*$ are the actual and predicted ratings of u_j for i_k, and $|TS|$ is the number of user–item pairs in TS, such that u_j has rated i_k. A smaller value of MAE represents more precise estimation.

We also evaluate the quality of the recommendations using the mean squared error (MSE):

$$\text{MSE} = \frac{\sum_{(u_j, i_k) \in TS} (r_{j,k} - r_{j,k}^*)^2}{|TS|}.\tag{15}$$

In the same manner as MAE, smaller values of MSE indicate more precise estimation.

Furthermore, we use PCC to evaluate the correlation between the actual and predicted ratings:

$$PCC = \frac{\sum_{(u_p,i_k) \in TS}(r_{p,k} - \overline{r_p}) \times (r^*_{p,k} - \overline{r^*_p})}{\sqrt{\sum_{(u_p,i_k) \in TS}(r_{p,k} - \overline{r_p})^2} \times \sqrt{\sum_{(u_p,i_k) \in TS}(r^*_{p,k} - \overline{r^*_p})^2}},$$ (16)

where $\overline{r_p}$ and $\overline{r^*_p}$ are the mean values of all actual and predicted ratings of u_p, respectively. PCC values that are closer to 1 denote a stronger correlation between the actual and predicted ratings.

7.2 Parameter Settings

For all experiments, the number of neighbor users contributing to the prediction of the target user *num* is set to 35, the safety threshold *ST* is set to 0.9, the number of clusters *K* is set to 10, the population size *S* is set to 20, the maximum number of iterations t_{max} is set to 100, and the balance factor λ is set to 0.9. These parameters were set by first outlining their approximate ranges and then determining their optimal values by progressive refinement. As an example, we describe the determination process for *num* on Movielens 100K. We compared the results of four algorithms in terms of the three evaluation metrics as *num* was varied from 5 to 40. The results are shown in Figs. 1, 2 and 3.

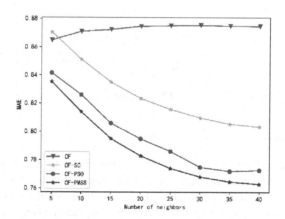

Fig. 1. Comparison of the MAE with different numbers of neighbors.

From the comparison results for MAE and MSE, the performance of CF-SC, CF-PSO, and CF-PMSS improves as *num* increases; when *num* exceeds 35, however, the MAE of CF-PSO becomes worse, and the MSE of these three algorithms increases. In terms of correlation, the PCC values of the same three algorithms increase as *num* increases, but start to decrease once *num* exceeds 35. Hence, we set *num* to 35 in the experiments. The other insight from these three figures is that CF-PMSS always outperforms the other three algorithms under different numbers of neighbors.

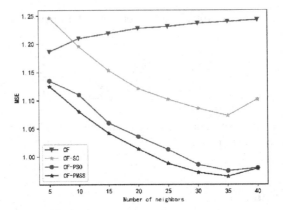

Fig. 2. Comparison of the MSE with different numbers of neighbors.

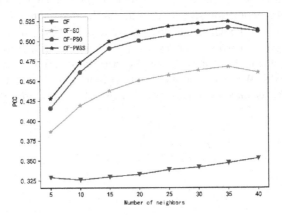

Fig. 3. Comparison of the PCC with different numbers of neighbors.

7.3 Comparison Results

The results given by the four algorithms are compared in Table 2. For each dataset, the underlined entries in each column represents the optimal values for that measure/dataset.

Compared with the other three algorithms, it is clear that the proposed CF-PMSS model provides more accurate estimations of user ratings and produces a high correlation between the actual and predicted ratings on both datasets. This is because the ratings for both specific items and item types are considered in CF-PMSS. For ratings on specific items, refinement is performed through the preference model. Furthermore, when calculating the predicted ratings, only users who are in the same cluster as the target user are considered. The clustering process is optimized through SSA. Consequently, unnecessary similarity calculations among irrelevant users are avoided, which enhances the performance.

Table 2. Comparison of the recommendation results given by different algorithms

Dataset	Method	MAE	MSE	PCC
MovieLens 100K	CF	0.875	1.243	0.353
	CF-SC	0.804	1.102	0.461
	CF-PSO	0.767	0.983	0.523
	CF-PMSS	0.766	0.977	0.525
MovieLens 1M	CF	0.876	1.244	0.329
	CF-SC	0.789	1.027	0.491
	CF-PSO	0.783	1.013	0.507
	CF-PMSS	0.782	1.011	0.511
FilmTrust	CF	0.763	0.966	0.431
	CF-SC	0.733	0.923	0.468
	CF-PSO	0.698	0.862	0.514
	CF-PMSS	0.684	0.844	0.532

8 Conclusions

We have studied the rating prediction problem of CF from two aspects. First, we attempted to optimize the rating calculation. To this end, we refined the original coarse-grained rating using a preference model. The second aspect involved accurately identifying neighbor users to participate in the rating calculation. For this, we used the SSA to determine the initial centers of the K-means clustering algorithm, and incorporated item type information into the calculation of similarity among users. Our experimental results demonstrate that the proposed method achieves high recommendation accuracy.

Acknowledgments. This work was partially supported by the National Natural Science Foundation of China (61977001), and Great Wall Scholar Program (CIT&TCD20190305).

References

1. Forestiero, A.: Heuristic recommendation technique in Internet of Things featuring swarm intelligence approach. Expert Syst. Appl. **187**, 115904 (2022)
2. Lee, J., Lee, D., Lee, Y.-C., Hwang, W.-S., Kim, S.-W.: Improving the accuracy of top-N recommendation using a preference model. Inf. Sci. **348**, 290–304 (2016)
3. Ofem, O.A., Agana, M.A., Felix, E.O.: Collaborative filtering recommender system for timely arrival problem in road transport networks using viterbi and the hidden Markov algorithms. Int. J. Softw. Innov. **11**(1), 1–21 (2023)
4. Peng, C., Zhu, L., Xu, Y., Li, Y., Guo, L.: Binary multi-modal matrix factorization for fast item cold-start recommendation. Neurocomputing **507**, 145–156 (2022)
5. Schafer, J.B., Frankowski, D., Herlocker, J., Sen, S.: Collaborative filtering recommender systems. In: Brusilovsky, P., Kobsa, A., Nejdl, W. (eds.) The Adaptive Web. LNCS, vol. 4321, pp. 291–324. Springer, Heidelberg (2007). https://doi.org/10.1007/978-3-540-72079-9_9

6. Song, W., Li, X.: A non-negative matrix factorization for recommender systems based on dynamic bias. In: Torra, V., Narukawa, Y., Pasi, G., Viviani, M. (eds.) MDAI 2019. LNCS (LNAI), vol. 11676, pp. 151–163. Springer, Cham (2019). https://doi.org/10.1007/978-3-030-26773-5_14

7. Song, W., Liu, S.: Collaborative filtering based on clustering and simulated annealing. In: Proceedings of the 3rd International Conference on Big Data Engineering, pp. 76–81 (2021)

8. Song, W., Liu, S.: Optimal user categorization from a hierarchical clustering tree for recommendation. In: Fujita, H., Fournier-Viger, P., Ali, M., Wang, Y. (eds.) IEA/AIE 2022. LNCS, pp. 759–770. Springer International Publishing, Cham (2022). https://doi.org/10.1007/978-3-031-08530-7_64

9. Song, W., Yang, K.: Personalized recommendation based on weighted sequence similarity. In: Wen, Z., Li, T. (eds.) Practical Applications of Intelligent Systems. AISC, vol. 279, pp. 657–666. Springer, Heidelberg (2014). https://doi.org/10.1007/978-3-642-54927-4_62

10. van der Merwe, D.W., Engelbrecht, A.P.: Data clustering using particle swarm optimization. In: Proceedings of the 2003 Congress on Evolutionary Computation, pp. 215–220 (2003)

11. Wang, J., Chen, Z., Qin, Y., He, D., Lin, F.: Multi-aspect co-attentional collaborative filtering for extreme multi-label text classification. Knowl. Based Syst. **260**, 110110 (2023)

12. Wu, C., et al.: Recommendation algorithm based on user score probability and project type. EURASIP J. Wirel. Commun. Netw. (2019)

13. Xue, J., Shen, B.: A novel swarm intelligence optimization approach: sparrow search algorithm. Syst. Sci. Control Eng. **8**(1), 22–34 (2020)

14. Zhou, Q., et al.: Hybrid collaborative filtering model for consumer dynamic service recommendation based on mobile cloud information system. Inf. Process. Manag. **59**(2), 102871 (2022)

Asynchronous Federated Learning Framework Based on Dynamic Selective Transmission

Ruizhuo Zhang[1], Wenjian Luo[1,2(✉)], Yongkang Luo[1], and Shaocong Xue[1]

[1] Guangdong Provincial Key Laboratory of Novel Intelligence Technologies, School of Computer Science and Technology, Harbin Institute of Technology, Shenzhen 518055, Guangdong, China
{zhangruizhuo,luowenjian}@hit.edu.cn,
{21S151078,22S051026}@stu.hit.edu.cn
[2] Peng Cheng Laboratory, Shenzhen 518055, Guangdong, China

Abstract. This paper proposes an asynchronous federated learning framework based on dynamic selective transmission to solve the communication efficiency problem in asynchronous federated learning. The framework first dynamically determines the network layer to be transmitted in each round according to the training progress, and then dynamically adjusts the ratio of parameters to be transmitted according to the training progress and model staleness for the selected network layers, which effectively reduces the negative impact of the local model staleness of the client on the performance of global model while significantly reducing the uplink communication cost of the asynchronous federated learning framework. On two datasets, this paper designs experiments to compare the proposed framework with the existing work. Through the analysis of the experimental results, the effectiveness of the proposed framework is verified.

Keywords: Federated learning · Asynchronous federated learning · Communication efficiency · Machine learning

1 Introduction

The earliest federated learning framework adopted synchronous communication protocols, such as FedAvg [7] and its improved variants [2,5,8,10,11]. Since the server and clients need to communicate a lot in federated learning system, the communication efficiency has always been one of the important research directions in the field of federated learning. Nishio *et al.* [8] proposed a federated

This study is supported by the National Key R&D Program of China (Grant No. 2022YFB3102100), Shenzhen Fundamental Research Program (Grant No. JCYJ20220818102414030), the Major Key Project of PCL (Grant No. PCL2022A03, PCL2021A02, PCL2021A09), Guangdong Provincial Key Laboratory of Novel Security Intelligence Technologies (Grant No. 2022B1212010005). *(Corresponding author: Wenjian Luo.)*

Y. Tan et al. (Eds.): ICSI 2023, LNCS 13969, pp. 193–203, 2023.
https://doi.org/10.1007/978-3-031-36625-3_16

learning framework FedCS based on client selection, which can reduce the waiting time of the server and improve the learning efficiency of the federated learning system by selecting clients with better computing resources and network quality as much as possible in each round of training. Konečný *et al.* [3] proposed two optimization methods for the uplink communication of federated learning, i.e., structural update and sketch update, which effectively reduced the communication cost of the federated learning system. Reisizadeh *et al.* [9] proposed a quantitative message delivery method which quantifies the client updates and then uploads them to the parameter server to improve the communication efficiency and scalability of federated learning.

In recent years, asynchronous federated learning has attracted more and more attention, because it is more suitable for real application scenarios and has more engineering feasibility [13]. However, there are still many problems to be solved in the field of asynchronous federated learning, and new frameworks also bring new challenges.

The asynchronous communication mode makes the model returned by the client with a slow update have a high degree of staleness, thus causing a noticeable impact to the performance of the global model [12]. Furthermore, the asynchronous federated learning framework often enables clients with better computing performance and network environments to communicate with the server more frequently, resulting in greater uplink traffic and greater network pressure and computing load on the server and such clients. Therefore, the above issues have become the key issues and major bottlenecks in asynchronous federated learning.

Focusing on the key issues of communication cost optimization in asynchronous federated learning, by hybridizing the ideas in [1] and [6], this paper proposes an asynchronous federated learning framework based on dynamic selective transmission. By dynamically determining the network layer to be transmitted in each round according to the training progress, and then dynamically adjusting the ratio of parameters to be transmitted in the network layer according to the training progress and model staleness for the selected network layer, the proposed framework effectively reduces the uplink communication and improves the communication efficiency of the asynchronous federated learning framework on the premise of ensuring the training efficiency and model performance of the global model. At the same time, the proposed framework also reduces the negative impact of the local model staleness on global model performance. Through comparative experiments with existing work on different data sets, the effectiveness of the proposed framework for communication cost optimization is verified.

The rest of this paper is organized as follows. First, we introduce some existing work about communication efficiency problems in federated learning in Sect. 2. Next, we describe the proposed asynchronous federated learning framework based on dynamic selective transmission in Sect. 3. Then, Sect. 4 shows experimental results. Finally, Sect. 5 draws a brief conclusion.

2   Related Work

The communication in federated learning can be generally divided into model parameter communication and gradient communication [14]. Using the model parameter communication mode, Chen et al. [1] proposed an asynchronous federated learning framework with layer wise update and temporally weighted aggregation (ASTW) to reduce the number of parameters transmitted between the server and the clients. Using the gradient communication mode, Li et al. [6] proposed a double-end sparse gradient compression algorithm (DESC) to reduce the gradient amount required for transmission and optimize the communication efficiency. The framework proposed in this paper is based on both ASTM [1] and DESC [6]. The details of these two existing techniques are as follows.

2.1   ASTW

To improve the communication efficiency of model parameters in federated learning, Chen et al. [1] proposed a temporally weighted aggregation asynchronous federated learning framework (ASTW), which aggregates and updates the parameters in the shallow and deep layers of the deep neural network with different frequencies, thus reducing the number of parameters transmitted between the server and the clients. Based on the work by Yosinski et al. [15] and Krizhevsky et al. [4], the deep neural network can be divided into shallow parts w_g and deep part w_s, while the parameters in the shallow layer of the network are more critical to the performance of the global model. At the same time, the parameter quantity S_g of the shallow layers is usually far less than the parameter quantity S_s of the deep layers. Therefore, the shallow layers' parameters should be updated at a higher frequency than the deep layers' parameters. The parameters in the shallow layers and deep layers can be updated asynchronously at different frequencies, thus reducing the communication cost between the server and the client.

However, although the method proposed by Chen et al. [1] transmits the deep layers and the shallow layers with different intervals, the transmission interval of each client is fixed. Therefore, the method has poor adaptability to different training tasks, which could have a negative impact on the performance improvement efficiency and the final performance of the global model. At the same time, the method proposed by Chen et al. [1] still uses all parameters transmission for each network layer, so there is still much room to reduce the communication cost.

2.2   DESC

Aiming at the optimization of gradient communication efficiency in federated learning, Li et al. [6] proposed a double-end sparse gradient compression algorithm (DESC), which uses the modified Top-K AllReduce sparse compression algorithm in the server and clients to reduce the gradient amount required for transmission and improve the communication efficiency. For the server, only the top p changes will be transmitted to the clients and participate in the update; for each client, only the top q gradients will be transmitted to the server and participate in the update. Generally, most of the gradients transmitted during the training process of federated learning is sparse, especially in the late stage of the training process, the significant change of the predicted value usually occurs on a very small amount of data in a batch of training. Therefore, this algorithm can significantly reduce the gradient transmission ratio in the training process of federated learning, and significantly improve the communication efficiency under the premise of ensuring that the loss of global model performance is small.

However, the method proposed by Li *et al.* [6] only considers the federated learning framework based on gradient communication, and does not consider the federated learning framework based on model parameter communication. More importantly, the method proposed by Li *et al.* [6] does not take into account the negative impact of model staleness. For clients with different staleness, they used the same strategy for partial gradient transmission, which could cause the gradient update of clients with higher staleness to damage the performance of the global model.

3   The Proposed Framework

In order to improve the communication efficiency in asynchronous federated learning, this paper improves and combines the methods proposed by Chen *et al.* [1] and Li *et al.* [6], and proposes an asynchronous federated learning framework based on dynamic selective transmission (AFL-DST). The framework first dynamically determines the network layer to be transmitted in each round according to the training progress, and then dynamically adjusts the ratio of parameters in the selected layer to be transmitted according to the training progress and model staleness, so as to effectively reduce the uplink communication cost and improve the communication efficiency of the asynchronous federated learning framework under the premise of ensuring the final performance of global model, and reduce the impact of model staleness on the performance of the global model.

3.1   AFL-DST

As shown in Fig. 1, the framework proposed in this paper mainly includes two components: *Server* and *Client*. Each client has one thread, while *Server* has two parallel threads, *Server-Scheduler* and *Server-Updater*.

(1) **Client**

The client performs asynchronous training according to the scheduling of the server, selects and returns the network layer and model parameters of the local model according to the dynamic selective transmission strategy (*DynamicSelectiveTransmission*). Each client needs to maintain the number of completed training round k, the latest round of deep layers' transmission g, and the deep layers' transmission interval θ, and initialize them to 0. When a client

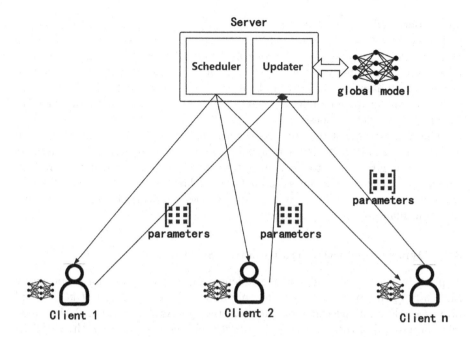

Fig. 1. The architecture of AFL-DST

receives the current round number t and the current global model w_t sent by the server, save as τ and w_τ, respectively, and then use its local dataset to train w_τ for E rounds. After training, in order to dynamically adjust the parameter selection ratio according to the model staleness, the client needs to establish a connection with the server and request the current number of global model training rounds t to calculate the local model staleness $t - \tau$. Then, according to the dynamic selective transmission strategy, the clients determine whether to carry out shallow layer transmission or full-layer transmission, and calculate the parameter selection ratio. Finally, the set $\mathbb{W}$ composed of the selected parameters in the layers to be transmitted is returned to the server. The detailed *DynamicSelectiveTransmission* algorithm will be described in the Sect. 3.2.

(2) Server-Scheduler

Server-Scheduler is responsible for scheduling all the clients participating in current training. Assume the number of total training round is T. In each training round, *Server-Scheduler* randomly selects the clients participating in the training from n clients according to the client scheduling ratio r and puts them into the set $\mathbb{C}$. Then, the current training round t and the current global model w_t is sent to each client in $\mathbb{C}$ and process a local training on these clients.

(3) Server-Updater

Server-Updater is responsible for updating the global model and establishing network connections with the clients to process the client's requests. When *Server-Updater* receives a model parameter set $\mathbb{W}$ returned by a client, for the layers contained in $\mathbb{W}$ and the updated model parameter $\mathbb{W}_i$, aggregate with the global model w to update the global model. That is, $w_i = (1-\alpha)w_i + \alpha \mathbb{W}_i$ where w_i is the global model parameter corresponding to $\mathbb{W}_i$ and α is a mixing hyperparameter. Additionally, when *Server-Updater* receives a client c_i's request to obtain the number of training rounds, it establishes a connection with the client and sends the current number of training rounds t to c_i, enabling the client to calculate its model staleness.

The framework first dynamically determines the layers to be transmitted in each round according to the training progress, and then dynamically adjusts the ratio of parameters to be transmitted according to the training progress and model staleness for the selected network layer. Each client has different transmission periods, effectively reducing the communication cost and ensuring the performance of the global model.

3.2   Dynamic Selective Transmission Strategy

Inspired by the double-end sparse gradient compression algorithm proposed by Li *et al.* [6], this paper improves the method proposed by Chen *et al.* [1], and proposes a dynamic selective transmission strategy. As shown in Fig. 2, the transmission interval of the deep layers of the model (the green part in the Fig. 2) is dynamically adjusted according to the local training progress of the clients, and then the parameter transmission ratio in the selected network layer is dynamically adjusted according to the local training progress of the client and the local model staleness, and only the parameters with larger changes are selected for transmission. Thus, the deep layers of the network is transmitted at a higher frequency in the early stage of training and the parameter selection ratio is higher, and the parameter selection ratio of the shallow layers is also higher in the early stage of training. The parameter transmission amount in each layer is dynamically reduced with the training progress, so as to effectively reduce the communication while ensuring the performance improvement efficiency of the global model, and at the same time reduce the impact of the model staleness on the performance of the global model.

The pseudo-code of the dynamic selective transmission strategy is given in Algorithm 1.

First, Algorithm 1 calculates the model parameter selection ratio of the client in this round according to the local staleness and local training progress. When $\tau == t$, that is, the local model is not stale, the calculation formula of the parameter selection ratio μ is

$$\mu \leftarrow \frac{1}{\lambda_k \cdot k + 1}, \tag{1}$$

where λ_k is the hyperparameter that determines the influence weight of the training progress on the parameter selection ratio. The bigger the λ_k is, the greater the impact of the training progress on μ. When $\lambda_k = 0$, the training progress has no effect on μ, and μ always equals to 1. That is, the client always

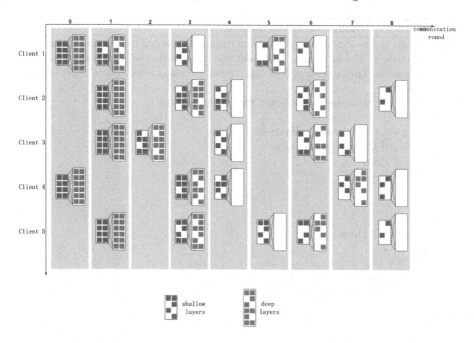

Fig. 2. Dynamic Selective Transmission

returns all the local model parameters. The calculation formula makes the initial parameter transfer ratio always 1. That is, all the parameters of the local model are transferred. As the number of training completed by the client increases, the transfer ratio of model parameters μ decreases. It not only ensures that the global model can rapidly improve its performance in the early stage of training, but also ensures that the transmission of model parameters in the later stage of training can be greatly reduced, thus reducing the uplink communication.

When $\tau \neq t$, that is, the local model is stale, the calculation formula of parameter selection ratio μ is

$$\mu \leftarrow \frac{1}{\lambda_s(t - \tau) + 1}, \tag{2}$$

where $(t - \tau)$ represents the staleness of the local model after this training, λ_s is the hyperparameter that determines the influence weight of staleness on the parameter selection ratio. The bigger the λ_s is, the greater the impact of staleness on μ. When $\lambda_s = 0$, the local model staleness has no effect on μ, and μ is always 1. That is, the client will return all the local model parameters regardless of the staleness. This calculation formula makes the transfer ratio of model parameters μ decrease with the increase of local model staleness, that is, the clients with higher local model staleness transmit less model parameters. It not only effectively reduces the negative impact of local model staleness on

---

Algorithm 1. DynamicSelectiveTransmission

Require: Global rounds before client local training τ; Global rounds after local training of client t; The local model of client w_τ; The number of rounds that the client
has trained k; The latest round of deep layers' transmission g; Deep layers transmit
interval θ

Ensure: BPS (Parameter set selected for transmission $\mathbb{W}$)

1: $\mathbb{W} \leftarrow \varnothing$
2: $\mathbb{L}_{shallow} \leftarrow$ All the shallow layers of w_τ
3: $\mathbb{L}_{deep} \leftarrow$ All the deep layers of w_τ
4: **if** $\tau == t$ **then**
5:    $\mu \leftarrow \frac{1}{\lambda_k \cdot k + 1}$
6: **else**
7:    $\mu \leftarrow \frac{1}{\lambda_s (t-\tau) + 1}$
8: **end if**
9: **if** $k == g + \theta$ **then**
10:    $g \leftarrow k$
11:    $\theta \leftarrow \theta + 1$
12:    **for** $\mathcal{L} \in \mathbb{L}_{shallow} \cup \mathbb{L}_{deep}$ **do**
13:       $\mathbb{W} \leftarrow \mathbb{W} \cup SelectLargerPart(\mathcal{L}, \mu)$
14:    **end for**
15: **else**
16:    **for** $\mathcal{L} \in \mathbb{L}_{shallow}$ **do**
17:       $\mathbb{W} \leftarrow \mathbb{W} \cup SelectLargerPart(\mathcal{L}, \mu)$
18:    **end for**
19: **end if**
20: **return** $\mathbb{W}$

---

global model's performance, but also reduces the model parameter transmission
of the client, and reduces the communication load of the system.

Second, after the calculation of the model parameter selection ratio, the
dynamic selective transmission strategy further selects the network layer that
the client's local model needs to transmit, that is, dynamically adjust the transmission interval for the shallow and deep parts of the network.

$\mathbb{L}_{shallow}$ is the set of all the shallow layers in w_τ, $\mathbb{L}_{deep}$ is the set of all the
deep layers in w_τ. The $SelectLargerPart(\mathcal{L}, \mu)$ in the Algorithm 1 is used to
select μ percent parameters with lager update changes from layer $\mathcal{L}$.

When $k == g + \theta$, the parameters with larger update changes are selected in
both shallow layers and deep layers according to the model parameter selection
ratio μ. When $k \neq g + \theta$, the parameters with larger update change are selected
for transmission in only the shallow layers according to the model parameter
selection ratio μ. In addition, when the deep layer transmission is carried out,
the deep transmission interval θ will increase, which makes the transmission
frequency of the deep layers higher in the early training period and lower in the
late training period. This not only ensures that the performance of the global
model can be rapidly improved in the early stage of training, but also effectively

reduces the transmission of model parameters in the later training stage, and improves the training efficiency of the system.

4   Experiments

In order to verify the performance improvement of the asynchronous federated learning framework based on dynamic selective transmission strategy proposed in this paper, we set up a comparative experiment based on two different data sets and a FC3 network model. FC3 is a neural network model with three fully connected layers consisting of 2-hidden layers (each has 200 units and uses ReLU activation) and an output layer. Our code is available at: https://github.com/ MiLab-HITSZ/2023ZhangAFLDST.

Set the number of clients $n = 100$, each client has 500 local data samples, the client scheduling ratio of each asynchronous training round $r = 0.1$, the local learning rate $\eta = 0.1$, the local learning batch size $B = 50$, the local training rounds $E = 1$, the global training rounds $T = 40$, and the model aggregation weight $\alpha = 0.2$. For the division of the shallow part and the deep part of the neural network, we take the input layer of the FC3 network model as the shallow part, and the hidden layer and the output layer as the deep part.

This paper compares the proposed dynamic selective transmission strategy with the strategy proposed by Chen et al. [1] and the benchmark asynchronous federated learning framework using the full parameter transmission strategy. Experimental results are shown in Fig. 3 . By analyzing the experimental results, it can be found that when $\lambda_k = 1$ and $\lambda_s = 1$, the dynamic selective transmission strategy proposed in this paper has almost no significant reduction in the global model performance compared with the layered asynchronous update strategy and the full parameter transmission strategy. When λ_k and λ_s are large, the global model performance is only slightly reduced under the dynamic selective transmission strategy proposed in this paper.

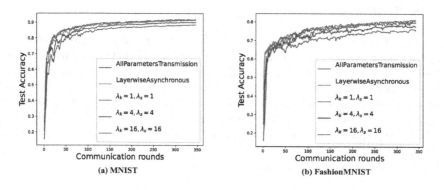

Fig. 3. Model performance in MNIST and FashionMNIST

In order to further analyze the impact of the asynchronous federated learning framework based on dynamic selective transmission, the traffic reduction rate of the framework in the whole training process $\mathcal{R}$ and the final accuracy rate of the global model Γ_{final} (the average test set accuracy of the global model in the last 10 rounds) is comprehensively compared, as shown in Table 1. From the experimental results, it can be found that with the increase of λ_k and λ_s, under the dynamic selective transmission strategy proposed in this paper, compared with the existing work, the total communication cost has been significantly further reduced, but the final accuracy of the global model Γ_{final} has been affected only a little. For the MNIST dataset, when $\lambda_k = 1$ and $\lambda_s = 1$, compared with the full parameter transmission, the proposed strategy further reduces the communication by 15.9%, while the final accuracy of the global model is only reduced by 0.3% on average. For the MNIST dataset, when $\lambda_k = 16$ and $\lambda_s = 16$, compared with the full parameter transmission, the proposed strategy further reduces 42.6% of the traffic, and the final accuracy of the global model is only reduced by 3.2% on average.

Table 1. Communication cost and final model performance

	MNIST		FashionMNIST	
	$\mathcal{R}$	Γ_{final}	$\mathcal{R}$	Γ_{final}
AllParametersTransmission	0.0%	91.8%	0.0%	80.7%
LayerwiseAsynchronous	34.8%	91.6%	35.2%	80.3%
$\lambda_k = 1, \lambda_s = 1$	50.7%	91.5%	51.3%	79.9%
$\lambda_k = 4, \lambda_s = 4$	70.2%	90.3%	69.3%	77.7%
$\lambda_k = 16, \lambda_s = 16$	77.4%	88.6%	77.6%	76.3%

5   Conclusion

This paper studies the communication efficiency optimization problem in the asynchronous federated learning framework and proposes the asynchronous federated learning framework based on dynamic selective transmission. We introduce the main components of the framework, and the dynamic selective transmission strategy in detail. In the experimental part, we compare the framework proposed in this paper with the existing work based on two different data sets, and analyzes the parameters of the proposed framework, thus showing the effectiveness of the proposed framework in improving the communication efficiency of asynchronous federated learning.

References

1. Chen, Y., Sun, X., Jin, Y.: Communication-efficient federated deep learning with layerwise asynchronous model update and temporally weighted aggregation. IEEE Trans. Neural Netw. Learn. Syst. **31**(10), 4229–4238 (2020)
2. Kamp, M., et al.: Efficient decentralized deep learning by dynamic model averaging. In: Berlingerio, M., Bonchi, F., Gärtner, T., Hurley, N., Ifrim, G. (eds.) ECML PKDD 2018. LNCS (LNAI), vol. 11051, pp. 393–409. Springer, Cham (2019). https://doi.org/10.1007/978-3-030-10925-7_24
3. Konečný, J., McMahan, H.B., Yu, F.X., Richtárik, P., Suresh, A.T., Bacon, D.: Federated learning: Strategies for improving communication efficiency. In: Proceedings of the NIPS Workshop Private Multi-Party Machine Learning, pp. 1–10. MIT Press (2016)
4. Krizhevsky, A.: One weird trick for parallelizing convolutional neural networks. arXiv preprint arXiv:1404.5997 (2014)
5. Leroy, D., Coucke, A., Lavril, T., Gisselbrecht, T., Dureau, J.: Federated learning for keyword spotting. In: Proceedings of the IEEE International Conference on Acoustics, Speech and Signal Processing, pp. 6341–6345. IEEE, Brighton, United Kingdom (2019)
6. Li, M., Chen, Y., Wang, Y., Pan, Y.: Efficient asynchronous vertical federated learning via gradient prediction and double-end sparse compression. In: International Conference on Control, Automation, Robotics and Vision (ICARCV), pp. 291–296. IEEE, Shenzhen, China (2020)
7. McMahan, B., Moore, E., Ramage, D., Hampson, S., y Arcas, B.A.: Communication-efficient learning of deep networks from decentralized data. In: Proceedings of the 20th International Conference on Artificial Intelligence and Statistics. vol. 54, pp. 1273–1282. PMLR, Fort Lauderdale, FL, USA (2017)
8. Nishio, T., Yonetani, R.: Client selection for federated learning with heterogeneous resources in mobile edge. In: IEEE International Conference on Communications (ICC), pp. 1–7. IEEE, Shanghai, China (2019)
9. Reisizadeh, A., Mokhtari, A., Hassani, H., Jadbabaie, A., Pedarsani, R.: Fedpaq: A communication-efficient federated learning method with periodic averaging and quantization. In: International Conference on Artificial Intelligence and Statistics, pp. 2021–2031. PMLR, Palermo, Sicily, Italy (2020)
10. Sahu, A.K., Li, T., Sanjabi, M., Zaheer, M., Talwalkar, A., Smith, V.: On the convergence of federated optimization in heterogeneous networks, vol, 3, p. 3. arXiv:1812.06127 (2018)
11. Wang, S., et al.: When edge meets learning: Adaptive control for resource-constrained distributed machine learning. In: Proceedings of the 2018 IEEE Conference on Computer Communications, pp. 63–71. IEEE, Honolulu, HI, USA (2018)
12. Xie, C., Koyejo, S., Gupta, I.: Asynchronous federated optimization. arXiv:1903.03934 (2019)
13. Xu, C., Qu, Y., Xiang, Y., Gao, L.: Asynchronous federated learning on heterogeneous devices: A survey. arXiv:2109.04269 (2021)
14. Yang, Q., Liu, Y., Chen, T., Tong, Y.: Federated machine learning: Concept and applications. ACM Trans. Intell. Syst. Technol. (TIST) **10**(2), 1–19 (2019)
15. Yosinski, J., Clune, J., Bengio, Y., Lipson, H.: How transferable are features in deep neural networks. In: Advances in Neural Information Processing Systems 27 (2014)

Data Mining

Small Aerial Target Detection Algorithm Based on Improved YOLOv5

TianLe Yang[1], JinLong Chen[1(✉)], and MingHao Yang[2]

[1] Guilin University of Electronic Technology, Guangxi 541004, Guilin, China
641577329@qq.com
[2] Research Center for Brain-inspired Intelligence (BII), Institute of Automation, Chinese Academy of Sciences (CASIA), Beijing 100190, China

Abstract. Small target detection is a difficult problem in the field of target detection. Small targets have many characteristics, such as less visual information, difficult to extract discriminative features, easy background confusion and so on, as a result, its detection accuracy is often only half of the large target. To solve these problems, we study small target detection in UAV aerial scene, and propose an improved YOLOv5-based small target detection algorithm. First, we performed two optimizations on the YOLOv5 prediction scale, reducing the missed detection of small targets and reducing model parameters and computational complexity, while reducing the negative impact of scale mismatches. Then, based on the idea of BiFPN, we introduce the skip connection from backbone network to PAN, and use Concat weighted fusion instead of the weighted fusion in BiFPN to improve the detection precision of the model. Finally, we do some experiments on the data set of Visdrone2019. The experimental results show that the mAP@0.5 and mAP@0.5:0.95 of the improved model are improved by 6.6% and 4.2%, respectively, the parameters of the model are reduced and YOLOv5 is light weight.

Keywords: small target detection · YOLOv5 · Visdrone2019

1 Introduction

With the development of artificial intelligence and the improvement of computing power, target detection algorithm based on depth learning has replaced the traditional target detection algorithm which needs to extract features manually. The deep learning model has stronger generalization ability, and shows better stability and higher detection accuracy in practical application. It has attracted a large number of scholars to join the tide of Deep Learning Research, at the same time, the application of target detection technology in city safety [1], industrial production [2], medical diagnosis [3], driverless and other fields is promoted. In recent years, UAV are widely used in military and civil fields because of their low cost and easy operation, and the object detection technology for UAV aerial images has become a hot research topic. The UAV has a wide field of vision at its flying altitude, and the images it takes contain a lot of scene information, however,

Y. Tan et al. (Eds.): ICSI 2023, LNCS 13969, pp. 207–219, 2023.
https://doi.org/10.1007/978-3-031-36625-3_17

ground targets such as vehicles and pedestrians can not be detected successfully because of their lack of pixels and obvious features. In addition, the noise generated by UAV equipment, image blur and occlusion caused by object motion have a greater impact on small targets in images, and small targets detection has become a key and difficult task in aerial target detection.

YOLO series [4–8] algorithm is a classical one-stage target detection algorithm, although there is a certain gap in the detection accuracy compared with two-stage target detection algorithm, but in the detection rate advantage is obvious. Among them, YOLOv5 is one of the most widely used target detection algorithms, with excellent flexibility and detection rate, suitable for UAV platform deployment. Therefore, we choose YOLOv5 as the basic detection model and validate the improved effect on the UAV aerial data set Visdrone2019 [9].

2 Related Work

In this section, first of all, the research status of small target detection technology is summarized, and then the basic detection model YOLOv5 is introduced in detail.

2.1 Small Object Detection

In the design of target detection algorithm, more and more scholars have integrated the performance of small target detection, and formed some effective solutions, these schemes can be divided into data enhancement, multi-scale learning, context learning and so on.

It is the simplest and most effective way to expand the small target sample size and improve the generalization ability and robustness of the model by using data enhancement strategy. Kisantal et al. increased the number and location of small targets by means of copy-and-paste and angle transformation in response to the problem of inadequate training due to the often low proportion of small targets in the dataset [10]. Chen et al. proposed stitcher method to solve the problem of small and medium-scale targets with too little information by using image zooming and stitching when the small target loss is too small [11]. However, it is difficult to achieve the optimal results in different application scenarios, so Zoph et al., proposed a learning-based data enhancement method [12]. The authors construct a set of substrategies that are randomly applied to each batch of images to find the optimal strategy through reinforcement learning.

The deepening of the network will lead to the partial loss of target information, which has a greater impact on the small target with less information. Multi-scale learning, which considers both deep semantic information and shallow location information, is an effective strategy to solve the above problems. The BiFPN [13] proposed by Google sets a learning weight for feature fusion nodes to weigh the importance of different input features, and adds jump links between features of the same level, more features are integrated at a lower cost, which provides a way for the design of the model in this paper. Objects are often closely related to the surrounding environment. The performance of small target detection can be improved by enriching the feature expression of small target through context information. The CoupleNet [14] model, proposed by Zhu et al., obtains local and global information through two branches formed by R-fcn and Roi Pooling, improving the detection accuracy of small targets. Lim et al. used higher-level features as a context to combine with features of small targets and increased the detection rate of small targets by reducing unnecessary background information through attention mechanisms [15].In the context reasoning method proposed by Fu et al., the semantic relation and spatial layout relation are modeled and inferred, which effectively reduces the problem of small target misdetection [16].

2.2 YOLOv5

When YOLOv5 was released in June 2020, its model was only 1/9 the size of YOLOv4, and its reasoning speed was twice that of YOLOv4. On the basis of YOLOv5L, five different network size models, such as YOLOv5N/s/m/l/x, are formed by scaling the width and depth. With the increase of network volume, YOLOv5 shows higher detection accuracy, but its reasoning speed gradually decreases. Compared with YOLOv5s, YOLOv5s is more suitable for the UAV environment in this paper.

The overall structure of YOLOv5 is similar to that of YOLOv4 and can be divided into four parts: Input, Backbone, Neck, and head. In the first part, the images entered into the network are randomly cropped, scaled, and arranged in a Mosaic-enhanced manner to form a single image, thus enriching the background for object detection, at the same time, the data volume of the small target can be expanded. In addition, YOLOv5 integratesthe

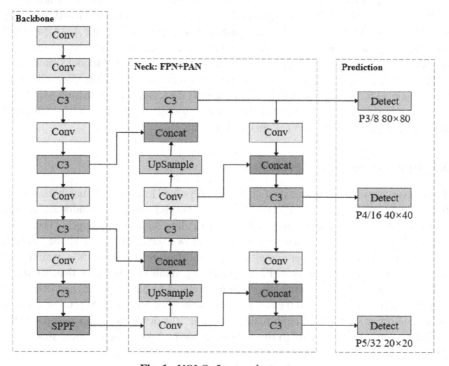

Fig. 1. YOLOv5 network structure

calculation of the adaptive anchor frame into this part, and the runtime automatically performs a clustering algorithm based on the data set's Ground Truth to obtain the optimal anchor frame size. YOLOv5 replaces the original SPP structure with a faster SPPF structure in Backbone and designs a CSP structure that can be used in Backbone and Neck, respectively. CSP structure can make the network lightweight, so that the network can reduce the amount of computing while learning more rich features. Neck is a feature fusion network composed of FPN and PAN. FPN transmits the deep semantic information to the shallow layer by adding paths, which improves the prediction effect on the shallow feature layer. In addition, Pan adds a feature fusion path to the deep layer of the network, which further improves the prediction ability of the feature layer at all scales. YOLOv5 has three different-scale detection heads at the head end, which are generated by convolution of the fused feature mAP at the Neck end. P3, P4 and P5 are used to detect small, medium and large scale objects in the input image, respectively. The loss function of YOLOv5 is composed of CIOU loss, confidence loss and classification loss, and NMS is used to output the final prediction result during post-processing. The network structure of YOLOv5 is shown in Fig. 1.

3   Methodology

Aiming at the difficulties of detecting small aerial target of UAV, an improved algorithm of small aerial target detection based on YOLOv5 is proposed. First, we optimize the YOLOv5 prediction scale. A 160×160 scale feature layer is obtained by adding an additional upper sampling layer to fuse the deep semantic information with the more shallow feature information in the backbone network. At the same time, the C3 module and the convolution module at the end of Pan are deleted to reduce the computation of the model. Secondly, we incorporate BiFPN into the optimized model structure. In order to integrate more features at lower cost, jump connection is introduced between the same level backbone nodes and PAN nodes. At the same time, in feature fusion, the use of weighted approach to enhance the proportion of important information. Figure 2 shows the improved YOLOv5 network architecture.

3.1   Optimize the Scale of Prediction

The target scale of the Visdrone data set is generally small and does not match the prediction scale set by YOLOv5. P3, P4 and P5 detection heads have low resolution and lack of shallow feature information, which makes it difficult to detect small targets. Therefore, we extracted the feature information of the shallower layer in the backbone network and added an additional upper sampling layer and feature fusion module, resulting in a 160×160-scale detector head P2 for predicting small targets. After that, we try to delete the C3 module and convolution module related to the P5 probe, and obtain the reduction of computation and the further improvement of accuracy. At present, another method to optimize the prediction scale is to reduce the down-sampling multiple of the backbone network directly to 16 times, but this method will produce more than twice our computation.

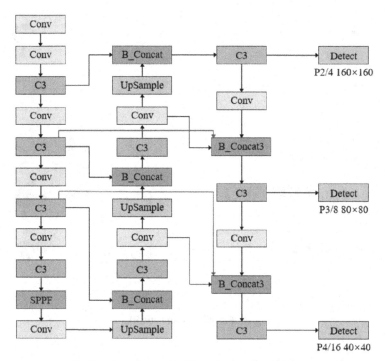

Fig. 2. The improved YOLOv5 network architecture.

3.2 Fusing BiFP

Feature fusion network is used to fuse the deep semantic information and the shallow position information. In YOLOv5, this part adopts the network structure of FPN and PAN. FPN transmits the deep information to the shallow information, which makes up for the lack of semantics in the shallow information. Pan transmits the fused information from the shallow layer back to the deep layer, so that each feature layer contains more comprehensive information, as shown in Fig. 3(a). BiFPN **N** introduces additional edges between the same hierarchical backbone nodes and PAN nodes, fusing more features at a lower cost while removing nodes with only one input, simplifying the network structure. BiFPN is shown in Fig. 3(b).

BiFPN takes into account the importance of different information, gives each feature a weight when features are fused, and introduces a fast normalization mechanism (1) to constrain the size of each weight.

$$Out = \sum_i \frac{\omega_i \times I_i}{\varepsilon + \sum_j \omega_j} \tag{1}$$

where ω_i denotes the weight parameter that the model can learn, it is guaranteed to be non-negative under the application of Relu function, I_i is the characteristic input of different levels, ε is a small quantity to guarantee numerical stability.

Based on the idea of BiFPN, a jumping connection from backbone network to PAN is added to the characteristic layer of 80×80 and 40×40, which reduces the repetition

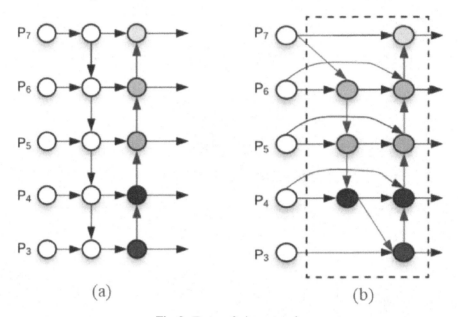

(a)   (b)

Fig. 3. Feature fusion networks

of C3 module in PAN, a richer blend of features at a lower cost. We also add the weight to the feature fusion, through the experiment, we find that the BiFPN's weighted fusion method will increase the computational complexity and produce a small benefit to the accuracy, so we still use Concat operation and introduce a small number of parameters that can be learned, and get a better detection result.

4   Experiments

4.1   Dataset

In order to verify the effectiveness of the improved YOLOv5-based small target detection algorithm, this paper selects a representative and influential data set of UAV aerial scene, VISDRONE2019. The dataset was provided by AISKYEYE team from the machine learning and Data Mining Laboratory at Tianjin University, China, data were collected by different types of uavs in 14 cities and villages in China, which are thousands of kilometers apart, taking into account factors such as scene density and weather conditions, increasing the diversity and representativeness of the data. Among them, Visdrone2019-det data set is mainly used for image target detection task, including 10,209 static images, divided into 6,471 training sets, 584 verification sets, 1,610 test sets and 1,580 game-specific test sets. The dataset contained 10 different categories of objects and more than 2 million tags were manually labeled. The number of tags in different categories is shown in Fig. 4(a), and it can be seen that the sample sizes of different categories vary too much, which can easily lead to inadequate learning of some categories by the model at the time of training. Figure 4(b) shows the distribution of objects of different sizes, showing that the axis is denser near the origin, reflecting the fact that the data set contains a large number of small objects, which is bound to make the model more difficult to detect.

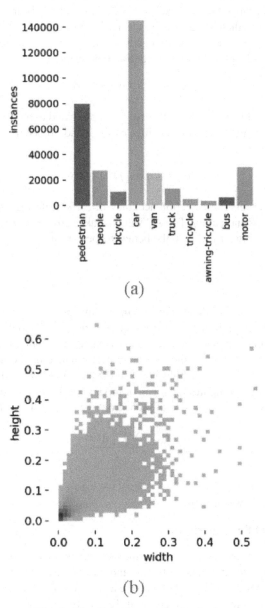

Fig. 4. Visualizes the sample distribution.

4.2 Evaluation Indicators

Precision, Recall, F1, AP, mAP, FPS and so on are the common evaluation indexes of target detection. In this paper, the use of several evaluation indicators and the determination of the results of the prediction methods made a brief introduction.

Precision, the ratio of the number of positive cases predicted correctly to the number of samples judged to be positive in the experimental result.

$$Precision = TP/(TP + FP) \tag{2}$$

Recall, the ratio of the number of positive cases predicted to be correct in the experimental result to the number of actual positive cases. It and Precision in general, the pros and cons, mutual constraints.

$$Recall = TP/(TP + FN) \tag{3}$$

AP, which represents the detection accuracy of a class of objects, can balance the opposition of P and R indicators. The AP value is taken from the area of PR curve and coordinate axis. The bigger the area, the better the performance of the model.

$$AP = \int_0^1 p(r)dr \tag{4}$$

mAP is an important index of model performance in the task of target detection. Because there are many categories in the task of target detection, the model will get AP value of each category and calculate average value of AP value of all categories, you get a mAP, so the higher the number, the better the performance of the model. mAP@0.5 is calculated when the IOU threshold is set to 0.5, and mAP@0.5:0.95 is calculated, mAP@0.5:0.95 further computes the mean of mAP at different IOU thresholds. Where the IOU threshold is set at a step size of 0.05, from 0.5 to 0.95 Table 1.

$$mAP = \frac{\sum_{i=1}^{k} AP_i}{k} \tag{5}$$

4.3 Experimental Environment

4.4 The Results of the Experiment

Based on YOLOv5s, this paper improves the experiment. The experiment adopts the same model hyperparameters and software and hardware environment to ensure the reliability of the experiment. The settings were as follows: Input Image Resolution 640 × 640, batchsize at training 8, Batchsize at Test 1, Epoch 200, initial learning rate 0.01, momentum size 0.937, and weight decay coefficient 0.0005. Using the default data enhancement method, the default pre-training weights are used.

We first trained the YOLOv5s and the model in this article, using data sets for VisDrone2019-DET-train and VisDrone2019-DET-val. Figure 5 shows the border regression loss, confidence loss, and classification loss for the YOLOv5s and our model

Table 1. Experimental environment.

number	equipment	model
1	CPU	Intel(R) Xeon(R) CPU E5–2630
2	Memory	32GB
3	The operating system	Ubuntu 16.04
4	GPU	NVIDIA Tesla V100
5	Programming language	C + +, Python3.9
6	Deep learning framework	PyTorch

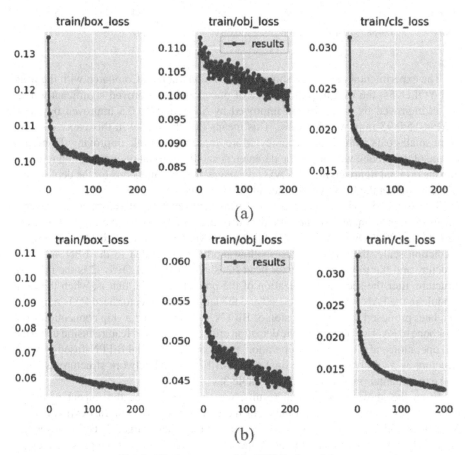

Fig. 5. The loss curve of the YOLOv5s and Ours

on the VisDrone2019-DET-train, respectively, as can be seen from the use of the official pre-training weights, at the beginning of the training, the losses of the two models presented a straight-line downward trend, gradually smoothed after 10 rounds and gradually saturated around 200 rounds.

Through training, we obtained the optimal weights for the YOLOv5s and the present model over 200 rounds, and compared the performance of the model using the optimal weights at VisDrone2019-DET-val, the results of which are shown in Table 2.

Table 2. Performance comparison of YOLOv5s and Ours.

Models	P	R	mAP@0.5	mAP@0.5:0.95	Parameters(M)	GFLOPs
YOLOv5s	0.446	0.347	0.34	0.189	7.04	15.8
Ours	0.488	0.404	0.406	0.231	5.47	17.6

The experimental results are shown in Fig. 5 and Table 2. Compared with the original YOLOv5s, the overall performance of our model is improved significantly. Precision improved by 4.2%, Recall improved by 5.7%, mAP@0.5 improved by 6.6%, mAP@0.5:0.95 improved by 4.2%. This means that the model in this paper captures more small goals with higher accuracy. More importantly, while improving the performance of the model, we control the parameters and the amount of computation to maintain the original lightweight YOLOv5s advantage, it is advantageous for the deployment of the model on the UAV with limited airborne resources.

We further designed a comparative experiment to verify the effectiveness of different stages of improvement. For the sake of illustration, we'll refer to the original network structure as YOLOv5s. YOLOv5s -1 is the network structure of the first optimized prediction scale, that is, when the shallow modules are added to the FPN and PAN structure and the tiny target detector head P2 is obtained. YOLOv5s -2 is the network structure after the second optimization of the prediction scale, that is, when the deep module in the PAN structure is deleted and the large-scale detector head P5 is removed. In a later proposed scheme for fusion of BiFPN, we introduced a skip connection from backbone to PAN and compared the detection effects of weighted feature fusion using the add operation and the CONCAT operation, we record the original BiFPN-fused network structure as YOLOv5-3 and the final concat-BiFPN-fused network structure as Ours. The experimental results are shown in Table 3.

From the above experimental results, we can directly see the impact of different improvements on the performance of the model. From the contrast experiment of YOLOv5s and YOLOv5s-1, we can see that adding small target detection layer can effectively capture more small targets, and it is an effective way to improve the effect of target detection. From the experimental results of YOLOv5s-2, it can be seen that removing the large-scale detection layer is beneficial to the lightweight of the model, and can reduce the negative influence of the scale mismatch on the detection. From the comparison experiment between YOLOv5s-3 and the final model, it can be seen that the fusion Concat-BiFPN produces less parameters and computation, and can bring higher recall rate and mAP than the fusion of BiFPN.

Table 3. Performance comparison of different stages of improvement.

Models	P	R	mAP	Parameters(M)	GFLOPs
YOLOv5s	0.446	0.347	0.34	7.04	15.8
YOLOv5s-1	0.491	0.386	0.393	7.18	18.7
YOLOv5s-2	0.491	0.396	0.401	5.39	17.2
YOLOv5s-3	0.494	0.396	0.403	5.65	18.6
Ours	0.488	0.404	0.406	5.47	17.6

Figures 6 and 7 visualize the detection effect of Yolov5s and our model on the Visdrone data set, respectively. It can be seen that Yolov5s can detect most of the near targets, but the detection effect of the near occluded targets is not good, and there are many missed detection of distant targets. Our model can improve the detection of near and far targets, but there is still room to improve the detection of far and low-contrast targets.

Fig. 6. Visualize the detection effect of YOLOV5s

Fig. 7. Visualize the detection effect of our model

5   Conclusion

In this paper, an improved YOLOv5-based small target detection algorithm is proposed, and the YOLOv5 prediction scale is optimized twice, and then the multi-scale feature fusion is carried out based on the idea of BiFPN, finally, Concat-BiFPN is used as a new feature fusion network. Experimental results show that the proposed algorithm model has excellent comprehensive performance, while improving the detection accuracy while maintaining the YOLOv5 lightweight advantages, it has a good application prospect in UAV platform and other small target scenarios.

Acknowledgements. The research work of this paper is supported by Guangxi Science and Technology Project (AB21220038), (AB21075004), (AB21220011).

References

1. Xie, Y., Zhu, J, Cao, Y., et al.: Efficient video fire detection exploiting motion-flicker-based dynamic features and deep static features. IEEE Access **8**(81), 904–81 917 (2020)
2. Sevo, I., Avramovic, A.: Convolutional neural network based automatic object detection on aerial images. IEEE Geosci. Remote Sens. Lett. 13(5), 740–744 (2016)
3. Xiong, F., He, D., Liu, Y., et al.: Pneumonia image classification based on convolutional neural network [J].Proc. SPIE **12057**, 120573C (2021)
4. Redmon, J., Divvala, S., Girshick, R., Farhadi, A.: You only look once:unified,real-time object detection. In: Proceedings of 2016 IEEE Conference on Computer Vision and Pattern Recognition. Washington D. C.,USA:IEEE Press, pp. 779–788 (2016)
5. Redmon, J., Farhadi, A.: YOLO9000:better, faster, stron-ger. IEEE Conf. Comput. Vision Pattern Recogn. (CVPR) 6517–6525 (2016)

6. Redmon, J, Farhadi, A.: YOLOv3:an incremental im-provement. ar Xiv Preprint, (2018)
7. Bochkovskiy, A., Wang, C.Y., Liao, H.Y.M.: YOLOv4:optimal speed and accuracy of object detection[EB/OL].[2021–08–10]. https://arxiv.org/abs/2004.10934
8. Jocher, G., Stoken, A., Borovec, J.,et al.: Ultralytics/YOLOv5:V3.1-bug fixes and performance improvements[EB/OL] (2020). https://doi.org/10.5281/zenodo.4154370
9. Du, D., et al.: VisDrone-DET2019: The vision meets drone object detection in image challenge results. In: Proceedings of the IEEE/CVF International Conference on Computer Vision Workshops, pp. 0–0 (2019)
10. Kisantal, M., Wojna, Z., Murawski, J.,et al.: Aug mentation for small object detection. arXiv: 1902.07296, (2019)
11. Chen, Y., Zhang, P., Li, Z., et al.: Stitcher:feedback-driven data provider for object detection. arXiv:2004.12432 (2020)
12. Zoph, B., Cubuk, E.D., Ghiasi, G., et al.: Learning data augmentation strategies for object detection. ArXiv Preprint arXiv:2019:1906.11172
13. Tan, M., Pang, R., Le, Q.V.: Efficient Det:scalable and efficient object detection. In: IEEE Conference on Computer Vision and Pattern Recognition (CVPR), pp. 10778–10787 (2020)
14. Zhu, Y., Zhao, C., Wang, J., et al.: Couplenet:coupling global structure with local parts for object detection. In: Proceedings of the IEEE International Conference on Computer Vision, pp. 4126–4134 (2017)
15. Lim, J.S., Astrid, M., Yoon, H.J, et al.: Small object detection using context and attention[EB/OL].(2019–12–13)[2019–12–16]. https://arxiv.org/abs/1912.06319
16. Fu, K., Li, J., Ma, L., et al.: Intrinsic relationship reasoning for small object detection[EB/OL].(2020–09–02)[2020–09–02]. https://arxiv.org/abs/2009.00833

Secondary Pulmonary Tuberculosis Lesions Detection Based on Improved YOLOv5 Networks

Haojie Xie[1,4], Mingli Lu[1(✉)], Jing Liu[2], Benlian Xu[3], Xianghang Shi[1,4], Chen Zhang[1,4], Jian Shi[1], and Jinliang Cong[1]

[1] School of Electrical and Automatic Engineering, Changshu Institute of Technology, Changshu, People's Republic of China
luml@cslg.edu.cn
[2] Department of Radiology, The Fifth People's Hospital of Suzhou, Suzhou, China
[3] School of Electronic and Information, Suzhou University of Science and Technology, Suzhou, People's Republic of China
[4] School of Electrical Engineering, Yancheng Institute of Technology, Yancheng, People's Republic of China

Abstract. Tuberculosis (TB) is an infectious disease caused mainly by Mycobacterium tuberculosis. It has been reported that the mortality rate of TB is extremely high if not timely detected and diagnosed in early stages. Currently, the diagnosis of secondary pulmonary tuberculosis (PTB) relies heavily on subjective analysis by specialized radiologists, which is time-consuming and inefficient. Thus, the application of neural networks augmented with intelligent algorithm rules holds significant value in detecting secondary PTB. This paper proposes an algorithm based on an attention mechanism and a loss function-improved YOLOv5 neural network to accurately detect secondary PTB lesions with four specific features. Furthermore, a multi-scale data augmentation method is proposed to expand the lesion dataset, enhancing the generalization ability and robustness of the trained model. Experimental results demonstrate that our proposed method can effectively improve the detection accuracy and speed of secondary PTB lesions, achieving accurate recognition of secondary PTB lesions with the four specific features.

Keywords: Lesion detection · Data augmentation · Attention mechanism · Loss function · Neural network · Pulmonary tuberculosis

1 Introduction

TB is a chronic infectious disease caused by Mycobacterium tuberculosis, which is primarily transmitted through respiratory droplets and is highly contagious through the air, dust, and aerosols [1]. China is one of the 30 countries with the highest incidence of TB in the world, ranking second in the world in terms of prevalence [2]. Therefore, it is crucial to have an accurate and effective diagnosis of Secondary PTB in the early stages and to begin early treatment.

Currently, the diagnosis of Secondary PTB lesions still relies mainly on radiologists and commonly used CT image-assisted software. The main functions

© The Author(s), under exclusive license to Springer Nature Switzerland AG 2023
Y. Tan et al. (Eds.): ICSI 2023, LNCS 13969, pp. 220–231, 2023.
https://doi.org/10.1007/978-3-031-36625-3_18

of the software are limited to viewing and managing CT images, and the main detection methods still rely on subjective judgment of radiologists. Due to the diverse and variable morphologies of Secondar PTB lesions in CT images, it is difficult to identify them. Medical staff need to read CT images among a large number of patients and determine the specific location and type of TB lesions. The long working hours and huge workload make medical staff more prone to fatigue and misdiagnosis. With the rapid development of technology, assisted detection systems for Secondary PTB lesions have begun to emerge in major cities, which can greatly help radiologists reduce their workload and assist in disease diagnosis [3,4]. However, the high cost of these detection systems makes it difficult for them to be widely applied. In recent years, the YOLO (You Only Look Once) series of networks has rapidly developed, with high accuracy and fast inference speed, surpassing many two-stage neural networks in performance. This paper aims to investigate the detection and classification of Secondary PTB lesions from CT images by improving the YOLOv5 neural network [5] and evaluate its performance in detecting Secondary PTB lesions.

2   Related Research

In recent years, many medical researchers have explored the capabilities of deep learning in the detection and classification of pulmonary TB lesions, resulting in the emergence of numerous deep neural network applications and algorithm improvements specifically for handling pulmonary lesions for disease diagnosis. Hwang et al. [6]proposed the first automatic TB detection system based on convolutional neural networks, which achieved promising results at that time. The team led by Andrew Ng proposed a deep learning algorithm called CheXNet [7], which is a 121-layer convolutional neural network capable of diagnosing 14 types of pulmonary diseases from frontal chest X-ray scans, outputting the probability of pneumonia. The model has a complex layer structure and a large amount of data, resulting in accurate model training. The classification performance can be further improved after correct interpretation by professional radiologists. Gao et al. [8] proposed a 3D block-based residual deep learning network (ResNet) [9]to address the issue of limited datasets for pulmonary TB and analyzed the features of lesions occupying limited regions in CT images. They injected depth information into each layer of the network (Depth-ResNet) and made progress in predicting lesion severity scores and analyzing severity probability. Li et al. [10] fully considered the design of Secondary PTB detection system and its interaction capability with radiologists. They combined the ideas of DenseVoxNet [11], 3D U-Net [12,13], and V-Net [14], and designed a network called VNET-IR-RPN by modifying the V-Net backbone, which implemented a Secondary PTB diagnosis system capable of generating quantitative CT reports.

However, despite the superior performance of deep learning-based methods in the detection of Secondary PTB lesions as demonstrated in the aforementioned studies, more efficient algorithms are still needed. This is because these methods often have complex architectures that prioritize accuracy over detection speed, overlooking the importance of timely diagnosis in radiology. To address the balance between detection accuracy and prolonged diagnosis time for radiologists, we have chosen YOLO (You Only Look Once) neural network as our base model. YOLO adopts a lightweight backbone network structure, enabling fast lesion detection while maintaining high accuracy.

3 Methods and Materials

Fig. 1 illustrates the overall workflow of our study experiments. CT scan images are preprocessed and annotated by expert radiologists, and the annotated training dataset is further augmented using multi-scale data augmentation techniques before being fed into the neural network for training. The trained model is then utilized for lesion localization and classification. In this section, we will discuss our research methodology, as well as the materials such as data, that we used in our study.

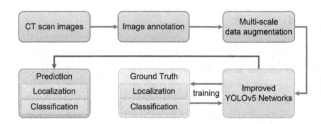

Fig. 1. Working flow of classification.

3.1 Dataset Description

In our study, we used CT lung scan images from five different CT scanning devices in three hospitals.

  i) Hospital A: Beijing Hospital CT Set
 ii) Hospital B: Suzhou Hospital CT Set
iii) Hospital C: Heilongjiang Hospital CT Set

The images from all three hospitals were exported in the required format for neural network training, with original size and containing 3 color channels, with dimensions of 512*512. The data from Hospital A and Hospital B were used as the training, validation, and internal testing sets, and were annotated by professional radiologists based on the four category features. The quantity of each lesion class in the dataset and the annotation file information are summarized in Table 1.

Table 1. The information on the number of lesions and annotation files.

Tag number	Type of lesion	Number of lesions	Number of annotated files
Nidus1	consolidation	1090	971
Nidus2	tree-in-bud	1552	1254
Nidus3	cavitation	463	415
Nidus4	nodules	449	375

We selected four categories of features related to secondary PTB for a retrospective study, as shown in Fig. 2. A total of 3015 annotation files were obtained. The data were divided into training set, validation set, and internal testing set in an 8:1:1 ratio. The data from Hospital C were used as an external testing set to evaluate the clinical applicability performance of the neural network model. As a result, there were 2412, 302, and 301 images in the training, validation, and testing sets, respectively. Hospital C contributed a total of 825 images.

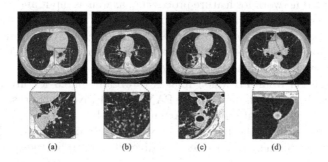

Fig. 2. The four categories of features related to secondary PTB: (a) Consolidation (b) Tree-in-bud (c) Cavitation (d) Nodules.

3.2 Data Augmentation

To address the issue of imbalanced data due to significant differences in the number of lesions in each category of pulmonary CT images, we proposed a data augmentation preprocessing strategy based on multiscale techniques to better utilize lesions from each category. This strategy helps to balance the training data set for different lesion categories and reduce overfitting while improving accuracy. We employed methods such as Gaussian blur, brightness/contrast adjustment, rotation, and flipping that do not alter the biological characteristics of the lesions for data augmentation. The augmentation factor for each lesion category was determined based on the maximum number of lesions multiplied by 5, resulting in a more balanced augmentation quantity. After augmentation, a total of 20,867 images containing Secondary PTB lesions were obtained. Fig. 3 shows the detailed statistical results of the arrangement of CT image data for each lesion category in the original training data set.

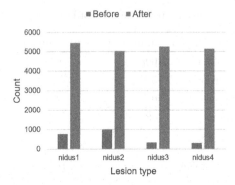

Fig. 3. Comparison of CT image count before and after Multiscale data augmentation.

3.3 Improved YOLOv5 Network

The original YOLOv5 neural network consists of a Backbone for feature extraction and a Head network for feature processing, as well as input and output ends. After improvement, the input end obtains data from multi-scale data augmentation operations, which are then processed through adaptive anchor box calculation and image-adaptive scaled input to the backbone network. The backbone network is constructed using CSPDarknet53 architecture and SPPF module for feature extraction from TB lesion images to obtain shared feature maps. The improved head network is built using PANet and FPNet networks, along with the embedded Normalization-based Attention Module (NAM) [15], for fusing the feature maps from the backbone network to obtain higher-level semantic information and further enhance feature diversity and robustness. The output end includes non-maximum suppression, IoU threshold determination, and the use of the improved SIoU loss function[16] to evaluate the object category confidence loss between the ground truth boxes and predicted boxes based on the fused high-level semantic features. Finally, the predicted bounding boxes and category information of TB lesions in the patients to be detected are obtained. The improved YOLOv5 neural network architecture is shown in Fig. 4.

Normalization-based Attention Module (NAM). NAM is a lightweight attention module that can be embedded into various convolutional neural networks (CNNs) and trained end-to-end. It adopts the architecture of the Convolutional Block Attention Module (CBAM) [17] and has been optimized for channel attention and spatial attention. The structure of the NAM module is shown in Fig. 5. Experimental results from the paper [15]demonstrate that incorporating NAM into different models significantly improves their performance on various classification and detection datasets, validating the effectiveness of this module. In CT images, the use of NAM module further suppresses less significant weights, and with channel and spatial attention, the channel attention focuses more on key features of the lesions, thereby enhancing the detection capability of Secondary PTB lesions. The spatial attention then refines the localization of lesions with more informative regions based on channel attention, improving the

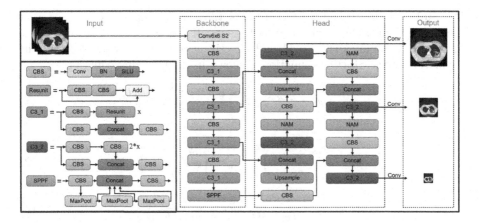

Fig. 4. The improved YOLOv5 network model structure.

classification capability for different categories of Secondary PTB lesions. This helps YOLOv5 to be more attentive to relevant lesion targets during training.

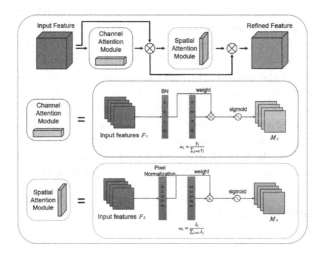

Fig. 5. NAM Module, refining feature maps using channel and spatial attention modules, along with residual pathways.

SCYLLA-IoU (SIoU) Loss Function. The loss function is primarily used as a measure of the inconsistency between the predicted values and the ground truth in a model. Therefore, it plays a critical role in the training process of neural networks, as choosing an appropriate loss function can not only improve the quality of the model but also enable faster convergence. In this study, we replaced the CIoU loss function used in YOLOv5 with the SIoU loss function.

The SIoU loss function further considers the vector angles between different regressions, which effectively reduces the overall degrees of freedom of the loss and greatly promotes the speed and effectiveness of model convergence. The calculation of the SIoU loss function is shown in the following equation:

$$Loss_{SIoU} = 1 - IoU + \frac{\sum_{t=x,y}(1-e^{-\gamma\rho_t}) + \sum_{t=w,h}(1-e^{-\omega_t})^\theta}{2} . \quad (1)$$

where:

$$\rho_x = \left(\frac{b_{c_x}^{gt} - b_{c_x}}{c_w}\right)^2 \quad \rho_y = \left(\frac{b_{c_y}^{gt} - b_{c_y}}{c_h}\right)^2 . \quad (2)$$

$$w_w = \frac{|w - w^{gt}|}{max(w, w^{gt})} \quad w_h = \frac{|h - h^{gt}|}{max(h, h^{gt})} . \quad (3)$$

$$\gamma = 2 - cos\left(2 * \left(arcsin\left(\frac{c_h}{\sigma}\right) - \frac{\pi}{4}\right)\right) . \quad (4)$$

where, $c_h = max\left(b_{c_y}^{gt}, b_{c_y}\right) - min\left(b_{c_y}^{gt}, b_{c_y}\right)$, $\sigma = \sqrt{(b_{c_x}^{gt} - b_{c_x}^{gt})^2 + (b_{c_y}^{gt} - b_{c_y}^{gt})^2}$. b^{gt} is the predicted bounding box, b is the ground truth bounding box, $(b_{c_x}^{gt}, b_{c_y}^{gt})$ are the center coordinates of the predicted box, (b_{c_x}, b_{c_y}) are the center coordinates of the ground truth box, (c_w, c_h) are the width and height of the minimum enclosing rectangle formed by the ground truth and predicted boxes, (w^{gt}, h^{gt}) and (w, h) are the width and height of the ground truth bounding box and the predicted box, c_h and σ are the height difference and distance between the center points of the ground truth and predicted boxes, respectively. In equation $\frac{c_h}{\sigma} = sin(\alpha)$, when α is equal to $\frac{\pi}{2}$ or 0, the angle loss is 0. During the training process, if $\alpha < \frac{\pi}{4}$, then minimize σ, otherwise minimize β. θ controls the degree of attention to the shape cost, with the parameter range of $[2, 6]$. Due to the introduction of vector angles between the ground truth and predicted boxes in SIoU, further improvements in convergence speed and regression accuracy in the neural network training process are achieved.

4   Experimental Results

4.1   Evaluation Metrics

The detection and classification performance of the trained network is evaluated using four performance metrics: recall, precision, mean average precision (mAP), and F1 score. The mean average precision represents the average precision of the four categories of lesion detection, and is used to evaluate the overall performance of the model. The formulas for the calculation of these metrics are as follows:

$$Precision = \frac{TP}{TP + FP} . \quad (5)$$

$$Recall = \frac{TP}{TP + FN} . \quad (6)$$

$$mAP = \frac{\sum_{m}^{i=0} AP_i}{m} . \tag{7}$$

$$F1 = 2 * \frac{Precision * Recall}{Precision + Recall} . \tag{8}$$

Where TP stands for true positive, TN for true negative, FP for false positive, and FN for false negative. AP represents the average formula of Precision corresponding to each recall:

$$AP = \sum_{i=1}^{n-1} (x_{i+1} - x_i) P_{inter}(x_i + 1) . \tag{9}$$

where x_1 , x_2 ... x_n is the recall corresponding to the first interpolation in the precision interpolation segment arranged in ascending order.

4.2 Model Training

To obtain comparative results, we utilized a five-fold cross-validation method to evaluate the models and to acquire efficient hyperparameter combinations. We separately trained both YOLOv5 and the improved YOLOv5 network with the same data and parameters. We then plotted the comparison of loss curves for the two networks based on the training results data, as presented in Fig. 6, where the horizontal and vertical axes denote iterations and loss values, respectively. From the figure, it can be observed that our improved network exhibits a faster and more stable training process with better convergence compared to YOLOv5, as evidenced by the lower loss values.

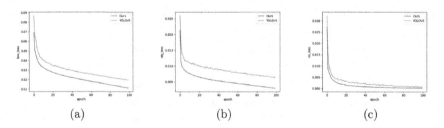

(a)   (b)   (c)

Fig. 6. Loss comparison: (a) box bounding loss (b) objectness loss (c) classification loss.

Fig. 7 shows the average classification results of our network using 5-fold cross-validation on the validation data, presented in confusion matrices for performance measurement. From the figure, it can be seen that the two networks perform similarly on nidus1 and nidus4, while our network demonstrates better results on nidus2 and nidus3.

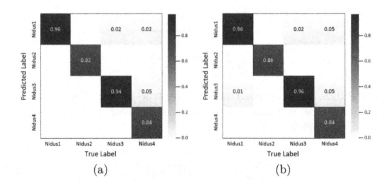

Fig. 7. Confusion matrix: (a) YOLOv5 (b) Ours.

4.3   Comparisons of Various Neural Networks

To validate the effectiveness of the proposed model, we compared the results obtained from training the same dataset with other neural network methods for pulmonary lesion detection published in recent years. Table 2 shows the performance evaluation on the internal testing set compared to five other networks, including two-stage Faster R-CNN[18], one-stage SSD[19], RetinaNet[20], YOLOv5, and YOLOv8[21]. The results demonstrate that the improved YOLOv5 outperforms the existing classical networks slightly in terms of performance.

Table 2. Comparison with performance of various Neural Networks.

Methods	P(%)	R(%)	mAP@.5(%)	F1-score(%)
Faster RCNN	53.1	89.9	78.2	66.8
SSD	86.8	57.2	82.1	68.9
RetinaNet	88.0	83.5	90.6	85.7
YOLOv5	85.5	88.4	92.4	86.9
YOLOv8	87.9	88.8	93.3	88.4
Ours	89.3	89.7	93.8	89.5

4.4   Performance Analysis on Cross-Hospital Dataset

To validate the cross-dataset performance of our trained model using data from Hospital A and Hospital B, additional external testing experiments were conducted using data from collaborating Hospital C. Table 3 summarizes the performance of our improved YOLOv5 model on the external testing set, based on

Table 3. Performance comparison of YOLO Networks on external test data.

Methods	P(%)	R(%)	mAP@.5(%)	F1-score(%)
YOLOv5	62.3	57.7	59.9	58.9
YOLOv8	62.3	59.7	60.9	59.1
Ours	68.7	72.4	70.5	70.1

the calculated detection results. A comparison with two other YOLO networks is also provided for reference.

Evaluation metrics are a valuable tool for assessing network models, but they are not the only means. Based on the detection results of the model, we conducted a separate analysis of the visualized results for four categories of lesions. Fig. 8 shows a comparison of the detection capabilities of our trained model relative to the original YOLOv5 and YOLOv8 neural network models, using data from Hospital C, and the results were reviewed and annotated by professional radiologists. The neural network model automatically annotated the areas of secondary PTB lesions with bounding boxes of different colors for different categories. The results showed that our neural network model was able to detect more lesion areas and provided higher confidence in the detected lesions. Overall, the detection results of our improved YOLOv5 model were superior to the original YOLOv5 and YOLOv8 networks, and were comparable to the

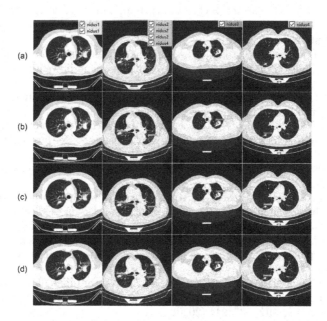

Fig. 8. Visual comparison of detection results with annotated results by radiologists (**a**) Radiologists (**b**) YOLOv5 (**c**) YOLOv8 (**d**) Ours.

radiologists' annotations, demonstrating its potential for assisting radiologists in the diagnosis of secondary PTB in practical applications.

5   Conclusion

In this study, we proposed and implemented an improved YOLOv5 scheme. The proposed scheme incorporated multi-scale data augmentation, channel and spatial attention mechanisms, and an improved loss function. The data augmentation improved the network's generalization ability, making it capable of handling more complex datasets. The attention mechanisms effectively suppressed false positives and improved recognition ability for secondary PTB. The improved loss function considered the angle relationship between the ground truth boxes and predicted boxes, accelerating network convergence. The experimental results showed that our improved network outperformed current classical neural networks in terms of detection performance. In the internal test set, the accuracy, recall rate, mean average precision (mAP), and F1 score were 89.3%, 89.7%, 93.8%, and 89.5%, respectively. In the external test set, the accuracy, recall rate, mAP, and F1 score were 68.7%, 72.4%, 70.5%, and 70.1%, respectively. These results demonstrate the potential of our improved network in assisting radiologists in screening secondary PTB lesions from pulmonary CT images.

References

1. Rangaka, M.X., Cavalcante, S.C., Marais, B.J., Thim, S., et al.: Controlling the seedbeds of tuberculosis: diagnosis and treatment of tuberculosis infection. The Lancet. **386**, 2344–2353 (2015)
2. Chakaya, J., Khan, M., Ntoumi, F., Aklillu, E., et al.: Global Tuberculosis report 2020-reflections on the global TB burden, treatment and prevention efforts. Int. J. Infect. Dis. **113**, S7–S12 (2021)
3. Zhang, G., Jiang, S., Yang, Z., Gong, L., et al.: Automatic nodule detection for lung cancer in CT images: a review. Comput. Biol. Med. **103**, 287–300 (2018)
4. Wang, Y., Wu, B., Zhang, N., Liu, J., et al.: Research progress of computer aided diagnosis system for pulmonary nodules in CT images. J. X-Ray Sci. Technol. **28**, 1–16 (2020)
5. Glenn jocher et al. Ultralytics/yolov5 v6.0 (2021). https://github.com/ultralytics/yolov5/releases/tag/v6.0.
6. Hwang, S., Kim, H.E., Jeong, J., Kim, H.J.: A novel approach for tuberculosis screening based on deep convolutional neural networks. In: Medical imaging 2016: Computer-aided Diagnosis, vol. 9785, pp. 750–757. Proc. SPIE, California (2016)
7. Lakhani, P., Sundaram, B.: Deep learning at chest radiography: automated classification of pulmonary tuberculosis by using convolutional neural networks. Radiology **284**, 574–582 (2017)
8. Gao, X.W., James-Reynolds, C., Currie, E.: Analysis of tuberculosis severity levels from CT pulmonary images based on enhanced residual deep learning architecture. Neurocomputing **392**, 233–244 (2020)
9. He, K., Zhang, X., Ren, S., Sun, J.: Deep residual learning for image recognition. In: Proceedings of the IEEE Conference on Computer Vision and Pattern Recognition, pp. 770–778. IEEE Press, Las Vegas (2016)

10. Li, X., Zhou, Y., Du, P., Lang, G., et al.: A deep learning system that generates quantitative CT reports for diagnosing pulmonary Tuberculosis. Appl. Intell. **51**, 4082–4093 (2021)
11. Yu, L., et al.: Automatic 3D cardiovascular MR segmentation with densely-connected volumetric ConvNets. In: Descoteaux, M., Maier-Hein, L., Franz, A., Jannin, P., Collins, D.L., Duchesne, S. (eds.) MICCAI 2017. LNCS, vol. 10434, pp. 287–295. Springer, Cham (2017). https://doi.org/10.1007/978-3-319-66185-8_33
12. Çiçek, Ö., Abdulkadir, A., Lienkamp, S.S., Brox, T., Ronneberger, O.: 3D U-Net: learning dense volumetric segmentation from sparse annotation. In: Ourselin, S., Joskowicz, L., Sabuncu, M.R., Unal, G., Wells, W. (eds.) MICCAI 2016. LNCS, vol. 9901, pp. 424–432. Springer, Cham (2016). https://doi.org/10.1007/978-3-319-46723-8_49
13. Chen, J., Yang, L., Zhang, Y., Alber, M., et al.: Combining Fully Convolutional and Recurrent Neural Networks for 3D Biomedical Image Segmentation. In: Advances in Neural Information Processing Systems. 29 (2016)
14. Milletari, F., Navab, N., Ahmadi, S.A.: V-Net: Fully Convolutional Neural Networks for Volumetric Medical Image Segmentation. In: 2016 Fourth International Conference on 3D Vision, pp. 565–571. IEEE Press, Stanford (2016)
15. Liu, Y., Shao, Z., Teng, Y., Hoffmann, N.: NAM: Normalization-based Attention Module. arXiv preprint arXiv. 2111.12419 (2021)
16. Gevorgyan, Z.: SIoU Loss: More Powerful Learning for Bounding Box Regressiond. arXiv preprint arXiv. 2205.12740 (2022)
17. Woo, S., Park, J., Lee, J.-Y., Kweon, I.S.: CBAM: convolutional block attention module. In: Ferrari, V., Hebert, M., Sminchisescu, C., Weiss, Y. (eds.) ECCV 2018. LNCS, vol. 11211, pp. 3–19. Springer, Cham (2018). https://doi.org/10.1007/978-3-030-01234-2_1
18. Ren, S., He, K., Girshick, R., Sun, J.: Faster R-CNN: towards real-time object detection with region proposal networks. In: Advances in Neural Information Processing Systems 28 (2015)
19. Liu, W., et al.: SSD: single shot multibox detector. In: Leibe, B., Matas, J., Sebe, N., Welling, M. (eds.) ECCV 2016. LNCS, vol. 9905, pp. 21–37. Springer, Cham (2016). https://doi.org/10.1007/978-3-319-46448-0_2
20. Lin, T.Y., Goyal, P., Girshick, R., He, K., et al.: Focal loss for dense object detection. In: 2017 IEEE International Conference on Computer Vision, pp. 2999–3007. IEEE Press, Venice (2017)
21. Glenn jocher et al. Ultralytics/YOLOv8 (2023). https://github.com/ultralytics/ultralytics.

Abnormal Traffic Detection Based on a Fusion BiGRU Neural Network

Lin Jiang[(✉)], Da-hua Zhang, Ya-yun Zhu, and Xiao-juan Zhang

China Electric Power Research Institute, Beijing 100192, China
{jianglin,zhangdh,zhuyayun,zhangxiaoj}@epri.sgcc.com.cn

Abstract. As network security is getting more and more attention, methods for anomalous traffic detection are proposed. However, the methods for anomalous traffic detection have problems such as low detection rate and high false alarm rate, so this paper proposes a two-branch neural network based on BiGRU network as the backbone. The model uses the dual-branch and BiGRU networks to extract and analyze the spatial and temporal features of the data, and finally discriminates the abnormal traffic using a Softmax classifier. Experimental results on the ISCX-IDS-2012 dataset and CIC-IDS-2017 dataset show that the model has a low false alarm rate and high accuracy rate, which is better than existing methods (This work was supported by the Research and development project of China Electric Power Research Institute, (NO. 5242002000QT)).

Keywords: Abnormal traffic detection · Deep Learning · Neural network

1 Introduction

With the advancement of technology, the Internet has developed rapidly and has become an indispensable infrastructure in people's lives, with wide applications in work, education and other areas. However, enjoying the convenience of the Internet also brings security problems. In order to effectively deal with security threats on the Internet, effective dynamic management of network traffic and abnormal traffic monitoring is required. Therefore, as a very important defense tool, abnormal network traffic detection is becoming more and more important in network security protection.

The main current detection methods are divided into four categories: payload feature-based detection techniques, traffic feature-based detection methods, statistical-based detection methods, with machine learning-based detection methods [1] The first three traditional detection methods each have certain limitations: it is difficult to handle encrypted traffic, difficult to cope with unknown anomalous traffic, and difficult to accurately determine the type of anomalous traffic. Therefore, the traditional security defense system based on feature detection and statistical rules has been difficult to meet modern network security needs. And deep learning-based anomalous network traffic detection technology came into being.

In this paper, we propose a new neural network-based anomalous traffic detection method with BiGRU network as the backbone, in which the neural network does not need to manually identify features in anomalous traffic detection. In order to minimize the missing feature information due to pooling operation, this paper uses maximum pooling and global pooling fusion to compensate for the missing feature information. In addition, the model in this paper uses a two-branch network to improve model performance.

2 Related Work

2.1 Abnormal Traffic Detection

Abnormal traffic refers to the abnormal data flow that occurs in the network. It can be caused by hacker attacks, virus infections, malware, misuse, and other factors, thus posing a threat to the security and stability of the network. Network traffic anomalies can be classified into many different types, some of the common ones include: Attack traffic: This type of traffic may contain malicious codes that cause harm to the network system, for example: Denial of Service Attack (DoS): This is a malicious attack that makes the network or service unavailable by sending a large number of requests or data streams. Distributed Denial of Service (DDoS): This is an attack that uses multiple devices to send attack traffic at the same time to pose a greater threat. Network Spoofing (Spoofing): This is an attack in which a malicious attacker controls network traffic by spoofing network devices. Scanning attack: This is an attack in which a malicious attacker attempts to gather information or find vulnerabilities by scanning the network. Worm: This refers to malware that spreads automatically through the network.

Spam traffic: This type of traffic may contain large amounts of spam, junk messages, etc., which take up network bandwidth and reduce the efficiency of the network. Abnormal traffic: This type of traffic may be caused by network equipment failure, misoperation, etc. And does not conform to the normal traffic patterns in the network, e.g., network paralysis, network data leakage, etc. Therefore, abnormal traffic detection is of great importance in the field of network security. By detecting and identifying abnormal traffic, network administrators can take appropriate protection measures to prevent the network from security threats and protect the integrity of the network system.

2.2 Deep Learning

Deep learning algorithms use multilayer network models to learn the features of data to identify anomalous traffic more accurately based on large amount of network traffic data. Zhang Y et al. [2] proposed a parallel cross-convolutional neural network (PCCN) algorithm that improves the detection performance of unbalanced anomalous flows by fusing the learned features of two branching convolutional neural networks (CNNs) using deep learning techniques. Kim TY et al. [3] proposed the C-LSTM model, which can effectively model the spatial and temporal information in flow data. The model integrates convolutional neural network (CNN), long short-term memory (LSTM), and deep neural network (DNN) to improve the anomaly detection performance on flow data. Wei et al. [4] proposed a deep learning-based hierarchical spatio-temporal feature learning (HAST-NAD) network anomaly detection method by fusing convolutional neural networks (CNN) with recurrent neural networks (RNN) for the purpose of automatically learning traffic features and improving the efficiency of network traffic anomaly detection, addressing the problem of feature dependency and high false alarm rate in network anomaly detection. Koh et al. proposed a deep learning-based self-encoder-based deep learning model to identify anomalies in network traffic by unsupervised learning of the data. Dong [1] proposed a semi-supervised dual-depth network-based anomaly detection method, which can be detected by identifying features in network packets. In addition, Li [5] et al. proposed a new anomaly detection method based on BiIndRNN network, which has greater advantages in parallel computation and gradient adjustment compared to previous recurrent neural networks (RNN). The model extracts bidirectional structural features of network traffic by positive and negative inputs and is able to capture spatial influences in the data stream. To establish the dependencies on the positive and negative directions of network traffic, the study also proposes a BiIndRNN model incorporating global attention (GA) to focus more on moments containing important information. Experiments are conducted on the UNSW-NB15 dataset to perform GA representation of packet feature vectors of the network, feature fusion, and loss computation of multilayer fully connected layers. The experimental results show that the GA-BiIndRNN model converges faster, with accuracy, precision and F1 scores exceeding 99%, and a false positive rate (FPR) close to 0.36%, compared to traditional deep and shallow machine learning and other state-of-the-art techniques, and is able to effectively identify normal and malicious network activities. These results provide a theoretical basis for rapid implementation of protection measures. Marir [6] et al. proposed a new distributed method for detecting anomalous behavior in large-scale networks. The model finds anomalous behaviors from large-scale network traffic data in a distributed environment through a combination of deep feature extraction and multilayer integrated support vector machines (SVMs). First, nonlinear dimensionality reduction is performed on large-scale network traffic data by distributed deep belief networks, and then the obtained features are fed into multilayer integrated SVMs. The construction of the integration is done by an iterative reduction paradigm based on Spark. Zhong [7] et al. proposed a new framework for network traffic anomaly detection to address three problems of existing machine learning anomaly detection algorithms: data obsolescence, inability to relearn new models, and inapplicability of a single detection algorithm to complex network attack environments. The framework is implemented through the organic integration of multiple deep learning techniques, extracting features by the damped incremental statistical algorithm, labeling anomaly scores by self-encoder, then training by LSTM, and finally using a weighting method to obtain the final anomaly scores. Experimental results show that this HELAD algorithm has better adaptability and accuracy than other existing algorithms. Chen [8] proposed a network traffic classification model called ArcMargin, which incorporates metric learning into convolutional neural networks (CNNs) to make CNN models more discriminative. ArcMargin maps network traffic samples of the same category more closely, while samples of different categories are mapped as far away as possible. The metric learning regularization feature is called additive angular edge loss and is embedded in the objective function of the traditional CNN model. The ArcMargin model is validated on three datasets and compared with other related algorithms. Based on a set of classification metrics, it is demonstrated that the ArcMargin model performs better in both network traffic classification tasks and open-ended tasks. In addition, in the open-ended task, the ArcMargin model can cluster unknown data classes that do not exist in the previous training dataset.

3   Method

In this section, we design a new neural network model. The model incorporates a Bidirectional GRU [9] network to improve the detection performance of anomalous traffic data. Because the dataset is composed of multiple bytes, the continuous features of anomalous traffic are more suitable to be processed using a neural network with processing and preservation. The model consists of spatial features, temporal features, bilinear fusion, and soft-max classifier. These four modules form the end-to-end model, with two parallel networks as the backbone, namely the large and small receptive field layers. The basic flow of the model is converted the packets into image form, use multilayer 2-dimensional convolution to feature the output image features, fully analyze the temporal features of the packets after two stages of forward and backward GRU networks, then use bilinear pooling to fuse the features, and finally use a dichotomous classifier to learn the temporal features and perform the detection of abnormal traffic. The two modules, spatial features and temporal features, are described in detail below. The overall network model is shown in Fig. 1.

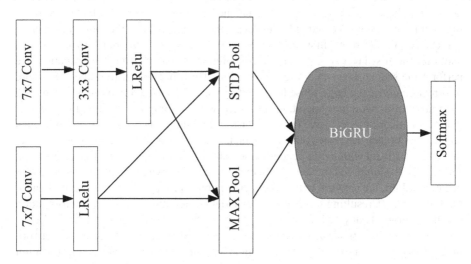

Fig. 1. Network structure

3.1   Data Pre-processing

Not all data are credible data, and untrustworthy data will eventually lead to bias in the model representation. In addition, preprocessing the data can better express the characteristics of the traffic, so this paper preprocesses the data set before training the model. A data package exists with different continuous features, and in order to unify these features in a fixed range of values, this paper uses data normalization. The data normalization formula is as follows, where x is the sample feature value, $\max(x)$ denotes the maximum value in the sample, and $\min(x)$ denotes the minimum value in the sample. From Eq. 1 the data can be normalized and mapped to the range [0, 1]. The normalization of the data set improves the consistency of the data.

$$y = \frac{x - \min(x)}{\max(x) - \min(x)} \tag{1}$$

The pre-processed data is in the form of network streams, which are further processed into one-hot data form. One-hot encoding separates each category into separate features. In one-hot encoding form, each traffic byte is encoded as an n-dimensional vector. Zhang [10] et al. verified the performance effectiveness of different encodings in deep learning and determined that One-hot encoding gives a more stable performance.

3.2   Spatial Feature Extraction

Spatial pooling is essentially a feature extraction. The size of the convolutional layer in the model is denoted as $w \times h \times c$, where w, h, c are the width, height, and a number of channels of the convolutional kernel, respectively. The size of the convolution kernel determines the size of the receptive field, and the larger the receptive field range the more image information is received [11] The larger the receptive field, the more information is received [11]. Here we first set the convolution kernel of Top branch to $7 \times 7 \times 1$, and after the first layer of 7×7 large convolution to extract feature information, we perform the second layer of 3×3 convolution. This operation is to refine the feature information under the large sense field, and a layer of maximum pooling is added after each layer of convolution operation. In recent years, to improve the ability to extract features, network models have been designed with dual branches [12, 13] Top Branch and Bottom branch can effectively complement each other in feature extraction. To make the network model with better generalization ability, we use the Leaky ReLu activation function [14]. Compared to the ReLu activation function, when the data is close to 0 or negative values, the network will cannot perform back propagation when the gradient of the function is 0, resulting in the network model not learning the relevant information. On the contrary, Leaky ReLu solves the above problem. On the other hand, the Leaky ReLu activation function can map the data to a specified range, which makes the network model less prone to the divergence problem in learning [15]. The Leaky ReLu activation function is as follows.

$$\text{Leaky Relu} = \max(0, x) + \text{leak} * \min(0, x) \tag{2}$$

Where Leak is a constant and x is the input data feature. The specific steps of spatial feature learning are as follows.

Spatial feature learning

Input: Network traffic converted into image form

Step 1: Top Branch

    1. The first layer is a 7×7 convolutional layer with 50 channels, after Leaky ReLu activation function, add a layer of maximum pooling layer, average pooling layer

    2. The second layer is a 3×3 convolutional layer with 50 channels, after Leaky ReLu activation function, add a layer of maximum pooling layer, average pooling layer

Step 2: Bottom Branch

    1. The first layer is a 7×7 convolutional layer with 50 channels, after Leaky ReLu activation function, add a maximum pooling layer, average pooling layer

Convolution: After the convolution kernel operates on the image, a new feature is an output, and the feature is represented by the following equation C. where w is the convolution kernel, x denotes the image information, b denotes the bias, and f() denotes the Leaky ReLu activation function.

$$c = f(w * x + b) \tag{3}$$

Pooling: The maximum pooling operation is performed on the features to get the maximum value in the feature area.

$$C = Max(c)(i = 1 \ldots \ldots n) \tag{4}$$

3.3 Timing Feature Extraction

The two-way GRU network performs further temporal learning on the features passed from the upper layer. The two-way GRU network is shown in Fig. 2. GRU is a variant of RNN and has a simpler structure compared to other RNN networks. Each GRU unit consists of an update gate and a reset gate. xt is the video frame feature input at t time step, h_{t-} is the hidden layer information at $t - 1$ time step, and ht is the hidden layer information at t time step. $\tilde{h}_{t-1}$ is the candidate hidden state at the t time step, which is used to calculate the hidden layer information for the next frame. The BiGRU network is calculated as follows

$$r_t = \sigma \left(w_r * \left[h_{t-1}, x_t \right] \right) \tag{5}$$

$$z_t = \sigma \left(w_z * \left[h_{t-1}, x_t \right] \right) \tag{6}$$

$$h_t = \tanh \left(w * \left[r_t * h_{t-1}, x_t \right] \right) \tag{7}$$

$$\tilde{h}_t = (1 - z_t) * h_{t-1} + z_t * h_t \tag{8}$$

Where w_z, w_r are the input weights of the previous moment and the current moment, respectively; z_t, z_r are the state information and reset information of t time step, respectively σ is the sigmoid function. The timing feature x is processed by the update gate and reset gate, respectively. Meanwhile, if the reset gate information is close to 0, then the hidden information of the previous frame is discarded, if the reset gate information is close to 1, the hidden information of the previous frame is kept, and the result of multiplying the information of the previously hidden layer with the reset gate result is connected with the input, and then the candidate hidden information is calculated by the activation function. At this point, the hidden information of the previous frame already contains all the historical information up to t time step. Finally, the update gate of t time step combines the hidden information h_{t-1} of the previous time step and the candidate hidden information of the current time step.

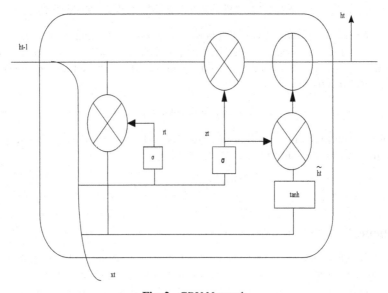

Fig. 2. GRU Network

The detailed steps of the time-series feature learning are as follows

Time-series feature learning
Input: Spatial features
Step 1: Top Branch
1.   Adding a bidirectional GRU network layer to the upper layer features
Step 2: Bottom Branch
1.   Adding a bidirectional GRU network layer to the upper layer features
Step 3: Bilinear pooling
1.   The features of the two branches processed by the bidirectional GRU network layer are fused for maximum pooling and output to a fully connected layer
Step 4: Training the model
Calculate the binary cross-entropy loss function: -log(yt|yp)=-(ytlog(yp)+(1-yt)log(1-yp))
Adam Optimizer Optimization Model

3.4 Soft-Max Classifier

At the end of the model, we use the Soft-max classifier to output the final classification results. Among the commonly used classifiers Soft-max classifier can widen the gap in confidence scores and make the classification effect more obvious. Soft-max classifier is formulated as follows, where xi is the output probability of the i th category and the sum of the exponential functions is the probability value of the i th category calculated. The total sum of all the category probabilities j is 1.

$$softmax(X_i) = \frac{\exp(x_i)}{\sum_{i=1}^{j} \exp(x_i)} \tag{9}$$

4 Experiment

In this section, we first present two existing datasets and then compare the proposed model to the ISCX-IDS-2012 [16] dataset and CIC-IDS-2017 [17] and performed ablation experiments, and finally analyze the experimental results derived from the model.

4.1 Dataset

The ISCX-IDS-2012 dataset is an intrusion detection dataset released by the University of New Brunswick, Canada, which contains normal and abnormal traffic for 7 days. The abnormal traffic is (Brute Force SSH, Http Dos, DDos, Infiltrating) respectively. The CIC-IDS-2017 dataset contains normal traffic and common attacks with a total of 1168671 stream data. The data time starts at 9:00 am on July 3, 2017, and ends at 5:00 pm on July 7, 2017. Among the abnormal traffic include: FTP-Parator, SSH-Parator, Hearbleed, Web Attack, Infiltration, PortScans, Dos Hulk, Dos Slowhttp, Dos Slowloris, DDos, BotNet, Goldeneye, a total of 12 types, as shown in Table 1. These two datasets are similar in distribution, so this paper uses 80% as the training set and 20% as the test set.

4.2 Evaluation Metrics and Model Training

To analyze the anomaly testing performance of the proposed model, we use overall accuracy (OA-ACC), detection rate (DR), and false alarm rate (FAR). The specific formulas are as follows

$$OA - ACC = \frac{\sum_{i=1}^{K} TP_i}{N} \tag{10}$$

$$DR_i = \frac{TP_i}{TP_i + FN_i} \tag{11}$$

$$FAR = \frac{FP_i}{TN_i + FP_i} \tag{12}$$

Table 1. ISCX-IDS dataset detail

Type	Number	Percentage
FTP-Parator	19941	1.71%
SSH-Parator	27545	2.36%
Hearbleed	9859	0.84%
Web Attack	10537	0.90%
Infiltration	5330	0.46%
BotNet	2075	0.18%
PortScan	319636	27.35%
Dos Hulk	474656	40.62%
Dos Slowloris	10537	0.90%
Dos Slowhttp	6789	0.58%
DDos	261226	22.35%
Goldeneye	20543	1.76%

Where TP_i denotes the number of correctly identified target traffic, TN_i denotes the number of correctly detected non-class i traffic, FP_i denotes the number of incorrectly detected non-class i target traffic, and FN_i denotes the number of incorrectly detected class i traffic.

The hyperparameters of the training model are epoch, learning rate and Epoch: the network is trained by one epoch for the whole training set, if the epoch is set too small, the network model may not have enough data to train well. If the epoch is set too large, it may cause the overfitting problem of the network model and make the network model deviate in the test set. Here we set the epoch to 40, and the learning rate is the degree of convergence of the loss function to the minimum value. When the learning rate is set too large, it tends to cause the model to oscillate around the gradient minimum. If the learning rate is set too small, the model will converge slowly. In this paper, the learning rate is set to 0.001. For the determination of the hyperparameters, see the ablation experiment section. For hardware, the CPU I7-13700K is used and the memory is 32G. The graphics card is NVIDIA RTX3080 with 12G of video memory. The model's deep learning framework is Pytorch [18], running on a Win10 64-bit system.

5   Experimental Results and Analysis

In order to verify the performance of the model in this paper, the model is compared with existing methods. Table 2 shows the comparison results. It can be seen that this paper's method achieves better results in DR metrics and OA-ACC metrics, 99.8% and 99.5%, respectively. Compared with the AD-H1CD method, this paper's method is 4.3% higher in the DR metrics and equal in the OA-ACC metrics. Compared with the KMGNBC method, it was 0.1% higher in the DR metrics and 4.1% higher in the OA-ACC metrics.

Overall, the method in this paper achieved relatively good results in all three evaluation indexes, ranking first among the five methods, and most of the indexes were higher than those in recent years. As Table 3 shows, the method in this paper achieves relatively high results in all three metrics of the four traffic types, among which the OA-ACC metrics reach 99.68% and 99.72% on DDos type and HttpDos type, respectively. In addition, the model in this paper achieves a good result of 99.65% on the OA-ACC metric, but on the FAR metric, the result is less satisfactory, reaching 0.03%.

Table 4 shows the comparison of the methods on the CIC-IDS-2017 dataset. In the DR metric, the method in this paper is 1.1% higher than the second-ranked method, and the effect is not significant in the FAR metric, which is due to the low performance of the FAR metric because the CIC-IDS-2017 dataset is not uniform in features, and thus the network model has difficulty in obtaining statistical features. In addition, it is observed that the OA-ACC metric reaches 99.6%, which is an encouraging result. Overall, the method in this paper has good performance and is highly competitive.

Table 2. ISCX2012 dataset experimental results comparison

Method	DR	FAR	OA-ACC
ALL-AGL [19]	93.2%	0.30	95.4%
MHCVF [20]	68.2%	0.03	99.5%
KMGNBC [21]	99.7%	0.30	95.4%
AMG2NB [22]	94.7%	7.0	94.5%
AD- H1CD [23]	95.5%	0.02	99.3%
Our	99.8%	0.03	99.5%

Table 3. ISCX2012 dataset DR, FAR, ACC metrics performance

Class	DR	FAR	OA-ACC
DDos	83.96%	0.02	99.68%
BFSSH	96.71%	0.01	99.52%
Infiltrating	91.70%	0.01	99.69%
HttpDos	92.53%	0.02	99.72%
Overall	90.78%	0.03	99.65%

Table 4. CIC-IDS-2017 dataset experimental results comparison

Method	DR	FAR	OA-ACC
ALL-AGL	92.0%	0.20	90.4%
MHCVF	73.8%	0.01	98.7%
KMGNBC	94.5%	0.81	91.4%
AMG2NB	90.7%	0.18	94.5%
AD- H1CD	93.2%	0.02	99.5%
Our	95.6%	0.04	99.6%

6 Ablation Experiments

We did ablation experiments to verify the effect of double branching and hyperparameters on the model performance.

6.1 Dual Branch Comparison

In this section, we have compared the performance of the Top branch, the Bottom branch and the Double branch respectively. As shown in Fig. 3. The OA-ACC metrics of the Top branch performance is around 90%, and the OA-ACC index of the Bottom branch is the lowest, which is because the Top branch has a finer feature extraction ability on the data compared to the Bottom branch. The Dual branch combines the advantages of the Top branch and the Bottom branch, and the dual branch complements the lack of performance of a single branch.

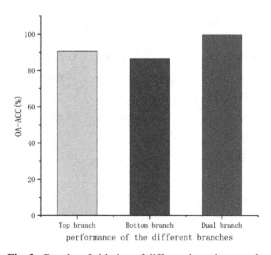

Fig. 3. Results of ablation of different branch networks

6.2 Hyperparameter Epoch

As Fig. 4 shown, this paper investigates the performance of OA-ACC metrics in epochs from 0 to 40, and it can be seen that the OA-ACC metrics grow gradually as epochs increase. When the epoch reaches 20, the OA-ACC growth gradually smooths out. In addition, the training curve and the validation curve are gradually fitted, and it can be concluded that no overfitting phenomenon occurs. As Fig. 5 shown in the figure, with the increase of epoch, the Loss function gradually decreases, and when epoch reaches about 30, the Loss function remains stable with a value less than 0.02, and with the increase of epoch the Loss oscillates around 0.02. Based on the above comprehensive experimental determination, epoch is chosen to be 40.

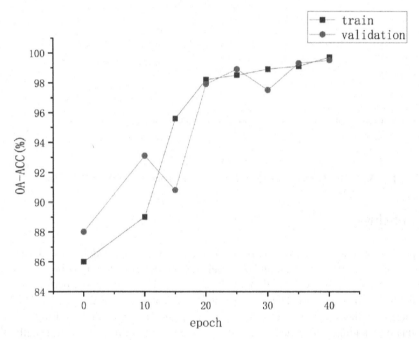

Fig. 4. Comparison of OA-ACC metrics for different epochs on ISCX2012 dataset

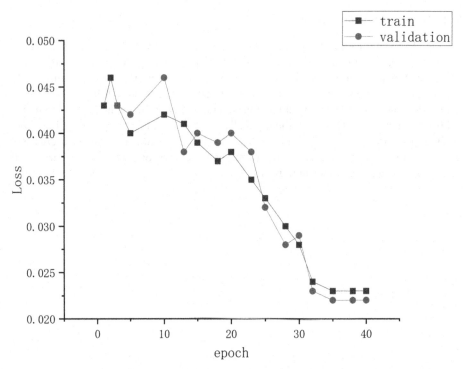

Fig. 5. Comparison of loss function for different epochs on ISCX2012 dataset

7 Conclusion

We propose a new dual-flow neural network-based network model to extract spatial features and temporal features, and the dual-flow branches of the model can compensate for the shortcomings of single branches. The model improves the performance of anomalous traffic detection. The experimental results on CIC-IDS-2017 and ISCX2012 datasets show that the model in this paper has individual anomalous traffic detection and has outstanding performance for overall anomalous traffic detection. Finally, the performance of different branches for anomalous traffic detection and the impact of hyperparameters on the performance of the model is verified.

In future work, we will try to use novel network models to improve anomalous traffic detection performance and propose more complete anomalous traffic datasets, which are important for network security.

References

1. Dong, S., Xia, Y., Peng, T.: Network abnormal traffic detection model based on semi-supervised deep reinforcement learning. IEEE Trans. Netw. Serv. Manage. **18**(4), 4197–4212 (2021)

2. Zhang, Y., Chen, X., Guo, D., et al.: PCCN: parallel cross convolutional neural network for abnormal network traffic flows detection in multi-class imbalanced network traffic flows. IEEE Access **7**, 119904–119916 (2019)

3. Kim, T.Y., Cho, S.B.: Web traffic anomaly detection using C-LSTM neural networks. Exp. Syst. Appl. **106**, 66–76 (2018)

4. Wei, G., Wang, Z.: Adoption and realization of deep learning in network traffic anomaly detection device design. Soft. Comput. **25**(2), 1147–1158 (2021)

5. Li, H., Ge, H., Yang, H., et al.: An abnormal traffic detection model combined BiIndRNN with global attention. IEEE Access **10**, 30899–30912 (2022)

6. Marir, N., Wang, H., Feng, G., et al.: Distributed abnormal behavior detection approach based on deep belief network and ensemble SVM using spark. IEEE Access **6**, 59657–59671 (2018)

7. Zhong, Y., Chen, W., Wang, Z., et al.: HELAD: a novel network anomaly detection model based on heterogeneous ensemble learning. Comput. Netw. **169**, 107049 (2020)

8. Chen, M., Wang, X., He, M., et al.: A network traffic classification model based on metric learning. CMC Comput. Mater. Continua **64**(2), 941–959 (2020)

9. Dey, R., Salem, F.M.: Gate-variants of gated recurrent unit (GRU) neural networks. In: 2017 IEEE 60th International Midwest Symposium on Circuits and Systems (MWSCAS), pp. 1597–1600. IEEE (2017)

10. Zhang, X., LeCun, Y.: Which encoding is the best for text classification in Chinese, English, Japanese and Korean? arXiv preprint arXiv:1708.02657 (2017)

11. Luo, W., Li, Y., Urtasun, R., et al.: Understanding the effective receptive field in deep convolutional neural networks. In: Advances in Neural Information Processing Systems 29 (2016)

12. Gao, F., Huang, T., Wang, J., et al.: Dual-branch deep convolution neural network for polarimetric SAR image classification. Appl. Sci. **7**(5), 447 (2017)

13. Goel, N., Kaur, S., Bala, R.: Dual branch convolutional neural network for copy move forgery detection. IET Image Proc. **15**(3), 656–665 (2021)

14. Xu, B., Wang, N., Chen, T., et al.: Empirical evaluation of rectified activations in convolutional network. arXiv preprint arXiv:1505.00853 (2015)

15. Xu, J., Li, Z., Du, B., et al.: Reluplex made more practical: leaky ReLU. In: 2020 IEEE Symposium on Computers and Communications (ISCC), pp. 1–7. IEEE (2020)

16. Shiravi, S., Shiravi, A., Shiravi, H., Tavallaee, M., Ghorbani, A.A.: Toward developing a systematic approach to generate benchmark datasets for intrusion detection. Comput. Secur. **31**(3), 357–374 (2012)

17. Sharafaldin, I., Lashkari, A.H., Ghorbani, A.A.: Toward generating a new intrusion detection dataset and intrusion traffic characterization. ICISSp **1**, 108–116 (2018)

18. Imambi, S., Prakash, K.B., Kanagachidambaresan, G.R.: PyTorch. In: Prakash, K.B., Kanagachidambaresan, G.R. (eds.) Programming with TensorFlow. EICC, pp. 87–104. Springer, Cham (2021). https://doi.org/10.1007/978-3-030-57077-4_10

19. Sallay, H., Ammar, A., Saad, M.B., et al.: A real time adaptive intrusion detection alert classifier for high speed networks. In: 2013 IEEE 12th International Symposium on Network Computing and Applications, pp. 73–80. IEEE (2013)

20. Hacibeyoğlu, M., Karlik, B.: Design of multilevel hybrid classifier with variant feature sets for intrusion detection system. IEICE Trans. Inf. Syst. **99**(7), 1810–1821 (2016)

21. Yassin, W., et al.: Anomaly-based intrusion detection through k-means clustering and Naives Bayes classification, pp. 298–303 (2013)

22. Tan, Z., Jamdagni, A., He, X., et al.: Detection of denial-of-service attacks based on computer vision techniques. IEEE Trans. Comput. **64**(9), 2519–2533 (2014)

23. Ma, C., Du, X., Cao, L.: Analysis of multi-types of flow features based on hybrid neural network for improving network anomaly detection. IEEE Access **7**, 148363–148380 (2019)

A Fabric Defect Detection Model Based on Feature Extraction of Weak Sample Scene

Maosen Wang[1], Mengtian Wang[1(✉)], Jun Liu[1], Shaozhang Niu[1], Wen Zhang[1], and Jiaqi Zhao[2]

[1] Southeast Digital Economic Development Institute, Quzhou 324100, China
19845175463@163.com
[2] Quzhou College of Technology, Quzhou 324100, China

Abstract. Defect detection of fabrics is a necessary step for quality control in industries related to fabrics such as clothing and tents. Traditional fabric inspection relies on visual inspection, which is inefficient and inaccurate. The abnormal detection model based on PatchCore is based on the feature extraction of a pre-trained model on the general ImageNet large data set and performs well in industrial abnormal detection tasks. However, it is difficult to adapt to the noise problem of factory fabrics and scene adaptability. Therefore, in this paper, feature extraction is scenarized, and a small amount of real fabric data is used to fine-tune the pre-trained feature extraction network guided by object detection. This allows it to adapt to real industrial fabric abnormal detection scenes, and the scoring function is optimized to improve segmentation accuracy for noise problems. This solves the problem of insufficient fabric defect samples and the speed and accuracy requirements of defect detection in industrial scenes. The deployment and testing in the factory have effectively solved the problem of fabric detection in industrial scenes.

Keywords: Fabric Defect Detection · Feature Extraction · Unsupervised

1 Introduction

Fabric has always been one of the most closely related products to human life, our daily clothing, towels, sheets, and tents, most of them are directly or indirectly produced by fabrics as raw materials, and fabric production is currently one of the most important industries in the world. In the process of fabric production, due to the generation of equipment or human factors, the produced fabric may have a variety of defects, and the textile industry has defined more than 70 kinds of fabric defects [1]. This article focuses on two types of defects in cloth: dyeing defects and structural defects. If the defect shown in Fig. 1(a) is a dyeing defect, such defect is a defect in the appearance of the cloth due to uneven dyeing or dirt on the cloth during the production process. The defects shown in Fig. 1(e) are structural defects, which are problems such as missing threads and holes in the fabric production process due to weaving reasons or the drawing of the fabric and the production machine. Structural defects are more damaging to the quality of the fabric and more difficult to detect than dyeing defects.

Y. Tan et al. (Eds.): ICSI 2023, LNCS 13969, pp. 246–259, 2023.
https://doi.org/10.1007/978-3-031-36625-3_20

Defect detection is an indispensable part of the fabric production process, in reality, most of the company's fabric defect detection work is mainly completed manually, limited by human eye fatigue and lack of concentration, and other factors, manual inspection can hardly provide reliable and stable results [2], In addition, these results are often subjective and cannot be quantified. How to replace manual labor, relying on machines to achieve robust and accurate automatic cloth defect detection is of great significance.

By setting up an industrial camera on the cloth inspection machine instead of manual labor to detect the cloth, the image processing algorithm is used to process and identify the surface defects of the cloth and whether there are defects.

Vision-based inspection systems typically consist of three steps: image acquisition, defect detection, and post-processing. The image acquisition program is mainly responsible for the digital image capture of the defect sample, the defect detection program is used to locate and segment the defect area, and post-processing refers to the subsequent processing of the image after the defect detection, such as defect type classification and defect level evaluation. In this article, we mainly focus on the research of key technologies for defect detection.

There are three main challenges in the defect detection task, first, there are many types of defects, as shown in Fig. 1, Stain, Pleats, Coating, Knot, Hole, and Coarse yarn, etc., the same defects on different fabrics, due to different patterns and textures also show different characteristics, the diversity of fabrics and defects increases the difficulty of defect detection. Second, the same imperfections appear differently due to different lighting and fabrics. Third, in industrial scenarios, it is extremely difficult to collect a large number of fabric defect samples, and data imbalance makes model training difficult.

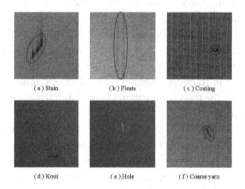

Fig. 1. Different types of defects in fabric.

Based on the above task challenges, a new unsupervised defect detection network is proposed, and the normal samples are used to select the network, and the following work is done. 1. For the feature extraction network, transfer learning [3] uses the pretrained feature extraction network based on ImageNet and uses the labeled cloth scene dataset to embed it into the target detection scene to optimize the feature representation ability; 2. Aiming at the problem of weak noise immunity of abnormal detection, corresponding post-processing is designed to improve the accuracy under the condition of ensuring recall; 3. Lightweight compression optimization of network parameters to improve speed under balanced accuracy. The network structure is divided into three modules: image feature extraction based on CNN [4], anomaly score calculation and defect region segmentation module, and lightweight compression module of the network.

2 Related Work

In recent years, there have been many studies on visual cloth defect detection algorithms, which are generally divided into two categories according to image processing methods: traditional image processing methods and image processing methods based on deep learning.

2.1 Traditional Cloth Defect Detection

Traditional fabric detection methods are generally based on image processing methods, through artificial construction of defect features, the fabric defects are detected from the image, traditional fabric defect detection methods mainly have two categories: image-based spatial features and frequency-domain-based methods.

At present, there are many different feature options for fabric defect detection, and the grayscale co-occurrence matrix [5] (GLCM), local binary mode [6] (LBP) feature, and difference histogram [7] (SDH) feature are used as features, and fractal dimensions are used as features. Among them, GLCM is used to describe the joint distribution of all pixel grayscales separated by a certain spatial relationship in the image, which can be used to extract recurring texture information in the image, and LBP is used to count various edge features in the image, such as straight edges and corners. The sum histogram (SDH) feature is calculated in a similar way to GLCM, but only the gray and difference distributions of all pixels in the image that are separated by a certain spatial relationship are counted. The idea of fractal dimensions as features is to use fractal geometry methods to describe cloth images in fractional dimensions. After obtaining the fractal dimension of the image, the value of the fractal dimension is used as a feature to distinguish whether the image contains defects.

The method based on the frequency domain is to use the frequency domain transformation of the image to achieve defect detection because most of the fabric contains rich texture features, these textures show certain frequency characteristics, and the position of the fabric defect has mutations in the texture of the image, and the normal texture frequency characteristics of the fabric will be very different, the use of the frequency domain characteristics of the fabric defect texture to achieve the defect detection of the cloth. Frequency-domain detection transformation methods include fast Fourier transform, discrete cosine transform, wavelet transform, and Gabor transform [8]. Since the defects of the fabric exhibit high-frequency characteristics in the frequency domain, the fast Fourier transform and discrete cosine transform are used to detect the defect information using a high-pass filter after the image is transformed. Because wavelet transform and Gabor transform can obtain frequency position information, widely used in cloth defect detection, the defect information of the image is usually included in these high-frequency components, wavelet transform decomposes the image in the horizontal and vertical directions, calculates the high-frequency components of the image horizontal, vertical and diagonal, and small through wave transformation can achieve defect detection. Gabor transform and wavelet transform can extract high-frequency components in any direction and can also achieve defect detection.

2.2 Fabric Defect Detection Based on Deep Learning

CNN (Convolutional Neural Network) is a very important and widely used branch of DNN (deep neural network), CNN since it was proposed, because of its strong feature extraction ability and can greatly reduce the number of parameters, it is widely used in image feature extraction, and in a variety of image recognition applications to achieve far better than traditional feature extraction methods. The existing methods can be roughly divided into the following categories: methods based on distance measurement, methods based on classification polygon construction, and methods based on image reconstruction.

Liu C et al. [9] proposed that Deep SVDD) (Deep support vector data description) is a common method based on distance measurement, which artificially specifies a point as the feature center in the feature space, and then uses the sum of the distance from the normal sample to the point as the main body of the loss function to train the feature extraction network. The trained network can map the normal samples in the original image space near the center point of the feature, while the features corresponding to the abnormal sample may be far from the center point, so the sample to be measured can be judged whether the sample to be measured is an abnormal sample according to the distance. Based on the classification surface construction method, the core idea is to convert a single class normal sample into a multi-class sample to train the classifier, so as to construct a classification surface in image space to realize the classification of normal samples and potentially abnormal samples.

Liang S et al. proposed ODD (Out-of-distribution) [10] based on the dim sum method constructed on the classification surface, the original single-class samples are obtained by geometric transformation, and the confidence method is used to carry out anomaly detection, and the classification model may also produce high probability values for abnormal samples, which affects the process of anomaly detection, which is also a content worthy of further research and improvement. Oza et al. trains a classification network by extracting the features of normal images as positive samples by using the pre-trained neural network as positive samples, and using random Gaussian noise vectors centered on the origin as negative samples in the feature space. However, this method only uses Gaussian noise as negative samples, which are relatively single and easy to classify, and it is easy to overfit and cause the network to fail to detect new abnormal samples.

Based on the image reconstruction method, the core idea is to encode and decode the normal image of the input, train the neural network to learn the distribution pattern of the normal image with the goal of reconstructing the input, and then perform anomaly detection by analyzing the difference between the images before and after reconstruction in the detection stage. According to the training mode, the method based on image reconstruction includes two types: autoencoder-based [11] and generative adversarial networks [12] (GAN).

Based on the most commonly used network structure of autoencoder (AE), the autoencoder trained by normal samples can reconstruct normal images well in the test stage, and for abnormal images, there will be a large difference from normal images in the image encoding and subsequent reconstruction process, and the size of the difference is an indicator to measure the abnormality of the sample to be measured.

The method of GAN reconstruction takes into account that the original GAN only creates a mapping relationship from hidden space to image space, so it adopts iterative optimization to obtain the reconstructed image. Fujioka et al. [13] proposed AnoGAN (Anomaly detection with a generative adversarial network), an anomaly detection with a generative adversarial network, starts from a random variable, calculates the difference between the image generated by the variable and the image to be measured, and iteratively optimizes the random variable through gradient descent so that the generated image gradually approaches the image to be measured. Since the generator only uses normal samples for training, only normal samples can theoretically be generated. When there are anomalous areas in the image to be measured, the generator generates an image that is as close to it as possible but belongs to the normal category as a reference, and anomaly detection is performed by calculating the difference between the image to be measured and the generated image.

3   Methods

Improving the speed and accuracy of fabric defect detection is an important goal of defect detection task, and the overall detailed detection of 5 million pixel shooting images takes a lot of time, while the overall rough detection is difficult to meet the area positioning of small defects such as 0.05 mm defects. In order to achieve this, defect detection is divided into two stages: initial screening of defect areas and fine-grained detection of defects. The first stage of the initial screening model ensures high recall and speed requirements, through a lightweight classification model, sliding windows into small figures for classification, quickly screening out possible defect areas from the whole cloth map, and training a model with high recall and low accuracy by synthesizing defect data to roughly locate defect areas; The second stage of defect fine-grained detection model, based on pixel-level defect area detection, although the image processing time is relatively long, the recall and accuracy can meet the requirements, the proposed method is mainly optimized for the second stage algorithm.

3.1   Structure

Figure 2 is the overall structure that concludes the feature extraction module based on object detection, the training process based on the patch feature pool referring to PatchCore, the loss function, and the anomaly score calculation module to generate the result. The image feature extraction module based on CNN is mainly used for feature extraction of flawless dataset samples and is used for feature comparison in the defect detection stage by collecting a feature pool with comprehensive information and a small amount of data. The pre-trained CNN network structure based on ImageNet, COC.

O, and other large datasets often has good feature extraction network parameters, which can achieve a good model for conventional unsupervised scenarios, but the transfer learning method cannot better meet the high-precision requirements of specific clothes through the feature extraction of the pre-trained network structure of large datasets, so the feature extraction is optimized in the training stage of the object detection architecture.

The anomaly score calculation and defect region segmentation module is used to calculate the anomaly score of the newly entered samples in the inference stage, and finally used for the regression abnormal heat value map for regional segmentation, in order to prevent the noise map from causing the regional segmentation false detection, the score regression is optimized.

The lightweight compression module of the network is used to accelerate the defect detection process and ensure the efficiency of inference time without losing some accuracy.

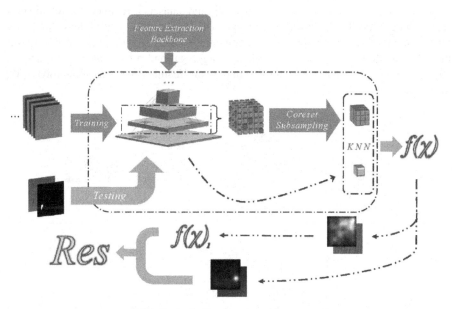

Fig. 2. Overall network architecture.

3.2 Defective Feature Extraction Backbone

In the industrial fabric defect detection scenario, the acquisition of defect sample maps is difficult, the proportion of positive and negative samples is very different, the defect situation has no corresponding law or paradigm, and the randomness is large, so the algorithm in this paper extracts image features based on the flawless picture dataset, supplemented by its own data for optimization. Use X_n to represent all flawless picture collections, X_d means all flawed picture collections, The corresponding label $y_i(i \in X)$ is 0 for flawless and 1 for defective. Use a transfer learning pre-trained network Φ to extract features X_{ij} in X_n. i represents the i-th image, j represents the output features of different feature levels of the pre-trained network, representing the feature maps at different resolutions.

Different convolution levels of CNN can obtain feature maps of different scales, and the feature parameters obtained at different resolutions can better characterize feature perception, and the shallow features are susceptible to noise interference due to the small receptive field. Because the task bias is not conducive to the comparison and differentiation of defective features, feature extraction only extracts middle-level features. Such as 2, 3 level features that are $j = 2$ and $j = 3$. Assuming that the original input map is changed to 224 * 224 * 3 size, taking Wideresnet-50 as an example, after several pooling of the features of layers 2 and 3, the feature map sizes of the 2nd and 3rd layers are 512 * 28 * 28 and 1024 * 14 * 14, respectively. In order to stack features of different scales, refer to [14], use embedded block connection, and connect the features after upsampling and alignment, and the size of the aligned patch is batchsize * 1536 * 28 * 28. If the feature pool obtained by all the flawless graph datasets is retained, the feature pool will be large, and then the inference stage will lead to a large time loss due to the huge amount of data in the pool to be compared, so the greedy algorithm proposed in the reference literature performs sub-feature aggregation on the feature pool, and the large feature pool is characterized by a smaller sub-feature pool to save the time of the inference stage [15].

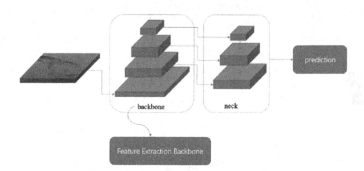

Fig. 3. Backbone through object detection.

In order to better adapt to the cloth data of the factory scene, for the feature extraction module, this paper takes Wideresnet-50 as the feature extraction network backbone, embeds it in the target detection scene, as shown in Fig. 3 of the common network structure of object detection, first the input image is extracted through the backbone, then the neck is used to optimize the feature representation ability such as feature fusion, and finally, the final position regression and category regression are carried out at the head layer. Therefore, this paper conducts simple training oriented to object detection for some labeled data, so as to optimize the feature extraction ability in the training stage, bias part of the feature learning ability for cloth patterns, and use the optimized feature extraction network for pre-order feature extraction.

3.3   Anomaly Score Calculation

The training stage sends the flawless sample training set to the CNN feature extraction network or feature pool, and the inference stage extracts the feature map of the current test sample with the same CNN network. X_t To represent the current test graph, use Φ to extract the same block feature data in the same way, and fuse features of different scales through the same feature aggregation method. The loss function is used to calculate the feature gap between the feature map obtained in this test picture and the feature pool of the flawless sample, and determine whether the whole map is a defective sample according to the size of the loss value of the whole picture. Suppose the aggregate feature pool obtained in the training stage is M, the feature pool obtained by greedy coreset is M_c, the feature pool obtained by greedy coreset is m_t, first compare the feature map of the online stage with the feature pool in a similar KNN way, and obtain the nearest K feature map m_i, $i = 1, ...k$ from the feature pool, and then calculate the anomaly score based on the following formula, where S represents the maximum difference between the adjacent K feature map and the test feature map.

$$S = \max(\|m_t - m_i\|_2), i = 1, 2, ...k \tag{1}$$

After the anomaly score is calculated, the image that meets the threshold S can be considered as a suspicious defect, and if the fixed S is set too large, it may cause less obvious defects to be missed, and if the setting is too small, it may cause many abnormalities caused by noise to be regarded as defects. The normal defect location often shows a proportional relationship between the outlier and the abnormal area in the heat value map generated by the abnormal contrast distance. When the outliers are normalized, if the grayscale map has a large area of anomalies but the anomaly score is very low, it is often based on the abnormal presentation under noise interference. As shown in Fig. 4:

(a) Normal defect.   (b) Noisy defective.

Fig. 4. Origin image (left) and heatmap (right) for normal defect and noisy.

Therefore, in this paper, the abnormal calorific value map and the anomaly score are discriminated twice before the region is segmented. Firstly, an anomalous calorific value map is generated according to the outlier size of the feature block, the size is 28 * 28, and it is upsampled to the original size of the image 224 * 224, and the Gaussian kernel of sigma = 4 is used for filtering, and secondly, in order to deepen the size gap between the abnormal defective area value and the non-anomalous flawless value, the abnormal calorific value map area is subjected to quasi-cross entropy treatment, assuming that the outlier value of a certain location point of the feature map is assumed as, through:

$$xnew = \exp(x) \qquad (2)$$

Amplify the anomaly gap to deepen the difference between different positions, and then obtain the difference pixel size relationship based on min-max regularization, and project the probability value to the grayscale range between 0 and 255.

$$xnew_i = \frac{xnew_i - \min(xnew)}{\max(xnew) - \min(xnew)} \qquad (3)$$

$$gray_i = 255 * xnew_i \qquad (4)$$

Where $gray_i$ indicates the brightness and darkness of the pixel after normalization, represents the specific gray value of the pixel after transformation. The larger the outlier, in principle, the brighter the grayscale plot, and the smaller the outlier, in principle, the darker the grayscale plot. The anomaly map caused by noise interference has the characteristics of large outliers but a low overall score and the grayscale map of abnormal calorific value shows a large bright area but a low score. Therefore, this paper establishes the relationship between grayscale luminance intensity and anomaly score.

$$S_{gray} = \sum_{i=1}^{w*h} gray_i \qquad (5)$$

$$score = S_{gray} - \alpha * (S - r) \qquad (6)$$

Among them, S_{gray} is considered to be the grayscale intensity of the abnormal calorific value grayscale map which is the sum of the grayscale values of each calorific value image prime point to characterize the brightness degree, S which is the sum of the grayscale values of each calorific value image prime point to characterize the brightness degree, r and α is a hyperparameter, which needs to be compared by testing. The obtained greater than 0 is considered to be a defective map, and after filtering by this method, it can better filter out the abnormal situation under the influence of noise, and ensure that more shallow defects are also detected.

3.4 Acceleration

The first-stage defect classification screening model is a high-recall, low-precision model designed to provide rough location information for subsequent high-precision detection models. As this classification model is a simple CNN network structure transfer, it has not been optimized for model accuracy. However, due to the traversal of the entire image, the time for processing large images is directly proportional to that of small images. Therefore, considering the needs of actual factory deployment, this screening model needs to be quantitatively accelerated. In this article, TensorRT is used to quantize the float32 model to int8, with ONNX as the intermediate format, in order to achieve the goal of speeding up inference.

In the detection model part, fabric data captured in a factory environment typically have large sizes, such as 2048 * 2448. As the first-stage screening model has high recall and low precision, the defect images filtered by the screening model are often still large. Therefore, sliding window processing is also used to optimize the processing time. As a compromise, large sliding windows are used for image traversal. The large sliding windows need to be scaled to the target size, such as 224 * 224, which may result in a loss of some details. However, the defect score threshold can be appropriately relaxed. If a defect is detected, small sliding windows are used to restore the original details of the large image area for detection. The sliding window size of this scheme is based on the anchor box clustering selection in object detection, which can bring some speed improvements while maintaining accuracy.

4 Experiments

Dataset: The experimental data set comes from the factory cloth data, divided into untextured cloth and textured cloth, of which 500 test pictures are selected in each of the three categories, and 200 pieces of untextured cloth and 200 original large pictures of textured cloth are selected for the training data, and the sliding window of 224 * 224 size and the data difference is compared, and as many kinds of flawless pictures are selected as much as possible. For the optimization of the feature extraction network, 1000 pieces of original data were prepared, and the data augmentation was expanded to 10000 images.

Benchmark: In this paper, PatchSVDD, PatchCore, and CutPaste are used as the benchmark models, and PatchCore is the optimal model based on the industrial scene anomaly detection dataset MVTec AD.

Evaluation metric: The current popular evaluation method in the field of industrial image-level anomaly detection is to use AUROC as the evaluation index, that is, the area under the ROC curve. For finer defect region models to evaluate segmentation capabilities, it is represented by pixel-level AUC, i.e. the AUC between the predicted image and the mask of the true defect region is calculated on a pixel-by-pixel basis. In this paper, two metrics, image-level AUROC, and pixel-level AUROC, are used for comparison.

The experiment in this paper is based on the Linux operating system dual 2080ti GPUs for training, the BatchSize is set to 64, the image transformation process converts it into a grayscale map for training, makes it compatible with more kinds of fabric, and trains to extract the features of the middle level 2–3 layers.

4.1 Accuracy

In this paper, since the optimization scheme is mostly based on the actual fabric of the enterprise, the results of the factory fabric dataset are tested. Due to the inability of supervised models such as object detection models to achieve high accuracy for small defects and few samples, this article only compared unsupervised models. The test compared two additional models trained on ImageNet, as well as PatchCore, and an improved version of the PatchCore model, fine-tuned on its own dataset. The results were compared on the enterprise test set, and the I/P AUROC metrics of each model are shown in Table 1.

Table 1. I/P AUROC of different models.

Model	I-AUROC	P-AUROC
PatchSVDD	93.8	93
CutPaste	95.5	95
PatchCore	99.6	96.5
Ours	99.8	96.9

The comparison shows that the original PatchCore has achieved high model accuracy at the image level, but the P-AUROC results are not as good due to the presence of noise, mainly due to the influence of noise and shooting interference on the textured fabric. In this paper, after fine-tuning the self-cloth dataset, the I-AUROC of the model was improved by 0.2%, while the P-AUROC was improved more due to the post-processing, reaching a 0.4% improvement, which achieved a good effect. Therefore, applying feature extraction based on weakly supervised samples to unsupervised models can yield better results. For both types of cloth with and without texture, P-AUROC was tested under PatchCore and the improved model in this paper, and the results are shown in Table 2:

Table 2. P-AUROC comparison of our model and PatchCore model for different fabrics.

	PatchCore	Ours
Cloth A	97.3	97.3
Cloth B	96.9	97.0
Cloth C	95.2	96.3

The image of the cloth map after post-processing in this paper is compared with the original PatchCore as shown in Fig. 5, the imaging map obtains the spatial resolution of the input sample by interpolation method through the filtered abnormal heat value map, and then generates the final heat value map, as shown in Fig. 5, it can be seen from the figure that the algorithm proposed in this paper has obtained good results on the three cloths regardless of whether it is noisy or not.

4.2 Time Elapse

To analyze the time changes of small images with a size of 224 * 224 before and after quantization in this article, the first-stage defect screening model is processed through TensorRT inference, quantized to int8 format, and compared with inference

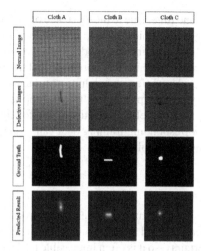

Fig. 5. Defect detection results of enterprise datasets

without processing. The comparison table of classification time for defect existence in this article's algorithm is shown in Table 3 below:

As the actual scenario involves data with a size of 2448 * 2048, in order to test the efficiency and accuracy changes of the large image inference, this paper calculates the frames per second (FPS) and I-AUROC of the model. The changes in accuracy and overall inference efficiency of the model based on quantization and sliding window optimization are shown in the following table (Table 4):

After being quantized and optimized for large images, the average FPS for large images is 2.1. The efficiency without large image optimization takes about 2 s per image, which is much higher than that of supervised defect detection models such as object detection models, and cannot meet the real-time requirements of the factory. However, after quantization and large image optimization, it can meet the real-time requirements of enterprise fabric inspection, with a certain loss of accuracy, achieving a

Table 3. Comparison table of defect screening model time before and after quantization.

	Before quantization	After quantization
Time(s)	0.019	0.011

Table 4. Comparison of FPS and I-AUROC before and after quantization and optimization.

	Before optim	After optim
FPS	0.5	2.1
I-AUROC	99.4	99.3

balance between accuracy and time. Due to the existence of the high recall rate first-stage classification model, there is no significant drop in I-AUROC after time optimization.

Regarding inference efficiency, since the model's time complexity is positively correlated with the amount of features obtained by normal model training data acquisition, the matching time can be affected by different matching accuracies and different amounts of data according to different algorithm principles. In actual application, if further efficiency improvement is required, more models can be subdivided for similar fabrics, and different models with smaller corresponding training data can reduce the time consumption of a single model in engineering applications.

5   Conclusion

Fabric defect detection is an important step in the process from cloth production to fabric processing, which has been achieved through manual verification in the past, limited by labor costs and labor uncertainty, so it is often impossible to effectively assess the quality of fabric suppliers, and directly affect the subsequent production of finished products. In order to solve the problem of real-time detection of fabric defects, aiming at the problem of sample imbalance, this paper first performs transfer learning based on the unsupervised anomaly detection scheme and trains a basic model based on the normal sample data of enterprise cloth. Based on the defect data of previous iterations, the feature extraction capability is improved and the model is iteratively optimized. For the noise data of textured cloth, post-processing is designed to make up for the lack of feature extraction ability and better segment the abnormal area. In view of the model speed problem, a strategy is designed on the large graph to accelerate. Finally, the improved model based on PatchCore is better in speed and accuracy than the original PatchCore model under the optimization of its own enterprise cloth dataset, which can effectively solve the problem of artificial detection of enterprise fabric defects.

Although the method in this paper can effectively solve the problem of enterprise fabric defects, based on PatchCore to adapt its own fabric data, it is still not as effective as PatchCore on various datasets such as MVTecAd, and future research still needs to solve the problem of detection model redundancy and model accuracy from the general method. The main research directions in the next step are: for the feature extraction network, a better network is needed to eliminate the influence of image noise; Aiming at the problem of model redundancy, reduce the detection accuracy of the network to achieve basically similar detection. For different data types, unify into one large model for all-category fabric adaptive defect detection.

Acknowledgments. This work was supported by the project "Research on Key AI Visual Technologies Based on Deep Learning and Their Industrial Application in Industrial Scenarios" of Quzhou Science and Technology Bureau.

References

1. Ngan, H.Y.T., Pang, G.K.H., Yung, N.H.C.: Automated fabric defect detection—a review. Image Vis. Comput. **29**, 442–458 (2011)

2. Zhou, J., Wang, J.: Unsupervised fabric defect segmentation using local patch approximation. J. Text. Inst. **107**, 800–809 (2016)
3. Kirchler, M., et al.: TransferGWAS: GWAS of images using deep transfer learning (2021). https://doi.org/10.1101/2021.10.22.465430
4. Heaton, J.: Ian Goodfellow, Yoshua Bengio, and Aaron Courville: Deep learning. Genet. Program Evolvable Mach. **19**(1–2), 305–307 (2017). https://doi.org/10.1007/s10710-017-9314-z
5. Honeycutt, C.E., Plotnick, R.: Image analysis techniques and gray-level co-occurrence matrices (GLCM) for calculating bioturbation indices and characterizing biogenic sedimentary structures. Comput. Geosci. **34**, 1461–1472 (2008)
6. Guo, Z., Zhang, L., Zhang, D.: Rotation invariant texture classification using LBP variance (LBPV) with global matching. Pattern Recogn. **43**, 706–719 (2010)
7. Unser, M.: Sum and difference histograms for texture classification. IEEE Trans. Pattern Anal. Mach. Intel. (PAMI) **8**, 118–125 (1986)
8. Gonzalez, R.C., et al.: Digital image processing, third edition. J. Biomed. Opt. **14**(2), 029901 (2009). https://doi.org/10.1117/1.3115362
9. Liu, C., Gryllias, K.: A deep support vector data description method for anomaly detection in helicopters. In: PHM Society European Conference, p. 9 (2021)
10. Liang, S., Li, Y., Srikant, R.: Enhancing the reliability of out-of-distribution image detection in neural networks. arXiv: Learning (2017)
11. Vincent, P., et al.: Extracting and composing robust features with denoising autoencoders. Presented at the (2008). https://doi.org/10.1145/1390156.1390294
12. Choi, Y., et al.: StarGAN: unified generative adversarial networks for multi-domain image-to-image translation. Presented at the (2018). https://doi.org/10.1109/cvpr.2018.00916
13. Fujioka, T., et al.: Efficient anomaly detection with generative adversarial network for breast ultrasound imaging. Diagnostics **10**(7), 456 (2020). https://doi.org/10.3390/diagnostics10070456
14. Defard, T., Setkov, A., Loesch, A., Audigier, R.: PaDiM: a patch distribution modeling framework for anomaly detection and localization. In: Del Bimbo, A., et al. (eds.) ICPR 2021. LNCS, vol. 12664, pp. 475–489. Springer, Cham (2021). https://doi.org/10.1007/978-3-030-68799-1_35
15. Roth, K., et al.: Towards Total Recall in Industrial Anomaly Detection. Presented at the (2022). https://doi.org/10.1109/cvpr52688.2022.01392

Intrusion Detection Method Based on Complementary Adversarial Generation Network

Lixiang Li🆔, Yuxuan Liu(✉)🆔, and Haipeng Peng🆔

Information Security Center, State Key Laboratory of Networking and Switching Technology, Beijing University of Posts and Telecommunications, Beijing 100876, China
{lixiang,lixiang,penghaipeng}@bupt.edu.cn

Abstract. How to effectively identify the attack behavior in the network environment is an urgent problem in the field of network security. In this paper, an intrusion detection method based on complementary generative adversarial network is proposed. The algorithm is One Class Classification algorithm based on generative adversarial network architecture. The algorithm uses normal data in training, normal and abnormal data in testing phase, and calculates the accuracy rate. In this algorithm, complementary samples are generated by complementary generator and input to discriminator as exception samples for training. The discriminator in this paper uses stacked Non-Symmetric autoencoders to recognize normal samples from training data and abnormal samples generated by generators through training. Experimental results show that the algorithm proposed in this paper is superior to other algorithms of the same kind in performance, with higher accuracy and excellent time performance.

Keywords: Intrusion Detection · Deep Learning · Complementary GAN · Auto-Encoder

1 Introduction

The rapid progress of science and technology, the Internet has been popularized to thousands of households. Internet technology continues to innovate, our way of life has also undergone earth-shaking changes. Nowadays, network technology has penetrated into our daily life, and everything we say and do is closely connected with the network. But cyberspace is not a perfect utopia for us ordinary netizens or the country's secret service. It is a virtual world full of tricks and deceit.

With the development of information technology, the means of network attack are constantly improving. The traditional defense methods such as firewall, virtual network, user identity encryption technology have been unable to effectively deal with a variety of attacks. As a result, more reliable defense methods have emerged. Intrusion detection is a Proactive Defense technology, which can be used as a supplement to traditional defense methods to help the network detect attacks in time. Compared with the previous passive defense methods such as firewalls, intrusion detection can find attacks faster and more effectively.

Y. Tan et al. (Eds.): ICSI 2023, LNCS 13969, pp. 260–271, 2023.
https://doi.org/10.1007/978-3-031-36625-3_21

2 Related Work

In this section, we introduce the related work about the description of intrusion detection and Generative adversarial network (GAN).

2.1 Intrusion Detection

Intrusion Detection Systems (IDS) is a security tool which is used to guarantee the security of information and network system. The research on intrusion detection originated from James, who first proposed the concept of intrusion detection in 1980. In 1987, Denning and Neumann proposed the first intrusion detection system model IDES (Intrusion Detection Expert System), which explained the structure of intrusion detection system completely for the first time. With the development of network attacks, the intrusion modes become more complex and various, and the scope of intrusion is expanded. In order to adapt to the deteriorating network environment security, the International Internet Engineering Task Force set up IDWG special group to design the intrusion detection system model. After a series of researches, Denning et al. proposed a general Intrusion Detection model called CIDF (Common Intrusion Detection Framework). CIDF model has strong extensibility and has been widely approved by the industry [1].

Intrusion detection algorithms can be divided into intrusion detection algorithms based on data mining and machine learning and intrusion detection algorithms based on deep learning. Data mining and machine learning algorithms have been widely used in various fields since the last century. Examples of applying classical data mining and machine learning algorithms to intrusion detection include: association rules [2], support vector machine (SVM) [3], K-nearest neighbor (KNN) [4], etc. As a branch of machine learning, deep learning has been proved to be very effective in various neighborhood in recent years. Some scholars have applied this deep learning algorithm to intrusion detection. A. Javaid et al. proposed an algorithm based on deep neural network for stream-based intrusion detection, and the experimental results show that this algorithm can be used for intrusion detection of software-defined networks [5]. T. A. Tang et al. proposed A self-learning method based on deep learning based on the NSL-KDD dataset and proved the effectiveness of this method [6].

From the perspective of detection methods, intrusion detection can be divided into misuse detection and anomaly. Misuse detection requires that the attack library of known attacks be specified in advance and the characteristics and patterns defined in the information be analyzed according to the attack library. Common algorithms for misuse detection include expert systems and pattern matching. This method has high detection efficiency for known attacks. The idea of the anomaly detection method is to protect the "normal" behavior of the system by generating an abnormal alarm when the gap between the observed value and the outlier reaches a threshold value. The advantage of anomaly detection is that previously unknown intrusion behaviors can be detected [7]. The intrusion detection method studied in this paper is based on anomaly detection.

2.2 Generative Adversarial Network

Generative Adversarial Network (GAN) was first proposed by Goodfellow et al., which uses the competition between generative and discriminant models to implement the functions of generative models [8]. The difference between GAN model and traditional neural network model lies in the introduction of antagonistic factors in model training, and the generator and discriminator have different optimization directions and compete with each other. Compared with traditional generation algorithms, GAN model can generate samples of higher quality. GAN package contains two modes: Generative Model and Discriminative Model. The generative model is used to generate fake samples of "very-like" real samples to fool the discriminator, whose task is to identify whether a given sample is real or fake.

After GAN model was proposed, some improved models were proposed successively. F-divergence is a method to measure the similarity between two distributions. f-GAN expands the objective function of GAN model by using any given function as F-divergence [9]. Wasserstein GAN (WGAN) algorithm proposed by Arjovsky et al. proposed an effective method for measuring the distance between and [10]. Instead of learning probability distribution itself directly, the algorithm learns by generating functions to convert input into. The viewpoint of WGAN improvement points out that the weight cutting of key parts will make the discriminator appear ill behavior when training WGAN. Therefore, it is suggested to use a gradient standard as a penalty term to replace weight clipping [11]. LS-GAN and WGAN use Lipschitz constraints, but they are used differently [12]. LS-GAN makes the loss of a real sample smaller than that of a generated sample by learning the loss function instead of the key factor. Another related improvement is the RWGAN, where a loose Wasserstein distance is obtained by combining Bregman divergence and Wasserstein distance [13].

Researchers have applied many of these algorithms to intrusion detection. These algorithms basically pay attention to the imbalance of network traffic data and the problem that the data is difficult to be effectively labeled, and propose corresponding improved algorithms based on this problem. SGAN introduces a new training architecture to improve the performance of intrusion detection classifier [14]. DGAN introduces Earth-Mover (EM) distance to capture the distribution of low-dimensional space and use the encoder structure catch latent space representation [15]. Efficient-GAN based on BiGAN, and designed a new anomaly score function for model training [19]. The model is a one-class classification model. GAN-RF used GAN to solve the problem of data imbalance [16]. In this algorithm use Random Forest for classification. GANs-ECNN propose a new architecture for training for improve accuracy for multi-attack detection and use CNN to stabilize the training process [17]. IGAN-IDS add an imbalanced data filter and convolutional layers based on typical GAN for data imbalance problem [18].

3   Method

In this paper, a network model based on complementary adversarial generation network and Non-Symmetric stack autoencoder is proposed, which can be used for network traffic intrusion detection. In this chapter, we first introduce the concept of complementary generation algorithm, and then introduce the application of Non-Symmetric stack autoencoder in network model, and then give the overall network architecture and the specific process of intrusion detection.

3.1 Complementary Confrontation Generates Adversarial Networks

With the development of adversarial learning, how to generate effective adversarial samples has become a hot topic. The most direct way is model-based generation, that is, it is assumed that the distribution of samples conforms to a certain model distribution, such as Gaussian mixture model, and the parameters of the model are calculated by means of maximum likelihood estimation to obtain a reasonable generation model [20]. The idea of generating adversarial network is to use game theory model, use generator to generate pseudo-true samples, and use discriminator to discriminate. The two sub-models compete with each other and iterate successively, and it is better to obtain a generator that can generate "realistic" samples.

The generator used in this paper is based on the idea of complementary adversarial generation network. The complementary distribution of normal samples can be obtained through generator learning to assist the training of discriminator [21]. Based on the idea of game theory, the generative adversarial network is trained by generators and discriminators to obtain generators that can generate realistic samples. Since the samples generated by the generator are set as abnormal samples, the generated abnormal samples and normal samples need to have a certain degree of differentiation to ensure that the abnormal samples contain enough information [21]. Therefore, a complementary GAN generator is proposed to generate exception samples. The generator of complementary GAN is a feedforward neural network whose output layer has the same dimension as the sample x. Formally, define the generated sample as. Conventional GAN can learn the sample distribution similar to the normal sample distribution, but the generator G of complementary-GAN can learn the distribution p^* close to the complementary distribution of normal sample, which is exactly the goal of generating abnormal samples. The complementary distribution p^* can be defined as:

$$p^* = \begin{cases} \frac{1}{\varepsilon} \frac{1}{p_{data}} & if \ p_{data}(\tilde{x}) > \tau \ and \ \tilde{x} \in B_x \\ C & if \ p_{data}(\tilde{x}) \ll \ and \ \tilde{x} \in B_x \end{cases} \tag{1}$$

Where, τ is the threshold value used to judge whether the generated sample is located in a high-density area. ε is the regularized parameter; C is the smaller constant; B_x is the sample space, and we need to make sure that all the generated $\tilde{x}$ are in the sample space.

In order to make the generated samples conform to the p^* distribution, that is, make the generated distribution p_G close to the complementary distribution p^*. f-divergence is used to represent the gap between two distributions. f-divergence is a general function used to calculate the gap between two distributions. The expression of f-divergence is:

$$D_f(P\|Q) = \int q(x)f\left(\frac{p(x)}{q(x)}\right)dx \tag{2}$$

Use f-divergence to describe the difference between p_G and p^*:

$$D_f(p_G\|p^*) = \int p^*(\tilde{x})f\left(\frac{p_G(\tilde{x})}{p^*(\tilde{x})}\right)d\tilde{x} \tag{3}$$

Where f is the hyperparameter of the representation $D_f(p_G\|p^*)$, as a function satisfies 1) The function is a concave function; 2). $f(1) = 0$.

In addition, $f(x) = x \log x$ uses *KL* divergence to calculate the gap between two distributions, then the generator G needs to be trained to minimize the *KL* divergence between p_G and p^*. The complete objective function of the generator is defined as:

$$\min_{G} - \mathcal{H}(p_G) + E_{\tilde{x} \sim p_G} \log p_{data}(\tilde{x}) I \left[p_{data}(\tilde{x}) > \tau \right]$$
$$+ \left\| E_{\tilde{x} - p_G} f(\tilde{x}) - E_{x \sim p_{data}} f(x) \right\|^2 \qquad (4)$$

Where $\mathcal{H}(\cdot)$ is entropy and $I[\cdot]$ is the indicator function. The latter item is used to ensure that the generated samples are confined to sample space B_x. The objective function of the complementary generator aims to make the generated distribution p_G close to the complementary sample p^*, i.e. $p_G = p^*$, and make the generated sample come from a different region from the normal sample.

In the early stages of generation, the generator may not be able to generate enough exception samples around normal samples. But over enough iterations, the generator gradually learns the generation mechanism and generates more and more potential outliers that occur within or near the real data. It's an active learning process. The iterative effect of the complementary generator is shown in Fig. 1.

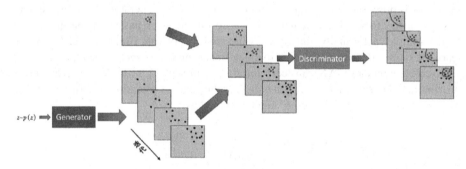

Fig. 1. The training process of the complementary generator

The $\mathcal{H}$ term is not easy to calculate directly, so a pull-away term (*PT* term) is used instead of entropy [22]. This term is used to measure the orthogonality of the data, which increases the diversity of the samples. Optimizing this term is equivalent to optimizing the entropy. Define the *PT* term as follows:

$$L_{PT} = \frac{1}{N(N-1)} \sum_{i}^{N} \sum_{j \neq i}^{N} \left(\frac{f(\tilde{v}_i)^T f(\tilde{v}_j)}{\|f(\tilde{v}_i)\| \|f(\tilde{v}_j)\|} \right)^2 \qquad (5)$$

Where represents the size of the batch when training in batch. Update the overall optimization function of the generator as follows:

$$\min_{G} L_{PT} + E_{\tilde{v} - p_G} \log p_{data}(\tilde{v}) \mathbb{I} \left[p_{data}(\tilde{v}) > \varepsilon \right]$$
$$+ \left\| E_{\tilde{v} - p_G} - E_{v - p_{data}} f(v) \right\|^2 \qquad (6)$$

The discriminator can be directly used to classify the samples after the training is completed, and the final output layer has the same dimension as the samples, which realizes the generation of samples.

The architectures of regular GAN and complementary GAN are similar, but the complementary GAN has stronger anomaly detection ability. The generator of a regular GAN is used to generate data with the same distribution as normal samples. Therefore, the abnormal samples generated after the generator has been trained to a certain extent are located in the same region as the normal samples. The probability of the discriminator of the conventional GAN to judge whether the sample is normal or abnormal is close to 0.5, so that the discriminator of the conventional GAN cannot normally identify normal or abnormal data. But the generator using the complementary GAN network can generate complementary samples of normal samples. Since the generated samples have the same distribution as the abnormal samples, the discriminator of the complementary GAN can be abnormal samples.

3.2 Stacked Non-symmetry Autoencoder

In this paper, the discriminator of the generative adversarial network uses the stacked non-symmetric autoencoder structure, which can reduce the amount of calculation without affecting the accuracy.

Autoencoders are self-supervised learning models that are commonly used for data compression [23, 24]. After the rise of neural networks, scholars used generative adversarial networks to implement autoencoders [26]. Autoencoders are special neural network models that are often used to extract features. Autoencoder models typically consist of three layers: (1) input layer; (2) output layer; (3) Hidden layers. Compared to the input and output layers, the number of neurons in the hidden layer is small. The operation process of the model is to first convert the input into a form with fewer dimensions through the encoder, and then use the decoder to restore the data to reconstruct the original input, which is called encoder-decoder process. The original autoencoder emphasizes data compression, that is, the compressed data of the original data is obtained through the output of the hidden layer, and the function is similar to PCA. In this paper, the feature extraction ability of autoencoder is used to distinguish normal samples and abnormal samples [25].

The stacked autoencoder is a neural network nested by multiple autoencoders, where each autoencoder contains a certain number of layers of neural network for encoding and decoding [23]. The idea of this model construction comes from deep learning [27]. The training of such deep networks is achieved by training each layer of autoencoders in turn. The increase in depth will reduce the amount of data required for training, as well as reduce the computational cost and improve accuracy. The output of each hidden layer is used as input to the next layer. The first layer of the stacked autoencoder can be learned as raw input. The second layer learns complex feature categories from the first layer, and higher layers learn more advanced features. The deterministic function represents the neural network operation used to map the input to the latent space, and the mapping function then reconstructs the original input from the latent variables. The obtained from the first autoencoder is used to train the second autoencoder so that the mapping function tries to transform into the hidden representation named as. Another mapping function can reconstruct the original from.

The stacked Non-Symmetric autoencoder used in this paper stacks two Non-Symmetric autoencoders on top of each other to create a deep learning hierarchy. Unlike regular autoencoders, Non-Symmetric autoencoders compute the hidden representation directly without a decoder [24]. Compared with the symmetric autoencoder, the Non-Symmetric autoencoder used in this paper can reduce the computational cost and has almost no impact on the accuracy when the network structure is reasonably designed. Non-Symmetric autoencoders can perform unsupervised feature extraction structure on input high-dimensional data, and their training mode is similar to that of regular autoencoders.

The Non-Symmetric autoencoder takes the input and progressively maps it to a hidden layer whose activation function is:

$$h_i = \sigma(W_i \cdot h_{i-1} + b_i); i = \overline{1, n}, \tag{7}$$

Where, $h_0 = x$, σ is an activation function and n is the number of hidden layers. In this study, an autoencoder is used without the decoder, and the output result is calculated by the function:

$$y = \sigma(W_{n+1} \cdot h_n + b_{n+1}) \tag{8}$$

Use the minimum squared error as the loss function:

$$E(\theta) = \sum_{i=1}^{m} \left(x^{(i)} - y^{(i)} \right)^2 \tag{9}$$

For the loss function, the reconstruction loss of normal samples should be as small as possible, while the reconstruction loss of abnormal samples should be as large as possible.

This model can distinguish samples according to the complex relationship between different features. Each Non-Symmetric encoder uses a $14 \times 28 \times 28$ network structure, and uses ReLU as the activation function. Figure 2 shows the stacked Non-Symmetric autoencoder network structure.

The advantage of autoencoder is that it can better extract the features in the sample, but the pure classification ability is weak, and the classification effect of directly using softmax function is not as good as other machine learning algorithms. Therefore, this paper combines autoencoder and shallow classifier to achieve classification together. In this paper, SVM is selected as the classifier, which has a better effect in binary classification [23]. Therefore, after calculating the hidden representation, the result is input into SVM for classification to realize anomaly detection. SVM implements classification based on the maximum margin principle and uses the maximum margin decision function. SVM based on kernel function can realize nonlinear classification. Soft Support Vector Machine (SSVM) and Radial Basis function (RBF) kernel function are often used together in network intrusion detection system [26], so this paper uses soft support

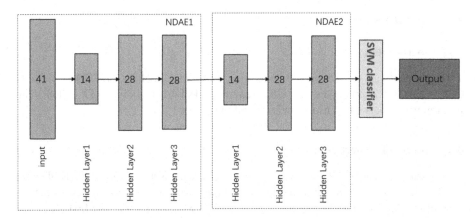

Fig. 2. Stacked Non-Symmetric Autoencoder Network architecture

vector machine based on RBF kernel function as a shallow classifier. In the training phase, the hidden representation calculated by the stacked Non-Symmetric autoencoder and the labels of normal and abnormal are used to train the SVM to obtain the classifier. The classifier is used for classification in the testing phase. Figure 3 shows the overall structure of the model.

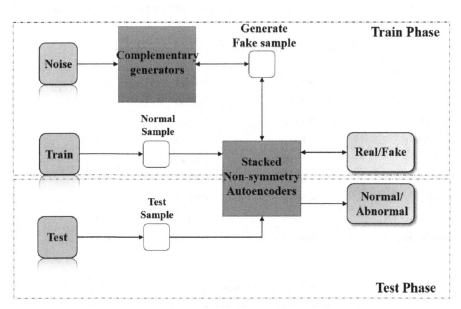

Fig. 3. Complementry GAN structure

4 Experiment

4.1 Environment

This chapter uses windows10 operating system, Intel(R)Core(TM)i7-10750H CPU@2.90 GHz processor, 32 GB memory, NIVDIA GeForce RTX 2060. The programming tools are TensorFlow2.7.2 and python3.6.

4.2 Process of Experiment

In this paper, the KDD99 dataset is used as the test dataset. The KDD99 (KDD Cup 1999) dataset was developed to train and validate intrusion detection systems relevant to traditional information technology systems [27, 28]. In this paper, we first clean and normalize the data before using the dataset to remove unwanted features and fill in missing values. Convert some string-based features (e.g., usage protocol, service type) to numeric representation. The dataset is split into training and test sets before use.

Experiment 1. Compare the Accuracy of Different Models

Firstly, we compare the proposed model with the machine learning algorithms Isolation Forest, OC-SVM and other GAN-based algorithms. We tested the accuracy, recall and F1 score of the selection algorithm, and compared the results of the model of this paper with the results of other algorithms. The obtained results are shown in Table 1 and Fig. 3:

Table 1. Accuracy comparison of different models

Model	accuracy	recall	F1-score
Isolation Forest	0.4415	0.3260	0.3750
OC-SVM	0.7457	0.8523	0.7954
DSEBM-r	0.8521	0.6474	0.7328
DSEBM-e	0.8619	0.6446	0.7328
AnoGAN	0.8786	0.8297	0.8865
BiGAN	0.8698	0.9523	0.9058
Efficient-GAN	0.9324	0.9473	0.9398
Our Model	**0.9330**	**0.9456**	**0.9315**

As can be seen from the Fig. 4, Our model performs the best in accuracy, second only to Efficient-GAN in recall and F1 score, and far better than other algorithms in performance. So, from the perspective of accuracy, the algorithm proposed in this paper is better.

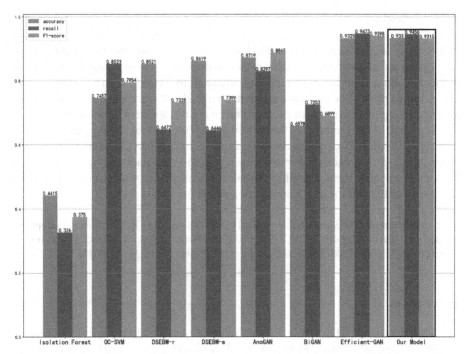

Fig. 4. Bar graph comparing the performance of different algorithms

Experiment 2. Time Performance Comparison of Different Models
Another important metric is speed. In the test, the algorithm used in this paper only uses
the discriminator for anomaly classification, and the computational complexity is less
than that of the algorithm that needs the discriminator and the generator to calculate
the anomaly score, such as AnoGAN. Therefore, the proposed algorithm is compared
with AnoGAN and BiGAN on the same machine under this paper. In this test, 100000
samples (50000 normal samples and 50000 abnormal samples) are randomly selected as
the test set, and the time consumed by each test is calculated for 10 times. The average
time consumed is taken as the standard for comparison. The results of the calculated
time performance are shown in Table 2.

Table 2. Comparison of time performance of different algorithms

Model	Time(ms)	Base time
AnoGAN	132443	4.42
BiGAN	35986	1.20
Our Model	**29986**	**1.0**

5 Conclusion

In this paper, a generative adversarial network based on complementary generator and asymmetric stacked autoencoder discriminator is presented. The algorithm can be trained using only normal samples during training, and abnormal samples can be identified during the testing phase. The algorithm makes use of the idea of complementary distribution, and the training generator generates complementary distribution of sample distribution to assist training, thereby improving the performance of the operation. And use asymmetric encoders to improve the performance of calculations. Compared with traditional algorithms, this algorithm greatly improves the recognition accuracy and greatly improves the operation speed. The algorithm is suitable for network data with no or lack of known anomalous traffic and for real-time intrusion detection.

Acknowledgements. This work was supported by the National Key Research and Development Program of China (Grant No. 2020YFB1805403), the National Natural Science Foundation of China (Grant Nos. 62032002, 61932005) and the 111 Project (Grant No. B21049).

References

1. Khraisat, A., Gondal, I., Vamplew, P., et al.: Survey of intrusion detection systems: techniques, datasets and challenges. Cybersecurity **2**(1), 1–22 (2019)
2. Azeez, N.A., Ayemobola, T.J., Misra, S., et al.: Network intrusion detection with a hashing based apriori algorithm using Hadoop MapReduce. Computers **8**(4), 86 (2019)
3. Kuang, F., Xu, W., Zhang, S.: A novel hybrid KPCA and SVM with GA model for intrusion detection. Appl. Soft Comput. **18**, 178–184 (2014)
4. Li, W., Yi, P., Wu, Y., et al.: A new intrusion detection system based on KNN classification algorithm in wireless sensor network. J. Electr. Comput. Eng. **2014**, 1–8 (2014)
5. Javaid, A., Niyaz, Q., Sun, W., et al.: A deep learning approach for network intrusion detection system. In: Proceedings of the 9th EAI International Conference on Bio-inspired Information and Communications Technologies (formerly BIONETICS), pp. 21–26 (2016)
6. Tang, T.A., Mhamdi, L., McLernon, D., et al.: Deep learning approach for network intrusion detection in software defined networking. In: 2016 International Conference on Wireless Networks and Mobile Communications (WINCOM), pp. 258–263. IEEE (2016)
7. Garcia-Teodoro, P., Diaz-Verdejo, J., Macia-Fernandez, G., et al.: Anomaly-based network intrusion detection: techniques, systems and challenges. Comput. Secur. **28**(1–2), 18–28 (2009)
8. Goodfellow, I., Pouget-Abadie, J., Mirza, M., et al.: Generative adversarial nets. In: Advances in Neural Information Processing Systems, pp. 2672–2680 (2014)
9. Nowozin, S., Cseke, B., Tomioka, R.: f-GAN: training generative neural samplers using variational divergence minimization. In: Proceedings of the 30th International Conference on Neural Information Processing Systems, pp. 271–279 (2016)
10. Arjovsky, M., Chintala, S., Bottou, L.: Wasserstein generative adversarial networks. In: International Conference on Machine Learning, pp. 214–223. PMLR (2017)
11. Gulrajani, I., Ahmed, F., Arjovsky, M., et al.: Improved training of wasserstein GANs. In: Proceedings of the 31st International Conference on Neural Information Processing Systems, pp. 5769–5779 (2017)
12. Qi, G.J.: Loss-sensitive generative adversarial networks on Lipschitz densities. Int. J. Comput. Vis. **128**(5), 1118–1140 (2020)

13. Guo, X., Hong, J., Lin, T., et al.: Relaxed Wasserstein with applications to GANs. In: ICASSP 2021–2021 IEEE International Conference on Acoustics, Speech and Signal Processing (ICASSP), pp. 3325–3329. IEEE (2021)
14. Yin, C., Zhu, Y., Liu, S., et al.: Enhancing network intrusion detection classifiers using supervised adversarial training. J. Supercomput. **76**(9), 6690–6719 (2020)
15. Hao, X., Jiang, Z., Xiao, Q., et al.: Producing more with less: a GAN-based network attack detection approach for imbalanced data. In: 2021 IEEE 24th International Conference on Computer Supported Cooperative Work in Design (CSCWD), pp. 384–390. IEEE (2021)
16. Lee, J.H., Park, K.H.: GAN-based imbalanced data intrusion detection system. Pers. Ubiquit. Comput. **25**, 121–128 (2021)
17. Soleymanzadeh, R., Kashef, R.: A stable generative adversarial network architecture for network intrusion detection. In: 2022 IEEE International Conference on Cyber Security and Resilience (CSR), pp. 9–15. IEEE (2022)
18. Huang, S., Lei, K.: IGAN-IDS: an imbalanced generative adversarial network towards intrusion detection system in ad-hoc networks. Ad Hoc Netw. **105**, 102177 (2020)
19. Chen, H., Jiang, L.: Efficient GAN-based method for cyber-intrusion detection. arXiv preprint arXiv:1904.02426 (2019)
20. Zheng, P., Yuan, S., Wu, X., et al.: One-class adversarial nets for fraud detection. In: Proceedings of the AAAI Conference on Artificial Intelligence, vol. 33, no. 01, pp. 1286–1293 (2019)
21. Dai, Z., Yang, Z., Yang, F., et al.: Good semi-supervised learning that requires a bad GAN. In: Advances in Neural Information Processing Systems, vol. 30 (2017)
22. Shone, N., Ngoc, T.N., Phai, V.D., et al.: A deep learning approach to network intrusion detection. IEEE Trans. Emerg. Top. Comput. Intell. **2**(1), 41–50 (2018)
23. Wang, W., Du, X., Shan, D., et al.: Cloud intrusion detection method based on stacked contractive auto-encoder and support vector machine. IEEE Trans. Cloud Comput. **10**(3), 1634–1646 (2020)
24. Hinton, G.E., Salakhutdinov, R.R.: Reducing the dimensionality of data with neural networks. Science **313**(5786), 504–507 (2006)
25. Imran, M., Haider, N., Shoaib, M., et al.: An intelligent and efficient network intrusion detection system using deep learning. Comput. Electr. Eng. **99**, 107764 (2022)
26. Liu, C., Yang, J., Wu, J.: Web intrusion detection system combined with feature analysis and SVM optimization. EURASIP J. Wirel. Commun. Netw. **2020**, 1–9 (2020)
27. Ferrag, M.A., Maglaras, L., Moschoyiannis, S., et al.: Deep learning for cyber security intrusion detection: approaches, datasets, and comparative study. J. Inf. Secur. Appl. **50**, 102419 (2020)
28. KDD99 Dataset [EB/OL]. http://kdd.ics.uci.edu/databases/kddcup99/kddcup99.html. Accessed 15 Dec 2022

EEG-Based Subject-Independent Depression Detection Using Dynamic Convolution and Feature Adaptation

Wanqing Jiang[1,2], Nuo Su[2,3], Tianxu Pan[2], Yifan Miao[2], Xueyu Lv[4],
Tianzi Jiang[2], and Nianming Zuo[2(✉)]

[1] School of Artificial Intelligence, University of Chinese Academy of Sciences,
Beijing 100049, China
jiangwanqing2020@ia.ac.cn
[2] Brainnetome Center and NLPR, Institute of Automation,
Chinese Academy of Sciences, Beijing 100190, China
nmzuo@nlpr.ia.ac.cn
[3] College of Computer Science and Technology, Ocean University of China, Qingdao
266100, China
[4] Psychology and Sleep Medicine Department of Guang'anmen Hospital,
China Academy of Chinese Medical Sciences, Beijing 10053, China

Abstract. Depression is a debilitating condition that can seriously impact quality of life, and existing clinical diagnoses are often complicated and dependent on physician experience. Recently, research on EEG-based major depressive disorder (MDD) detection has achieved good performance. However, subject-independent depression detection (i.e., diagnosis of a person never met) remains challenging due to large inter-subject discrepancies in EEG signal distribution. To address this, we propose an EEG-based depression detection model (DCAAN) that incorporates dynamic convolution, adversarial domain adaptation, and association domain adaptation. Dynamic convolution is introduced in the feature extractor to enhance model expression capability. Furthermore, to generalize the model across subjects, adversarial domain adaptation is used to achieve marginal distribution domain adaptation and association domain adaptation is used to achieve conditional distribution domain adaptation. Based on experimentation, our model achieved 86.85% accuracy in subject-independent MDD detection using the multimodal open mental disorder analysis (MODMA) dataset, confirming the considerable potential of the proposed method.

Keywords: Depression detection · Electroencephalogram (EEG) · Domain adaptation · Subject-independent

1 Introduction

According to the World Health Organization (WHO), depression is a serious illness that affects approximately 350 million people worldwide [1]. At present,

W. Jiang, N. Su—Contribute equally to this work.

© The Author(s), under exclusive license to Springer Nature Switzerland AG 2023
Y. Tan et al. (Eds.): ICSI 2023, LNCS 13969, pp. 272–283, 2023.
https://doi.org/10.1007/978-3-031-36625-3_22

diagnosis of depression is based on clinical interviews and psychiatric question-naires conducted by physicians on patients. However, given the current lack of objective criteria [2], diagnosis is highly influenced by physician experience as well as the authenticity and subjectivity of patient self-description. In contrast, the electroencephalogram (EEG) is an objective and reliable depression assess-ment method, with the advantages of high temporal resolution, relatively low cost, easy recording, and non-invasiveness. EEG signals provide a direct repre-sentation of neural activity in the brain and are highly correlated with a person's emotional state [3]. An increasing number of scholars have used EEG-based data to construct machine learning models for detecting depression. Support vector machine (SVM), linear discriminant analysis (LDA), naive Bayes (NB), and k-nearest neighbor (KNN) algorithms have been used to make predictions with better results when effective feature selection is performed on EEG. Bashir et al. [4] proposed an EEG-based major depressive disorder (MDD) detection model, which achieved 99.70% and 99.60% accuracy with the KNN and convolutional neural network (CNN) classifiers for a dataset of 34 MDDs and 30 healthy sub-jects, respectively. Song et al. [5] used 0.5–50 Hz SincFIR filters to preprocess EEG signals and fed them into a combined CNN and long short-term memory (LSTM) classification model. They obtained 93.98% and 94.69% accuracy using 30 depressed subjects and 40 healthy subjects as the dataset, using gamma-band and full-band EEG signals, respectively. Although previous studies using machine learning models to detect depression have achieved high accuracy, most have not considered individual independence and are therefore of limited value for clinical use. Furthermore, current studies exploring individual independent depression diagnoses have achieved relatively modest results. For example, in the context of individual independence, Gulay et al. [6] achieved 76.08% classi-fication accuracy using the novel twin pascal's triangles lattice pattern model, while Chen et al. [7] achieved an 84.91% two-class correct rate using the self-attention graph pooling with soft label (SGP-SL) model based on the multi-modal open mental disorder analysis (MODMA) dataset. To further improve depression diagnosis and reduce the adverse effects of variability, transfer learn-ing may be a useful approach [8], known as domain adaptation when the data distribution of the source and target differ but the two tasks are the same [9]. Zhao et al. [10] proposed a plug-and-play domain adaptive approach to deal with inter-subject variability, a common challenge. However, few studies have focused on domain adaptation methods in MDD detection. Thus, we propose a novel network, called DCAAN, which uses dynamic convolution as a feature extractor to improve model representation, while adapting to both global and local categories using dynamic convolution, adversarial domain adaptation, and association domain adaptation. Experiments and comparisons were performed to validate the effectiveness of our model.

2   Methods

2.1   Overview

The EEG dataset is defined as $\left\{\left(x^1, y^1\right), \ldots, \left(x^i, y^i\right), \left(x^N, y^N\right)\right\}$, where N represents the number of EEG samples. Let $x^i \in \mathbb{R}^{C \times L}$ denote an EEG sample with C electrodes and L sampling points and $y^i \in \mathbb{R}^K$ denote the corresponding label, where K represents the number of categories. EEG-based depression diagnosis is modeled as a domain adaptation problem, with $D_s = \left\{x_s^1, x_s^2, \cdots, x_s^i, \cdots, x_s^{n_s}\right\}$, and $D_t = \left\{x_t^1, x_t^2, \cdots, x_t^i, \cdots, x_t^{n_t}\right\}$ representing the samples of the source and target domains in training, respectively, where n_s is the number of samples in source domain D_S and n_t is the number of samples of target domain D_T. Our primary objective is to construct a model capable of better predictions for never-before-seen newly collected target domain subjects by using all labeled source data and unlabeled target data, i.e., achieve subject-independent depression detection. As illustrated in Fig. 1, the proposed DCAAN algorithm empowers the dynamic perception of the feature extractor and combines it with domain alignment to improve the accuracy of depression diagnosis. The DCAAN model consists of two modules, i.e., dynamic feature extractor based on dynamic convolution and domain adaptation.

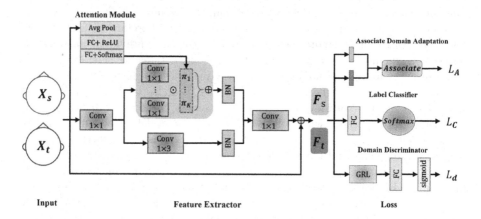

Fig. 1. Framework of proposed DCAAN

2.2   Dynamic Feature Extractor

The main architecture of the feature extractor is a variant of the residual network, called a bottleneck residual block [11], which uses a stack of three convolutional layers. These three layers consist of 1×1, 1×3, and 1×1 convolutions, with the two 1×1 convolutions at the beginning and end used to reduce channel dimensions of channels and the 1×3 convolution used to form a bottleneck structure. The purpose of this is to reduce the number of parameters and matrix multiplications while increasing depth. 1D-CNN is a powerful tool for feature extraction of time series data, and is widely used in audio signal recognition [12], behavioral detection [13], and other fields [14]. As EEG signals are time series data with high temporal resolution and multiple channels, we used 1D-CNN to extract spatiotemporal information, where the input data are the original EEG signals. Considering that dynamic networks can be difficult to learn [15], we only added the dynamic convolution block to the middle 1×3 of the bottleneck block, making a small number of parameters vary dynamically with the sample. Dynamic convolution consists of an attention mechanism and parallel convolution kernels (Let M denote the number of parallel 1D-convolution kernels). After dynamically integrating the parallel convolution blocks using sample-dependent attention weights, in line with the classical approach [11], we employed batch normalization and rectified linear unit (ReLU) activation functions.

Attention Module. In the attention module, squeeze-and-excitation [16] is employed to generate attention weights for kernels. Squeeze-and-excitation is widely used in the image field and can be adapted to extract important features. The squeeze-and-excitation network (SENet) [16] improves the representational capability of the network through dynamic feature recalibration of channels. However unlike SENet, our model does not generate attention scores over the channels of 2D-images, but over the M 1D-convolution kernels. The squeeze function compresses the temporal information by global average pooling, such that 1D feature map of each channel is "squeezed" into a single numeric value.

$$\text{Avgpooling}(x^i) = \frac{1}{L} \sum_{n=1}^{L} x^i(n) \tag{1}$$

where $x^i(n)$ represents the nth column vector of x^i. The excitation function consists of a fully connected layer, a ReLU, a fully connected layer, and a softmax in turn, to yield sample-specific normalized attention weights $\pi\left(x^i\right) \in \mathbb{R}^{1 \times M}$ for the M convolution kernels. Moreover, the computational cost of attention is very low, much smaller than that of the convolution operation [17].

$$\pi(x^i) = \text{Softmax}\left(\boldsymbol{W}_2\left(\text{ReLU}\left(W_1\,\text{Avgpooling}(x^i) + b_1\right) + b_2\right)\right) \tag{2}$$

Parallel Convolution Kernels. The M convolution kernels share the same kernel size and dimensions. In general, kernel size and number of parallel M convolution kernels are set to be small, so the computational complexity of the integration operation is low. Let $\{(W_m, b_m)\}$ represent the weight matrix and bias vector of the mth kernel. The attention calculated by the attention module is then used for dynamic integration. Let π_m denote the attention weight of the

m^{th} kernel. The formulae are as follows.

$$\Delta W(x^i) = \sum_{m=1}^{M} \pi_m(x^i)W_m$$

$$\Delta b(x^i) = \sum_{m=1}^{M} \pi_m(x^i)b_m \tag{3}$$

where ΔW is the aggregated weight and Δb is the aggregated bias. Therefore, ΔW and Δb are sample-dependent and dynamic convolution increases non-linearity.

2.3 Domain Adaptation Scheme

The training objective consists of three parts, i.e., depression classification loss L_c, domain classifier loss L_d, and loss L_A of the association domain adaptation part. Thus, the overall training objective is shown as:

$$L = L_c + L_d + L_A \tag{4}$$

We adopted the cross-entropy function as the classification loss:

$$L_c = H(\tilde{y}^i, y^i) \tag{5}$$

where H represents cross-entropy loss, $\tilde{y}^i$ is the output of the label classifier for x^i and y^i is the depression label.

Transfer learning involves the use of general patterns and features acquired by a neural network from the source domain to enable it to make predictions on new target domain data. The domain adaptation algorithm aims to minimize the distribution of potential features in the source and target domains. We selected adversarial domain adaptation and association domain adaptation to perform marginal and conditional distribution alignment.

Adversarial Domain Adaptation. The adversarial domain adaptation approach was initially applied to image classification [18], but has since shown efficacy in the analysis of EEG signals [19–21]. This approach primarily consists of two components, i.e., gradient reversal layer (GRL) [18], and domain classifier. The domain classifier determines whether a given sample is from the source or target domain, with the label $d \in \{0, 1\}$. Each sample is first passed through the feature extractor to obtain feature F and subsequently through the domain classifier, where θ_f represents the network parameters of the feature extractor and θ_d represents the network parameters of the domain classifier. The principal aim of adversarial domain adaptation is to induce the feature extractor to capture domain-invariant features. This is achieved by inserting the GRL between the feature extractor and domain classifier, whereby the GRL multiplies the gradient by a negative value only during backpropagation. The GRL enables the θ_f feature extractor parameter to be continually adjusted during training to maximize the loss of domain binary classification, while the θ_d domain classifier parameter is adjusted to minimize the loss of domain classification. The domain classifier cannot correctly separate the source domain from the target domain i.e., domain-invariant features are extracted. The loss function of this part can be written as:

$$L_d = \frac{1}{n_s + n_t} \sum_{x^i \in D_s \cup D_t} H(\tilde{p}_i, d_i) \tag{6}$$

where $\tilde{p}_i$ is the output of the domain classifier for x^i and d_i is the domain label. The optimization objectives of θ_f and θ_d are as follows:

$$\tilde{\theta}_f = \arg\max_{\theta_f} L_d$$

$$\tilde{\theta}_d = \arg\min_{\theta_d} L_d \tag{7}$$

Association Domain Adaptation. To achieve a more fine-grained distribution alignment, the association domain adaptation(ADA) [22] method is used to draw the source and target domains according to different categories. As shown in Fig. 1, F_s^i and F_t^j are high-level feature representations of the source domain sample x_s^i and target domain sample x_t^j extracted by the feature extractor, respectively. The dot product of F_s^i and F_t^j is used as a similarity measurement denoted as $sim(i,j) = <F_s^i, F_t^j>$. Thus, the transition probability from F_s^i to F_t^j (i.e., $F_s^i \to F_t^j$) can be defined as:

$$P_{ij}^{st} = P\left(F_t^j \mid F_s^i\right) = \frac{\exp\left(sim\left(i,j\right)\right)}{\sum_{j'} \exp\left(sim\left(i,j'\right)\right)} \tag{8}$$

Furthermore, P_{ij}^{sts} can be defined as the circular transfer probability $F_s^i \to F_t^j \to F_s^j$, i.e., the probability of a sample randomly wandering from source domain F_s^i to target domain F_t^j and then back to source domain F_s^j, which can be written as:

$$P_{ij}^{sts} = \left(P_{ij}^{st} P_{ij}^{ts}\right) \tag{9}$$

To constrain F_s^i circular wandering to F_s^j belonging to the same class with all circular wanderings of the same class having the same probability, walker loss L_w is defined so that the circular transfer probability is close to the distribution of U using the following equation:

$$L_w = H(U, P^{sts})$$

$$U = \begin{cases} \dfrac{1}{|class(D_s^i)|} & , class\left(D_s^i\right) = class\left(D_s^j\right) \\ 0 & , else \end{cases} \tag{10}$$

Furthermore, L_v is introduced to constrain the association of the two domains to cover as many samples of the target domain as possible.

$$L_v = H(V, P^v)$$
$$V_j = \frac{1}{|D_t|}, P_j^v = \sum_{x_s^i \in D_s} P_{ij}^{st} \qquad (11)$$

Therefore, the loss of ADA can be written as Equation(12), and the weights of the two parts can be dynamically adjusted using λ.

$$L_A = L_v + \lambda L_w \qquad (12)$$

3   Experiment

3.1   Datasets

The public MODMA dataset [23] was used in our experiment. The MODMA dataset includes EEG and audio data from clinically depressed patients and normal controls, with EEG signals recorded from a total of 24 MDD patients and 29 HCs. The patients included were carefully diagnosed and selected by professional psychiatrists. Participants were asked to remain awake and still, without unnecessary eye movements or any body movements. The data set was acquired using 129 channels (128 electrodes, E1 to E128; and a reference electrode located at the central midline position Cz). MDD was confirmed by psychiatric assessment. All subjects were assessed with questionnaires on psychosocial and general health and sleep quality. A detailed description of the dataset is presented in the table below. We use the raw EEG signal as input.

Table 1. Details of MODMA dataset

Information	MDD		HC	
Sex	Female	Male	Female	Male
Number	11	13	9	20
Age (years)	27.54 ± 9.13	33.69 ±10.85	29 ± 7.8	32.55 ± 9.65
Education (years)	14.9 ± 3.44	11.84 ± 3.57	16.44 ±2.87	15.75 ± 2.07
PHQ9 score	19 ± 4.40	17.65 ±2 .55	1.55 ± 1.94	3.15 ± 1.53

3.2   State-of-the-Art Models

In the following experiments, we compared and analyzed the performance of depression detection with the following methods. Domain adversarial neural networks (DANN) [18] uses an adversarial domain classifier to assist the feature extractor to extract domain invariant features. Deep adaptation network

(DAN) [24] explicitly narrows the distance between the feature distributions of the source and target domains extracted by multiple task-specific layers with the help of multi-kernel maximum mean discrepancy to enhance model transferability. Self-attentive graph embedding method (SAGE) [25] is a graph classification framework that embeds graph-level instances into a fixed-length vector, which is used to implement graph classification tasks. SST-Emotionnet [26] maps EEG signals into 2D data according to electrode locations and integrates spatial-spectral-temporal features simultaneously. Self-attention graph pooling with soft label (SGP-SL) [7] constructs graph structures based on local and global connections between EEG channels, while avoiding information loss through a self-attention graph pooling module and improving feature discriminability with soft labels.

3.3   Experimental Settings

In our experiments, 10-fold cross-validation assessment was applied. In detail, we randomly selected four people as the test set (target domain), including two HCs and two MDD patients. Samples of the remaining subjects were used as the training set (source domain). The above 10-fold cross-validation experiment was repeated 10 times, with the mean, standard deviation, precision, recall, and F1-score of the classification results then compared and analyzed. For the dynamic feature extractor, the main bottleneck residual block adopted 1×1, 1×3, and 1×1 convolutions with 128, 62, and 128 filters, respectively. The output vector is then downscaled to a 32-dimensional vector by a fully connected layer. The label classifier and domain classifier were both composed of a fully connected layer, reshaping the feature from 32D to 2D. The number of 1×1 parallel M convolution kernels was set to 2. The training process was optimized by adaptive moment estimation (Adam), and we set the initial learning rate to 0.001 with a weight decay of 1e-3. The batch size was 128, with training for 100 epochs.

3.4   Evaluation Metrics

In depression diagnostic binary classification tasks, accuracy (Acc), precision (Pre), recall (Rec), and F1-score are typically used as evaluation metrics to measure the effectiveness of a model, defined as:

$$Acc = \frac{TP + TN}{TP + FN + FP + TN}$$
$$Pre = \frac{TP}{TP + FP}$$
$$Rec = \frac{TP}{TP + FN} \tag{13}$$
$$\text{F1-score} = \frac{2 \times Pre \times Rec}{Pre + Rec}$$

where TP, FN, FP, and TN represent true positive, false negative, false positive, and true negative, respectively.

Table 2. Subject-independent classification performance on MODMA dataset

Model	Acc	Pre	Rec	F1-score
SAGE [25]	67.92	64.00	66.67	65.30
SST-Emotionnet [26]	73.58	69.23	75.00	72.00
SGP-SL [7]	84.91	80.77	87.50	84.00
DAN [24]	64.77	55.72	64.88	55.81
DANN [18]	85.08	90.01	85.06	84.09
DCAAN (ours)	86.85	91.14	86.81	85.97

4   Results

We compared our model with several state-of-the-art MDD diagnosis methods based on subject-independent experiments, as shown in Table 2. SAGE [25], SST-Emotionnet [26], and SGP-SL [7], which do not use transfer learning, achieve MDD diagnosis by fully capturing valid spatiotemporal high-dimensional features. The results in the Table 2 were obtained from subject-independent experiments conducted by Chen et al. [7], with SGP-SL showing the highest accuracy of 84.91%. Given the limited research on the use of transfer learning in MODMA, we used the DANN and DAN approaches according to the same experimental paradigm. Results showed that DCAAN achieved the highest classification accuracy (86.85%), and best Pre, and F1-score metrics. DCAAN exceeded SGP-SL by 2% and outperformed both classical domain adaptation methods (DAN and DANN). Our approach, which uses a relatively simple but efficient dynamic feature extractor combined with powerful marginal and conditional domain adaptation, achieved the highest subject-independent MDD diagnosis. As shown in Fig. 2, we drew confusion matrices of the first- and fourth-fold cross-validations and 10-fold average results, with DCAAN showing the high diagnostic accuracy for both HCs and MDDs.

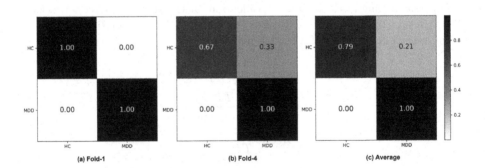

Fig. 2. Confusion matrices of depression detection on MODMA dataset

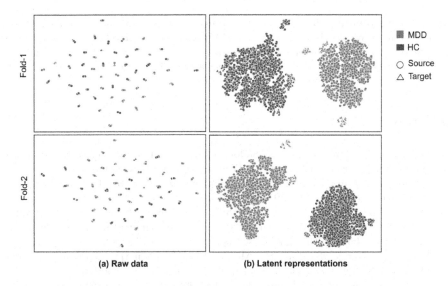

Fig. 3. Visualization of DCAAN using t-SNE

To explain the effectiveness of our algorithm, we used t-SNE [27] to visualize changes in the feature distributions, as shown in Fig. 3. We randomly selected 3000 samples in the source domain and 200 samples in the target domain in each cross-validation experiment, with the first- and fourth-fold results shown in Fig. 3. Figure 3(a) shows the original distribution of the source and target domains, and Fig. 3(b) shows the feature distribution of the output of the feature extractor. Clearly, our method distinctly separated the source and target domains based on classes, resulting in reduced intra-class distance and a well-defined inter-class boundary.

5   Conclusions

The proposed DCAAN model applies domain adaptation to EEG-based MDD diagnosis, which can better attenuate the effects of individual variability in EEG signals on diagnostic accuracy. This model, trained on limited known data, showed good diagnostic results for unknown individuals, thereby offering an objective and scientific approach for detecting depression. Based on a series of experiments, we demonstrated the feasibility and usefulness of dynamic convolution for adaptive feature extraction. On the publicly available 128-channel MODMA dataset, our model achieved 86.85% binary classification accuracy using a 10-fold cross-validation strategy. Although we only applied the DCAAN model to the field of EEG depression diagnosis in this study, our approach could be widely applied to other EEG-based classification problems, such as emotion recognition, motor imagery, epilepsy detection, and sleep stage classification.

Acknowledgements. This work was partially supported by the National Natural Science Foundation of China (Grant No. 61971420), the Science Frontier Program of the Chinese Academy of Sci-ences (Grant No. QYZDJ-SSW-SMC019) and the Science and Technology Innovation 2030 - Brain Science and Brain-Inspired Intelligence Project (Grant No. 2021ZD0200200).

References

1. Sadock, B,J., et al.: Kaplan & Sadock's synopsis of psychiatry: behavioral sciences/clinical psychiatry, vol. 2015. Wolters Kluwer Philadelphia, PA (2015)
2. Edition, F., et al.: Diagnostic and statistical manual of mental disorders. Am. Psychiatric. Assoc. **21**(21), 591–643 (2013)
3. Sharma, M., Achuth, P.V., Deb, D., Puthankattil, S.D., Acharya, U.R.: An automated diagnosis of depression using three-channel bandwidth-duration localized wavelet filter bank with eeg signals. Cognit. Syst. Res. **52**, 508–520 (2018)
4. Bashir, N., Narejo, S., Naz, B., Ali, A.: EEG based major depressive disorder (MDD) detection using machine Learning. In: Djeddi, C., Siddiqi, I., Jamil, A., Ali Hameed, A., Kucuk, İ (eds.) MedPRAI 2021. CCIS, vol. 1543, pp. 172–183. Springer, Cham (2022). https://doi.org/10.1007/978-3-031-04112-9_13
5. Song, X.W., Yan, D., Zhao, L., Yang, L.: Lsdd-eegnet: An efficient end-to-end framework for eeg-based depression detection. Biomed. Signal Process. Control **75**, 103612 (2022)
6. Tasci, G., et al.: Automated accurate detection of depression using twin pascal's triangles lattice pattern with eeg signals. Knowl.-Based Syst. **260**, 110190 (2023)
7. Chen, T., Guo, Y., Hao, S., Hong, R.: Exploring self-attention graph pooling with eeg-based topological structure and soft label for depression detection. IEEE Trans. Affect. Comput. **13**(4), 2106–2118 (2022)
8. Zhuang, F., et al.: A comprehensive survey on transfer learning. Proc. IEEE **109**(1), 43–76 (2020)
9. Farahani, A., Voghoei, S., Rasheed, K., Arabnia, H.R.: A brief review of domain adaptation. In: Advances in Data Science and Information Engineering: Proceedings from ICDATA 2020 and IKE 2020, pp. 877–894 (2021)
10. Zhao, L.-M., Yan, X., Lu, B.-L.: Plug-and-play domain adaptation for cross-subject eeg-based emotion recognition. In: Proceedings of the AAAI Conference on Artificial Intelligence, vol. 35, pp. 863–870 (2021)
11. He, K., Zhang, X., Ren, S., Sun, J.: Deep residual learning for image recognition. In Proceedings of the IEEE Conference on Computer Vision and Pattern Recognition, pp. 770–778 (2016)
12. Chowdhury, A., Ross, A.: Fusing mfcc and lpc features using 1d triplet cnn for speaker recognition in severely degraded audio signals. IEEE Trans. Inf. Forensics Secur. **15**, 1616–1629 (2019)
13. Trelinski, J., Kwolek, B.: Embedded features for 1d cnn-based action recognition on depth maps. In: VISIGRAPP (4: VISAPP), pp. 536–543 (2021)
14. Kiranyaz, S., Avci, O., Abdeljaber, O., Ince, T., Gabbouj, M., Inman, D.J.: 1d convolutional neural networks and applications: a survey. Mech. Syst. Signal Process. **151**, 107398 (2021)
15. Li, Y., Yuan, L., Chen, Y., Wang, P., Vasconcelos, N.: Dynamic transfer for multisource domain adaptation. In: Proceedings of the IEEE/CVF Conference on Computer Vision and Pattern Recognition, pp. 10998–11007 (2021)

16. Hu, J., Shen, L., Sun, G.: Squeeze-and-excitation networks. In: Proceedings of the IEEE Conference on Computer Vision and Pattern Recognition, pp. 7132–7141 (2018)

17. Chen, Y., Dai, X., Liu, M., Chen, D., Yuan, L., Liu, Z.: Dynamic convolution: Attention over convolution kernels. In: Proceedings of the IEEE/CVF Conference on Computer Vision and Pattern Recognition, pp. 11030–11039 (2020)

18. Ganin, Y., et al.: Domain-adversarial training of neural networks. J. Mach. Learn. Res. **17**(1), 2030–2096 (2016)

19. Li, H., Jin, Y.-M., Zheng, W.-L., Lu, B.-L.: Cross-subject emotion recognition using deep adaptation networks. In: Cheng, L., Leung, A.C.S., Ozawa, S. (eds.) ICONIP 2018. LNCS, vol. 11305, pp. 403–413. Springer, Cham (2018). https://doi.org/10.1007/978-3-030-04221-9_36

20. Hang, W., et al.: Cross-subject eeg signal recognition using deep domain adaptation network. IEEE Access **7**, 128273–128282 (2019)

21. Jin, Y.-M., Luo, Y.-D., Zheng, W.-L., Lu, B.-L.: Eeg-based emotion recognition using domain adaptation network. In: 2017 International Conference on Orange Technologies (ICOT), pp. 222–225. IEEE (2017)

22. Haeusser, P., Frerix, T., Mordvintsev, A., Cremers, D.: Associative domain adaptation. In: Proceedings of the IEEE International Conference on Computer Vision, pp. 2765–2773 (2017)

23. Cai, H., et al.: Modma dataset: a multi-modal open dataset for mental-disorder analysis. arXiv preprint arXiv:2002.09283 (2020)

24. Long, M., Cao, Y., Wang, J., Jordan, M.: Learning transferable features with deep adaptation networks. In: International Conference on Machine Learning, pp. 97–105. PMLR (2015)

25. Li, J., Rong, Y., Cheng, H., Meng, H., Huang, W., Huang, J.: Semi-supervised graph classification: A hierarchical graph perspective. In: The World Wide Web Conference, pp. 972–982 (2019)

26. Jia, Z., Lin, Y., Cai, X., Chen, H., Gou, H., Wang, J.: Sst-emotionnet: Spatial-spectral-temporal based attention 3d dense network for eeg emotion recognition. In: Proceedings of the 28th ACM International Conference on Multimedia, pp. 2909–2917 (2020)

27. Van der Maaten, L., Hinton, G.: Visualizing data using t-sne. J. Mach. Learn. Res. **9**(11) (2008)

Multi-label Adversarial Defense Scheme Based on Negative Correlation Ensemble

Hongwei Zhang[1], Wenjian Luo[1,2($\boxtimes$)], Zhijian Chen[1], and Qi Zhou[1]

[1] Guangdong Provincial Key Laboratory of Novel Security Intelligence Technologies,
School of Computer Science and Technology, Harbin Institute of Technology,
Shenzhen 518055, Guangdong, China
{20s151127,21B951010,22s051036}@stu.hit.edu.cn, luowenjian@hit.edu.cn
[2] Peng Cheng Laboratory, Shenzhen 518055, Guangdong, China

Abstract. Adversarial examples have become an important issue in the field of deep learning security. There have been many studies on adversarial example attack and defense algorithms for single-label classification models. However, in the real world, multi-label classification models are also widely used. There are only a few studies on adversarial example attack and defense algorithms for multi-label classification models. In this paper, we propose a negative correlation ensemble defense scheme against multi-label adversarial examples (ML-NCEn). The fundamental principle of ML-NCEn is to make the gradient directions and magnitudes of member models negatively correlated with those of other members, respectively, in the positive and negative label sets. Experimental results show that ML-NCEn has good adversarial robustness.

Keywords: Adversarial examples · Ensemble · Negative correlation

1 Introduction

In recent years, Deep Neural Networks (DNNs) have grown rapidly and been used in many areas [3,4,12,13,21,24]. Although artificial intelligence (AI) systems based on deep learning techniques perform well in processing complex problems such as speech, images, and text, a considerable amount of research demonstrates that deep learning models are vulnerable to adversarial examples, which are generated by adding well-designed and imperceptible perturbations to original examples. As a result, deep learning models would produce incorrect classification results with high confidence, posing a significant security risk [22]. The emergence of adversarial examples poses a serious challenge to the secure deployment of AI systems.

This study is supported by the National Key R&D Program of China (Grant No. 2022YFB3102100), Shenzhen Fundamental Research Program (Grant No. JCYJ20220818102414030), the Major Key Project of PCL (Grant No. PCL2022A03, PCL2021A02, PCL2021A09), Guangdong Provincial Key Laboratory of Novel Security Intelligence Technologies (Grant No. 2022B1212010005).

© The Author(s), under exclusive license to Springer Nature Switzerland AG 2023
Y. Tan et al. (Eds.): ICSI 2023, LNCS 13969, pp. 284–294, 2023.
https://doi.org/10.1007/978-3-031-36625-3_23

Currently, the majority of research on generating adversarial examples focuses on single-label classification. In a single-label classification, the model takes the label which has the highest prediction confidence as the prediction result. In a single-label adversarial attack, the attacker only needs to make the confidence score of any negative label beyond that of the positive label, then the model will misclassify and the attack will be successful [10]. Unlike single-label classification models, multi-label classification models rely on a specified threshold for prediction, i.e., if the confidence score of a label exceeds this threshold, the example is considered to contain this label. Due to the complexity of multi-label classification models and the correlation between labels [2], existing single-label classification attack techniques cannot be directly applied to multi-label classification models. Song et al. [20] proposed two targeted attack frameworks for attacking multi-label classification and ranking tasks. They developed two ranking attack algorithms, i.e., ML-RankI and ML-RankII, and two classification attack algorithms, i.e., ML-CW and ML-DeepFool. Since multi-label classification tasks are very common in real-world scenarios, the research on generating and defense of multi-label adversarial examples is of great significance, which is important for improving the robustness of deep learning models and the security of deep learning systems.

Existing techniques of defending against adversarial examples could be divided into detecting adversarial perturbations and improving model robustness [25]. The ensemble learning has a natural adversarial robustness due to the different initial weights, model structures, and training methods of each member in the ensemble. This means that an adversarial example which can deceive one model may not successfully deceive the other members in the ensemble [5]. By ensemble learning, the vulnerability of a single model could be avoided for the purpose of defending against adversarial examples. Recently, the ensemble training scheme proposed by Luo et al. [17] based on the negative correlation principle could effectively defend against the attack of adversarial examples. However, the negative correlation ensemble method in [17] is designed for single-label classification models. In this paper, we adapt the negative correlation ensemble method in [17] for multi-label classification models.

Specifically, the contributions of this paper are described as follows.

- In this paper, we propose a multi-label adversarial defense scheme based on negative correlation ensemble (ML-NCEn).
- Experimental results demonstrate that ML-NCEn has better adversarial robustness compared to typical ensemble defense schemes.

The rest of this paper is organized as follows. We introduce the related research work on negative correlation and single-label ensemble defense algorithms in Sect. 2. We describe the principle of a multi-label ensemble defense algorithm based on negative correlation in Sect. 3. Then, Sect. 4 gives the validation process and results of the experiments. Finally, Sect. 5 draws a brief conclusion.

2   Related Work

2.1   Ensemble Defense

The model ensemble is an effective way to improve model robustness [1,5,11, 18]. After training, different members in the ensemble would naturally exhibit diversity due to their distinct model structures, initial weights, and training methods. Based on this idea, Kariyappa et al. [11] proposed Gradient Alignment Loss (GAL) to improve the adversarial robustness of the ensemble by considering the diversity of gradient directions across the members. However, GAL does not consider the optimal geometric boundary for the diversity of gradient directions within the ensemble, nor does it balance the magnitude of the gradient for each member.

Therefore, Dabouei et al. [7] proposed Gradient Phase and Magnitude Regularization (GPMR). GPMR aims to raise the lower bound of perturbation by considering the optimal geometric bound which diversifies the gradient directions and by balancing the magnitude of the gradients of the members in the ensemble. Thus, the first-order interaction defense of the members in the ensemble is constructed. However, GPMR does not sufficiently consider the interactions among the members of the ensemble.

Pang et al. [18] proposed Adaptive Diversity Promoting (ADP). ADP redefines the diversity of individual members in the ensemble, i.e., the diversity among non-maximal predictions of individual members in the ensemble, to provide better adversarial robustness.

2.2   Negative Correlation

The negative correlation ensemble was originally proposed by Liu and Yao in 1999 [14,15]. Liu and Yao [15] proposed a Cooperative Ensemble Learning System (CELS) to improve the prediction accuracy of the ensemble by negative correlation learning. Liu et al. [16] also attempted to estimate the optimal number of neural networks in the ensemble by negative correlation learning and evolutionary learning, while maintaining good interaction of individual members. Chan et al. [6] proposed Negative Correlation Learning via Correlation-Corrected Data (NCCD). Wang et al. [23] proposed a new negative correlation learning algorithm: AdaBoost.NC. AdaBoost.NC has better generalization performance and the training time cost is significantly lower than CELS and NCCD.

Recently, Luo et al. [17] proposed a negative correlation ensemble (NCEn), where the gradient direction and gradient magnitude of each member in the ensemble are negatively correlated at the same time, and experimental results showed that NCEn has good adversarial robustness in single-label classification scenarios.

3   The Proposed Method

This section will introduce the multi-label adversarial defense scheme based on negative correlation ensemble (ML-NCEn) in detail. ML-NCE is adapted from NCEn in [17].

3.1   Gradient Direction

In a multi-label classification model, a label is classified as positive if its confidence score exceeds a specific threshold, and negative otherwise. ML-NCEn uses the principle of negative correlation to make the gradient directions of ensemble members on the positive and negative label sets negatively correlated with other members, thereby increasing diversity among ensemble members.

When the attacker adds an adversarial perturbation ϵ along the gradient direction of the model, the loss function of the model changes the most. Assuming that each member in the ensemble has very different gradient directions, an adversarial perturbation ϵ added by the attacker in a certain gradient direction will only significantly change the output of member models whose gradients are similar to attacking gradient direction, and the adversarial perturbation ϵ is difficult to change the output confidence of other members. In this paper, we still use cosine similarity to evaluate the similarity between model gradients, and make each member of the ensemble negatively correlate in the gradient direction on the positive and negative label sets, respectively. The regularization terms proposed are shown in the formulas (1) and (2).

$$\text{Loss}_{\cos}^{\Omega^+} = \text{CS}\left(\nabla J_i^{\Omega^+}, \nabla J_{\text{ensemble}}^{\Omega^+}\right) \sum_{\substack{j=1 \\ j\neq i}}^{k} \text{CS}\left(\nabla J_j^{\Omega^+}, \nabla J_{\text{ensemble}}^{\Omega^+}\right), \qquad (1)$$

$$\text{Loss}_{\cos}^{\Omega^-} = \text{CS}\left(\nabla J_i^{\Omega^-}, \nabla J_{\text{ensemble}}^{\Omega^-}\right) \sum_{\substack{j=1 \\ j\neq i}}^{k} \text{CS}\left(\nabla J_j^{\Omega^-}, \nabla J_{\text{ensemble}}^{\Omega^-}\right), \qquad (2)$$

where,

$$\nabla J_{ensemble}^{\Omega^+} = \frac{1}{k} \sum_{i=1}^{k} \nabla J_i^{\Omega^+}, \qquad (3)$$

$$\nabla J_{ensemble}^{\Omega^-} = \frac{1}{k} \sum_{i=1}^{k} \nabla J_i^{\Omega^-}, \qquad (4)$$

$\nabla J_i^{\Omega^+}$ represents the gradient of the loss value of the positive label set of the ith member relative to the input. $\nabla J_i^{\Omega^-}$ is the gradient of the loss value of the negative label set of the ith member relative to the input. $\nabla J_{ensemble}^{\Omega^+}$ and $\nabla J_{ensemble}^{\Omega^-}$ is the mean of the gradients of all members in terms of positive and negative labels, respectively. k is the number of members in the ensemble.

3.2   Gradient Magnitude

The gradient magnitude of the loss function relative to the input indicates the extent to which the adversarial perturbation along the gradient direction could alter the classifier's output confidence. The larger the gradient magnitude, the greater the same perturbation could change the classifier's loss function value, thereby changing the label confidence of the classifier's output.

To address this issue, this paper constrains the gradient magnitude of each member in the ensemble through the principle of negative correlation. Through the negative correlation training, the gradient ranges of all members in the ensemble are negatively correlated in both positive and negative label sets, which means it is difficult for an attacker to change a positive label to a negative label or a negative label to a positive label by attacking a member with a larger gradient magnitude. Therefore, this paper suggests regularization terms on the gradient magnitude as shown in the formulas (5) and (6).

$$\text{Loss}_{\text{norm}}^{\Omega^+} = \frac{1}{g^{\Omega^+2}} \left(\left\| \nabla J_i^{\Omega^+} \right\|_2 - g^{\Omega^+} \right) \sum_{j \neq i}^{k} \left(\left\| \nabla J_j^{\Omega^+} \right\|_2 - g^{\Omega^+} \right), \qquad (5)$$

$$\text{Loss}_{\text{norm}}^{\Omega^-} = \frac{1}{g^{\Omega^-2}} \left(\left\| \nabla J_i^{\Omega^-} \right\|_2 - g^{\Omega^-} \right) \sum_{j \neq i}^{k} \left(\left\| \nabla J_j^{\Omega^-} \right\|_2 - g^{\Omega^-} \right), \qquad (6)$$

where,

$$g^{\Omega^+} = \frac{1}{k} \sum_{i=1}^{k} \left\| \nabla J_i^{\Omega^+} \right\|_2, \qquad (7)$$

$$g^{\Omega^-} = \frac{1}{k} \sum_{i=1}^{k} \left\| \nabla J_i^{\Omega^-} \right\|_2, \qquad (8)$$

g^{Ω^+} and g^{Ω^-} represent the mean of the gradient magnitude of all members in terms of the positive and negative label sets, respectively. This paper uses L_2 norm to represent the magnitude of the gradient.

Given that all members are trained at the same time, different members can learn different characteristics from the training dataset and have different gradient sizes on the same input, because it is not easy to always have very small gradient magnitudes throughout the input space. This prevents an attacker from altering the label properties of the input examples by attacking a few models with large gradient magnitudes in the ensemble.

3.3   ML-NCEn

Constraining the ensemble training from either the gradient direction or the gradient magnitude of the members in the ensemble will improve the robustness of the ensemble, but constraining from a single aspect has its drawbacks. Similar to [17], this paper considers both the direction and the magnitude of the gradient, making them negatively correlated in the terms of the positive and negative label sets, so as to improve the robustness of the ensemble. The loss function for ensemble training is shown in the formula (9):

$$Loss = BCE + \lambda_{\cos} \text{Loss}_{\cos}^{\Omega^+} + \lambda_{\cos} \text{Loss}_{\cos}^{\Omega^-} + \lambda_{\text{norm}} \text{Loss}_{\text{norm}}^{\Omega^+} + \lambda_{\text{norm}} \text{Loss}_{\text{norm}}^{\Omega^-},$$
(9)

$\lambda_{\cos}$ and λ_{norm} are hyperparameters. BCE (Binary Cross-Entropy) is the general loss function of the multi-label classification model, which can be called directly from the Pytorch library [19]. Its calculation formula is as follows.

$$BCE(X,Y) = \frac{1}{NM} \sum_{i=1}^{N} \sum_{c=1}^{M} loss_{i,c},$$
(10)

$$loss_{i,c} = -w_i * [y_{i,c} * \log (\text{sigmoid}(x_{i,c})) + (1 - y_{i,c}) * \log (1 - \text{sigmoid}(x_{i,c}))],$$
(11)

where N is the number of samples, M is the number of classes, $loss_{i,c}$ is the loss function of the i-th sample to the class c, $x_{i,c}$ is the model's output for the i-th sample corresponding to class c, $y_{i,c}$ is the label of the i-th sample corresponding to class c, and w_i is the hyper-parameter used to address the issue of imbalanced samples between classes.

Algorithm 1. ML-NCEn

Require: Dataset: X, the correct label set for the dataset X: label, list of all members in the ensemble: model_list, number of members in the ensemble: model_num
Ensure: Trained ensemble
1: **for** $epoch \leftarrow begin\_epoch$ **to** end_epoch **do**
2:    $loss = 0$
3:    $loss\_P = 0$
4:    $loss\_N = 0$
5:    **for** $model_i \leftarrow model\_list$ **do**
6:       $loss+ = BCE(output, target)/model\_num$
7:       $loss\_P+ = BCE(output[target == 1], target[target == 1])/model\_num$
8:       $loss\_N+ = BCE(output[target == -1], target[target == -1])/model\_num$
9:    **end for**
10:    $gradients\_P \leftarrow autograd.grad(loss\_P)$
11:    $gradients\_N \leftarrow autograd.grad(loss\_N)$
12:    $NCEn\_loss\_P \leftarrow \lambda_{cos} \cdot get\_cos(gradients\_P) + \lambda_{norm} \cdot get\_norm(gradients\_P)$
13:    $NCEn\_loss\_N \leftarrow \lambda_{cos} \cdot get\_cos(gradients\_N) + \lambda_{norm} \cdot get\_norm(gradients\_N)$
14:    Back-propagation using the sum of $loss$, $NCEn\_loss\_P$ and $NCEn\_loss\_N$
15:    Update the model parameters for each model
16: **end for**

The negative correlation training algorithm for the multi-label classification ensemble is shown in Algorithm 1. In each training epoch, this paper calculates the cross-entropy *loss* of all members in the ensemble on all labels, the cross-entropy *loss_P* over the set of positive labels and the cross-entropy *loss_N* over the set of negative labels, respectively. Then the gradients of the input examples for each member in the ensemble are calculated using *loss_P* and *loss_N*, and the corresponding regulations are calculated, respectively. Finally, the sum of *loss*, $NCEn\_loss\_P$ and $NCEn\_loss\_N$ is back-propagated and the parameters of each member in the ensemble are updated.

4   Experiments

This section describes the experimental setup and results of ML-NCEn. The source codes of ML-NCEn are available at https://github.com/MiLab-HITSZ/2023ZhangML-NCEn.

4.1   Experimental Setup

Dataset. Two multi-label datasets, VOC2007 [9] and VOC2012 [8], are used in this paper. The number of examples in the training set, validation set and test set are shown in Table 1. Since the test set is not included in VOC2012, the original validation set is used as the test set in this paper. The 80% of the original training set is used as the new training set, and the remaining 20% is used as the validation set. In experiments, the size of all images is unified as $299 * 299$.

Table 1. Dataset

Dataset	Training set and validation set	Test set	Label Category
VOC2007	7521	4952	20
VOC2012	5017	5821	20

Comparison Method. Five typical methods are used in the experiments to compare the performance of ML-NCEn. The first one is a single-model scheme with only one classifier, and the classifier used is ML-LIW from [20]. The second comparison method is a simple ensemble baseline without any regularization terms, i.e., three ML-LIW models as ensemble members. The third comparison method is a multi-label version ML-GAL of GAL [11], using three ML-LIW models as ensemble members. The fourth comparison method is ML-GPMR, which is a multi-label version of GPMR [7], using three ML-LIW models as ensemble members. ML-GAL and ML-GPMR are not different from GAL and GPMR except for the loss function. The fifth comparison method is ML-ADP, a multi-label porting scheme for ADP [18], also using three ML-LIW models as ensemble members. ML-ADP does not differ from ADP except that it replaces the unique prediction labels in ADP with a collection of positive labels.

Attack Parameter Setting. In this experiment, we mainly use the attack strategy of hiding a single label. We select examples with at least two positive labels from the examples that can be correctly classified by the ensemble, and then choose to hide one of the positive labels as a negative label. To shorten the attack time, for each attack, this paper experimentally selects the first 600 examples that satisfy the condition as the adversarial examples. This paper adopts four multi-label adversarial example attack algorithms to fool the ensembles, namely ML-CW, ML-DeepFool, ML-RANK1, ML-RANK2 [20].

4.2 Experimental Results and Analysis

This paper uses two general metrics, i.e., the accuracy of clean examples (ACE) and the attack success rate of adversarial example (ASA), to measure the performance of the ensemble defense scheme.

The clean example test accuracy is the ratio of the test examples which could be correctly identified by the ensemble. The ACE is calculated as follows.

$$success(F(x), y_{true}) = \begin{cases} 1, F(x) == y_{true} \\ 0, F(x)! = y_{true} \end{cases}, \tag{12}$$

$$ACE = \frac{\sum_{i=1}^{N} success(F(x_i), y_{i,true})}{N}, \tag{13}$$

where x is the input example, $F(x)$ is the ensemble prediction result, y_{true} is the ground truth label, $y_{i,true}$ is the ground truth label of the i-th sample, and N is the number of examples in the test set.

The definition of ASA is shown below.

$$success(F(x'), y_{target}) = \begin{cases} 1, F(x') == y_{target} \\ 0, F(x')! = y_{target} \end{cases} \tag{14}$$

$$ASA = \frac{\sum_{i=1}^{M} success(F(x'_i), y_{i,target})}{M} \tag{15}$$

where x' is the adversarial example generated by the adversarial example generation algorithm, y_{target} is the target label of the attack, $y_{i,target}$ is the target label of the i-th sample, and $F(x')$ is the prediction given by the ensemble. M is the number of adversarial examples used to attack the ensemble. In this paper, the first 600 adversarial examples are selected for each attack method against the ensemble in this paper.

Tables 2 and 3 show ACE and ASA of the single model and the five ensemble defense schemes on two different datasets, where the baseline is a simple ensemble without any regularization term. The data in the table show that the clean example test accuracies of the single model and the five ensemble defense schemes on two different datasets remain around 90%, all of which can produce correct prediction results for clean examples with a high confidence level.

In Tables 2 and 3, RANK denotes the sum of the rankings of each defense scheme under the four different attack methods, and a lower RANK indicates the better overall performance of the defense scheme. The bold part in the table indicates the best performance. When using dataset VOC2007, the attack success rate of ML-RANK1 on ML-NCEn is only 0.17, which is two percentage points lower than ML-ADP; the attack success rate of ML-DeepFool on ML-NCEn is 0.29, which is five percentage points lower than ML-ADP. When using dataset VOC2012, the attack success rate of ML-RANK1 on ML-NCEn is 0.15; ML-RANK2 on ML-NCEn is 0.82; ML-DeepFool on ML-NCEn is 0.13, which is nineteen percentage points lower than ML-ADP. ML-NCEn has the lowest RANK value, which indicates that ML-NCEn has the best overall performance under four different attack methods. This indicates that it is difficult for an attacker to change the confidence level of a positive label to below the threshold by hiding a single label.

From the experimental results, it can be concluded that ML-NCEn has stronger robustness compared to other ensemble schemes. This means that by making the gradient direction and gradient magnitude of each member negatively correlated on the set of positive and negative labels, respectively, the overlap area between the adversarial subspaces of each member in the ensemble

Table 2. Performance comparison of each ensemble defense algorithm on VOC2007

	ACE	ML-CW	ML-RANK1	ML-RANK2	ML-DeepFool	RANK
Single Model	**0.9201**	0.9998/6	0.2192/4	0.9256/5	0.4519/6	21
Baseline	0.9152	0.8783/4	0.2706/6	0.9418/6	0.4479/5	21
ML-GAL	0.9025	0.8000/2	0.1962/3	**0.8145**/1	0.3478/3	9
ML-GPMR	0.9122	0.9915/5	0.2273/5	0.8873/4	0.3748/4	18
ML-ADP	0.9054	**0.7998**/1	0.1900/2	0.8640/3	0.3445/2	8
ML-NCEn	0.8916	0.8755/3	**0.1750**/1	0.8383/2	**0.2983**/1	**7**

Table 3. Performance comparison of each ensemble defense algorithm on VOC2012

	ACE	ML-CW	ML-RANK1	ML-RANK2	ML-DeepFool	RANK
Single Model	**0.9213**	0.9966/6	0.1598/3	0.8769/4	0.4269/6	19
Baseline	0.9109	0.8735/4	0.2017/6	0.8786/5	0.3880/5	20
ML-GAL	0.9011	0.8085/2	0.1783/5	0.8914/6	0.3412/4	17
ML-GPMR	0.9029	0.9615/5	0.1504/2	0.8308/3	0.3334/3	13
ML-ADP	0.9096	**0.7894**/1	0.1640/4	0.8285/2	0.3264/2	9
ML-NCEn	0.8999	0.8533/3	**0.1550**/1	**0.8200**/1	**0.1333**/1	**6**

could be effectively reduced. Thus, the diversity of each member in the ensemble can be improved.

5   Conclusion

In this paper, we propose an adversarial example defense scheme based on the negative correlation ensemble for multi-label classification scenarios, i.e., ML-NCEn. ML-NCEn is a multi-label version of NCEn in [17]. In the ML-NCEn defense scheme, we make the gradient direction and gradient magnitude of each member in the ensemble as negatively correlated as possible on the positive label set and negative label set, respectively, thereby increasing the diversity of ensemble members, and interactive training for all members in the ensemble at the same time. We compare the ML-NCEn defense scheme with other typical defense schemes through experiments on the mainstream multi-label classification datasets, i.e., VOC 2007 and VOC 2012, and a typical multi-label classification model, i.e., ML-LIW from [20]. Experimental results show that ML-NCEn can provide good adversarial robustness.

References

1. Abbasi, M., Gagné, C.: Robustness to adversarial examples through an ensemble of specialists. In: Proceedings of the 5th International Conference on Learning Representations. OpenReview (2017)
2. Akhtar, N., Mian, A.S.: Threat of adversarial attacks on deep learning in computer vision: a survey. IEEE Access **6**, 14410–14430 (2018)
3. Akhtar, Z., Mouree, M.R., Dasgupta, D.: Utility of deep learning features for facial attributes manipulation detection. In: Proceedings of the IEEE International Conference on Humanized Computing and Communication with Artificial Intelligence, pp. 55–60. IEEE (2020)
4. Al-Qizwini, M., Barjasteh, I., Al-Qassab, H., Radha, H.: Deep Learning Algorithm for Autonomous Driving Using GoogLeNet. In: Proceedings of the IEEE Intelligent Vehicles Symposium, pp. 89–96. IEEE (2017)
5. Bagnall, A., Bunescu, R., Stewart, G.: Training Ensembles to Detect Adversarial Examples. CoRR abs/ arXiv: 1712.04006 (2017)
6. Chan, Z.S.H., Kasabov, N.K.: A preliminary study on negative correlation learning via correlation-corrected data (NCCD). Neural Process. Lett. **21**(3), 207–214 (2005)
7. Dabouei, A., Soleymani, S., Taherkhani, F., Dawson, J.M., Nasrabadi, N.M.: Exploiting joint robustness to adversarial perturbations. In: 2020 IEEE/CVF Conference on Computer Vision and Pattern Recognition, pp. 1119–1128. IEEE, Seattle, WA, USA (2020)
8. Everingham, M., Eslami, S., Van Gool, L., Williams, C.K., Winn, J., Zisserman, A.: The Pascal Visual Object Classes Challenge: A Retrospective. Int. J. Comput. Vision **111**(1), 98–136 (2015)
9. Everingham, M., Van Gool, L., Williams, C.K., Winn, J., Zisserman, A.: The Pascal Visual Object Classes (VOC) Challenge. Int. J. Comput. Vision **88**(2), 303–338 (2010)

10. Goodfellow, I.J., Shlens, J., Szegedy, C.: Explaining and harnessing adversarial examples. In: Proceedings of the International Conference on Learning Representations, OpenReview (2015)
11. Kariyappa, S., Qureshi, M.K.: Improving Adversarial Robustness of Ensembles with Diversity Training. arXiv e-prints pp. arXiv-1901 (2019)
12. Krizhevsky, A., Sutskever, I., Hinton, G.E.: ImageNet classification with deep convolutional neural networks. In: Proceedings of the Advances in Neural Information Processing Systems, pp. 1106–1114 (2012)
13. LeCun, Y., Bengio, Y., Hinton, G.: Deep Learning. Nature **521**(7553), 436–444 (2015)
14. Liu, Y., Yao, X.: Ensemble learning via negative correlation. Neural Netw. **12**(10), 1399–1404 (1999)
15. Liu, Y., Yao, X.: Simultaneous training of negatively correlated neural networks in an ensemble. IEEE Trans. Syst. Man Cybern. **29**(6), 716–725 (1999)
16. Liu, Y., Yao, X., Higuchi, T.: Evolutionary ensembles with negative correlation learning. IEEE Trans. Evol. Comput. **4**(4), 380–387 (2000)
17. Luo, W., Zhang, H., Kong, L., Chen, Z., Tang, K.: Defending adversarial examples by negative correlation ensemble. In: Data Mining and Big Data: 7th International Conference, DMBD 2022, Beijing, China, November 21–24 2022, Proceedings, Part II. pp. 424–438. Springer (2023). https://doi.org/10.1007/978-981-19-8991-9_30
18. Pang, T., Xu, K., Du, C., Chen, N., Zhu, J.: Improving adversarial robustness via promoting ensemble diversity. In: Proceedings of the 36th International Conference on Machine Learning vol. 97, pp. 4970–4979. PMLR, Long Beach, California, USA (2019)
19. Paszke, A., et al.: Pytorch: An imperative style, high-performance deep learning library. In: Advances in Neural Information Processing Systems 32 (2019)
20. Song, Q., Jin, H., Huang, X., Hu, X.: Multi-label Adversarial Perturbations. In: Proceedings of the IEEE International Conference on Data Mining, pp. 1242–1247. IEEE, Singapore (2018)
21. Sun, Y., Chen, Y., Wang, X., Tang, X.: Deep learning face representation by joint identification-verification. In: Proceedings of the Advances in Neural Information Processing Systems, pp. 1988–1996 (2014)
22. Szegedy, C., et al.: Intriguing properties of neural networks. arXiv preprint arXiv:1312.6199 (2013)
23. Wang, S., Chen, H., Yao, X.: Negative correlation learning for classification ensembles. In: Proceedings of the International Joint Conference on Neural Networks, pp. 1–8. IEEE, Barcelona, Spain (2010)
24. Yan, S., Xiong, Y., Lin, D.: Spatial temporal graph convolutional networks for skeleton-based action recognition. In: Proceedings of the AAAI Conference on Artificial Intelligence, pp. 7444–7452 (2018)
25. Yuan, X., He, P., Zhu, Q., Li, X.: Adversarial examples: Attacks and defenses for deep learning. IEEE Trans. Neural Netw. Learn. Syst. **30**(9), 2805–2824 (2019)

Analysis of the Impact of Mathematics Courses on Professional Courses in Science and Engineering Majors

Zhihong Sun[✉], Zhigang Zhang, and Danling Wang

School of Mathematics and Physics, University of Science and Technology, Beijing, China
szh15613295837@163.com, wang_dan_ling@ustb.edu.cn

Abstract. In order to explore the influence of mathematics courses on professional courses, we studied the course scores of students majoring in four science and engineering majors. Firstly, we counted the distribution of students' grades in professional courses under each grade band of mathematics courses. Then we identified the correlation coefficient between mathematics courses and professional courses and explored the degree of influence of every course in liberal studies courses on professional courses with canonical correlation analysis. As a result, we found that mathematics courses have a strong correlation with professional courses and this correlation cannot be affected by external factors such as grade level. In addition, mathematics courses have a greater correlation with professional courses compared with other liberal studies courses.

Keywords: Mathematics courses · Pearson correlation coefficient · Canonical correlation analysis

1 Introduction

The mathematics courses in university are based on the foundation of secondary school mathematics, mandatory basic courses for science and technology majors, and provide students with sufficient knowledge reserves and theoretical guidance for their subsequent professional courses [1, 2]. "Mathematics courses are significant liberal studies courses for university teaching and play a pivotal role in talent development." [3] Crucially, most courses of study at the university level require mathematics, and the ability to master mathematical skills is an essential indicator of a student's potential in various academic areas [4]. Li [5] et al. analyzed the relationship between mathematics courses and the core knowledge points of subsequent professional courses in computer science to verify the importance of mathematical knowledge for computer science students. Liu [6] analyzed the importance of offering mathematics courses in art majors in higher education institutions from four aspects such as: the need for social and technological development, the need for students' thinking development, the need for higher education in art, and feasibility analysis. Then, it is a common concern for students, teachers and teaching administrators how to quantitative the relevance between mathematics courses and subsequent professional courses.

Y. Tan et al. (Eds.): ICSI 2023, LNCS 13969, pp. 295–306, 2023.
https://doi.org/10.1007/978-3-031-36625-3_24

"When course content is correlated, the grades for taking those courses are also correlated." [7] Philip [8] et al. used the grades of foundational science and mathematics courses taken in high school as the independent variables and course performance in introductory college science courses as the dependent variables to explore the relationship between courses taken in high school and performance in introductory college science courses in biology, chemistry, and physics. Kyle [9] et al. used course grades to analyze the relationships between math and science courses and early engineering courses in two large majors to gain insight into which math and science courses are most predictive of later engineering courses performance.

Some researchers have used descriptive statistics to verify the correlation between courses. Lu [10] counted the number of students who passed four professional foundation courses under each grade band of mathematical analysis course grades and obtained that mathematical analysis courses directly influenced the learning of professional foundation courses. Sheng [11] et al. counted the number of students in the class of 2009 who achieved passing and excellent grades in the CET-4, divided the CET-4 grades into excellent, passing and failing exams according to the CET-4 grades, and calculated the average College English courses grades and the average CET-4 grades of each category of students, finally verified that learning English well is the basis for passing the CET-4 exam. There are also statistical methods used to verify the correlation between courses. Peng [12] et al. found the correlation coefficients between physics, other public basic courses and four professional basic courses and obtained the conclusion that the correlation coefficients of physics and four professional courses were much higher than those of other public basic courses, which verified the basic position of physics in the public basic courses of science and engineering institutions. Xia [13] et al. used the course grades of the first three years of applied physics students to find the correlation coefficients between two courses and obtained strong correlations between natural science courses, professional foundation courses and professional courses. Liu [14] et al. found the Pearson correlation coefficients between calculus and other courses and concluded that the correlation with calculus grades was significant for mathematical and scientific subjects such as line algebra, probability and economics, while the correlation with subjects such as university language and Introduction to the Principle of Marxism was not significant. Wang [15] et al. used Pearson correlation coefficients to conduct the correlation between school grades and total grades in the college entrance examination and between grades in each subject in the college entrance examination and grades in each semester correlation analysis and got different correlations between the grades of each subject in the college entrance examination and the grades of courses at different stages of undergraduate studies. Cheng [16] et al. conducted a typical correlation analysis between the basic and professional courses of statistics in a school. They got the influence of the grades of basic courses on the grades of professional courses. Li [17] used canonical correlation analysis to study the interrelationship between basic and professional mathematics courses and between basic and professional computer courses.

This paper uses students' course grades to explore the degree of correlation between mathematics courses and professional courses. The following article will discuss and analyze the degree of correlation between mathematics courses and professional courses from three aspects: the distribution of professional courses' grades under each grade band of mathematics courses to see the degree of correlation between mathematics courses and professional courses; quantifying the degree of correlation between mathematics courses and professional courses based on the correlation coefficients between mathematics courses and professional courses; and highlighting the high correlation between mathematics courses and professional courses based on the canonical correlation coefficients between various types of courses in the liberal studies courses and professional courses.

2 The Theoretical Basis of the Algorithm Used

2.1 Pearson Correlation Coefficient [18]

The Pearson correlation coefficient measures the degree of linear correlation between two variables and has a value between -1 and 1. If there are two sets of data: X : $\{X_1, X_2, \cdots X_n\}$ and $Y : \{Y_1, Y_2, \cdots Y_n\}$, Then the Pearson correlation coefficient of X and Y is:

$$\rho_{XY} = \frac{cov(X, Y)}{\sigma_X \sigma_Y} \tag{1}$$

$$cov(X, Y) = \frac{\sum_{i=1}^{n}(X_i - E(X))(Y_i - E(Y))}{n} \tag{2}$$

$$\sigma_X = \sqrt{\frac{\sum_{i=1}^{n}(X_i - E(X))^2}{n}} \tag{3}$$

$$\sigma_X = \sqrt{\frac{\sum_{i=1}^{n}(Y_i - E(Y))^2}{n}} \tag{4}$$

2.2 Canonical Correlation Analysis

Canonical correlation analysis is an analytical method to study the interrelationship between two sets of variables. This algorithm first performs principal component analysis on the two sets of variables and extracts the principal components. Each set of variables is combined linearly separately so that the newly obtained composite variables can represent most of the information in the original variables. Here the obtained composite variables are called canonical correlation variables. Each pair of canonical correlation variables reflects the linear correlation between two variables. The correlation coefficient between each pair of canonical correlated variables is called the canonical correlation coefficient. Then the pair of canonical correlation variables with the largest canonical correlation coefficient is found, called the first pair of canonical variables. The second pair of canonical variables, and so on, until the correlation between the two groups of variables can be extracted entirely. Finally, the significance test of canonical correlation coefficients is performed, and the canonical pairs of variables with significant correlations are extracted for analysis.

3   Data Processing and Analysis of Results

In this paper, the four-year course grades of 2015 and 2016 undergraduate students from four majors (Communication Engineering, Computer Science & Technology, Internet of Things Engineering and Information Security) in the School of Computer and Communication Engineering of the University of Science and Technology Beijing are used as research data, and the courses are divided into four categories according to the attributes of the courses: liberal studies courses, disciplinary foundation courses, professional compulsory courses and professional elective courses. Among them, there are many kinds of professional elective courses, and students' elective courses are different, so professional courses used in this study are disciplinary foundation courses and professional compulsory courses. In addition, the liberal studies courses are roughly divided into ideological and political courses, foreign language courses, physical education courses, physics courses, mathematics courses and related disciplinary foundation courses. Since the related disciplinary foundation courses undoubtedly lay the foundation for the subsequent professional courses and are closely associated with the professional courses, the correlation between the related disciplinary foundation courses and professional courses is not discussed and analyzed in this paper. Table 1 below shows Communication Engineering major courses used in this paper.

3.1   Statistical Descriptive Analysis of Course Grades

In order to explore the correlation between mathematics courses and professional courses, it is necessary to synthesize mathematics courses' grades and professional courses' grades in the process of descriptive statistical analysis and extract the index that can represent the learning situation of students in mathematics courses and professional courses --- the courses' composite score. Therefore, the four types of course grades of each student are weighted, i.e., they are weighted according to the credits of the courses to obtain the comprehensive scores of each type of course, where the four types of courses include mathematics courses, disciplinary foundation courses, professional compulsory courses and total professional courses. In addition, the grades of the four types of courses are processed in segments: those below 60 are failing grades, those from 60 to 80 are passing grades, and those above 80 are excellent grades.

The following is the distribution of the 2015 and 2016 Communication Engineering students' performance in their professional courses under each grade band in mathematics courses, as shown in Fig. 1 and 2.

The following features can be obtained by observing Fig. 1.

The number of the Communication Engineering students in the class of 2015 who got excellent grades in mathematics courses was 78, among which the number of students who got excellent grades in the disciplinary foundation courses was 59, accounting for 78%. The number of students who got excellent grades in the professional compulsory courses was 39, accounting for 50%. The number of students who got excellent grades in the total professional courses was 46, accounting for 58.9%. In addition, the number of students who received excellent grades in the professional compulsory courses was slightly smaller than the number of students who received passing grades, and one student received a failing grade in the professional compulsory courses. The results showed that

Table 1. Specific classification of liberal studies courses for Communication Engineering major.

Course Catalogue		Course Name
liberal studies courses	ideological and political courses	I1) Chinese Contemporary History Summary, I2) Thought Morals Tutelage and Legal Foundation, I3) Introduction to the Principle of Marxism, I4) Introduction to Mao Zedong Thought I, I5) Introduction to Mao Zedong Thought II
	foreign language courses	E1) College English I, E2) College English II, E3) College English III
	physical education courses	P1) Physical Education I, P2) Physical Education II, P3) Physical Education III, P4) Physical Education IV
	physics courses	PS1) College Physics AI, PS2) College Physics AII
	mathematics courses	M1) Mathematical Analysis for Technology I, M2) Linear Algebra, M3) Mathematical Analysis for Technology II, M4) Functions of Complex Variables & Integral Transformations, M5) Probability Theory & Mathematical Statistics, M6) Stochastic Process
professional courses	disciplinary foundation courses	DF1) Graphing of Engineering, DF2) Basis of Circuit Analysis I, DF3) Basis of Circuit Analysis II, DF4) Analog Electronics Technique, DF5) Digital Electronic Technique, DF6) Principle & Interface Technique of Micro-computer, DF7) Signal & Systematic, DF8) Principle of Communication
	professional compulsory courses	PC1) Communication Electronic Circuits, PC2) Electromagnetic Fields and Antennas, PC3) Digital Signal Processing, PC4) Basis of Fiber Optical Communication, PC5) Modern Switching Technology, PC6) Digital Communication Systems, PC7) Modern Communication Technology, PC8) Theory of Communication Network

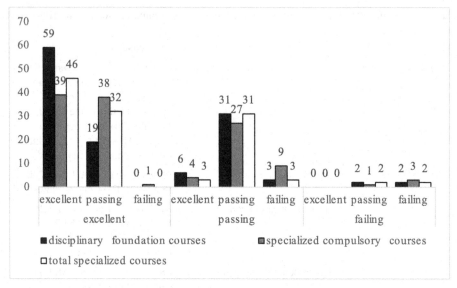

Fig. 1. The distribution of the 2015 Communication Engineering students' performance in their professional courses under each grade band in mathematics courses.

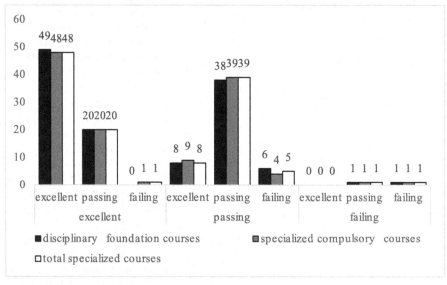

Fig. 2. The distribution of the 2016 Communication Engineering students' performance in their professional courses under each grade band in mathematics courses.

when students get excellent grades in mathematics courses, there is a high probability that they will also get excellent grades in their subsequent professional courses; in addition, compared with the professional compulsory courses, mathematics courses have more influence on the disciplinary foundation courses. This conclusion verifies that disciplinary foundation courses are part of the basic knowledge category in the major and are a transition and bridge from liberal studies courses to professional courses [14].

The number of the Communication Engineering major's students in the class of 2015 who achieved passing grades in mathematics courses was 40, of which 31 students (77.5%) achieved passing grades in the disciplinary foundation courses, 27 students (67.5%) achieved passing grades in the professional compulsory courses and 31 students (77.5%) achieved passing grades in the total professional courses. In addition, the maximum probability that students with passing grades in mathematics courses will achieve excellent grades in professional courses is 15%, and the maximum probability of achieving failing grades is 22.5%. It can be concluded that when a student achieves passing grades in mathematics courses, i.e., his or her ability to master mathematics is at an intermediate level, this student is likely to achieve passing grades or even better grades in the subsequent professional courses, but there is no shortage of risk of failing grades.

There are four students who received failing grades in mathematics courses in the class of 2015. When students received failing grades in mathematics classes, their professional courses grades were also unsatisfactory.

Observing Fig. 2, we can get that the grade distributions of different types of professional courses under each segment of grades in mathematics courses of the Communication Engineering students in the class of 2016 are similar. The same characteristics as the grade distribution of the Communication Engineering students in the class of 2015 are: when students get excellent grades in mathematics courses, the probability of their professional course grades are also excellent; when students get passing grades in mathematics courses, the probability of their professional courses is also passing, but there is no shortage of risk of failing grades.

Through the above statistical analysis, we came to an obvious conclusion: mathematics courses are the tools and foundation for learning the discipline foundation courses and professional compulsory courses, and the good or bad grades of mathematics courses directly affect the good or bad learning of professional courses.

3.2 Correlation Coefficient Between Courses

In this paper, we next use the Pearson correlation coefficient in statistics to quantify the conclusions obtained from the statistical analysis above, i.e., the degree of correlation between mathematics courses and professional courses. Figure 3 below shows the heat map obtained from the correlation coefficient analysis between mathematics courses and disciplinary foundation courses in Communication Engineering major in the classes of 2015. Figure 4 below shows the heat map obtained from the correlation coefficient analysis between mathematics courses and professional compulsory courses in Communication Engineering major in the classes of 2015. Also, the names of the courses corresponding to the variables on the left and upper side in Fig. 3 and 4 are shown in Table 1.

The correlation coefficients of mathematics courses and professional courses of the Communication Engineering students in 2015 can be observed. The correlation coefficients of mathematics courses and disciplinary foundation courses are concentrated

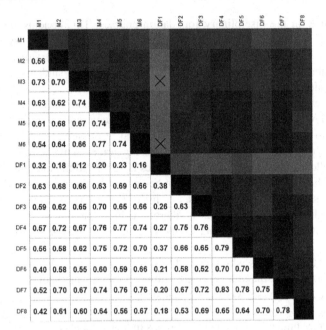

	M1	M2	M3	M4	M5	M6	DF1	DF2	DF3	DF4	DF5	DF6	DF7	DF8
M1														
M2	0.56													
M3	0.73	0.70												
M4	0.63	0.62	0.74											
M5	0.61	0.68	0.67	0.74										
M6	0.54	0.64	0.66	0.77	0.74									
DF1	0.32	0.18	0.12	0.20	0.23	0.16								
DF2	0.63	0.68	0.66	0.63	0.69	0.66	0.38							
DF3	0.59	0.62	0.65	0.70	0.65	0.66	0.26	0.63						
DF4	0.57	0.72	0.67	0.76	0.77	0.74	0.27	0.75	0.76					
DF5	0.56	0.58	0.62	0.75	0.72	0.70	0.37	0.66	0.65	0.79				
DF6	0.40	0.58	0.55	0.60	0.59	0.66	0.21	0.58	0.52	0.70	0.70			
DF7	0.52	0.70	0.67	0.74	0.76	0.76	0.20	0.67	0.72	0.83	0.78	0.75		
DF8	0.42	0.61	0.60	0.64	0.56	0.67	0.18	0.53	0.69	0.65	0.64	0.70	0.78	

Fig. 3. Correlation coefficients for Class of 2015 mathematics courses with disciplinary foundation courses

between 0.5 and 0.8, among which the correlation coefficients of two disciplinary foundation courses, Analog Electronics Technique and Signal & Systematic, are relatively high with mathematics courses. The correlation coefficients of mathematics courses and professional compulsory courses are concentrated between 0.1 and 0.5, among them, the two professional compulsory courses, Digital Signal Processing and Digital Communication Systems have a relatively high correlation with the mathematics courses. Similarly, we conducted a Pearson correlation analysis of the mathematics courses and professional courses of Communication Engineering students in 2016. The correlation coefficients between the mathematics courses and disciplinary foundation courses are concentrated between 0.5 and 0.7, among which the correlation coefficients of two disciplinary foundation courses, Analog Electronics Technique and Signal & Systematic, are relatively high with mathematics courses; the correlation coefficients between mathematics courses and professional compulsory courses are concentrated between 0.2 and 0.6, among them, the two professional compulsory courses, Digital Signal Processing and Digital Communication Systems, have a relatively high correlation with mathematics courses.

From the above analysis, the degree of correlation between mathematics courses and professional courses is not affected by external factors such as grade level, and the four professional courses of analog electronics, signal and system, digital signal processing and digital communication system have a higher correlation with mathematics courses. Also, compared with professional compulsory courses, mathematics courses correlate

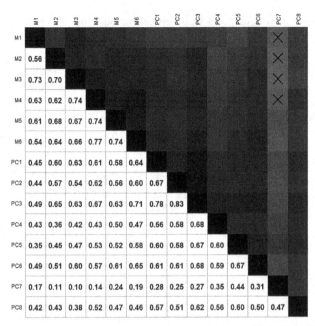

	M1	M2	M3	M4	M5	M6	PC1	PC2	PC3	PC4	PC5	PC6	PC7	PC8
M1													X	
M2	0.56												X	
M3	0.73	0.70											X	
M4	0.63	0.62	0.74										X	
M5	0.61	0.68	0.67	0.74										
M6	0.54	0.64	0.66	0.77	0.74									
PC1	0.45	0.60	0.63	0.61	0.58	0.64								
PC2	0.44	0.57	0.54	0.62	0.56	0.60	0.67							
PC3	0.49	0.65	0.63	0.67	0.63	0.71	0.78	0.83						
PC4	0.43	0.36	0.42	0.43	0.50	0.47	0.56	0.58	0.68					
PC5	0.35	0.45	0.47	0.53	0.52	0.58	0.60	0.58	0.67	0.60				
PC6	0.49	0.51	0.60	0.57	0.61	0.65	0.61	0.61	0.68	0.59	0.67			
PC7	0.17	0.11	0.10	0.14	0.24	0.19	0.28	0.25	0.27	0.35	0.44	0.31		
PC8	0.42	0.43	0.38	0.52	0.47	0.46	0.57	0.51	0.62	0.56	0.60	0.50	0.47	

Fig. 4. Correlation coefficients for Class of 2015 mathematics courses with professional compulsory courses

more with disciplinary foundation courses, this conclusion is more consistent with the conclusions of course correlations obtained from the statistical analysis above.

3.3 Canonical Correlation Analysis Between Course Classes

As liberal studies courses of science and technology majors, mathematics courses have a solid foundation status. In order to explore the basic position of mathematics courses in science and technology majors, this paper uses MATLAB to conduct canonical correlation analysis for two groups of courses: liberal studies courses and professional courses, using a sig value less than 0.05 when conducting the significance test of the canonical correlation coefficient. Here the study of the overall correlation between various types of liberal studies courses and professional courses can be transformed into an analysis of the correlation between Canonical variables that pass the significance test. Table 2 below shows the canonical correlation coefficients between the first canonical correlated variable pairs obtained after the canonical correlation analysis between various courses of liberal studies courses and their professional courses in each major.

Taking the canonical correlation results of Communication Engineering as an example for analysis, the canonical correlation analysis results obtained from the course grades of students in different grades of the same major do not vary much, i.e., the canonical correlation coefficient between mathematics courses and professional courses is the largest in the Communication Engineering major, and the canonical correlation coefficient between physical education courses and professional courses is the smallest. The canonical correlation coefficient between the first canonical correlated variable

Table 2. Canonical correlation coefficients between various types of courses in liberal studies courses and professional courses

canonical correlation coefficient		ideological and political courses	foreign language courses	physical education courses	physics courses	mathematics courses
Communication Engineering	2015	0.773	0.663	none	0.850	0.916
	2016	0.754	0.689	0.596	0.890	0.897
Computer Science & Technology	2015	0.718	0.569	none	0.822	0.904
	2016	0.789	0.740	0.565	0.862	0.906
Information Security	2015	0.864	0.821	0.768	0.915	0.915
	2016	0.888	0.848	none	0.886	0.917
Internet of Things Engineering	2015	0.918	0.818	0.751	0.899	0.964
	2016	0.886	0.707	none	0.822	0.904

pairs of mathematics courses and professional courses in the courses grades of the class of 2015 was 0.916, which means that the correlation between mathematics courses and professional courses was 91.6%; similarly, the relevance of mathematics courses in the course grades of the class of 2016 to professional courses is 89.7%. The first canonical correlation variable between physical education courses and professional courses in the course grades of 2015 students did not pass the significance test, i.e., the correlation between physical education courses and professional courses was not significant, and the correlation between physical education courses and professional courses in the courses grades of the class of 2016 was 59.6%. The content setting of physical education courses is not related to the content of professional courses at all, thus confirming the authenticity of the results of canonical correlation analysis.

Looking at the canonical correlation coefficients between the various types of courses in these four majors' liberal studies courses and the professional courses as a whole, it can be concluded that the mathematics courses have a higher degree of correlation with professional courses compared with other liberal studies courses. This conclusion can provide a scientific explanation for the connection between the content of mathematics courses and the mathematical knowledge required for professional courses.

4   Conclusion and Awareness

In order to investigate the degree of correlation between mathematics courses and professional courses of science and technology majors, this paper processed and analyzed the course grades of students in four majors in the School of Computer and Communication Engineering through statistical descriptions of course grade distributions, Pearson correlation coefficients between courses and courses, and canonical correlation analysis between liberal studies courses and professional courses, and obtained the following conclusions:

(I) Mathematics courses impact disciplinary foundation courses more than professional compulsory courses. When students achieved excellent grades in mathematics courses, the probability of achieving excellent grades in subsequent disciplinary foundation courses for this category was 71%-78%. However, the probability of achieving excellent grades in subsequent professional compulsory courses for this category was 50%-69%. The Pearson correlation coefficients of mathematics courses and disciplinary foundation courses were 0.5–0.8, while the Pearson correlation coefficients of mathematics courses and professional compulsory courses were 0.1–0.6.

(II) Students who perform poorly in mathematics courses find it difficult to do well in subsequent professional courses. When students received excellent, passing, and failing grades in mathematics courses, the maximum probabilities of failing in subsequent professional courses for this category were 1.2%, 22.5%, and 75%, respectively.

(III) Mathematics courses and professional courses have different correlations. Taking the course grades of the 2015 class as an example, the two mathematics courses, Functions of Complex Variables & Integral Transformations and Stochastic Process, have a more significant impact on the subsequent professional courses. The two professional courses, Analog Electronics Technique and Signal & Systematic, are more influenced by mathematics courses.

(IV) Mathematics courses correlated more with professional courses than the other liberal studies courses. The correlation coefficients of mathematics courses with professional courses for the four majors ranged from 0.897 to 0.964. The analysis of the canonical correlation between each type of liberal studies courses and professional courses for students in each major showed that the canonical correlation coefficients of mathematics courses were greater.

These conclusions verify that mathematics courses, as liberal studies courses for science and technology majors, significantly influence professional courses. Students of these majors can better promote the learning of subsequent professional courses only if they have a solid foundation in mathematics. These conclusions can provide some help to students' learning in the future. At the same time, the teachers can also help teachers' teaching by appropriately providing students with the ability to catch up on some relevant knowledge from previous courses according to the relevance of the courses they teach to other courses. In addition, it is also an essential reference to assist teaching managers in making reasonable arrangements for courses and other decisions.

Acknowledgements. This research was supported by fond Teaching Research Fund Project (JG2019ZD05).

References

1. Wu, J.H., Jiang, S.H., Dai, Z.X., GE, G.Y., Zhang, J.: Investigation and analysis on the current university mathematics education. J. Math. Educ. 36–39 (2007)

2. Fu, M.: Reflections on the reform of mathematics education for science majors in higher education institutions. High. Educ. Sci. 59–61 (2002)
3. Wang, Y.G.: Reform and practice on the course system and teaching contents of college mathematics. J. MATH. EDUC., **19**(04), 88–91 (2010)
4. Tang, H.E., Voon, L.L., Julaihi, N.H.: A case study of 'High-Failure Rate' mathematics courses and its' contributing factors on UiTM Sarawak diploma students. In: Paper presented at the Conference on Scientific & Social Research, pp. 14–15 (2009)
5. Li, X.F., Zhou, S.S., Zhang, L.F.: To explore the mathematics teaching in undergraduate computer major at applied colleges. High. Educ. Sci. **125**(01), 120–125 (2016)
6. LIU, H. L.: The Revolutionary research od art mathematical lesson in the university. Coll. Math. **154**(04), 8–10 (2011)
7. Xiao, Z.X., Hu, H.Y.: Correlation analysis of undergraduate applied physics courses based on complex network. Phys. Eng. **31**(4), 54–63 (2021)
8. Sadler, P.M., Tai, R.H.: The two high-school pillars supporting college science. Science **317**(5837), 457 (2007)
9. Whitcomb, K.M., Kalender, Z.Y., Nokes-Malach, T.J., Schunn, C.D., Singh, C.: Engineering students' performance in foundational courses as a predictor of future academic success*. Int. J. Eng. Educ. **36**(4), 1340–1355 (2020)
10. Lu, J.: Example analysis of the relativity of the students' grades of the course of mathematical analysis and the courses of electrics basics. J. Beijing Inst. Technol. (Social Sciences Edition), pp. 92–94 (2007)
11. Sheng, P., Zhang, L.S.: A study of the correlation between academic performance and CET-4 scores. Educ. Res. Exper., 84–87 (2015)
12. Zhang, P., Dong, T., Zhong, S.X., Zhao, H.: Analysis of the basic position of physics courses in the teaching of science and technology universities. J. Inner Mongolia Normal Univ. (Education Science) **25**(07), 127–130 (2012)
13. Liu, P., Xu, H.B.: Statistical analysis in the empirical study of the impact that calculus has on other follow-up courses. Math. Pract. Theor. **41**(24), 20–24 (2011)
14. Wang, C.J., Tan, C.C., Wang, H.: Statistical analysis on the university students' grades and the college entrance examination' grades. Coll. Math. **167**(04), 79–86 (2013)
15. Cheng, L.F., Chen, W.: A statistical analysis on influence of college basic courses score on major courses score. J. Shan Xi Finance Econom. Univ. (higher education edition), pp. 35–38(2007)
16. Li, X.G.: Canonical correlation analysis in data mining application. Changchun university of technology (2011)
17. Ding, F., Miu, B.Q., Ye, D.P.: Correlation analysis of college entrance exam results and college grades. China University Teaching, pp. 29–31 (2008)
18. Xu, J., Tang, B., He, H., Man, H.: Semi supervised Feature Selection Based on Relevance and Redundancy Criteria. In: IEEE Transactions on Neural Networks and Learning Systems, pp. 1974–1984 (2017)

Intelligent System Reliability Modeling Methods

Jiao Li, Wensheng Peng, Zhaoyang Zeng, Chenxi Li$^{(\boxtimes)}$, and Haochen Wang

CPAE, Beijing, China
m15715196329@163.com

Abstract. In order to solve the problem of fault coupling of intelligent systems, this paper focuses on the influence law of fault modeling and reliability, and fully investigates the applicability of the method. Aiming at the process fault caused by the self-protection strategy triggered by the intelligent system and the mutual influence of various structural faults of the system, a method of modeling the time series change process of the intelligent system is proposed based on the causal theory framework, and the causal relationship between various factors is described. Finally, the time of abnormal execution or interruption of the task of the intelligent system is calculated by Monte Carlo simulation to obtain the ability to meet the reliability requirements during the task operation of the complex intelligent system. And output the reliability of the intelligent system in the corresponding time.

Keywords: Intelligent system · Reliability · Reliability modeling · Reliability evaluation

1 Introduction

A series of events, such as autonomous driving accidents, 'killer' robots, 'Face ID cracking', and the Boeing 737max Maneuvering Characteristic Augmentation System (MCAS) collapse have greatly reduced public confidence in AI technology. Users care about and want to understand how AI decisions are made. In ISO Focus 2020, ISO/IEC TC 1 SC 42 pointed out ways to build trust in intelligent systems from transparency, interpretability, controllability, etc. One of the most challenging topics in this field is 'credibility'. It is also the core of many concerns of AI. Only by fully proving the reliability level of AI system can intelligent technology be widely used. However, the traditional reliability modeling method simply resists external interference and cannot summarize its intelligent characteristics, such as strong data dependence (Distribution of AI training samples and test samples, proportion of noise data, etc.), correctness of prediction results or behaviors.

Cui Tiejun [1] proposed a failure analysis method of artificial intelligence system based on information ecological methodology, considering the comprehensive decision-making and response of basic failure consciousness, emotion and reason, and instant failure semantic information. However, the research object is the intelligent analysis of system failure, not the intelligent system. In [2], the overall framework design of intelligent remote sensing target reliability identification is proposed, and the key reliability influencing factors in the process are analyzed. However, the reliability modeling method for specific intelligent remote sensing system needs to be further studied.Pan Xing [3] proposed a research idea based on MBSE for modeling design and simulation verification of intelligent sensing systems.The research object is limited to intelligent sensing system.

In summary, although previous authors have carried out preliminary explorations on the reliability of intelligent systems, the difficulties mainly focus on: 1) Due to the addition of AI modules, data dependence, model generalization, and adversarial attacks have brought more uncertainties, many influencing factors, and more complex failure modes; 2) Considering the self-protection strategy triggered by the intelligent system, it is difficult to find out the process failure mechanism caused by abnormalities such as identification, prediction and decision-making. 3) How to model the changing process of intelligent system timing state. Considering the uncertainty brought by intelligence, accurately modeling and evaluating the reliability of intelligent systems is the only way to prove and improve the reliability of intelligent systems.

2   Failure Mechanism, Risk and Applicability Analysis of Traditional Modeling Methods

Intelligent system functions mainly include: perception, recognition, learning, knowledge, prediction, decision-making, execution and other functions. The above functional failure modes and causes are analyzed, as shown in the following Fig. 1.

Unpredictability, the cognitive mechanism makes rapid and almost unconscious judgments on the actions that people and objects around them may take based on experience, repetition, and contact with similar scenes. Even small changes in external behavior can lead to a certain degree of unpredictability, which is inconsistent with our experience.

Machine learning in intelligent systems is essentially a process of ' curve fitting ' without considering causality. The introduction of intelligent features will cause many behaviors of intelligent systems to be unexplainable, and the design, development and deployment of AI modules will introduce more or aggravate existing vulnerabilities and threats, including dependence on data, opacity and unpredictability of machine learning models

Traditional reliability modeling methods generally take fixed and static system structure and function as the object, represented by reliability block diagram and fault tree. Because AI technology leads to high coupling, dynamic evolution, complex causality and other manifestations of intelligent systems, resulting in variable states changeable. When the traditional reliability modeling method is used to model the fault logic of intelligent system, it is inevitable modeling scales and state explosion. Therefore, a new reliability analysis method is imperative.

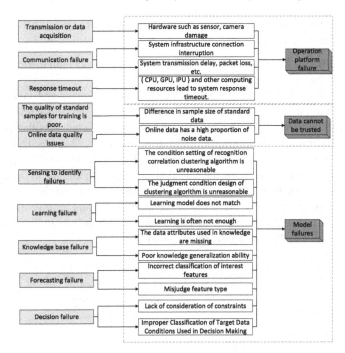

Fig. 1. Functional failure analysis

3 Intelligent System Fault Coupling Relationship Analysis Method

3.1 Overview of Causal Analysis Methods

Causal analysis theory provides a new idea for intelligent system reliability analysis modeling. Causation is an important basic concept of science, which can help engineers find out the causes and effects of intelligent system failures, and then understand the essence, laws and guidelines behind intelligent systems, and guide experts to make predictions, decisions, interventions and controls.

3.2 Quantitative Analysis of Fault Coupling Relationship Based on Causal Analysis

Based on the classical PC (Peter-Clark) algorithm in causal analysis, the quantitative analysis process of fault coupling relationship is as follows:

(1) Obtain all variables, add edges between variables to construct a complete graph. If the number of variables is N, the number of edges of the complete graph of the system is N * (N−1)/2;
(2) The independence or conditional independence test is used to determine whether there is independence between variables. If there is no relationship between variables, the edges between variables are cut off, and finally the skeleton of undirected graph is obtained.

(3) The direction of the local causality diagram is inferred by judging the V structure and other methods.

The key structure is characterized in a graphical mode, and a fault model considering the interaction between process faults is established.

Causal analysis and its related theories are used to discover and model the fault relationship in a specific complex intelligent system, so that the causal relationship can be found in the data of the observed system with less prior knowledge, without relying too much on the experience or prior knowledge of domain experts. The causal relationship of coupling faults between system faults is established through the causal graph structure.

Further, by reasonably limiting the causal mechanism, we find the asymmetric independence between the fault causality, so as to determine the causal direction. Based on the results of causal discovery, we simplify the fault model and remove the parameters unrelated to the causal chain in the model, which not only does not affect the effect of the model, but also makes the model more explanatory.

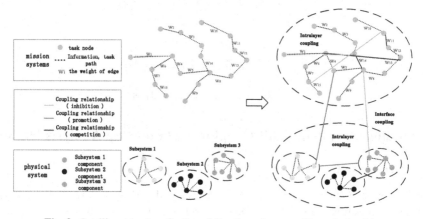

Fig. 2. Intelligent system fault modeling based on coupling relationship

Based on the theory of causal analysis, we find out one or more fault coupling relationships with the highest distribution frequency in a specific system. As shown in Fig. 2, the fault modeling of intelligent system based on coupling relationship is divided into two parts: physical system and task system. The physical system here represents the system composed of intelligent infrastructure such as software and hardware of intelligent system. The task system is a system composed of the task flow of the intelligent system. In the intelligent system fault modeling based on coupling relationship, including the coupling relationship within the layer and the coupling relationship between the layers, the coupling relationship and the failure diffusion/dissipation of the intelligent system caused by the coupling relationship are analyzed, and the key structure is characterized in a graphical mode. A fault model considering the interaction between process faults is established.

3.3 Reliability Modeling Method Considering Fault Coupling Relationship

Based on the three elements of object, fault coupling relationship and network dynamics process, the evaluation model of complex intelligent system is established to evaluate and calculate the operation of complex intelligent system within the specified time and output the reliability of the system. Since the algorithm, data and operating environment all have an impact on its reliability, this section proposes an evaluation model N that combines the fault coupling relationship of intelligent complex system to model complex intelligent system objects and describe their dynamic rules. The model can be formally expressed as:

$$N = \{\mathbb{G}, \mathbb{EC}, R\}$$

$\mathbb{G} = \{g_1(t), \ldots g_{GSN}(t)\}$ is an intelligent system object, which describes the value of each state of the system at time t. $g_i(t)$, $i \in 1, \ldots GSN$ denotes each state variable; $\mathbb{EC} = \{EC_1, \ldots, EC_{ECN}\}$ is the external disturbance of the intelligent system, which means that the external disturbance from 1 to ECN causes the state of the intelligent system object to change. Where $EC_i = \{t_i, EC_i^1, \ldots, EC_i^{FN}\}$ denotes that the i th disturbance is composed of occurrence time t_i and influencing factor $EC_i^j, j \in 1, \ldots, FN$; R: $\mathbb{G}(t_i) \times EC_i \rightarrow \mathbb{G}(t_{i+1})$ represents the change process of the intelligent system state at time t_i due to the disturbance causality of B to the intelligent system state at time $EC_i^j, j \in 1, \ldots, FN$.

The above model gives a formal definition framework for the reliability evaluation model of complex intelligent system considering fault coupling relationship. In the specific use, it is also necessary to combine the system objects and protocols for specific analysis and implementation of each part. Firstly, it is necessary to distinguish the physical system and information system of complex intelligent system to describe the characteristics in detail. Then, the process of failure of infrastructure and different functions in the operation process is described by simulation. Finally, the operation mechanism of complex intelligent system is analyzed to design the dynamic process of the system, so as to model and characterize the dynamic process of reliability of complex intelligent system under the influence of its own and external conditions.

Its functional attributes must be considered in the modeling of evolutionary objects for complex intelligent system information systems. Based on the above analysis, the attribute information of the information system mainly includes: the source node and the destination node of the process, the type of the functional process (identification perception\autonomous decision-making), the reliability requirements of the functional process, the functional process environment, the data set characteristics of the functional process (sample data, training data) and so on (Fig. 3).

External Disturbance

According to the above analysis of intelligent system objects, combined with the external factors (environment, human factors, user needs, etc.) of complex intelligent system operation, the factors that induce the state change of complex intelligent system are analyzed and designed.

1) Changes in the state of intelligent system components: equipment failure, occlusion, high temperature, electromagnetic or equipment movement and other factors caused

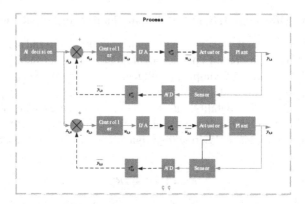

Fig. 3. Intelligent system function schematic diagram

by the state failure or interruption; state failure, base station failure, edge computing node failure etc.;

2) Changes brought by the state of sample data: standard data set non-compliance, data set interference points, noise points, outlier ratio non-compliance, sample data set interference etc.;

3) Insufficient algorithm design, unable to identify data with large interference, algorithm crash, algorithm response timeout etc.

4) The accuracy of the algorithm caused by functional faults in the intelligent system decreases, and the identification and decision-making faults caused by functional faults in the intelligent system.

5) System state changes brought by new technologies and frameworks (specifically, new influencing factors brought by new intelligent modules (such as AI) can be added to the existing framework system and foundation).

6) The change of system state caused by functional characteristics (specifically refers to the influencing factors of different intelligent system functional logic).

Temporal Variation Characteristics of Causality

The process of time series change describes the process in which the causal relationship between various factors changes dynamically with time. Specifically, the direction and strength of the causal arrow between the intelligent system rules of complex intelligent systems and AI intelligent perception, recognition, decision-making and other factors. According to the above analysis of external disturbances, the factors affecting the causality of the system include AI intelligent module, physical equipment failure of the system, deployment and configuration information of the intelligent system task flow (Fig. 4).

According to these conditions, the protocol and configuration and deployment mechanism between different intelligent system devices are combined. This project is mainly considered from the following aspects:

Causal changes caused by AI identification function, failure and recovery of physical system components.

Causal changes caused by deployment strategies such as priority allocation when multiple processes are deployed simultaneously.

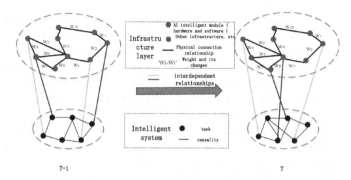

Fig. 4. Causal Relationship Over Time

Causal changes triggered by quantitative changes caused by process and its path generation and end.

Because the intelligent function fault of intelligent system involves many influencing factors and the state variability of AI at different times, as well as the mutual coupling between modules, it is difficult to quantify the weight of each part of the fault, and it is difficult to use the weighted sum of lower-level parameters to calculate its comprehensive reliability. Therefore, based on the running time of the intelligent system when the system stops running due to the above factors, this paper calculates its reliability by Monte Carlo simulation sampling method. The process and steps of the intelligent system reliability algorithm are as follows:

Step 1: Read in the intelligent system topology $G(V, E)$, task set *App*, external interference condition sequence S and causality rules related parameters;

Step 2: Calculate the fault type within the time T and add it to the external interference condition sequence;

Step 3: Analyze the causal relationship between faults based on causality, the length of causality state space H, and the interference condition state is initialized to $h = 1$;

Step 4: According to the occurrence time of the external interference conditions of the intelligent system, select the corresponding task path $AppPath_h$ to update;

Step 5: The repaired state in the set of external disturbance conditions of the intelligent system is added to the intelligent system topology, and the fault intelligent system components in the intelligent system topology are deleted from the intelligent system topology, and the current corresponding intelligent system topology is updated to G_h;

Step 6: Compare the fault intelligent system component set under the external interference condition state h of the current intelligent system and the task path set under the previous causal change time state $h - 1$, and count the task set of task failure under the current interference state h;

Step 7: If the fault task set is empty, jump to step 2; otherwise, the task set of the intelligent system is traversed, and the protection strategy of the intelligent system is selected according to the fault type.

Step 8: Obtain the attribute information under the change of the external interference state of the intelligent system, analyze the causality between the task protection strategy and the external interference factors of the intelligent system according to the causality rules set in the causality model, and judge whether there is still a fault;

Step 9: If there is a fault, the topology of the intelligent system is updated. According to the current external interference change state h, whether h is greater than H is judged. If it is less than h, $h + 1$ then updates and restores the topology of the intelligent system, records the unavailable time of the task under the current external interference state of the intelligent system and returns step 5.

Step 10: Count the unavailability time T_{off} of the state change process of the task under external interference in the intelligent system, so as to calculate the reliability of the task according to the task reliability $R = 1 - \frac{T_{off}}{T}$.

4   Case Study

In the process of intelligent inspection of UAV, the intelligent system is required to carry out remote control operation. The centralized control room sends instructions based on the training data of AI model, and transmits control instructions to UAV through the intelligent analysis and management module of inspection data and the trajectory tracking control module. Then, the intelligent control module of UAV system controls the flight power module and data acquisition module of UAV, so as to realize the intelligent inspection of UAV (Fig. 5).

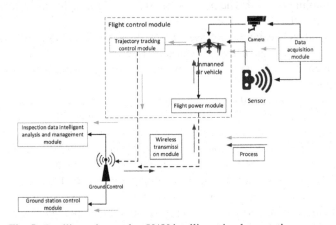

Fig. 5. Intelligent inspection UAV intelligent implementation process

This paper focuses on the analysis of intelligent modules. The topology of the case selection is shown in Fig. 6.

Among them, D represents the ground station control module, W represents the trajectory tracking control module, WX represents the intelligent transmission module, A and B represent the intelligent control system entrance, S represents the intelligent interaction module, C represents the camera, E represents the sensor, and P represents the UAV intelligent system controller. The object of UAV inspection intelligent system is shown in the following Table 1.

Based on the above analysis, the reliability of two UAV intelligent inspection control tasks (Client_1 and Client_2) is analyzed. The information and deployment of the two intelligent control tasks are as follows (Fig. 7).

Table 1. UAV inspection intelligent system information deployment

Task number	Task source node	Task host node	Tasks roadmap
Client1	C1	E1	C1—...—PC—...—E1
Client2	C1	E2	C2—...—PC—...—E2

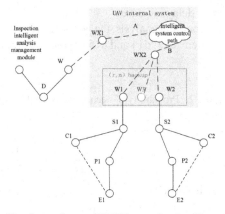

Fig. 6. Simple topology of UAV inspection intelligent system

Analysis 1: Quantitative analysis of coupling relationship between faults

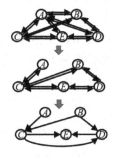

Fig. 7. Fault relationship analysis based on causal theory

Point A: AI computing timeout.
Point B: AI calculation deviation is too large.
Point C: E1 operation deviation.

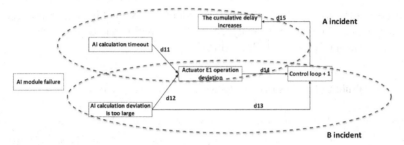

Fig. 8. Fault coupling relationship analysis of A and B events based on causal theory

Point D: Control loop + 1.
Point E: the cumulative delay increases.
According to the fault coupling analysis, the possible events are determined, namely, A, -A, B, and -B. Among them, d11, d14 and d15 are the corresponding causal influence coefficients.

Table 2. Probability of occurrence of each link

X	Y		
	C	D	E
A	0.85	0.15	0.15
-A(Not timed out)	0.35	0.16	0.02
B	0.80	0.30	0.15
-B(Not too large deviation)	0.4	0.04	0.02

Then, in the AI module, the impact of the A event 'AI calculation timeout' on the reliability of the Client1 service is:

$$R_{AI|A} = f(d11, d14, d15 \,|\, do\,(A = 1))$$
$$d11 = (P(C|A) - P(C|-A))$$
$$= (0.85 - 0.35)$$
$$= 0.5$$

Similarly, the values of d14 and d15 are calculated.

$$d11 = 0.5$$
$$d14 = 0.15$$
$$d15 = 0.07$$

So, RAI|A = a1xd11 + a2xd14 + a3xd15.
RAI | A = 0.36, where a is the weight and the weight of ownership is 0.5.

Similarly, calculate RAI|B = 0.47.

Because the B event (AI calculation deviation is too large) will lead to an additional increase in the number of loops, its impact on the final job delay and even business reliability RAI | B is greater than the A event RAI | A in this case.

Based on the causal theory, it can better explain the relationship between business and its faults, as well as the impact of faults on the overall business reliability. It has considerable prospects.

Analysis 2: Intelligent system reliability quantitative evaluation

Based on the above analysis, the reliability of the intelligent system is the nature that the intelligent system does not fail during use. According to the model constructed in Fig. 8 and the reliability evaluation process in Sect. 3.3, the Monte Carlo simulation sampling method is used to calculate the number and duration of abnormal operation caused by the above factors.

Through Monte Carlo simulation, the number of faults per unit time of the system is calculated to study the system reliability. Firing is the number of excitations in each Monte Carlo simulation, and the number of repetitions is the number of Monte Carlo simulations. Random Firing20000 simulation, a single simulation time 5h, simulation of continuous system operation process, repeat 5 times. According to the above settings, the 95% confidence interval of the number of abnormal operation of the system is shown in Table 2. The reliability calculation results are as follows (Table 3):

Table 3. Simulation result

simulation time	Total duration(h)	Abnormal times	95% confidence interval (±)
1	100000	0.016789	0.016539
2	100000	0.001949	0.0018626
3	100000	0.21108	0.020829
4	100000	0.029958	0.029628
5	100000	0.030959	0.030621

The observation point is set for the top-level task failure of the system. When the condition is satisfied, it is recorded as one failure of the system, and the failure probability of the system task F = the total number of failures n/the total number of simulations N.

According to the above simulation results, the task reliability evaluation result of the case system is: R = 1 − n/T = 1−0.086728 = 0.91327.

5   Conclusion and Foresight

The unreliability of the intelligent system is mainly attributed to the introduction of uncertainty factors in the deep learning process. The AI module and its faults will have a complex relationship with other modules in the system. In the process of autonomous decision-making of the intelligent system, it is linked to perception, recognition, prediction, etc. The evaluation of the combination of mechanism and data can be closer to the real system operation process than the traditional evaluation model. The main innovations of this paper include:

1) A method is proposed to identify the key fault coupling relationship in intelligent systems according to the fault mechanism of process faults.
2) A fault coupling relationship analysis method is proposed to quantify the influence of intelligent autonomous behavior of intelligent systems on reliability.
3) Based on the coupling relationship of intelligent systems, a modeling method for the reliability of intelligent systems is proposed.

References

1. Tiejun, C., Shasha, L.: Research on fault analysis principle of artificial intelligence system. J. Intell. Syst. **16**(4), 785–791 (2021)
2. Min, Z.: Artificial intelligence for reliability identification of remote sensing targets: overall framework design, current situation analysis and prospect. According J. Surv. Mapp. **50**(8), 1049–1058 (2021). 10.11947/j
3. Shi, W., Zhang, M.: Artificial intelligence for reliable object recognition from remotely sensed data: overall framework design, review and prospect. Acta Godaeticaet Crtoe a graphica Snicai **50**(8), 1049–1058 (2021). https://doi.org/10.11947/j.AGCS.2021.20210095
4. Wang, Y.X., Hou, M., Plataniotis, K.N., et al.: Towards a theoretical framework of autonomous systems underpinned by intelligence and systems sciences. IEEE-CAA J. Automatica Sinica **8**(1), 52–63 (2021)
5. Zhu, X., Wang, H., You, H.: Review of research on autonomous driving intelligent system testing. J. Softw. **32**(7), 2056–2077 (2020)
6. Mankad, K.B.: An intelligent process development using fusion of genetic algorithm with fuzzy logic. In: Artificial Intelligence: Concepts, Methodologies, Tools, and Applications, pp. 245–281. IGI Global (2017)
7. Stranieri, A., Sun, Z.: Only can AI understand me?: Big data analytics, decision making, and reasoning. In: Intelligent Analytics With Advanced Multi-Industry Applications, pp. 46–66. IGI Global (2021)
8. Köse, U., Arslan, A.: Chaotic systems and their recent implementations on improving intelligent systems. In: Handbook of Research on Novel Soft Computing Intelligent Algorithms: Theory and Practical Applications, pp. 69–101. IGI Global (2014)
9. Wu, G., Jiang, C., Zeng, G.: Several important questions about the research and development of autonomous intelligent systems. Electro-optics Control **9**(4), 27–31, 35 (2002). https://doi.org/10.3969/j.issn.1671-637X.2002.04.007
10. Wu, C., Zhang, T.: Autonomous intelligent system: the engine of economic development and scientific and technological progress. China's Sci. Technol. Wealth (8), 44–45 (2017). https://doi.org/10.3969/j.issn.1671-461X.2017.08.017
11. Yang, Z., Yang, C., Chen, F., et al.: Parameter estimation of machining center reliability model based on PSO algorithm and SVR model. J. Jilin Univ. Eng. Edn. (3), 8 (2015)
12. Qingbing, Z., Liu, Y., Ming, H., et al.: Online evaluation method of CPS system reliability based on machine learning. Comput. Eng. Appl. **50**(10), 128–130 (2014)
13. Quality elements and test methods of information technology artificial intelligence machine learning models and systems: T/CESA 1036-2019 (2019)
14. Dong, Z.L.: Quality elements and test methods of machine learning systems. Electron. Test. (9), 92–94,103 (2021). https://doi.org/10.3969/j.issn.1000-8519.2021.09.037

15. Li, D., Li, T., Liu, J., Jia, J.: Research on power communication reliability evaluation algorithm for large-scale complex networks. Shaanxi Electr. Power **42**(07), 14–17 + 22 (2014)
16. Di, W., Jing, H.: Reliability modeling of ship power system based on reliability block diagram. Ship Power Technol. **39**(S1), 28–31 (2019). https://doi.org/10.13632/j.meee.2019.s1.006
17. Sun, H.Z.: AFDX network performance reliability analysis based on dynamic fault tree. Comput. Sci. **43**(10), 53–56, 62 (2016)
18. Mo, H., Wang, W., Xie, M., Xiong, J.: Modeling and analysis of the reliability of digital networked control systems considering networked degradations. IEEE Trans. Autom. Sci. Eng. **14**(3), 1491–1503 (2017)
19. Khuntia, P., Hazra, R., Chong, P.: An efficient actor-critic reinforcement learning for device-to-device communication underlaying sectored cellular network. Int. J. Commun. Syst. **33**(10) (2020)
20. Wu, Z.-T., Huang, N., Wang, X.-W., et al.: Analysis of end-to-end delay on AFDX based on stochastic network calculus. Syst. Eng. Electron. **35**(1), 168–172 (2013)
21. Chen, R., Yang, C., Han, S., Zheng, J.: Dynamic path planning of USV with towed safety boundary in complex ocean environment. In: Proceeding of the 33rd China Conference on Control and Decision Making, pp. 444–449. IEEE (2021)
22. Xiao, J., Lu, J., Li, X.: Davies Bouldin Index based hierarchical initialization K-means. Intell. Data Anal. **21**(6), 1327–1338 (2017)
23. Liu, X., Li, Y., Zhang, J., et al.: Self-adaptive dynamic obstacle avoidance and path planning for USV under complex maritime environment. IEEE Access (2019)

Sensitivity of Clustering Algorithms to Sparseness of One Correlation Network

Anastasiia Timofeeva[1]([✉]) [iD] and Natalya Gorbacheva[2] [iD]

[1] Novosibirsk State Technical University, 20, Karl Marx Avenue, Novosibirsk, Russia
a.timofeeva@corp.nstu.ru

[2] Institute of Economics and Industrial Engineering, Siberian Branch of the Russian Academy of Sciences, 17, Academician Lavrentyev Avenue, Novosibirsk, Russia

Abstract. Correlation networks are popular in applications of bioinformatics, in the analysis of society and economic systems. Their analysis makes it possible to find clusters (modules) of highly correlated variables (genes, economic indicators). By constructing an adjacency matrix a hard threshold can be set. Edges with correlation values below the threshold are excluded, thereby increasing the sparseness of the graph. This allows us to simplify the structure of relations, but the results of network clustering are sensitive to sparseness. In this paper, using internal and external cluster validity indices, we study the sensitivity of clustering results to network sparseness. Algorithms of greedy and multi-level modularity optimization, using random walk and by label propagation are compared. The correlation network was built for indicators describing the socio-economic system of the Nordic region. As a result, it is shown that greedy and multi-level modularity optimization algorithms are the most sensitive to network sparseness. However, the optimal cluster partitions obtained are more suitable for interpretation. For external validation, it is proposed to find a compromise between the purity measure and the symmetric uncertainty.

Keywords: Correlation Network · Clustering Algorithm · Sparseness · Threshold

1 Introduction

With the development of digitalization, graph theory plays an important role in solving many problematic problems, including the economic field of activity. Graph theory methods are used not only at the micro level (for planning production processes, reducing transport costs, etc.), but also in the processing and analysis of macroeconomic data [1, 2].

The national economic system can be represented as a graph reflecting the relationship between the key indicators of the country's development. This approach has its advantages. Processing data using data mining methods requires that an array of objects and their properties be formed. In this case, for time series analysis it is required to bring them to the same period. This is a very difficult task because statistics for different indicators may be available for different periods. The ability to calculate pairwise similarity measures and use a correlation graph greatly simplifies this task.

However, when building a network, a number of problems arise. Socioeconomic data tends to be highly noisy, resulting in weak correlations. Therefore, not all algorithms succeed in detecting a well-interpreted community structure [3]. In addition, the results are sensitive to sparseness of the network. As a rule, graph edges corresponding to insignificant correlations are excluded. The level of significance is chosen by the researcher. Of interest is the study of the influence of the threshold value on the results of clustering in terms of their internal and external validity.

Further, Sect. 2 describes some theoretical points: a correlation network is introduced, compared clustering algorithms, and cluster validity measures are presented. Section 3 describes the software used. Subsection 3.2 describes in detail the collection and preprocessing of data for building a network. Section 3.3 presents the results of a comparison of clustering algorithms based on cluster validity measures with varying the distance matrix binarization threshold. Section 4 summarizes the work done.

2 Theoretical Background

2.1 Correlation Network

To build a correlation network [4], it is necessary to determine the association measure that characterizes the strength of relation between two variables. The association between variables can be defined through a measure of similarity or a measure of difference, between which there is a negative relation. Measures of similarity s_{ij} between variables i and j can be various correlation coefficients. At the same time, unsigned and signed similarity measures are distinguished. If the sign of the correlation is ignored, similarity means the presence of any relationship (positive or negative). However, this can confuse meaningful information. In contrast, in signed networks the similarity between variables takes into account the direction of the relation.

The dissimilarity matrix D consists of elements d_{ij} showing the dissimilarity between variables i and j. They have the following properties: $d_{ij} \geq 0$; $d_{ij} = d_{ji} \forall i, j$; $d_{ii} = 0 \forall i$. Distance measures can be used as dissimilarity measures.

The similarity and distance matrices must be transformed to build the network. The main transformation in network analysis is the discretization of these matrices and obtaining a binary adjacency matrix A based on arbitrary or probability cutoffs τ (thresholds). By 'hard' thresholding a network adjacency is defined to be 1 if $s_{ij} > \tau$ $\left(d_{ij} \leq \tau \right)$ and 0 otherwise. Adjacent edges can be assigned a weight equal to the distance between the corresponding variables. Such a transformation simplifies the graph, but leads to the loss of some information about weak relations. The results may be sensitive to the choice of the threshold.

2.2 Graph Clustering Algorithms

A number of graph clustering algorithms maximize network modularity. Network modularity is a measure of functional segregation that determines how well a network can be divided into non-overlapping groups of nodes or modules (clusters, communities). To calculate it, the graph is divided into clusters using the selected algorithm. Then modularity will characterize the concentration of relations in selected clusters (in the sense of the fraction of the edges) in comparison with the random distribution of relations between all nodes without taking into account clusters. If the fraction of edges within the given groups is no different from expected if edges were distributed at random, then the modularity will be zero. Nonzero values represent deviations from randomness, and in practice a value above about 0.3 is good enough to indicate a significant community structure in a network.

Optimization of modularity theoretically leads to the best possible division, however, it is an NP-complete problem, therefore, heuristic algorithms are commonly used.

The FastGreedy algorithm [5] is a greedy optimization of modularity. The initial number of clusters corresponds to the number of vertices, each vertex is placed in a separate cluster. Next, each node moves to the community that maximizes the modularity function. These aggregated communities are combined until the modularity function can no longer be increased. The method is computationally easy $((m \log n))$, easily applicable to large graphs, and, despite being greedy, often does a good work.

The Louvain algorithm [6] implements the multi-level modularity optimization based on a hierarchical approach. It is a greedy algorithm with time complexity in $((n \log n))$. It differs from the FastGreedy algorithm in that it iteratively exchanges nodes between communities and evaluates the change in modularity until it stops improving. The algorithm then collapses the communities into hidden nodes and identifies edge weights with other observable and hidden nodes, which provides a "multi-level" structure [7]. The final solution depends on the node order making the algorithm non-deterministic. For these reasons, in the course of the study below, the Louvain algorithm was run 100 times and the solution with the maximum modularity value was chosen.

Other algorithms use a random walk and updating the labels of the vertices by majority voting in the neighborhood of the vertex.

The WalkTrap algorithm [8] begins by computing a transition matrix where each element represents the probability (based on node strength) of one node traversing to another. Random walks are then initiated for a certain number of steps (typically 3 to 8), using the transition matrix for probable destinations. Each node starts as its own community and merges with adjacent communities (based on squared distances between each community) in a way that minimizes the sum of squared distances between other communities. Modularity is then used to determine the optimal partition of communities.

The Infomap algorithm [9] find community structure that minimizes the expected de-scription length of a random walker trajectory. Different from the Walktrap algorithm, Infomap is derived from information theory with the idea of 'compressing' the conditional information of a random walk on the network into Huffman codes. The major difference between these two algorithms is that Infomap captures the conditional flow of information across the network in a way that maximizes the information (e.g., bits) of the random walk process.

The Label propagation algorithm [10]. Each vertex in the graph is assigned to the community that owns the majority of its neighbors. If there are several such communities, then one of them is chosen randomly. At the initial moment of time, a separate community is assigned to all vertices. Then there are redistributions of communities. The algorithm terminates when all vertices belong to the same communities as most of their neighbors. It makes sense to run the algorithm several times and choose the best result. In the study below, we ran 100 times. This is a fast, nearly linear time algorithm. However, on noisy graphs, all vertices are often combined into one cluster.

2.3   Cluster Validity Indices

Distinguish between internal and external clustering validity indices. Internal ones try to characterize the quality of clustering only on the basis of data structure features. External indexes compare the obtained results with the a priori given partition into classes.

Since when building a network, there is information only on the matrix of distances between instances, and there is no data set typical for cluster analysis about the features of instances, therefore, we will consider validation methods based only on the distance matrix.

Silhouette index [11]. The silhouette value indicates how similar an object is to its cluster compared to other clusters. It takes values from -1 to 1. The closer the values are to 1, the better the clustering quality.

Index C Hubert & Levin [12]. Calculated from all within-cluster dissimilarities normalized to the sum of the largest and smallest dissimilarities in distance matrix. The minimum value corresponds to the best partition into clusters.

Index G2 is Baker & Hubert adaptation of Goodman & Kruskal's Gamma statistics [13]. All within-cluster dissimilarities and all between-cluster dissimilarities are compared. Comparisons are considered concordant (discordant) if a within-cluster dissimilarity is strictly less (strictly greater) than a between-cluster dissimilarity. The difference between the number of concordant and discordant comparisons is normalized by their sum. Therefore, the greater the value of the G2 index, the better the quality of clustering.

Index G3 Hubert & Levine is similar to index C, but for normalization, largest and smallest within-cluster dissimilarities are multiplied by number of within-cluster dissimilarities. Similarly, the smaller its value, the higher the validity of the cluster analysis results.

The connectivity indicates the degree of connection of the clusters, as determined by the k-nearest neighbors. The number of neighbors must be specified. The connectivity has non-negative values and should be minimized.

The Purity measure is used to compare the results of clustering with a given a priori partitioning. This measure is suitable for external validation of the quality of clustering. It shows to what extent the clusters contain one class. For each cluster, the number of observations from the most common class in the given cluster is counted. Next is the sum of all clusters divided by the sample size. A value of 1 corresponds to optimal clustering. However, if the given partition includes only one class, then the purity will always be 1. Therefore, along with the usual definition, let's call it Purity1, we will consider Purity2, which shows the extent to which the classes contain one cluster.

The SU index is a symmetrical uncertainty calculated as the harmonic mean of the uncertainty coefficients. The uncertainty coefficient is calculated as the ratio of mutual information to the entropy of the response variable. It shows the degree of relationship between features. If we consider the resulting clustering and the reference one as such features, then the values of the SU index close to 1 indicate a high external validity of the clustering results.

3 Research

3.1 Data Processing Software

For working with the website we used the rvest, httr, jsonlite packages in the R environment. The stringr and tokenizers packages were used for text processing.

Graph analysis was performed with igraph package [14] for R.

For validation of clustering results, the R-package clusterSim [15] was used. It was used to calculate silhouette indices, C, G2, G3. The clValid package [16] was used to calculate the connectivity validation measure.

3.2 Data Description

To identify clusters in the socio-economic system of the Nordic region (Denmark, Finland, Iceland, Norway, Sweden), its representation in the form of a network (graph) was chosen. The vertices of the graph correspond to the main socio-economic indicators. To build graphs, data for the Nordic region were collected from the website [17]. For automatic data collection, a JSON request posted on the website was used, the URL for the request was generated automatically based on the section name and indicator code. In total, there were 232 indicators with unique ciphers. Of these, 88 key indicators were identified. Indicator values were averaged or summed (depending on the meaning of the indicator) for all countries of the Nordic region. For each of the indicator, annual time series of different lengths are formed, on average 21 years (minimum 4 years, maximum 41 years).

Since socio-economic indicators are presented in dynamics, in order to exclude spurious correlation between time series, the analysis was performed in relative growth rates:

$$r_t = \frac{x_t - x_{t-1}}{x_{t-1}} = \frac{x_t}{x_{t-1}} - 1. \tag{1}$$

where x_t is the indicator value per year t. However, this approach only works when $x_t > 0$. If the indicator takes both positive and negative values (for example, net migration), then the relative growth rates (1) do not adequately show the direction of change in the indicator values. In such cases, the absolute growth rates are taken, normalized to the standard deviation:

$$a_t = \frac{x_t - x_{t-1}}{\sqrt{\frac{1}{N} \sum_{i=T_1}^{T_N} (x_i - \bar{x})^2}}. \tag{2}$$

where $\bar{x} = \frac{1}{N} \sum_{i=T_1}^{T_N} x_i$, T_1, T_N are first and last years of observation. Several indicators (COMP23: Economic growth; NAAC02: Real gross domestic product annual growth) were not subjected to the transformations described, since they are initially presented in terms of growth rates.

The next difficulty is related to the fact that when selecting pairwise intersecting time periods, many time series became quite short. The median length of such series was 17, minimum 2, maximum 31. A study was made of methods of correlation analysis and distance metrics from the point of view of the possibility of analyzing such short time series. As a result, the following approach is proposed: go to binary data (positive or non-positive relative growth rate) and calculate the Jaccard distance and Jaccard similarity measure. If there is not enough data to correctly estimate the relationship (row lengths are less than 5), the values of the similarity measure were set to zero (distance values were equated to one). There were 2.7% of such cases (103 out of 3828). This made it possible to exclude high correlations calculated from a small number of observations.

Thus, matrices of similarities (correlations) and distances between socio-economic indicators were obtained. When using a hard threshold, the value τ was set. The elements of the adjacency matrix of the network are given as

$$a_{ij} = \begin{cases} 1, & \rho_{ij} \leq \tau, \\ 0, & \rho_{ij} > \tau. \end{cases}$$

where ρ_{ij} is Jaccard distance between indicators i and j. The graph edges are weighted by the Jaccard distance between the corresponding vertices divided by the maximum pairwise distance between the vertices. A small value (0.001) has been added to all weights to avoid the problem with zero weights.

The closer τ to zero, the sparser the network is. The case when $\tau = 1$ corresponds to the original graph including all possible edges. We will consider it as a baseline for comparing clustering algorithms using cluster validity indices. It should be borne in mind that when obtaining a sparser network, isolated vertices appear that do not have common edges with other vertices. They, as a rule, stand out in separate clusters, thereby the number of clusters grows strongly with a decrease in τ.

3.3   Results

In the study, the threshold values ranged from 0.005 to 1 with a step of 0.005. At each threshold value, a graph was constructed, communities were identified using the selected algorithms, and clustering validity indices were calculated based on the original Jaccard distance matrix and cluster partition obtained by various methods.

Figure 1 shows the optimal modularity values corresponding to the cluster partitions obtained using the algorithms under study for various threshold values. It can be seen that on a large range of threshold values, the Louvain algorithm gives results no worse than other algorithms. In some cases, the FastGreedy algorithm performs better than Louvain algorithm. At $\tau \in [0.045, 0.055]$ and 0.08 the best result is given by the Infomap and Label Propagation algorithms. Infomap algorithm outperforms other algorithms at $\tau \in [0.06, 0.07]$. In general, when the network is very sparse, the Infomap and Label Propagation algorithms work better. In other cases, Louvain and FastGreedy algorithms are preferable if the goal is to optimize modularity.

Table 1 shows the values of the internal validity indices for the original network ($\tau = 1$). Baseline clustering is preferred by Infomap, Label Propagation and WalkTrap

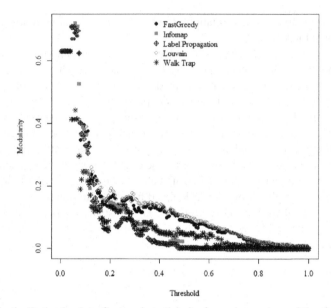

Fig. 1. Optimal value of network modularity for various values of threshold

algorithms. It provides the best values of all considered indices of internal validity. From a practical point of view, however, such a partition is of little interest, since it distinguishes two clusters, one of which contains only one node (SOCI07: Total number of pensioners). This is due to the fact that the length of the time series of this economic indicator is 4 years (2014–2017), which did not allow calculating correlations with other socio-economic indicators.

Clustering using the FastGreedy and Louvain algorithms is more interesting in terms of interpretation. There are two fairly large clusters. But these algorithms give clearly worse values of internal validity indices. Only by index G3 are all methods close to each other.

Table 1. Baseline values of indices of internal validity.

Index	FastGreedy	Infomap	Label Propagation	Louvain	WalkTrap
Silhouette	−0.021	**0.369**	**0.369**	−0.029	**0.369**
C	0.482	**0**	**0**	0.485	**0**
G2	0.035	**1**	**1**	0.021	**1**
G3	0.630	0.627	0.627	0.632	0.627
Connectivity	133.670	**2.929**	**2.929**	112.645	**2.929**

To characterize the variability of the results of clustering when the threshold changes, Table 2 shows the deviations of the values of the internal validity indices from the baseline. In terms of silhouette and connectivity indices, the Fast-Greedy and Louvain algorithms provide the least variation, that is, they are more resistant to network sparseness. However, for the rest of the indices of internal validity and for all indices of external validity, they lose. The clustering partitions obtained by the Infomap and Label Propagation algorithms are closest to the baseline partitionings by external validity indices.

Table 2. Deviations of values of internal validity indices from the baseline.

Index	FastGreedy	Infomap	Label Propagation	Louvain	WalkTrap
Silhouette	**0.065**	0.175	0.204	**0.076**	0.264
C	0.228	**0.064**	**0.065**	0.241	0.105
G2	0.529	**0.118**	**0.120**	0.555	0.212
G3	0.260	0.258	0.244	0.268	0.293
Connectivity	**46.64**	108.07	104.48	**60.97**	127.40
Purity1	0.241	**0.000**	**0.000**	0.200	**0.000**
Purity2	0.695	**0.460**	**0.438**	0.708	0.522
SU	0.868	**0.789**	**0.764**	0.872	0.870

Let's compare the optimal values of the internal validity indices (see Table 3) with the values achieved for the original network (see Table 1). It can be seen that in some cases the use of a hard threshold makes it possible to obtain a better partition into clusters. For example, an improvement in the silhouette index is observed for the FastGreedy and Louvain algorithms. Nevertheless, the achieved value is still worse than for the Infomap, Label Propagation, WalkTrap algorithms. For them, the optimal partitioning includes two clusters with isolated nodes and one with other nodes. Another isolated node is the indicator - NAAC08: Employment by activity manufacturing. No positive dynamics was observed for this indicator, therefore, correlations with other socio-economic indicators cannot be calculated. Again, this partition is not very useful in terms of interpretation.

The optimal values of the C and G2 indices obtained using the FastGreedy and Louvain algorithms are already close to the best values achieved using other algorithms. Therefore, changing the threshold can significantly improve the internal validity of clustering, estimated by the C and G2 indices, for the FastGreedy and Louvain algorithms.

In general, when applying the FastGreedy and Louvain algorithms, it makes sense to vary the sparseness of the network to select the best clustering.

External validity was determined in comparison with clustering by the original network, so the case $\tau = 1$ in Table 3 is not taken into account (for it, all external validity indices are equal to 1). It also reveals the advantages of the Infomap, Label Propagation and WalkTrap algorithms, which add another isolated cluster to the original partition. Interestingly, with a very small decrease in the network sparseness (from $\tau = 1$ to

Table 3. Optimal values of indices of internal, external (when $\tau = 1$ are not taken into account) validity.

Index	FastGreedy	Infomap	Label Propagation	Louvain	WalkTrap
Silhouette	0.151	**0.373**	**0.373**	0.151	**0.373**
C	0.063	**0**	**0**	0.057	**0**
G2	0.933	**1**	**1**	0.936	**1**
G3	0.103	0.104	0.104	0.103	**0.006**
Connectivity	80.835	**2.929**	**2.929**	112.645	**2.929**
Purity1	0.932	**1**	**1**	0.966	**1**
Purity2	0.648	**0.989**	**0.989**	0.670	**0.989**
SU	0.226	**0.667**	**0.667**	0.230	**0.667**

$\tau = 0.995$), the FastGreedy and Louvain algorithms give very different clustering from the baseline one. This also indicates a strong sensitivity of the algorithms to the sparseness of the network.

Table 4 shows the threshold values at which the optimal values shown in Table 3 are achieved. It is worth noting that for the Infomap, Label Propagation, WalkTrap algorithms, in many cases, optimal values are achieved over a wide range of threshold values. This is due to the fact that the same partition into clusters is obtained. Thus, the results can be characterized as more stable compared to the FastGreedy and Louvain algorithms. In Table 4, bold marks are cases where the range of optimal threshold values includes 1, that is, clustering on the original network gives the best result. This applies to the Infomap, Label Propagation and WalkTrap algorithms and internal validity indices such as C, G2, connectivity, as well as Louvain and connectivity, i.e. (10 cases out of 25). Therefore, in most cases, the use of a hard threshold and a sparse network has improved the validity of clustering. For the Infomap, Label Propagation and WalkTrap algorithms, you can choose a threshold close to one. Whereas for the FastGreedy and Louvain algorithms, the situation is different. For most indicators of internal validity, the optimal thresholds are close to zero. This results in a very sparse network with a large number of single clusters. As with the three clusters detecting with Infomap, Label Propagation, and WalkTrap, this partitioning is not very helpful in terms of interpretation. The connectivity index behaves differently, the optimal values of which are achieved at sufficiently large threshold values. Using the FastGreedy algorithm, two large clusters (51 and 35 nodes) and two single ones are detected. In terms of interpretation, such a partition can be more interesting.

Table 4. Optimal Threshold Values Based on Indexes of Internal and External (when $\tau = 1$ not taken into account) Validity.

Index	FastGreedy	Infomap	Label Propagation	Louvain	Walk Trap
Silhouette	0.06	[0.735, 0.995]	[0.67, 0.995]	0.06	[0.865, 0.995]
C	0.11	**[0.735, 1]**	**[0.67, 1]**	[0.1, 0.105]	**[0.865,1]**
G2	0.13	**[0.735, 1]**	**[0.67, 1]**	0.17	**[0.865,1]**
G3	[0.065, 0.07]	[0.045, 0.055]	[0.045, 0.055]	[0.065, 0.07]	[0.045, 0.055]
Connectivity	0.715	**1**	**1**	**1**	**1**
Purity1	[0.005, 0.04]	[0.005, 0.995]	[0.005, 0.995]	[0.005, 0.075]	[0.005, 0.995]
Purity2	0.915	[0.735,0.995]	[0.67, 0.995]	0.965	[0.865, 0.995]
SU	[0.005, 0.04]	[0.735,0.995]	[0.67, 0.995]	0.075	[0.865, 0.995]

Interesting patterns are revealed in the change of the Purity2 and SU indices for the FastGreedy and Louvain algorithms. There is a negative relationship between them. This can be judged by the optimal threshold values, for Purity2 they are high, for SU, on the contrary, they are close to zero. Figure 2 shows a scatterplot. It can be seen that the relationship is somewhat non-linear. This means that it is impossible to reach the optimum for two indices at once, that is, to ensure that the partition is close to that built on the original network and strongly correlates with it. Based on the scatterplot, it would be possible to select the points closest to the upper right corner (marked with a large gray triangle and large black circle). They correspond to a division into 4 clusters: 2 single ones (indicators SOCI07, NAAC08) and 2 clusters, including 39 and 47 nodes for Louvain algorithm, 48 and 38 nodes for FastGreedy algorithm. This version of the partition is much more interesting from the point of view of interpretation. Thus, when choosing the optimal partition using the FastGreedy and Louvain algorithms, it makes sense to variate the network sparseness and focus on a compromise between the Purity and SU indices.

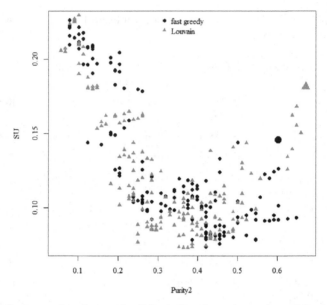

Fig. 2. Scatterplot of external validity indices for various values of threshold

4 Conclusion

On the example of one network describing the socio-economic system of the Nordic region, the work of clustering algorithms is compared when varying the threshold that determines the sparseness of the network. To assess the quality of clustering results, indices of internal and external validity are used. It is shown that Infomap, Label Propagation, WalkTrap algorithms behave more stable. However, their results are useless: almost all economic indicators are included in one cluster. The results of the FastGreedy and Louvain algorithms are more interesting in terms of interpretation.

For the Infomap, Label Propagation, WalkTrap algorithms, partitioning is optimal from the point of view of internal validity on a wide range of threshold values, usually close to 1, that is, for the original non-sparse network. While for the FastGreedy and Louvain algorithms, the optimal threshold values are in most cases close to zero, which corresponds to a highly sparse network. Such results are also not very suitable for interpretation, since they include many small and single clusters.

An interesting behavior of the external validity indices was found when using the FastGreedy and Louvain algorithms. Clustering on the original and weakly sparse network gives high values of the measure of purity, but a low relationship between partitions into clusters. On the contrary, when the network is highly sparse, the purity measure is lower, while the symmetric uncertainty increases. Therefore, it is not possible to optimize both of these indexes. It makes sense to find some compromise.

In general, in most cases, varying the threshold made it possible to obtain a better partition compared to the results of clustering on the original network. Therefore, it can be recommended to select a threshold based on a particular validity index based on the goals of the analysis.

Acknowledgments. The research is supported by the Russian Science Foundation and the Government of the Novosibirsk Oblast (project No. 22-28-20308).

References

1. Shkoda, M.: Economic indicators of development cluster partnership systems of the national economy. Eur. Sci. **1**, 85–92 (2021)
2. Skrabulakova, E.F., Gresova, E., Khouri, S.: Similarity assessment of economic indicators of selected countries by graph theory means. Transform. Bus. Econ. **18**, 333–347 (2019)
3. Golino, H., et al.: Investigating the performance of exploratory graph analysis and traditional techniques to identify the number of latent factors: a simulation and tutorial. Psychol. Methods **25**, 292–320 (2020)
4. Steinhauser, D., Krall, L., Müssig, C., Büssis, D., Usadel, B.: Correlation networks. In: Analysis of Biological Networks, pp. 305–333. Wiley, Hoboken, New Jersey (2008)
5. Clauset, A., Newman, M.E., Moore, C.: Finding community structure in very large networks. Phys. Rev. E **70**(6), 066111 (2004)
6. Blondel, V.D., Guillaume, J.L., Lambiotte, R., Lefebvre, E.: Fast unfolding of communities in large networks. J. Stat. Mech: Theory Exp. **10**, P10008 (2008)
7. Gates, K.M., Henry, T., Steinley, D., Fair, D.A.: A Monte Carlo evaluation of weighted community detection algorithms. Front. Neuroinform. **10**, 45 (2016)
8. Pons, P., Latapy, M.: Computing communities in large networks using random walks. In: Yolum, P., Güngör, T., Gürgen, F., Özturan, C. (eds.) ISCIS 2005. LNCS, vol. 3733, pp. 284–293. Springer, Heidelberg (2005). https://doi.org/10.1007/11569596_31
9. Rosvall, M., Axelsson, D., Bergstrom, C.T.: The map equation. Eur. Phys. J. Special Topics **178**(1), 13–23 (2009)
10. Raghavan, U.N., Albert, R., Kumara, S.: Near linear time algorithm to detect community structures in large-scale networks. Phys. Rev. E **76**, 036106 (2007)
11. Rousseeuw, P.J.: Silhouettes: a graphical aid to the interpretation and validation of cluster analysis. J. Comput. Appl. Math. **20**, 53–65 (1987)
12. Hubert, L.J., Levin, J.R.: A general statistical framework for assessing categorical clustering in free recall. Psychol. Bull. **83**(6), 1072–1080 (1976)
13. Baker, F.B., Hubert, L.J.: Measuring the power of hierarchical cluster analysis. J. Am. Stat. Assoc. **70**(349), 31–38 (1975)
14. Kolaczyk, E.D., Csárdi, G.: Statistical Analysis of Network Data with R. Springer, New York (2014)
15. Walesiak, M.: Cluster analysis with ClusterSim computer program and R environment Acta Universitatis Lodziensis. Folia Oeconomica **216**, 303–311 (2008)
16. Brock, G., Pihur, V., Datta, S., Datta, S.: clValid: an R package for cluster validation. J. Stat. Softw. **25**, 1–22 (2008)
17. Nordic Statistics. https://pxweb.nordicstatistics.org/pxweb/en/Nordic%20Statistics/. Accessed 1 Feb 2023

Routing and Scheduling Problems

Monte Carlo Tree Search with Adaptive Estimation for DAG Scheduling

Alexander Allahverdyan[1] , Anastasiia Zhadan[1] , Ivan Kondratov[1] ,
Vikenty Mikheev[1] , Ovanes Petrosian[1(✉)] , Aleksei Romanovskii[2] ,
and Vitaliy Kharin[2]

[1] St.Petersburg State University,
7/9 Universitetskaya Nab., 199034 Saint Petersburg, Russia
kondratov.ivan.vladimirovich@gmail.com, petrosian.ovanes@yandex.ru
kondratov.ivan.vladimirovich@gmail.com, petrosian.ovanes@yandex.ru
[2] Huawei, Saint-Petersburg Research Institute,
69/7 Marata, 191119 Saint-Petersburg, Russia

Abstract. Scheduling is important for improving performance in a distributed heterogeneous computing environment where workflows represented as a directed acyclic graph (DAG). The DAG task scheduling problem has been extensively studied, and many modern heuristic algorithms such as DONF and others have been proposed. Goal of this study is to propose an Artificial Neural Network (ANN) based scheduling scheme with Monte Carlo Tree Search (MCTS). Numerical representation of each DAG node are used as inputs to the ANN, and the outputs are the execution priority of the node. The MCTS method utilized to determine the actual scheduling policies; The proposed algorithm shows an improvement of up to 42% on certain graph topologies with average performance gain of 7.05% over the best scheduling algorithm. Algorithm managed to cover 40.01% of the proximity interval from the best scheduling algorithm to the global optimal solution obtained via the Mixed-Integer Linear Programming approach.

Keywords: Monte Carlo Tree Search · Neural networks · Scheduling · Directed Acyclic Graph

1 Introduction

Efficient task scheduling on heterogeneous computing resources is crucial for achieving high performance, particularly for dependent tasks in which workflows are represented as directional acyclic graphs, where node edges represent data dependencies between tasks. Finding an optimal solution to this problem is often an NP-complete problem [14].

This paper will consider static scheduling models in which information about the computational graph (adjacency matrix, node and edge weights) is known

Supported by Saint-Petersburg State University.

prior to execution at compile time. Static scheduling algorithms are classified as follows: mixed integer programming approach [2]; heuristic-based algorithms [5,7,11]; machine learning methods [1,8,9,13].

Mixed Integer Programming Approach (MILP). The MILP approach offers a global optimal solution to DAG scheduling, but it is often intractable for practical problems due to high dimensionality and computational complexity. The DAG scheduling problem is first formulated as a MILP, and a modern solver like Gurobi [4] can be used to solve it. This paper uses the MILP approach to compare the proposed algorithm to the global optimum.

Heuristic-Based Algorithms. Although these heuristic algorithms have good versatility and stability and have polynomial time complexity instead of exponential complexity [5], planning strategies are mostly based on greedy strategies and there is still much room for improvement, and it is difficult to adapt to rapidly changing requirements. According to the results in [7,11] the most effective and the frequently used algorithms for static DAG scheduling (in the problem of minizing the makespan) are the algorithms Critical Path on a Processor (CPOP), Heterogeneous Critical Parent Tree (HCPT), High Performance Task Scheduling (HPS), Performance Effective Task Scheduling (PETS) and Degree of Node First (DONF). These algorithms were selected for comparison with the algorithm proposed in this research.

Machine Learning is a logical response to the growing interest in adaptive data-driven scheduling algorithms. Reinforcement learning (RL) [1] and Monte Carlo Tree Search (MCTS) [13] are two examples of machine learning-based graph scheduling methods. RL requires significant computational resources to train, while MCTS is less resource-intensive. Orhan [9] proposed a reinforcement learning-based method for heterogeneous distributed system scheduling, but it has some unrealistic assumptions. Loth M [8] proposed using MCTS for job shop scheduling, which is simpler than DAG task scheduling in terms of data dependency.

This paper proposes a novel approach of using neural network-based Monte Carlo tree search for DAG, using an adaptive metric. Figure 1 illustrates the proposed algorithm's structure, with the first level being a Monte Carlo tree search using a neural network to estimate node quality and determine execution queue. The second level assigns problems to computational resources based on the Earliest Finish Time (EFT) algorithm. Node input features are defined by modern graph metrics (Sect. 3.2), which determine task priority from ready-to-run tasks.

The remainder of the paper is organized as follows. Section 2 describes the abstracted model of heterogeneous systems and mathematical problem statement. Section 3 describes the proposed MCTS-based scheduling algorithm and the algorithms considered for comparison. Section 4 evaluates the MCTS-based scheduling algorithm in comparison with state-of-the-art scheduling algorithms for a different workspaces and comparison with the exact MILP solution. Section 5 contains the conclusion.

2   DAG Scheduling Problem Description

The problem addressed in this paper is the scheduling of a single DAG in distributed heterogeneous system. The scheduling model consists of three parts:

The Problem Definition Directed Acyclic Graph $G = (J, E)$, where J is a set of nodes and E is the set of edges.

- Edge $(v_i, v_j) \in E$ denotes the precedence constraint such that node v_j must wait until task v_i finishes its execution ($S_j > S_i$, where $S_i, S_j \in R^n$ is a scheduled start time of nodes v_i, v_j)
- Cost of communication $b_{j,m}$ between nodes v_j and v_m, $(v_j, v_m) \in E$ should be taken into account if node v_j and v_m assigned to different executors, otherwise there is no cost of communication. If node v_m has several parents $(v_j,..,v_k)$ that were performed on executors other than the executor assigned to node v_m, then the cost of communication are taken into account from each parent $b_{j,m} +..+ b_{k,m}$.
- Each node v_j has a type that denotes on what type of executor should this node be executed.
- Set of immediate predecessors of node v_j in a DAG is expressed as $pred(v_j)$. The set of immediate successors of node v_j is expressed as $succ(v_j)$.

A Distributed Heterogeneous Computing System, which consists of a set P, $p_i \in P$ heterogeneous executors with a fully connected topology: with D computation cost matrix, where $D^{j,i} = w_{j,i}$ is the execution time for executor $p_i \in P$ to process node $v_j \in J$; where each executor p_i has a type that denotes what type of tasks can be defined for this executor. In order to assign a node $v_j \in J$ to a executor p_i, it is necessary that the type of node v_j and the type of executor p_i match. $B^{j,i} > 0$ – is a matrix with 1 if performer $p_i \in P$ can complete node $v_j \in J$, 0 otherwise.

Performance Criterion for Scheduling. Before presenting the final scheduling objective function, we first define the Makespan, Earliest Start Time (EST), Earliest Finish Time (EFT) attributes.

- Makespan. Is the finish time of the last node in the scheduled DAG. It is defined by $makespan = max\{AFT(v_{exit})\}$ where $AFT(v_{exit})$ is the actual finish time (AFT) of exit node v_{exit}. In the case where there are multiple exit nodes, the makespan is the maximum AFT of all exit nodes.
- EFT. $EFT(v_j, p_i)$ denotes the earliest finish time of node v_j on executor p_i and is defined as $EFT(v_j, p_i) = EST(v_j, p_i) + w_{j,i}.$, where $EST(v_j, p_i) = max\{T_{Ava}(p_i), max_{\{v_m \in pred(v_j)\}}\{AFT(v_m) + b_{m,j}\}\}$. $T_{Ava}(p_i)$ is the earliest ready time of executor p_i.

2.1   Mathematical Problem Statement

The objective function in Eq. 1 minimizes the makespan (C_{max}) over all nodes. Makespan is always longer than the completion time of any node: constraint 2, where X is variables matrix allocation of node to executor, where 1 if executor $p_i \in P$ is assigned to complete v_j, 0 otherwise. Constraint need Eq. 3 to minimize the number of irrelevant binary variables $X_{j,i}$. Constraint Eq. 4 specifies that each node must be assigned once to exactly one executor. Constraints Eqs. 5–8 define the nodes that are assigned to the same executor by using support variables $Z_{m,j}$ – 1 if node $v_m \in J$ and node $v_j \in J$ assigned to the same executor, 0 otherwise and $K_{i,j,m}$ is the product of binary variables $X_{j,i}$ and $X_{m,i}$. Since the product between variables violates the linearity of the model variable $K_{i,j,m}$, assumes its value through inequalities of model. Constraints Eqs. 9–10 the constraints state that a node cannot start until its predecessors are completed and data has been communicated to it if the preceding jobs were executed on a different executor, if required. Collectively, the constraints 11 – 12 specify that an executor may process at most one node at a time and that if two nodes v_j and v_k are assigned to the same executor, their execution times may not overlap. Where $Q^{j,m}$ is a full node precedence graph, 1 if node $v_j \in J$ comes any time before node $v_m \in J$, 0 otherwise. θ is a matrix support variable used to determine whether two nodes overlap, $\theta_{m,j}$ – 1 if node $v_m \in J$ is started before node $v_j \in J$, 0 otherwise. This approach to formulating the graph scheduling problem is common in the relevant literature, for example in [10] and [6].

$$Minimize : C_{max} \tag{1}$$

$$C_{max} \geq S_j + D^{j,i} \cdot X_{j,i}, \ \forall v_j \in J, \ \forall p_i \in P : D^{j,i} > 0. \tag{2}$$

$$X_{j,i} = 0, \ \forall v_j \in J, \ \forall p_i \in P : B^{j,i} = 0. \tag{3}$$

$$\sum_{p_i \in P} X_{j,i} = 1, \forall v_j \in J. \tag{4}$$

$$Z_{j,k} = \sum_{p_i \in P} K_{i,j,k}, \ \forall (v_j, v_k) \in E, \tag{5}$$

$$K_{i,j,k} \leq X_{j,i}, \ \forall (v_j, v_k) \in E, \ p_i \in P, \tag{6}$$

$$K_{i,j,k} \leq X_{k,i}, \ \forall (v_j, v_k) \in E, \ p_i \in P, \tag{7}$$

$$K_{i,j,k} \leq X_{k,i} + X_{j,i} - 1, \ \forall (v_j, v_k) \in E, \ p_i \in P. \tag{8}$$

$$S_k \geq S_j, \ \forall (v_j, v_k) \in E, \tag{9}$$

$$S_k \geq S_j + \sum_{p_i \in P} X_{j,i} \cdot D^{j,i} + \sum_{v_l \in pred(v_k)} b_{l,k} \cdot (1 - Z_{l,k}), \ \forall (v_j, v_k) \in E. \tag{10}$$

$$S_k - \sum_{p_i \in P} D^{j,i} \cdot X_{j,i} - S_j - \sum_{v_l \in pred(v_k)} b_{l,k} \cdot (1 - Z_{l,k}) \geq M \cdot \theta_{j,k},$$

$$S_k - \sum_{p_i \in P} D^{j,i} \cdot X_{j,i} - S_j - \sum_{v_l \in pred(v_k)} b_{l,k} \cdot (1 - Z_{l,k}) < M \cdot (1 - \theta_{j,k}),$$

$$\forall v_j, v_k \in J, \ v_j \neq v_k, \ Q^{j,k} \neq 0, \tag{11}$$

$$X_{j,i} + X_{k,i} + \theta_{j,k} + \theta_{k,j}, \ \forall v_j, \ v_k \in J, \ p_i \in P, \tag{12}$$

3   Algorithm

This section describes the proposed MCTS-based scheduling algorithm, graph scheduling metrics and the algorithms considered for comparison.

3.1   Algorithm Architecture

The first level of the algorithm is the Monte Carlo tree search, which uses a neural network to evaluate the quality of nodes using their metrics and determines the execution queue of nodes following these steps:

- Selection: it starts with the root node R and then selects the child node R according to the default policy. The newly selected node will be the root node and then repeat the above process until the leaf node L is reached
- Extension: create one or more child nodes L and select one node N until the game is over
- Simulation: select at node N and do with a random strategy such as uniform random walk until the game ends
- Back propagation: update node information on the path from node N to node R using the result of the random game

The neural network output is a single neuron that computes the Q, quality of a given node, based on metrics for the graph are described in Sect. 3.2. The output of the neural network represents the relevance of a node from the ready queue, which informs the Monte Carlo tree search as to which node to execute next. This relevance value is used to select the node with the highest value for execution. In the next step of the algorithm, a task is assigned to a computing resource of the same type using the EFT method. Once assigned, the task is marked as complete, and the readiness queue is updated. This process continues until all tasks have been completed. Figure 1 illustrates the overall system structure of the MCTS-based approach.

3.2   Graph Scheduling Metrics

The main stage of the algorithm development is the search of metrics that could be extracted on the local scale to describe of each node in the graph. A comprehensive overview of graph metrics is presented in paper [3]. Based on this review, we use the following graph metrics:

1. Weighted Out-Degree (WOD) of node v_i: $WOD(v_i) = \sum_{v_j \in succ(v_i)} \frac{1}{ID(v_j)}$, where $ID(v_j)$ is the in-degree of node v_j.
2. 2-degree
   WOD: $WOD_2(v_i) = \sum_{v_j \in succ(v_i)} \left(\frac{1}{ID(v_j)} + \alpha \cdot \sum_{v_k \in succ(v_j)} \frac{1}{ID(v_k)} \right)$, where α is a factor of the second out-degree.
3. HEFT prioritizing tasks: $rank(v_i) = \overline{w_i} + \max_{v_j \in succ(v_i)} (\overline{b_{i,j}} + rank(v_j))$.
4. Computational complexity $C(v_i)$ of the node v_i.

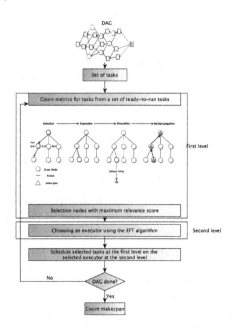

Fig. 1. Adaptive Neural Network with Monte Carlo Tree Search for DAG scheduling

5. Quantity of the predecessors $\|pred(v_i)\|$ node v_i.
6. Quantity of the successors $\|succ(v_i)\|$ node v_i.
7. Total weight of incoming edges TW_{in} in node v_i.
8. Total weight of outgoing edges TW_{ou} in node v_i.

3.3   State-of-the-Art DAG Scheduling Algorithms

In this section, we provide a detailed description of State-of-The-Art (SoTA) DAG scheduling algorithms, selected for comparison in accordance with the review paper for graph scheduling in heterogeneous environments [7,11].

Critical Path on a Processor (CPOP) Algorithm. Algorithm include of two phases. In the first phase, the priority of a node is defined by the upward rank $(rank_u(v_i) = \overline{w_i} + \max_{v_j \in succ(v_i)}(\overline{b_{i,j}} + rank_u(v_j)))$ and downward rank $(rank_d(v_i) = \max_{v_j \in pred(v_i)}(\overline{b_{i,j}} + \overline{w_j} + rank_d(v_j)))$. The final priority of node v_i is calculated by: $priority(v_i) = rank_d(v_i) + rank_u(v_i)$ In second phase, tasks are selected according to higher rank value and selects for scheduling to best suitable processor which minimizes the execution time of task.

Heterogeneous Critical Parent Trees (HCPT) Algorithm. It uses the method in which a graph is divided into two group of unordered parent tree for scheduling that is a critical path (CN). The zero variance among average latest start time and average earliest start time is specified as a critical node.

High-Performance Task Scheduling (HPS) Algorithm. The HPS algorithm has 3 execution stages. To sort the nodes, the traversing of DAG is being done in top down manner. In node prioritizing phase, computation of node priority is done by using the Up Link Cost, Down Link Cost, and Link Cost node priority attributes. In processor selection phase, the minimum EFT (given by the processor) is chosen for execution of that node.

Performance Effective Task Scheduling (PETS) Algorithm. The algorithm execute in three phases. In the first phase, algorithms try to grouping each task at each level for parallel executing. Also the second phase, assigns priority to each task based on its rank. In the processor selection phase, the processor is selected using EFT approach.

Degree of Node First (DONF) Algorithm. It is reasonable to maintain higher parallelism during the scheduling process in order to make full use of heterogeneous system resources. Thus the chosen node should have the property of enlarging parallelism as much as possible. Degree-of-node scheduling procedure can be shortly described as nodes with larger out-degree should be scheduled earlier. The weighted out-degree (WOD) of node v_i is defined by $WOD(v_i) = \sum_{v_j \in succ(v_i)} \frac{1}{ID(v_j)}$, where $ID(v_j)$ is the in-degree of node v_j. Next, a processor with minimizes EFT is selected.

Time Complexity shown in Table 1 with the parameters v, p where v denotes the tasks and where p is the count of processors, o is the maximum out-degree of any node.

Table 1. List Scheduling Algorithms for Heterogeneous System.

Algorithms	Complexity	Source
CPOP, HCPT	$O(v^2 \times p)$	[11]
HPS, PETS	$O(v^2 \times p \times \log v)$	[11]
DONF	$O(v \times o \times p)$	[7]
MCTS	$O(k \times o \times p \times d)$	this paper

Our proposed MCTS-based DAG scheduling algorithm has a time complexity of $O(k \times o \times p \times d)$, with k representing the number of nodes in the ready queue, d being the maximum tree depth, and o being large as $v - 1$. The algorithm's optimal depth was determined to be 30 through testing. While the algorithm may be slower than other heuristic algorithms, increasing the depth by a factor of d allows for better performance. The worst-case time complexity of our algorithm is $O(v^2 \times p \times d)$, whereas DONF has a worst-case complexity of $O(v^2 \times p)$, which is similar to other algorithms.

4   Numerical Simulation

In this section, the design of an experiment is explained and the performance of the proposed MCTS-based algorithm is evaluated. The hardware that was used in experiment is a PC with Intel Core i5-8600 with 6 cores and 3.10 Ghz base frequency, 16 gb of RAM and NVIDIA GeForce GTX 1060 with 6 gb of vRAM.

4.1   Simulation Environment

The working environment is determined by different heterogeneous configurations from three to nine executors are used. These executors differ in computational capabilities and types:

1. Workspace 1: 3 executors of various types are defined, which have differ in computational capabilities: 26, 134, and 34 GFlops respectively for the first, second and third types. Tested on dimensions of 30, 60, 90 DAG nodes.
2. Workspace 2: 6 executors, previous configuration has been expanded by an additional three performers, which have differ in computational capabilities: 50, 70, and 20 GFlops respectively for the first, second and third types. Tested on dimensions of 30, 60, 90 DAG nodes.
3. Workspace 3: 9 executors, here as well the second configuration is expanded by computing capacities: 125, 40, 60 GFlops respectively for the first, second and third types. Tested on dimensions of 30, 60, 90 DAG nodes.

For all test cases in the simulation, the data transfer starts only after the completion of all previous ones. The speed of data exchange between the executors is 1085 MB/s.

4.2   Training and Testing Graph Description

We use open-source project DAGGEN to generates random DAGs [12]. This solution was chosen in order to make a comparison conditions more equal: it was used as a benchmark for DONF, CPOP and PETS algorithms in respective papers [7,11]. This tool is based on a layer-by-layer approach with six parameters (the values for generation used in this paper are indicated in the intervals): number of nodes: $n \in \{30, 60, 90\}$, maximum number of tasks that can be executed concurrently: $f \in \{0.2, 0.5\}$, regularity of the distribution of tasks between the different levels of the DAG: $r \in \{0.2, 0.8\}$, numbers of dependencies between task of two consecutive DAG levels: $d \in \{0.1, 0.4, 0.8\}$, maximum number of levels spanned by inter-task communications: $j \in \{2, 4\}$, ratio of the sum of the edges weights to the sum of the node weights: $c \in \{0.2, 0.8\}$.

At a fixed dimensionality there are 48 topologies, for each topology 3 graphs are generated for training (total 144 for each dimension 30, 60, 90) and 10 graphs for evaluation (total 480 for each dimension 30, 60, 90). The total number of DAGs tested was 1440. A comparative analysis is presented in Sect. 4.4. Examples of DAGs are shown in Fig. 2.

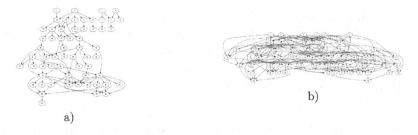

Fig. 2. Examples of DAGs with 60 nodes: a) fat 0.5, density 0.4, regular 0.2, jump 4, ccr 0.8, b) fat 0.5, density 0.8, regular 0.8, jump 4, ccr 0.8.

4.3   Training of Neural Network

We use a genetic algorithm to train the neural network. This algorithm requires five components: chromosome Encoding, where weights and biases are encoded as real numbers; evaluation Function, which calculates the total makespan for training graphs; initialization procedure, where initial member weights are randomly chosen between -1 and 1; operators, including TPI and SBX crossovers, mutation operators, and crossover operators; and parameter Settings, including a mutation probability of 80%, a population size of 150, 150 generations, and a total of 95 weights.

The neural network architecture includes 5 Dense layers. This means that each layer contains a set of neurons that are completely connected to the neurons of the previous and next layer. The number of neurons in each layer is 8, 6, 4, 2 and 1, respectively.

Activation functions are mathematical functions that are applied to the output data of neurons to control the rate of their activation. In this case, two activation functions are used: relu and sigmoid. ReLU (Rectified Linear Unit) is a function that returns the zero value for all negative input values and the value itself for all non-negative input values. Sigmoid is a function that converts an input value in the range from 0 to 1.

The total number of parameters in the neural network is 95. These parameters include neuron weights and offsets, which determine how signals are transmitted between neurons in different layers. The number of parameters can affect the performance of a neural network and its ability to learn from new data.

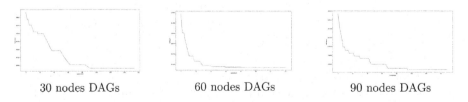

| 30 nodes DAGs | 60 nodes DAGs | 90 nodes DAGs |

Fig. 3. The learning curve of the MCTS training algorithm.

The training process of the MCTS-based algorithm is visualized in Fig. 3. The generations of the genetic algorithm are located on the X-Axis, the average makespan for all graphs is located on the Y-Axis.

4.4   Result Compared with SoTA

In this subsection, we will be comparing the MCTS solution that has been presented with the algorithms discussed in Sect. 3.3. The basis of this comparison will be the percentage of time saved by using the constructed MCTS solution as opposed to the corresponding algorithm. For each tested graph $p = \frac{(makespan_{sota} - makespan_{MCTS}) \cdot 100\%}{makespan_{sota}}$ where $makespan_{sota}$ is the finish time of the last node in the scheduled DAG using one of the State-of-the-Art (SoTA) algorithms from the list above, $makespan_{MCTS}$ is the finish time of the last node in the scheduled DAG using the MCTS-based algorithm. Obtained values are summarized and averaged for each workspace and DAG dimension DAG. The proposed MCTS-based algorithm is tested to measure its generalization and robustness:

- depending on different system configurations (workspaces) and different DAG dimensions,
- analyzing the closeness of the makespan for a provided schedule to the global optimal makespan (defined using the MILP solution, more details in the Sects. 2.1, 4.5).

Table 2. Average percentage of improvement MCTS-based algorithm compared to SoTA.

Workspace	Dimension	DONF	CPOP	HCPT	HPS	PETS
Workspace 1: 3 executors	30	**5.01**	17.96	15.23	18.51	18.61
	60	**10.01**	24.44	22.31	25.41	25.33
	90	**12.74**	26.67	25.46	28.19	28.21
Workspace 2: 6 executors	30	**2.42**	14.37	10.61	13.58	13.77
	60	**6.66**	23.81	18.89	22.82	22.71
	90	**10.04**	26.47	22.41	27.19	27.00
Workspace 3: 9 executors	30	**1.75**	10.91	6.56	8.28	8.31
	60	**5.40**	20.32	14.76	17.97	18.13
	90	**6.33**	22.11	17.15	20.74	20.91

Comparison results in Table 2 for each workspace show that the MCTS-based algorithm provides an average improvement of 5.01% compared to DONF at 30 node dimensions, and an average improvement of 17.7% compared to other algorithms. However, in workspaces 2 and 3, MCTS is competitive with DONF and does not provide significant improvements compared to DONF, but still outperforms the other baselines. It's worth noting that at DAG dimensions of 90 and 60, DONF is the closest heuristic to MCTS, while MCTS outperforms DONF at other DAG dimensions. The MCTS-based algorithm shows the biggest improvement in workspace 1 (on average, at all dimensions is 9.03%), with diminishing effectiveness in workspace 2 (on average, at all dimensions is 6.4%) and workspace 3 (on average, at all dimensions is 4.5%). Simulation results also indicate that the percentage improvement of MCTS-based algorithms over state-of-the-art algorithms increases as the DAG dimension increases. On average, for all workspaces, the improvement is 3.05% at dimension 30, 7.35% at dimension 60, and 9.75% at dimension 90.

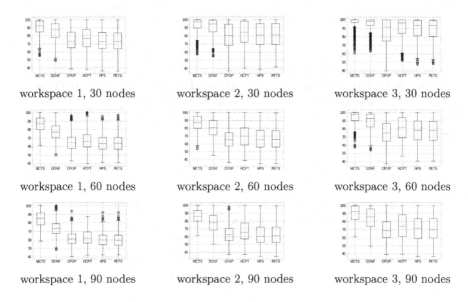

Fig. 4. Box plot with the comparison of proximity to the MILP solution for state-of-the-art DAG scheduling algorithms for each workspace and each DAG dimension.

Figure 4 compares different approaches for DAG scheduling to the global optimal solution obtained using the MILP approach, showing their proximity in terms of makespan. The MCTS-based algorithm is found to be closest to the global optimal solution on average, with a higher median than all other algorithms. The lower performance bound is also higher for the MCTS-based algorithm in all conditions, indicating better performance even under worst (low probability) conditions. The algorithm's results are more stable with fewer outliers, as shown by the shortest confidence interval length. See Sect. 4.5 for details on the exact solution implementation.

4.5   Detailed Comparison

In this subsection a details comparison of the proposed MCTS-based algorithm and the best state-of-the-art algorithm (DONF algorithm) is presented. The comparison is based on the proximity of the considered algorithms to the global optimal solution calculated by the MILP approach. The aim of this comparison is to determine the deviation of the resulting solution from the globally optimal solution, expressed as a percentage. This metric indicates the extent to which the solution can theoretically be further improved. The duration of the solution obtained by MILP is taken as 100%. For instance, a result of 96% signifies that the obtained solution is 4% longer than the MILP solution. The Mixed Integer-Linear Programming formulation presented in Sect. 2.1 is modeled and solved using Python 3.6 and MILP solver Gurobi 9.1.2 [4]. Since the mathematical model has a parameter M – upper bound on the makespan (set by default to a very large the same for all problems), then, in rare cases, this parameter can affect the convergence process of the gap. Therefore, the following parameters were chosen experimentally: time limit of 15 minutes (average time to solve problems of high dimensionality), relative MILP optimality gap of 0.03 (significantly reduce solution time with high scheduling accuracy), and deterministic concurrent method are used each MILP solution(significantly reduce solution time).

Table 3. Proximity of DONF and MCTS-based solutions to MILP solution.

Dimension	Workspace 3		Workspace 2		Workspace 1	
	DONF	MCTS	DONF	MCTS	DONF	MCTS
30	94.89	**96.58**	91.13	**93.39**	85.85	**90.57**
60	89.27	**94.33**	80.66	**86.51**	77.48	**86.48**
90	84.36	**89.96**	77.82	**86.48**	73.37	**84.67**

Table 3 shows the proximity results for MCTS-based and DONF algorithms to the MILP global optimal solution. For all the experiments MCTS-based algorithm provides higher proximity to the MILP solution. It is important to notice that with the help of the proposed MCTS-based algorithm it was possible to cover 40.01% of the proximity interval from the best DAG scheduling algorithm (DONF) to the global optimal solution obtained using the MILP approach: average proximity of best state-of-the-art algorithm DONF is 73.37% ; MCTS-based algorithm proximity is 84.67%.

Figure 5 shows the average proximity of MCTS-based and DONF solutions for each of the topologies under consideration. DAG topologies are located along the X-Axis, where the first letter denotes the generation parameter and the second number is the value of this parameter. Proximity percentages for MCTS-based and DONF solutions to the global optimal solution based on MILP are distributed along the Y-Axis. For workspace 1, MCTS-based algorithm provides

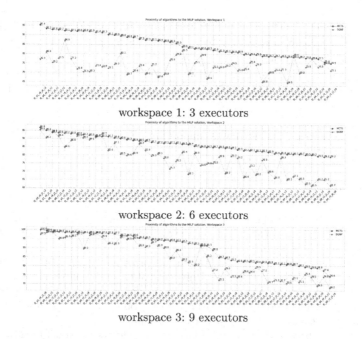

workspace 1: 3 executors

workspace 2: 6 executors

workspace 3: 9 executors

Fig. 5. Proximity of MCTS-based and DONF solutions to MILP solution for DAG topology: 90.

an improvement in the result for all DAG topologies. In other cases, one can observe some topologies where MCTS-based is quite close to the DONF algorithm. It is reasonable to interpret these results as features of training on a heterogeneous sample of data. Thus, having a similar computational complexity, the MCTS-based algorithm provides greater efficiency in almost all the DAG topologies.

5   Conclusion

This paper proposes a scheduling scheme for solving DAG static scheduling problems in heterogeneous environments using Artificial Neural Networks and Monte Carlo Tree Search methods. The proposed algorithm architecture is based on an novel adaptive DAG scheduling metric, where state-of-the-art graph scheduling metrics are aggregated by the neural network.

The algorithm has two levels. The first level is the Monte Carlo Tree Search algorithm, which uses a neural network to evaluate node quality and determine the order in the ready-to-execution queue of nodes. The second level is the assignment of a task to a computing resource of the same type as the task in accordance with the EFT algorithm. The proposed MCTS-based algorithm outperforms state-of-the-art heuristic rules, with an average improvement of 7.05% compared to DONF on a large number of random DAGs, and an average improvement of 24.1% over CPOP, HCPT, HPS, PETS algorithms. In small-scale DAGs, the MCTS-based algorithm provides proximity to the global optimal solution from 84.67% to 96.58%, covering 40.01% of the proximity interval from DONF to the global optimal solution.

Future work includes extending support for multi-dimensional executor requirements of DAGs and studying the DAG scheduling problem in online streaming settings.

Acknowledgements. This work was supported by Saint-Petersburg State University, project ID: 94062114.

References

1. Chen, S., Fang, S., Tang, R.: A reinforcement learning based approach for multi-projects scheduling in cloud manufacturing. Int. J. Prod. Res. **57**(10), 3080–3098 (2019)
2. Clausen, J.: Branch and Bound Algorithms-principles And Examples, pp. 1–30. Department of Computer Science, University of Copenhagen pp (1999)
3. Flint, C., Bramas, B.: Finding new heuristics for automated task prioritizing in heterogeneous computing. HAL-Inria preprint, pp. 1–43 (Nov 2020). https://hal.inria.fr/hal-02993015, working paper or preprint
4. Gurobi Optimization, LLC: Gurobi Optimizer Reference Manual (2023). https://www.gurobi.com
5. Hwang, J.J., Chow, Y.C., Anger, F.D., Lee, C.Y.: Scheduling precedence graphs in systems with interprocessor communication times. Siam J. Comput. **18**(2), 244–257 (1989)
6. Li, J., et al.: Path: Performance-aware task scheduling for energy-harvesting non-volatile processors. IEEE Trans. Very Large Scale Integr. (VLSI) Syst.**26**(9), 1671–1684 (2018). https://doi.org/10.1109/TVLSI.2018.2825605
7. Lin, H., Li, M.F., Jia, C.F., Liu, J.N., An, H.: Degree-of-node task scheduling of fine-grained parallel programs on heterogeneous systems. J. Comput. Sci. Technol. **34**, 1096–1108 (2019)
8. Loth, M., Sebag, M., Hamadi, Y., Schoenauer, M., Schulte, C.: Hybridizing Constraint Programming and Monte-Carlo Tree Search: Application to the Job Shop Problem. In: Nicosia, G., Pardalos, P. (eds.) LION 2013. LNCS, vol. 7997, pp. 315–320. Springer, Heidelberg (2013). https://doi.org/10.1007/978-3-642-44973-4_35
9. Orhean, A.I., Pop, F., Raicu, I.: New scheduling approach using reinforcement learning for heterogeneous distributed systems. J. Parallel Distrib. Comput. **117**, 292–302 (2018)
10. Singh, J., Mangipudi, B., Betha, S., Auluck, N.: Restricted duplication based milp formulation for scheduling task graphs on unrelated parallel machines. In: 2012 Fifth International Symposium on Parallel Architectures, Algorithms and Programming, pp. 202–209 (2012). https://doi.org/10.1109/PAAP.2012.37
11. Suman, C., Kumar, G.: Analysis of process scheduling algorithm for multiprocessor system. In: 2018 7th International Conference on Reliability, Infocom Technologies and Optimization (Trends and Future Directions)(ICRITO), pp. 564–569. IEEE (2018)
12. Suter, F., Hunold, S.: Daggen: A synthetic task graph generator (2013). https://github.com/frs69wq/daggen

13. Świechowski, M., Godlewski, K., Sawicki, B., Mańdziuk, J.: Monte carlo tree search: A review of recent modifications and applications. Artificial Intelligence Review, pp. 1–66 (2022)
14. Ullman, J.D.: Np-complete scheduling problems. J. Comput. Syst. Sci. **10**(3), 384–393 (1975)

Resource Allocation in Heterogeneous Network with Supervised GNNs

Qiushi Sun[(⊠)] [iD], Yuyi Zhang[iD], Haitao Wu[iD], and Ovanes Petrosian[iD]

St. Petersburg State University, Saint Petersburg 198504, Russia
{st059656,st088518,st082606}@student.spbu.ru, o.petrosyan@spbu.ru

Abstract. Device-to-device (D2D) transmission has become an essential form of wireless communication due to the rise of 5G and Internet of Things (IoT). Unfortunately, most currently available techniques for allocating resources are extremely time-consuming or computationally expensive. Graph neural networks (GNNs) have recently been proposed as a way to improve the efficacy of many network-related tasks. To tackle this issue, we propose a GNN-based method in a supervised manner. We denote heterogeneous networks as directed graphs and model various communication links as nodes and interference between links as edges. The proposed algorithm has two phases: (1) Offline supervised learning obtains the optimal solution through the particle swarm optimization (PSO) algorithm and uses the solution as a sample label to train GNN; (2) Online prediction by well-trained GNN. The simulation results indicate that the proposed method outperforms the state-of-the-art GNN-based benchmarks and achieves substantial speedups over conventional benchmarks.

Keywords: Resource allocation · Heterogeneous networks · Swarm intelligence · Graph neural networks

1 Introduction

Effective resource allocation (RA) is essential for optimizing the performance of wireless networks in light of the rapid increase in data traffic caused by the proliferation of wireless devices and the development of the internet of things. However, conventional RA problems (such as power control and beamforming) present computational difficulties. Because the wireless channel states change over time, these problems must be solved in real-time. Moreover, the mutual interference between connections and the strong coupling of optimization variables render numerous problems non-convex and NP-hard.

This work was supported by Saint Petersburg State University (project ID: 94062114).

In the past few decades, numerical optimization has been a critical part of solving problems with managing wireless resources. Nevertheless, optimization algorithms are typically quite complicated, which creates a severe gap between theoretical design and real-time processing. Optimal or near-optimal solutions can be obtained by applying specific search methods, such as genetic algorithms [1] and branch-and-bound techniques [2]. These algorithms are iterative algorithms that take as input a given set of real-time network parameters (such as channel realizations and signal-to-noise ratio specifications), run multiple iterations, and produce an optimal RA policy as output. Because of this, these methods are not suitable for large-scale heterogeneous D2D networks with many connections and different resources (such as channels, antennas, and transmit power). Suboptimal solutions obtained based on techniques such as iterative distributed optimization [3,4], heuristic algorithms [5], and game theory [6] are also typically computationally intensive or infeasible for large D2D networks due to high signal overhead. Moreover, these sub-optimal solutions may be far from the optimal ones, and their convergence properties and optimality gaps may still need to be discovered.

Deep learning (DL) has emerged as a revolutionary technique for effectively resolving various difficult problems in wireless communication systems. Learning the relationship between policies and their influencing parameters, a well-trained deep neural network (DNN) enables real-time RA. There are two distinct methods to develop approaches based on deep learning. The first is a data-driven approach [7,8] that replaces conventional building elements with neural networks that directly discover the optimal input-output mapping for the problem. The second is a model-driven approach [9,10] that substitutes neural networks for specific classical algorithm strategies. Both approaches use neural architectures inherited from applications such as computer vision, such as fully connected multilayer perceptrons (MLP) or convolutional neural networks (CNN). While these classical architectures can perform quickly and with near-optimal performance for optimal control on confined networks, their performance degrades significantly as the number of users increases.

Graph neural networks (GNN) have been developed to process non-Euclidean structured data for issues in communication networks because they can effectively exploit domain knowledge and capture spatial information hidden in the network topology. In [11], link scheduling was done with graph embedding instead of perfect channel state information (CSI), and good performance was achieved with small dataset. Message passing graph neural networks (MPGNN) were proposed to solve RA problem. In [12], the WMMSE algorithm improves convergence by inserting a learning GNN module. Investigators have created heterogeneous graph neural network (HetGNN) models for the RA problem in heterogeneous networks due to the inevitable heterogeneity of contemporary networks. In [13], A heterogeneous interference graph neural network (HIGNN) is constructed to learn the PC/BF policy. By treating as communication links between transmitters and receivers as different types of nodes, HetGNN was used in [14] to learn the PC policy. But most existing deep models ignore node features and edge features, and their performance is limited when the size of hidden layers in the network is larger than the size of node and edge features.

In this paper, we propose a supervised GNN-based resource allocation algorithm to achieve near-optimal performance while maintaining reasonable running time. It uses both node and edge features to show how deep components interact and make the model more expressive. This paper's primary contribution is:

- Wireless D2D systems are represented by oriented graphs, with nodes and edges standing in for transmission and interference connections, respectively. The resource allocation issue in wireless networks with integer parameters is addressed using a supervised GNN framework in which each node repeatedly aggregates feature data from its adjacent nodes and edges and merges the information with its own features. Furthermore, we focus on the relationship between node attributes and edge attributes.
- Employing a simulation system, the proposed framework is evaluated against four benchmark schemes: PSO, WMMSE, conventional DNN, and unsupervised GNN. The simulation results show that pre-trained GNN can achieve 97% performance of PSO, and at the same time, the calculation time for online power allocation is shortened by 99%. It can be concluded that the framework outperforms the benchmark design and maintains stable ultimate performance across a range of system parameters, network capacities, and network topologies.

The remaining sections are organized as follows. Section 2 describes the system model and assumptions. Section 3 presents the framework of supervised GNNs for resource allocations in heterogeneous networks. Section 4 presents experimental results to validate the proposed framework's performance. Conclusions are stated in Sect. 5.

2   System Model

Consider a heterogeneous D2D network with K transmit links between transmitters and receivers, where the quantity of antennas on each transmitter varies. The weighted sum rate optimization problem aims to design beamformers in each active connection. Various types of transceiver pairs share the same bandwidth W of the spectrum. The link type is indicated by set $M \triangleq \{1, ..., m\}$, and the antenna number of transmitters for link type m is N_m. The i-th link of type m is denoted as $i_m \in K$. Let $x_{i_m} \in \mathbb{C}^{N_m}$ denote the beamforming vector of link i_m. We presume complete knowledge of the channel state information (CSI). Consequently, the signal received at receiver i_m is the linear function of signals from various types of transmitters, as shown below:

$$y_{i_m} = h_{i_m i_m}^H x_{i_m} + \sum_{j_n \neq i_m} h_{i_m j_n}^H x_{j_n} + n_{i_m}, \tag{1}$$

where $h_{i_m i_m} \in \mathbb{C}^{N_m}$ describe the communication channel response, $h_{i_m j_n} \in \mathbb{C}^{N_m}$ indicate the interference channel response between the transmitter of link j_n and the receiver of link i_m. And $n_{i_m} \sim \mathcal{N}\left(0, \sigma_{i_m}^2\right)$ is the additive white Gaussian noise (AWGN). $X = \{x_{i_m} \,|\, \forall i \in K, m \in M\}$ denote the set of all beamforming vectors. The data rate to receiver i of type m is formulated as:

$$R_{i_m}(X) = W \log \left(1 + \frac{\left\| h_{i_m i_m}^H x_{i_m} \right\|_2^2}{\sum_{j_n \neq i_m} \left\| h_{i_m j_n}^H x_{j_n} \right\|_2^2 + \sigma_{i_m}^2}\right). \tag{2}$$

Particularly on SISO networks, the beamforming design deteriorates into a problem with power control. P_{max} is power constraint. The non-negative weights w_{i_m} indicate the priority of link i_m. The objective function of optimization issue is computed as:

$$\max_X \sum_{i_m} w_{i_m} R_{i_m}(X)$$
$$\text{s.t. } \|x_{i_m}\|_2^2 \leq P_{\max}, \forall i, m. \tag{3}$$

When all weights $w_{i_m} = 1$, the above problem is simplified to the sum rate maximization issue.

3   Graph Neural Networks Based Approach

In this section, we convert the resource management issue into a graph optimization issue and model the D2D wireless network as a heterogeneous graph. The key features of the resource management issue will be determined, and an efficient neural network architecture will be created using these features. The near-global optimal solution can be achieved by PSO, which is numerically proven. However, as the number of user equipments (UEs) increases (high-dimensional optimization space), PSO requires more iterations and longer running time. To achieve near-optimal total rate performance while maintaining a reasonable running time, GNN based method which is trained in a supervised manner has two phases: (i) the first phase applies offline supervised learning through the optimal allocation values calculated by PSO as labels for the nodes in the GNN graph and trains the GNN network. (ii) the second phase runs the trained GNN algorithm in a real-time online application to predict the best allocation.

3.1   Dataset Generation

The GNN-based framework proposed in this paper is trained in a supervised manner, which requires sufficient labeled training samples. A series of RA techniques based on heuristic algorithms for finding optimal allocations were compared in [5], where the PSO algorithm achieved excellent performance in various application scenarios.

In the optimization problem considered in this paper, PSO first generates a set of candidate solutions, i.e., resource allocation schemes, which move in the search space according to specific rules, and the candidate solutions are

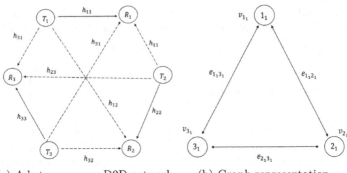

(a) A heterogeneous D2D network   (b) Graph representation
with three types of links

Fig. 1. A three-pair D2D network and its graph representation.

guided by their own known optimal positions in the search space and the known optimal positions of all candidate solutions. When improved positions are found, these positions guide the motion of neighboring candidate solutions. After several iterations, the PSO algorithm produces near-optimal solutions. These solutions can be used as labels for training data. The graph representation of nodes and edges in GNNs is described in the following sections.

3.2 Graph Representation

We formulate the sum rate optimization as learning over a direct heterogeneous graph. Tuples $G = (V, E, R)$ provide a formal description of the network, where V and E represent the set of vertices and edges. And the relation $r \in R$ is used to identify the type of node associated with the edge [15]. The index of each communication link and interference link is denoted as a vertex $i \in V$ and an edge $(i, j) \in E$. A vertex i is the intersection of an entity and an edge (i, j) specifies a directed relationship between vertices i and j. The neighboring set of vertex i is denoted by $\mathcal{N}_i = \{j \in V \,|\, (j, i) \in E\}$. v_i and e_{ij} define the attributes of vertex i and edge ij, respectively. A simple network and its graph representation is shown in Fig. 1.

Considering multiple types of vertices or edges in the heterogeneous graph. Denote the set of vertex types as S, the set of edge types as T. The graph can be represented by two mapping functions that map each vertex to its corresponding vertex type $\varphi : V \to S$ and each edge to its corresponding edge type $\psi : E \to T$. Denote the i-th vertex of type m as i_m, $\mathcal{N}_{i_m}^n = \{j \in V \,|\, (j_n, i_m) \in E\}$ is the set of i_m's adjacent vertices with type n. Node features are represented by $V \triangleq \{V_m\}_m$, where m specifies node types and $[V_m]_i = v_{i_m}$. Edge features are aggregated in $E \triangleq \{E_{mn}\}_{m,n}$, where $[E_{mn}]_{ij} = e_{i_m j_n}$ if edge (j_n, i_m) exists and 0 otherwise.

Each vertex's attributes contain its weight w_{i_m}, straight channel response $h_{i_m i_m}$ and noise variance $\sigma_{i_m}^2$. Each vertex's attributes contain its channel response $e_{i_m j_n} = [h_{i_m j_n}, h_{j_n i_m}]$ from the interfering transmitters to the inter-fered receivers. Furthermore, because there are various connection kinds with unique numbers of send and receive antennas, the size of the graph features may change.

3.3   Graph Neural Network

Our object is to determine a policy $p_\theta (\cdot)$ that maps the features of the hetero-geneous graph to an estimate of the optimal beamforming vector X. We param-eterize the policy by a learnable parameter θ, and the beamforming vector is estimated as $\hat{X} = p_\theta (F)$. GNNs have hierarchical neural network construction, like conventional MLPs. The GNNs aggregate the features of the edges and adja-cent nodes to update the representation of each node in each layer. The updating schema of the i-th node at layer l in GNNs is expressed as follows:

$$Aggregation : \alpha_v^{(l)} = \phi^\alpha \left(\left\{ \beta_u^{(l-1)} : u \in \mathcal{N}(v) \right\} \right), \tag{4}$$

$$Combination : \beta_v^{(l)} = \phi^\beta \left(\beta_v^{(l-1)}, \alpha_v^{(l)} \right), \tag{5}$$

where $\alpha_v^{(l)}$ represent the feature aggregated by node v from its adjacent node at layer l. $\beta_v^{(l)}$ denote the feature of node v at layer l. ϕ^α is a parameterized function that encodes vertex and edge attributes for each edge (j, i). Then each vertex i aggregates the updates of edges. ϕ^β represents the combination function used to derive the vertex update by combining the aggregated edge update $\alpha_v^{(l)}$ with the vertex's current attributes $\beta_v^{(l-1)}$. Message transmission is completed when each vertex's knowledge is embedded in edge updates and consequently assimilated by its adjacent vertices.

With permutation invariance operations (e.g., sum, mean, and maximum), neighborhood aggregation can capture the permutation invariance attributes of the interference channel. Since the size of edge features varies with the number of antennas, features from different relations cannot be directly processed by ordinary GNN and these features should be treated separately. Therefore, we assign separate update functions to each relation $r = (n, m)$, using MLPs for parameterization. First, message transmission is performed in each relation. The target vertex then samples and aggregates partial updates from multiple relations to achieve its final update. Define the update at vertex i_m in relation $r = (n, m)$ as:

$$e_{j_n i_m}^{(l)} = \phi_{(n,m)}^e \left(v_{j_n}^{(l-1)}, e_{j_n i_m}^{(0)} \right) \tag{6}$$

$$v_{(n,i_m)}^{(l)} = \phi_{(n,m)}^v \left(v_{i_m}^{(l-1)}, \max_{j \in \mathcal{N}_{i_m}^{(n)}} e_{j_n i_m}^{(l)} \right) \tag{7}$$

The initial edge attributes $e_{j_n i_m}^{(0)}$ are maintained in all edge update stages. This aids to maintain performance stability. The attributes of each vertex are used as initial inputs in forward calculation process, and then several iterations are performed to achieve the beamforming vector.

3.4 Unsupervised Learning

In an unsupervised manner, each graph takes attributes of all nodes and edges as input. The loss function of each graph is the sum-rate of all UEs, and the model parameters are trained to maximize the sum of loss function over all graphs in training set.

$$
L\left(\theta\right) = -E_H\left[\sum_{i,m} w_{i_m} \log\left(1 + \frac{\left\|h_{i_m i_m}^H x_{i_m}\left(\theta\right)\right\|_2^2}{\sum_{i_m \neq j_n}\left\|h_{i_m j_n}^H x_{j_n}\left(\theta\right)\right\|_2^2 + \sigma_{i_m}^2}\right)\right]. \tag{8}
$$

where θ represents the neural network's learnable parameters and the expectation is derived from all channel configurations. The channel state and the user's weights are inputs, and the beamforming vector is the output. Using this loss function iteratively eliminates sample labels.

3.5 Supervised Learning

In a supervised manner, each graph network takes attributes of nodes and edges as input and uses the best resource allocation obtained by the PSO algorithm as the output. The model parameters W are trained so that the empirical mean square error between the output of the GNN and the expected output of all training graphs is minimized. In our simulation experiments, sigmoid is used as the activation function for the output layer, which works just as well as the designed activation function.

4   Simulation Results and Analysis

This section presents the modeling configuration employed to generate the dataset, followed by the numerical results and computational analysis. The experimental procedure is based on the Deep Graph Library (DGL) and the PyTorch framework.

4.1 Simulation Setup

To evaluate the efficacy of the proposed supervised GNN, we consider a network with heterogeneous nodes. Transmitters and receivers are spread equally across a square area with side D. In the simulation experiment, assume there are two connection categories in the network: SISO and MISO with a ratio of 2:1. The channel gains possess a complex independent Gaussian distribution. System parameters are summarized in Table 1. The following benchmarks are considered for comparison.

- WMMSE [3]: An optimization-based algorithm that transforms the sum rate maximization issues in MIMO interfering broadcast channels to a weighted mean square error minimization problem.
- DNN [16]: A 4-layer classical supervised DNN that uses channel matrices as input. DNN's Hyperparameters are summarized in Table 2.
- PSO [17]: An iterative stochastic optimization technique based on swarm intelligence.
- GNN (unsupervised) [15]: It has the same framework and parameters as the supervised GNN that was proposed. GNN's Hyperparameters are summarized in Table 3.

Table 1. Simulation parameters

Square area radius	500 m
BS transmit power	20 dBm
D2D Pairwise distance	2–50 m
Path loss exponent	$\eta = 3.76$
Noise PSD	-174 dBm/Hz
Channel bandwidth	5 MHz

Table 2. DNN hyperparameters

Hidden layer size	(512-512-512-256)
Epoch size	30
Batch size	32
Learning rate	1×10^{-3}
Dropout rate	0.5
Optimizer	ADAM

Table 3. GNN hyperparameters

Hidden layer size	(512-512-256)
Epoch size	100
Learning rate	5×10^{-4}
Dropout rate	0.3
Optimizer	SGD

4.2   Performance Comparison

First, the training efficiency of the learning-based methods is tested. Figure 2
shows the performance results, the effectiveness of GNN increases gradually with
the sample size for training. The supervised GNN (3 layers) method outperforms
the benchmark solution generally. Since D2D beamforming is an issue of discrete
classification, supervised GNN surpasses unsupervised GNN after 1000 samples.
The possible reason is that the input features of our proposed method include
the priori knowledge of the near-global optimal allocation. In addition, classical supervised DNNs have the worst performance among all methodologies. Due
to the data-driven nature of supervised DNNs, large training datasets are typically necessary. Benefit from the superior sample efficiency, GNNs are often
recommended for real-world issues in wireless networks, where collecting enough
training data can be prohibitively costly or otherwise inconvenient.

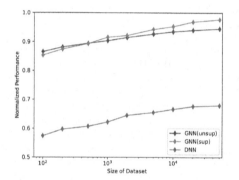

Fig. 2. Performance versus size of
dataset.

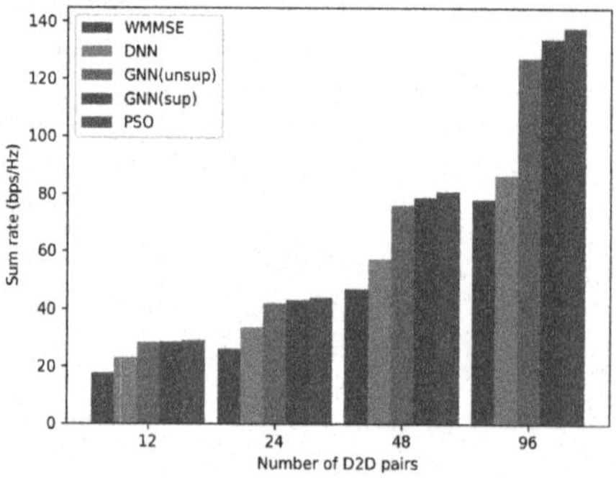

Fig. 3. Performance versus D2D pairs

Next, we test the scalability of the proposed method in networks of various
pairwise distances and densities. Figure 3 shows the performance of the proposed
method and the benchmarks with varying network size and link density. When
there are 12 pairs in the network, the performance of supervised GNN and
unsupervised GNN is similar and only gap from the near-global optimal solution
of PSO by 0.01. In the same environment, the hyperlink count increases by a
factor of 2, 4, and 8 in that order, and the area of the region increases by the
same multiple with the network size and link density, the normalization rate of
the supervised GNN stays above 0.97, while the unsupervised GNN can only stay
at 0.95. This indicates that the supervised GNN still has advanced performance
in this case.

Figure 4 demonstrates the performance of the supervised GNN and the
benchmark scheme for different D2D pairing distances, when the distribution
of pairing distances changes, the average normalization rate of the supervised

GNNcan be maintained above 0.96 of the PSO algorithm. The channel gain depends heavily on the distance, which can be embedded in the node characteristics as the geometric information of the wireless network. Therefore, the GNN-based optimization framework outperforms other benchmark schemes in all three tested parameter settings.

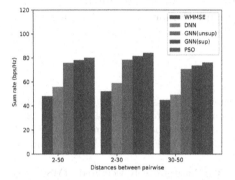
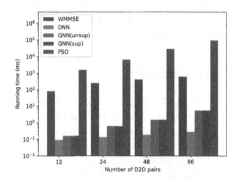

Fig. 4. Performance versus pairwise distances.

Fig. 5. Execution time of different benchmarks.

The learning-based methods are executed on GeForce GTX 2070 s s while the other methods on Intel(R) Core CPU i7-12700H. As shown in Fig. 5, PSO and WMMSE occupy a significant amount of time as the amount of D2D pairs rises, and the issue grows more complex as the network extends. Traditional RA algorithms are typically time-consuming and unsuitable for real-time applications. In contrast, the optimization issue in D2D networks is greatly aided by the supervised GNN method. It is more than 100 times faster than standard algorithms and over 1000 times quicker than heuristic algorithm PSO. This substantial acceleration of the proposed method augur well for its implementation in real time on wireless networks. Since the proposed supervised GNN method has the same network structure and input features as the unsupervised GNN, its runtime performance is comparable. Due to ignoring graph features, DNN obtains better runtime performance than GNN, but its terminal performance and sampling efficiency are inferior.

5 Conclusion

In this paper, we develop a scalable GNN-based neural network framework to address the issue of RA in heterogeneous D2D wireless networks. Instead of current learning-based approaches, we focus on designing neural structures to meet key performance requirements, such as low training cost, high computational efficiency, and good generalization. We furthermore integrate heuristic and learning-based optimization techniques. This paper suggests a supervised GNN model that is different from most other GNN approaches in that it uses both node and edge features to show how deep components interact and make the model more expressive. Simulations show that the proposed supervised GNN is better than other GNN-based methods at maximizing the sum rate. This is because it effectively uses a priori knowledge of the optimized solution the heuristic algorithm seeks. In the future, we will continue studying the optimal beamforming structures in a more general setting and explore more efficient neural network frameworks.

References

1. Zhao, J., Guan, X., Li, X.: Power allocation based on genetic simulated annealing algorithm in cognitive radio networks. Chin. J. Electron. **22**(1), 177–180 (2013)
2. Feiten, A., Mathar, R., Reyer, M.: Rate and power allocation for multiuser OFDM: an effective heuristic verified by branch-and-bound. IEEE Trans. Wirel. Commun. **7**(1), 60–64 (2008)
3. Shi, Q., Razaviyayn, M., Luo, Z.-Q., He, C.: An iteratively weighted MMSE approach to distributed sum-utility maximization for a MIMO interfering broadcast channel. IEEE Trans. Sig. Process. **59**(9), 4331–4340 (2011)
4. Shen, K., Wei, Yu.: Fractional programming for communication systems-part I: power control and beamforming. IEEE Trans. Sig. Process. **66**(10), 2616–2630 (2018)
5. Sun, Q., Haitao, W., Petrosian, O.: Optimal power allocation based on metaheuristic algorithms in wireless network. Mathematics **10**(18), 3336 (2022)
6. Palomar, D., Cioffi, J.M., Lagunas, M.A.: Uniform power allocation in MIMO channels: a game-theoretic approach. IEEE Trans. Inf. Theor. **49**(7), 1707–1727 (2003)
7. Sun, H., Chen, X., Shi, Q., Hong, M., Fu, X., Sidiropoulos, N.D.: Learning to optimize: training deep neural networks for interference management. IEEE Trans. Sig. Process. **66**(20), 5438–5453 (2018)
8. Liang, F., Shen, C., Wei, Yu., Feng, W.: Towards optimal power control via ensembling deep neural networks. IEEE Trans. Commun. **68**(3), 1760–1776 (2019)
9. He, H., Wen, C.-K., Jin, S., Li, G.Y.: Model-driven deep learning for MIMO detection. IEEE Trans. Sig. Process. **68**, 1702–1715 (2020)
10. He, H., Jin, S., Wen, C.-K., Gao, F., Li, G.Y., Xu, Z.: Model-driven deep learning for physical layer communications. IEEE Wirel. Commun. **26**(5), 77–83 (2019)
11. Eisen, M., Ribeiro, A.: Optimal wireless resource allocation with random edge graph neural networks. IEEE Trans. Sig. Process. **68**, 2977–2991 (2020)
12. Zhao, Z., Verma, G., Rao, C., Swami, A., Segarra, S.: Distributed scheduling using graph neural networks. In: ICASSP 2021–2021 IEEE International Conference on Acoustics, Speech and Signal Processing (ICASSP), pp. 4720–4724. IEEE (2021)
13. Chowdhury, A., Verma, G., Rao, C., Swami, A., Segarra, S.: Unfolding WMMSE using graph neural networks for efficient power allocation. IEEE Trans. Wirel. Commun. **20**(9), 6004–6017 (2021)
14. Zhang, X., Zhao, H., Xiong, J., Liu, X., Zhou, L., Wei, J.: Scalable power control/beamforming in heterogeneous wireless networks with graph neural networks. In: 2021 IEEE Global Communications Conference (GLOBECOM), pp. 01–06. IEEE (2021)

15. Wang, T.-Y., Zhou, H., Kannan, R., Swami, A., Prasanna, V.: Throughput optimization in heterogeneous MIMO networks: a GNN-based approach. In Proceedings of the 1st International Workshop on Graph Neural Networking, pp. 42–47 (2022)
16. Koc, A., Wang, M., Le-Ngoc, T.: Deep learning based multi-user power allocation and hybrid precoding in massive MIMO systems. In: ICC 2022-IEEE International Conference on Communications, pp. 5487–5492. IEEE (2022)
17. Koc, A., Le-Ngoc, T.: Swarm intelligence based power allocation in hybrid millimeter-wave massive MIMO systems. In: 2021 IEEE Wireless Communications and Networking Conference (WCNC), pp. 1–7. IEEE (2021)

Satellite Downlink Scheduling Under Breakpoint Resume Mode

Zhongxiang Chang[1,2]([✉]), Zhongbao Zhou[1,2], and Shi Cheng[3]

[1] School of Business Administration, Hunan University, Changsha 410082, China
zx_chang@hnu.edu.cn
[2] Hunan Key Laboratory of Intelligent Decision-Making Technology for Emergency Management, Changsha 410082, China
[3] School of Computer Science, Shaanxi Normal University, Xi'an 710119, China

Abstract. A novel problem called satellite downlink scheduling problem (SDSP) under breakpoint resume mode (SDSP-BRM) is studied in our paper. Compared to the traditional SDSP where an imaging data has to be completely downloaded at one time, SDSP-BRM allows the data of an imaging data be broken into a number of pieces which can be downloaded in different playback windows. By analyzing the characteristics of SDSP-BRM, we first propose a mixed integer programming model for its formulation and then prove the NP-hardness of SDSP-BRM. To solve the problem, we design a simple and effective heuristic algorithm (SEHA) where a number of problem-tailored move operators are proposed for local searching. Numerical results on a set of well-designed scenarios demonstrate the efficiency of the proposed algorithm in comparison to the general purpose CPLEX solver. We conduct additional experiments to shed light on the impact of the segmental strategy on the overall performance of the proposed SEHA.

Keywords: Scheduling · Satellite downlink scheduling problem · Segmental strategy · Breakpoint resume mode · Mixed integer programming · Heuristic algorithm and CPLEX

1 Introduction

Satellite mission planning and scheduling problem mainly contains two parts: data acquisition task scheduling and data downlink scheduling. The data acquisition is an imaging activity, while the data downlink is a playback activity as shown in Fig. 1. When the data acquisition activity is completed, the corresponding imaging data will be stored in the satellite, then which need to be transmitted to ground stations by data downlink activities. In theory, the data of each downlink activity could be a partial imaging data, a complete imaging data or even a combination of multiple complete or partial imaging data.

With the development of space technology, the imaging capability of satellites has been greatly enhancing, which causes a big explosion in the amount of imaging data. GAOFEN II [1], launched in 2014, marks the arrival of "submeter Era" for EOS in China.

While the downlink capacity of satellite antennas does not develop synchronously. The downlink rate of GAOFEN II antenna is only 2 × 450 Mbps. There is a big gap, the imaging data obtained by one-second observation will spend 4.5 s to play back, between the data acquisition capability and the data download capability. The big gap poses a new challenge to SDSP.

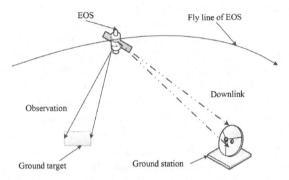

Fig. 1. Observation and Downlink

SDSP and its variations have been studied by many authors. Some of these works were focused on a single satellite [2] whereas others were more general purpose in the special satellite constellation or multi-satellite. On the other hand, some researchers saw SDSP as a time-dependent or resource-dependent problem and focused on the time-switch constraints between satellites and ground stations. In addition, other authors [3, 4] saw image data as an uncertain process, they considered satellite observation planning (SOP) and SDSP together or transformed SDSP as constraints for SOP. Chen [5] abstracted the possible positions of digital tasks in a scheduling sequence as nodes, and constructed a matrix solution construction graph of the pheromone distribution in the nodes. To solve it they proposed an ant colony algorithm. Chang [6] considered SDSP as a complex constrained combinatorial optimization problem, and proposed a particle swarm optimization algorithm with controllable speed, direction and size. Chen [7] regarded the data acquisition chain of satellite mission planning as a path planning problem with multi peak features, and established a framework for solving the problem based on the label a constrained shortest path method. Giuseppe [8] proposed a method to solve multiple satellites and multiple stations planning for automatically arranging tasks from the perspective of ground station. Chen [9] built a cooperative scheduling model considering observation and transmission for electronic reconnaissance satellites, and proposed a method based on the genetic algorithm for solving it. Li [10] established the descriptive model of satellite data transmission tasks and the satellite data transmission scheduling model. To solve the problem, they designed a hybrid genetic algorithm. Chen [11] established a satellite data transmission scheduling model which is suitable for practical application, and a quantum discrete particle swarm optimization algorithm was proposed to solve the problem. Maillard [12] presented a data transmission scheduling method for SDSP based on cooperation between satellite and ground in the presence of uncertain imaging data, and made contrast experiments with complete ground SDSP and Pure Onboard SDSP, results of which showed the method based on corporation had significant advantages. Li [3] regarded SDSP as a multi-constraint, multi-object and multi-satellite data transmission scheduling problem, and established a data transmission scheduling topology model. They solved the problem with the K-shortest path genetic algorithm.

Through a limited review about the researches according to SDSP, we find most of them are concerned about the allocation of playback window resources and resolving conflicts between playback windows. Especially, there is a similar assumption in their studies that one playback window can transmit multiple images and for a single image the data cannot be separated [13] and have to be transmitted in "First observed, First downlink (FOFD)" order [4]. Under these conditions, SDSP is equivalent to the satellite range scheduling problem (SRSP) [14]. But because of the above instruction that there is a big gap between the data acquisition capability and the data download capability, the assumption is no longer practical. In this paper, we make the first attempt to address the problem called satellite data downlink scheduling problem under breakpoint resume mode (SDSP-BRM). The breakpoint resume mode allows the transmission of an image data pause at some point and resume later on. Under this mode, the data of a single image can be divided into a number of small pieces. The purpose of the SDSP-BRM is to arrange the whole images or their small pieces into playback windows in order to maximize the total reward. Because of this additional dimension of complexity, SDSP-BRM is no longer a simple downlink request permutation problem (DRPP) [13]. SDSP-BRM is more complicated than SRSP [15] and SDSP [4, 14], both of which are NP-Hard [15, 16], so SDSP-BRM is NP-Hard, too.

The rest of this paper is organized as follows. In Sect. 2, we will present a mathematical formulation of SDSP-BRM and provide an analysis of the problem complexity. In Sect. 3, we will present a simple and effective heuristic algorithm (SEHA) for solving SDSP-BRM. Then experimental results and analysis are reported in Sect. 4 and concluding remarks are given in Sect. 5.

2 Problem Analysis

2.1 The Input and Output of the SDSP

The inputs, **image data** and **playback window**, and outputs, **playback task**, of SDSP(-BRM) is defined as,

Imaging Data. Imaging data (t) refers to the data acquired by the sensors of EOS, which can be represented by a six-element tuple below:

$$t = \langle n, p, os, oe, od, d \rangle \tag{1}$$

where n denotes the identity number of t. p reflects the priority of t. os, oe and od represents start time, end time and image duration of the image data t respectively. d denotes the duration for downloading t.

Since imaging data obtained by one-second observation will spend 4.5 s to play back as mentioned in Sect. 1. Therefore, the six-element tuple description can be simplified as a five-element tuple: $t = \langle n, p, os, oe, d \rangle$.

Playback Window. The playback window (w) refers to the visible time window between the satellite and receiving resources (like the ground stations and the relay satellites), which can be represented by a five-element tuple:

$$w = \langle m, ds, de, l \rangle \tag{2}$$

where m denotes the identity number of w. ds, de and l reflects start time, end time and duration of w respectively.

Playback Task. As mentioned above, the playback task (TS) is the output of SDSP-BRM, which is the action of satellites to playback the imaging data, and can be represented by a four-element tuple below:

$$TS = \langle m, ts, te, set \rangle \tag{3}$$

where m denotes the identity number of TS, which is directly inherited from the playback window. ts and te indicates the execution start time and end time of TS respectively. Set represents a set of the identity number of all imaging data transmitted in the playback window m.

2.2 Assumptions

Satellite has a file system to management all imaging data [1], so we assume that the imaging data can be transmitted discontinuously.

1) If only if an imaging data can be transmitted completely, there is possible to select it for scheduling. Otherwise the imaging data should be abandoned directly.
2) There is no any setup time for playback tasks.

2.3 Constraints

Service Constraint. A playback window can service an imaging data means that the window can be used to transmit the imaging data. We use service coefficient ($rij = 0$ or 1) expressing service of each window with each imaging data. If $dsj > oei$, then $ri,j = 1$, otherwise $ri,j = 0$.

Imaging Data Segmentable Constraint. The imaging data is stored in the on-board storage system as a continuous memory unit. In theory, the imaging data can be divided into several infinite pieces. But for the practical application, the infinite segmentation is unacceptable, therefore the minimum length (ld) of imaging data segmental is proposed.

2.4 Problem Complexity Analysis

If the imaging data can be segmented, SDSP becomes more challenging. Figure 2 illustrates the difference between SDSP with segmental strategy or not. Figure 2(a) shows that when the data download capability is comparable to the data acquisition capability. The focus of SDSP is how to schedule various imaging data and obtain more data under the limited playback windows, which is similar to the classic Knapsack problem that has been proved to be a NP-Hard problem [17].

However, with the synchronous development of the data acquisition capability and the data downlink capability, one playback window cannot transmit an imaging data completely. Therefore, considering the segmental strategy is imperative. Figure 2(b) shows SDSP considering segmental strategy, which includes two processes: segmenting the imaging data and allocating the playback windows to transmit the segmented imaging data.

SDSP-BRM is a completely new problem, which can decompose into three sub-problems:

(1) Whether an imaging data is transmitted or not;
(2) How to segment the imaging data;
(3) How to use limited playback windows to transmit all segmented imaging data.

The first sub-problem is a typical 0–1 integer programming problem; the second sub-problem is similar to the Cutting Stock problem (CSP), which is a NP-hard proved in theory [18]; the third sub-problem is similar to the Bin Packing problem (BPP) or the Knapsack problem, which has the typical characteristics of integer optimization. Moreover, they are closely related and interacted with each other. Therefore, the SDSP-BRM is more complex and is a NP-Hard problem.

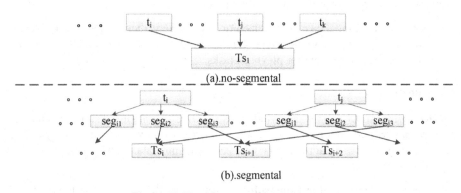

Fig.2. SDSP with segmental strategy or not

2.5 Mixed Integer Programming Model for SDSP-BRM

Let $T = \{ti \,|\, i = 1, 2, \cdots, N\}$ denotes a collection of imaging data needs to be transmitted, wherein N represents the total number of imaging data and $ti = \langle ni, pi, osi, oei, di \rangle$.

Let $W = \{wj \mid j = 1, 2, \cdots, M\}$ indicates the set of the playback windows, wherein M represents the total number of playback windows and $wj = \langle mj, dsj, dej, lj \rangle$.

Four decision variables should be considered in the model,

- xi A 0–1 integer variable. xi corresponds to the first sub-problem;
- yi,j A non-negative real number. yi,j corresponds to the second sub-problem;
- qj A 0–1 integer variable;
- gi,j A 0–1 integer variable. qj and gi,j correspond to the third sub-problem.

To transmit as much image data as possible is the original intention of SDSP. Without loss of generality, maximizing transmission revenue is our optimal objective. Therefore, the model is represented as follows:

$$\text{Maximize} \sum_{i=1}^{N} x_i \times p_i \tag{4}$$

S.T.

$$g_{i,j} \times ld \leq y_{i,j} \leq g_{i,j} \times d_i \tag{5}$$

$$\sum_{j=1}^{M} y_{i,j} = x_i \times d_i \tag{6}$$

$$\sum_{i=1}^{N} y_{i,j} \leq l_j \tag{7}$$

$$g_{i,j} \leq r_{i,j} \tag{8}$$

$$g_{i,j} \leq x_i \tag{9}$$

$$y_{i,j} \geq 0 \tag{10}$$

$$g_{i,j} = 0 \; or \; 1 \tag{11}$$

$$x_i = 0 \; or \; 1 \tag{12}$$

The right side of (6) represents the upper bound of segmentable pieces for each imaging data, and the left represents the lower bound.

The constraint Eq. (6) corresponds to the second assumption.

The constraint Eq. (7) means that the used time must less than or equal to length of the window.

The constraint Eq. (8) describes the relationship between the service coefficient(ri,j) and the decision variable (gi,j).

The constraint Eq. (10) means that yi,j is a non-negative continuous variable.

The constraint Eqs. (11) and (12) indicate xi and gi,j is a 0–1 integer variable.

3   A Simple and Effective Heuristic Algorithm (SEHA) for SDSP-BRM

As mentioned above, SDSP-BRM is a NP-hard problem, so there is not any exact algorithm can achieve the optimal solution in polynomial time generally. To deal with the large size practical problems, we design a simple and effective heuristic algorithm (SEHA) based on some heuristic rules according to the characteristics of SDSP-BRM and the pseudo-code of the algorithm as following:

Initialize parameters;
Construct a feasible solution based on a heuristic greedy algorithm;
While (Both *Max_Iter* and *Solve_Time* are not met) do
    While (*NoUp_Iter* is not met) do
        Apply the remove operator to constructed solution;
        Apply the insert operator to constructed solution;
        If (The value of the objective function increase)
            Update the solution;
            The time number of no improve reset;
        Else
            The time number of no improve add one;
        End if
    End while
End while

Then we want to explain how to construct an feasible solution, how to improve the solution and how to terminate the search algorithm.

3.1   Solution Construction

Two heuristic rules are designed according to the characteristic of the problem: prefer to select the imaging data with a greater contribution rate and prefer to use the playback window with a smaller service coefficient.

Rule 1 (Prefer to select the imaging data with a greater contribution rate). The contribution rate(ci) of an imaging data(ti) is defined as the Eq. (14). The rule 1 represents the higher value of ci is, the higher probability to select ti is.

$$c_i = \frac{p_i/l_i}{\max_{t_j \in T} p_j/l_j} \quad t_i \in T \tag{14}$$

Rule 2 (Prefer to use the playback window with a smaller service coefficient). The service coefficient (r_j) of a playback window (w_j) is defined as the Eq. (15). The rule 2 means that the smaller value of r_j is, the greater probability to adopt w_j is.

$$r_j = \sum_{t_i \in T} r_{ij} \tag{15}$$

3.2 Operators

The local search algorithm improves quality of a solution by changing the structure of the solution by a series of operators. We design two important operators for SEHA: remove operator and insert operator, as shown in Fig. 3.

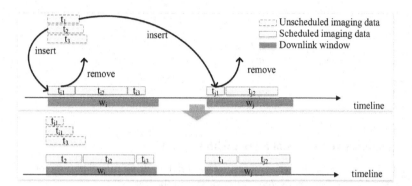

Fig. 3. Two types of operator

Use the remove operators to delete some scheduled imaging data from a given scheme randomly, which will produce more unused space in some data playback windows, and then use insert operator inserts to select some unscheduled imaging data randomly for adding in the scheme. After the process of move operators, the structure of the scheme will be changed and a new solution will be generated. If the value of the object improves, the new solution will be accepted.

3.3 Termination Criterions

We design three termination criterions for SEHA: 1) if the iterations of SEHA reach *Max_Iter*, SEHA will be terminated; 2) if the solution has no improvement after *NoUp_Iter* iterations, SEHA will be terminated; 3) if the run time of the algorithm reach *Solve_Time*, SEHA will be terminated.

4 Experimental Study

4.1 Experimental Setup

There is no any benchmark about SDSP in the literature. Therefore, considering on the actual project (GAOFEN II) and the universality of experiment, we design the several different scale scenarios randomly. Table 1 shows their principle to generate every element of the input data, the structure of which are proposed in the section above.

Since the lack of the data receiving resources, we set the relationship between the number of data window (M) and the number of imaging data (N) as $N = a * M$, wherein $a \sim U[1.5, 2.5]$. Using the above generation principle, we design several different size scenarios randomly. SEHA is coded in C++ and the version of CPLEX is 12.5. And both of them run on a PC with Intel i7-3520M (2.90 GHz) CPU and 12.0 GB RAM under Windows 7.

Table 1. Generation principle of scenario

Input data	Variable	Generation principle
t_i	p_i	$p_i \sim U[1, 10]$
	os_i	$os_i = oe_{i-1} + N(100, 1)$
	d_i	$d_i \sim U[2ld, 10ld]$
	oe_i	$oe_i = os_i + d_i/4.5$
w_j	ds_j	$ds_j = de_{j-1} + N(100, 1)$
	l_j	$l_j \sim U[ld, 5ld]$
	de_j	$de_j = ds_j + l_j$

4.2 Comparative Results of SEHA with CPLEX

After analyzing extensive experiments, we obtain the setting rule of parameters of SEHA (*Max_Iter*, *NoUp_Iter* and *Solve_Time*) shown in the Table 2.

Table 2. Parameters of SEHA

Max_Iter	*NoUp_Iter*	*Solve_Time(s)*
100000	5000	60

The comparative results of different scale simulation scenarios are shown in Table 3. Among them, *SN* denotes the serial number of each scenario. *N* indicates the number of imaging data in each scenario. *M* reflects the number of data playback window in each scenario. *Rmax*, *Rmin* and $\overline{R}$ denotes the best, worst and mean profit obtained by SEHA and CPLEX respectively. $\overline{T}$ indicates the mean running time of SEHA and CPLEX. Gap_R represents the difference between the best profit (*Rmax*) of CPLEX and SEHA. If Gap_R is a negative number, *Rmax* of SEHA is better than that of CPLEX, and vice versa. And $Gap_{\overline{R}}$ denotes the difference between the mean profit ($\overline{R}$) of CPLEX and SEHA. If $Gap_{\overline{R}}$ is a negative number, $\overline{R}$ of SEHA is better than that of CPLEX, and vice versa. From the experimental results in the Table 3, we can find some interesting phenomena.

- SEHA can obtain the optimal solution (i.e., scenario 1 and 2), and the running time (*Solve_Time*) of SEHA is significantly less than that of CPLEX.
- The profit of solution obtained by SEHA is approximate to that obtained by CPLEX (i.e., scenario 3 and 4), while the running time (*Solve_Time*) of SEHA is constantly better than that of CPLEX.
- Note that, much better solutions can be obtained by SEHA than that of CPLEX in acceptable running time (*Solve_Time* = 60 s) with the scale of task, like scenario 5 and scenario 6.
- With the size of task increases (i.e., scenario 7 and 8), especially, CPLEX already cannot obtain a feasible solution in acceptable running time (*Solve_Time*), while SEHA can still find satisfactory solutions or even near-optimal solutions.

Table 3. Comparative results of scenarios

SN	N	M	CPLEX				SEHA				Gap_R	$Gap_{\bar{R}}$
			Rmax	Rmin	$\overline{R_{MIP}}$	T	Rmax	Rmin	$\overline{R_{LS}}$	T		
1	20	8	**74**	**74**	**74**	2	**74**	**74**	74	**0.07**	**0**	0
2	30	15	**86**	**86**	**86**	4	**86**	84	85.21	**0.22**	**0**	0.79
3	50	24	148	148	148	60	146	142	144.47	**0.69**	2	1.53
4	100	70	286	284	285	60	282	277	279.19	**3.5**	4	5.81
5	200	85	443	443	443	60	452	444	447.23	**15.57**	*−9*	*−14.23*
6	500	220	564	564	564	60	1332	1319	1322.15	60	*−768*	*−758.15*
7	800	340	–	–	–	60	2105	2091	2096.78	60	–	–
8	1000	530	–	–	–	60	2903	2885	2891.49	60	–	–

– The value of *Rmax*, *Rmin* and $\bar{R}$ obtained by SEHA are very close for each scenario, which represents the robustness of SEHA is good.

Given all that above, comparing with CPLEX, SEHA can always obtain good quality solutions (satisfactory solutions or even near-optimal solutions) with fewer running time for different scale simulation scenarios for SDSP-BRM, especially with the scale of task increases.

4.3 The Impact of the Segmental Strategy

To evaluate the performance improvement brought by the segmental strategy, we perform contrast experiments and analysis results obtained by SEHA considering the segmental strategy and not considering it (shortly denoted by SG and Non-SG).

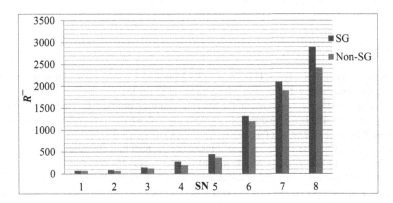

Fig. 4. The impact of the segmental strategy

As shown in Fig. 4, the segmental strategy significantly improves the profit of satellite playback. And the impact of segmental strategy is more obvious with the scale of problem increases. There are two reasons to explain the phenomenon. 1) Considering the segmental strategy ensures that some imaging data, which has a large contribution rate but cannot be transmitted by any single playback window, can be transmitted. 2) Some playback windows might have some residual space after scheduling without considering the segmental strategy, while considering the strategy we can divide some imaging data to fill those playback windows. Therefore, it is necessary for SDSP to consider the segmental strategy under the condition that the data acquisition capability and data download capability develops asynchronously.

5 Conclusions

Based on the analysis of the problem, we prove SDSP-BRM is a NP-Hard problem and establish a mixed integer programming model to formulate it, then we design a simple and effective heuristic algorithm (SEHA) to solve it. Extensive experiments demonstrate that SEHA can find optimal or near-optimal solutions in the acceptable running time for different scale simulation instances. While these two heuristic rules will improve the efficiency of SEHA apparently, and the segmental strategy can improve the profit of satellite downlink, and more significantly with the scale of problem increases.

The future work in our study is three-fold: 1) Considering more actual practical demands from engineering, add the time constraint of imaging data transmitted into the model. 2) In this paper, we only consider the mode of data playback, considering the various working modes is an in-depth research direction. 3) SDSP-BRM in this paper is an off-line scheduling, in another word, the imaging data acquisition schedule is given and not changed before downlink scheduling. However, with the development of satellite autonomy, the online scheduling considering the uncertainty of the imaging data acquisition schedule is an interesting direction.

Acknowledgements. The research of Zhongxiang Chang was supported by the science and technology innovation Program of Hunan Province (2021RC2048).

References

1. Huang, W., Sun, S.R., Jiang, H.B., et al.: GF-2 satellite 1 m/4 m camera design and in-orbit commissioning. Chin. J. Electron. **27**(6), 1316–1321 (2018)
2. Chang, Z., Chen, Y., Yang, W., et al.: Mission planning problem for optical video satellite imaging with variable image duration: a greedy algorithm based on heuristic knowledge. Adv. Space Res. **66**(11), 2597–2609 (2020)
3. Li, J., Li, J., Chen, H., et al.: A data transmission scheduling algorithm for rapid-response earth-observing operations. Chin. J. Aeronaut. **27**(2), 349–364 (2014)
4. Wang, P., Reinelt, G., Gao, P., et al.: A model, a heuristic and a decision support system to solve the scheduling problem of an earth observing satellite constellation. Comput. Ind. Eng. **61**(2), 322–335 (2011)
5. Xiao-Yue C X-G W. Ant colony algorithm for satellite data transmission scheduling problem. J. Syst. Eng. Electron. **24**, 451–398 (2009)
6. Chang, F., Wu, X.Y.: Satellite data transmission scheduling problem based on velocity controllable particle swarm optimization. J. Astronaut. **31**, 2015–2022 (2010)

7. Chen, H., Li, J., Jing, N., et al.: User-oriented data acquisition chain task planning algorithm for operationally responsive space satellite. J. Syst. Eng. Electron. **27**(5), 1028–1039 (2016)
8. Corrao, G., Falone, R., Gambi, E., et al.: Ground station activity planning through a multi-algorithm optimisation approach. In: Proceedings of the 2012 IEEE First AESS European Conference on Satellite Telecommunications (ESTEL) (2012)
9. Chen, H., Wu, J., Shi, W., Li, J.: Coordinate scheduling approach for EDS observation tasks and data transmission jobs. J. Syst. Eng. Electron. **27**(4), 822–835 (2016)
10. Li, Y.-F., Wu, X.-Y.: Application of genetic algorithm in satellite data transmission scheduling problem. Syst. Eng. Theor. Pract. **1**, 124–131 (2008)
11. Chen, H., Li, L., Zhong, Z., et al.: Approach for earth observation satellite real-time and playback data transmission scheduling. J. Syst. Eng. Electron. **26**(5), 982–992 (2015)
12. Maillard, A., Verfaillie, G., Pralet, C., et al.: Adaptable data download schedules for agile earth-observing satellites. J. Aerosp. Inf. Syst. **13**(8), 280–300 (2016)
13. Karapetyan, D., Mitrovic Minic, S., Malladi, K.T., et al.: Satellite downlink scheduling problem: a case study. Omega **53**, 115–123 (2015)
14. She, Y., Li, S., Li, Y., et al.: Slew path planning of agile-satellite antenna pointing mechanism with optimal real-time data transmission performance. Aerosp. Sci. Technol. **90**, 103–114 (2019)
15. Barbulescu, L., Watson, J.-P., Whitley, L.D., et al.: Scheduling space-ground communications for the air force satellite control network. J. Sched. **1**, 7–34 (2004)
16. Vazquez, A.J., Erwin, R.S.: On the tractability of satellite range scheduling. Optim. Lett. **9**(2), 311–327 (2014)
17. Garey, M., Johnson, D.: Computers and Intractability: A Guide to the Theory of NP-Completeness. WH Freeman (1979)
18. Delorme, M., Iori, M., Martello, S.: Bin packing and cutting stock problems: mathematical models and exact algorithms. Eur. J. Oper. Res. **255**(1), 1–20 (2016)

A Repetitive Grouping Max-Min Ant System for Multi-Depot Vehicle Routing Problem with Time Window

Wanfeng Chen[1], Ruochen Liu[1(✉)], Qi Guo[2], and Mengyi Niu[1]

[1] Key Lab of Intelligent Perception and Image Understanding of Ministry of Education, Xidian University, Xi'an 710071, China
`wfchen@stu.xidian.edu.cn, ruochenliu@xidian.edu.cn`
[2] 54th Research Institute of China Electronics Technology Group Corporation, Shijiazhuang, China

Abstract. The vehicle routing problem is a classic NP-hard problem in modern logistics industry. The aim of this paper is to propose a repetitive grouping Max-Min Ant System (MMAS) to solve multi-depot vehicle routing problem with time window. The whole algorithm adopts the framework of decomposition. Firstly, the algorithm defines boundary customers and groups them into corresponding depot groups repeatedly for sub-problem optimization. Secondly, the algorithm uses adaptive range technique to determine the size of boundary customers, so as to balance the convergence and resource consumption of the algorithm. Finally, local search operator is integrated into the sub-problem optimization to improve the search ability. The algorithm is tested on several benchmark problems. Experimental results show that the proposed algorithm can effectively improve the performance in most cases compared with several state-of-the-art evolutionary algorithms.

Keywords: Multi-depot vehicle routing problem with time window ·
Max-min ant system · Decomposition

1 Introduction

With the rapid development of science and technology, logistics industry has gradually become the basic industry of national economic development. The vehicle routing problem (VRP), as a key problem in logistics optimization system, has attracted extensive attention.

VRP is a well-known combinatorial optimization problem, which was first proposed by Dantzing and Ramser in 1959 [6]. The classic VRP has the following assumptions: there is a depot, each vehicle must leave the depot and return to the same depot, and each customer can only be served by one vehicle once. The current study focuses on variants of the traditional VRP, including capacitated VRP (CVRP) [23], VRP with time window (VRPTW) [12], multi-depot VRP (MDVRP) [3], and pick-up and delivery VRP (VRPPD) [16]. In real life,

Y. Tan et al. (Eds.): ICSI 2023, LNCS 13969, pp. 374–385, 2023.
https://doi.org/10.1007/978-3-031-36625-3_30

customers will impose restrictions on the item delivery time, that is, customers must be visited by vehicles within a specified time interval, and vehicles need to spend a specific service time on these customers. Therefore, this paper focuses on a combined model of multi-depot vehicle routing problem with time windows (MDVRPTW) [4], which takes into account both the time requirement of vehicles arriving at customers and the limitation of time windows.

Recently, Erdogan [9] developed an adaptive large neighborhood search to solve this problem. Mancini [15] and Rahim-Vahed et al. [21] considered an MDVRP problem which is closely related to the time period. Li et al. [14] considered the time window constraint, proposed a model with the minimum travel cost, and designed an HGA with adaptive local search to solve it. At the same time, the static time window problem and fuzzy time window problem have also received a lot of attention in the past few years [13]. Fan et al. [10] proposed a Multi-depot Open Vehicle Routing Problem based on the joint distribution mode of fresh food. MDVRPTW is a combined version of MDVRP and VRPTW, which is also a non-deterministic polynomial-time NP-hard problem. Desaulnier et al. [7] proposed MDVRPTW, where a fleet of vehicles is provided by different depots and the tasks performed by a group of vehicles are limited to start at a time interval. Bettinelli et al. [2] proposed a branch-price-pricing algorithm to find the exact solution for VRPTW variations. Bae and Moon [1] applied MDVRPTW to a practical and challenging logistics and supply chain management problem. Compared with the exact solution method, heuristic method is an effective technique for solving NP-hard problems. Ant colony optimization (ACO) algorithm is one of the most famous meta-heuristic algorithms [8]. Therefore, in order to solve the MDVRPTW problem effectively, this paper uses the heuristic algorithm Max-Min Ant System (MMAS), and proposes a repetitive grouping MMAS called RG-MMAS.

The contributions of the algorithm are summarized as follows:

1. This paper proposes the definition of the boundary customer. RG-MMAS considers not only the distance between the customers and the depots, but also the matching between the customers' time windows. And the roulette method is used to repeatedly divide boundary customers into the nearest depot and the second nearest depot according to the probability.
2. The boundary customers that need to be repeatedly grouped are selected based on the adaptive range, which makes the algorithm find a good balance between the accuracy of optimization and the consumption of time.
3. This paper introduces the λ-interchange operator. This operator uses multiple neighborhood structures, rather than a single neighborhood structure like other local search operators. Experiments have shown that it can better search more potential space.

The rest of this paper is organized as follows. Section 2 describes a detailed model for the MDVRPTW. The overall process and details of the algorithm will be given in Sect. 3. In Sect. 4, the results of RG-MMAS and the several comparison algorithms on the data set are discussed, and in Sect. 5, the paper is summarized and commented on possible directions for further research.

2 Problem Description

The objective of this study is to optimize the vehicle routing problem of multiple depots with time window through decomposition strategy and effective coordination scheme. Given a directed graph $G = (V, E)$, where V is the set of nodes containing all customers and depots with a certain capacity, and E is the arc set. Let $V = D \cup C$, where D represents the set of the depot nodes and the demand of each is 0, C represents the set of the customer nodes and the demand of each is $q_i (i \in C)$. E represents the set of d_{ij}, which is the Euclidean distance between i and $j (i, j \in V)$ in this problem. Each problem has a set of vehicle set R and every vehicle of them has a certain capacity limit, denoted as Q. k is the number of vehicles; K_m is the maximum number of vehicles at depot $m (m \in D)$.

Besides, some special symbols need to be explained in detail: $s_i (i \in C)$ is the customer's fixed service time, and the time window for open service is $[ET_i, LT_i](i \in C)$. If the vehicle arrives before the opening time ET_i, it will have to wait for a certain time. If the vehicle arrives after the end time LT_i, it will not be able to serve the customer, and the path of the vehicle will be determined as an infeasible path. T_i^k represents the time for vehicle k to arrive at i. And x_{ij}^{mk} is the decision variables for this model.

The integer programming mathematical model of MDVRPTW is as follows:

$$\min \sum_{i \in V} \sum_{j \in V} \sum_{k \in R} \sum_{m \in D} d_{ij} x_{ij}^{mk} + Overload + Tardiness. \tag{1}$$

$$s.t. \sum_{j \in C} \sum_{k \in R} x_{ij}^{mk} \leq K_m, \forall i = m \in D. \tag{2}$$

$$\sum_{j \in C} x_{ij}^{mk} = \sum_{j \in C} x_{ji}^{mk} \leq 1, \forall k \in R, i = m \in D. \tag{3}$$

$$\sum_{j \in V} \sum_{m \in D} \sum_{k \in R} x_{ij}^{mk} = \sum_{j \in V} \sum_{m \in D} \sum_{k \in R} x_{ji}^{mk} = 1, \forall i \in C. \tag{4}$$

$$\sum_{i \in C} q_i \sum_{j \in V} x_{ij}^{mk} \leq Q, \forall k \in R, m \in D. \tag{5}$$

$$T_i^k = 0, \forall i \in D, k \in R. \tag{6}$$

$$\sum_{m \in D} \sum_{i \in V} x_{ij}^{mk} \left(T_i^k + s_i + d_{ij} \right) \leq T_j^k, \forall k \in R. \tag{7}$$

$$ET_i \leq T_i^k \leq LT_i, \forall i \in V, k \in R. \tag{8}$$

$$\sum_{j \in D} x_{ij}^{mk} = \sum_{j \in D} x_{ji}^{mk} = 0, \forall i = m \in D, k \in R. \tag{9}$$

$$x_{ij}^{mk} \in \{0, 1\}. \tag{10}$$

where *Overload* and *Tardiness* in the function (1) represents the penalty received for a load violation and customer time window violation in this route, respectively. There are several constraints that are as follows: The objective (1) is to minimize the total vehicle travel cost. Constraint (2) limits the number of vehicles going out of the depot m less than K_m. Constraint (3) guarantees that each vehicle out from the depot and returns to it. Constraint (4) ensures that a customer is serviced exactly once by only one vehicle. Constraint (5) constrains the vehicle capacity. Constraint (6) initializes the beginning time of travelling from the depot. Constraint (7) and (8) are the time window constraints. Constraint (9) ensures that vehicles cannot travel from one depot to another. Constraint (10) is the decision variable constraint, if $x_{ij}^{mk} = 1$, then $T_j^k = \max\left\{T_i^k + s_i + d_{ij}, ET_i\right\}(i, j \in V)$.

3   The Proposed Algorithm RG-MMAS

This section proposes an efficient algorithm to MDVRPTW to meet the objectives and constraints. Firstly, the algorithm considered factors of distance and time window matching degree in the clustering phase, and then calculates the urgency of each customer, and normalized it to the probability required by roulette method. According to the probability, a certain number of customers are selected as boundary customers and are repeatedly grouped into the nearest and second nearest depot group. After decomposition, complex problems are transformed into a series of smaller sub-problems. Whereafter, routes are formed in each depot group. MMAS algorithm [18] is used to optimize the sub-solutions iteratively, and finally the sub-pheromone matrix is generated. At the same time, the local search operator λ-interchange is used for each generation of optimal solutions to further improve the quality of optimal solutions. After a period of optimization, the sub-pheromone matrix generated by the subproblem is fed back to the master pheromone matrix according to the rules follows to produce different guidance effects for the subsequent optimization.

The proposed algorithm is described carefully in Algorithm 1.

3.1   Grouping Phase

In this phase, an urgency is assigned to each customer, which can also be regarded as an assigned priority [19]. This urgency represents the extent to which the customer is at the boundary. If the urgency is high, it means that the customer is more suitable to be assigned to the nearest depot group. In this case, both depots are involved in the assessment of urgency in formula (11):

$$\mu_i = closeness\left(i, depot_i^1\right) - closeness\left(i, depot_i^2\right), i \in N. \tag{11}$$

where μ_i is the urgency of customer i, $depot_i^1$ is the closest depot to customer i, $depot_i^2$ is the second closest depot to customer i, N and M are the set of customers and depots respectively.

Algorithm 1 RG-MMAS

Input: The MDVRPTW data sets; Parameters of the algorithm;
Output: The best solution S;
1: **while** solution S is not feasible **do**
2:      calculate the urgency of each customer;
3:      calculate the number of boundary customers to be selected according to the current generation;
4:      roulette method was used to select boundary customers for repeated grouping;
5:      ordinary customers are assigned to the nearest depot group;
6:      **for** every group in all the depot groups **do**
7:          MMAS is used to optimize the subproblems and sub-solutions are generated;
8:          Use λ-interchange for the solutions of the subproblems;
9:      **end for**
10:     combine sub-solutions into a feasible solution s;
11:     update master pheromone according to the pheromone communication rules;
12:     **if** s is better than S **then**
13:         $S = s$;
14:     **end if**
15: **end while**

Because of the particularity of MDVRPTW, $closeness\,(i,j)$ is no longer refers to the Euclidean distance between node i and node j, but is a measure which considers both time window and distance. If the distance between node i and node j is further, and the time window matching between the two is lower, the value of $closeness\,(i,j)$ will be larger. The specific expressions are as follows (12):

$$closeness\,(i,j) = \frac{d_{ij}}{affinity\,(i,j)}, i \in N, j \in M. \tag{12}$$

where $affinity\,(i,m)$ measures the matching degree between customer i and the time windows of all customers that have been assigned to the depot m. The formula is as follows (13):

$$affinity\,(i,m) = \frac{n}{\sum\limits_{j \in S_N(m) \cup \{m\}} DTW\,(i,j) + t_{ij}}, m \in M; i, j \in N. \tag{13}$$

where $S_N(m)$ refers to all customers that have been assigned to depot m. It is worth mentioning that the calculation in the equation includes the time window matching degree between the customers and the depots. $DTW\,(i,j)$ represents the time window matching degree between customer i and customer j. The calculation method is as follow (14):

$$DTW\,(i,j) = \begin{cases} e_j - l_i, & l_i < e_j. \\ e_i - l_j, & l_j < e_i. \\ 0, & otherwise. \end{cases} \tag{14}$$

The calculated urgency was used as the probability of each customer to make a choice in roulette [17]. Then a certain range of boundary customers were repeatedly assigned to the nearest depot group and the second nearest depot group to participate in sub-problem optimization.

3.2 Adaptive Range

To overcome the resource consumption caused by repeated grouping, this paper proposes a method to adaptively set the range of boundary customers, which can find a balance between resource consumption and solution quality. The specific methods are as follows:

$$range\,(t) = \max \cdot \frac{T}{T+t}. \tag{15}$$

where max is the preset maximum range, T is the maximum number of generations, and the two parameters are a fixed value. The t represents the current number of generations.

3.3 Local Search

After grouping, each depot group uses MMAS for sub-problem optimization, and local search is also used in this process. The effectiveness of any iterative local search method is determined by the efficiency of the generation mechanism and the way to search the neighborhood. λ-interchange is a local search operator widely used in vehicle routing. The generation mechanism of λ-interchange for the MDVRPTW is described as follows.

Given the solution $S = \{r_1, r_2, \ldots, r_s\}$, where s is the number of routes contained in the solution and $r_i\,(i \in [1, \ldots, s])$ represents the set of customers and depots contained in path i. The exchange takes place between r_i and r_j $(i, j \in [1, \ldots, s])$. It is implemented by selecting subpaths from the two selected paths and exchanging them to form new paths, that is, a replacement of the sub-route $s_1 \subseteq r_i$ of size $|s_1| \cdot \lambda$ by another sub-route $s_2 \subseteq r_j$ of size $|s_2| \cdot \lambda$. So we can get two new routes $r_i^* = (r_i - s_1) \cup s_2$ and $r_j^* = (r_j - s_2) \cup s_1$, and a new neighboring solution $S^* = \{r_1, r_2, \ldots, r_i^*, \ldots, r_j^*, \ldots, r_s\}$. The neighborhood $N_\lambda(S)$ of a given solution S is the set of all neighbors S^* generated by the λ-interchange method.

The order in which the neighbors are searched is specified as follows. The permutation σ is the order of vehicle indices in a given solution:

$$S = \{r_1, r_2, \ldots, r_i, \ldots, r_j, \ldots, r_s\}(\sigma\,(p) = p, \forall p \in R). \tag{16}$$

An ordered search selects all possible combination of pairs (r_i, r_j) according to formula (17) and σ without repetition. A total number of $\nu \times (\nu - 1)/2$ different pair of routes (r_i, r_j) are examined to define a cycle of search in the following order:

$$(R_{\sigma(1)}, R_{\sigma(2)}), \ldots, (R_{\sigma(1)}, R_{\sigma(\nu)}), \ldots, (R_{\sigma(2)}, R_{\sigma(\nu)}), \ldots, (R_{\sigma(\nu-1)}, R_{\sigma(\nu)}). \tag{17}$$

Furthermore for a given group of paths (r_i, r_j) we must also define the search order for the customers to be exchanged. We consider the case of $\lambda = 1$ and $\lambda = 2$

for the neighboring search. The λ-interchange method between two routes results in customers either being shifted from one route to another, or customers being exchanged with other customers. The operator $(0,1)$ on routes (r_i, r_j) represents a shift of one customer from route i to route j. The operators $(1,0)$, $(2,0)$ and $(0,2)$ represent shifting of customers between two routes. The operator $(1,1)$ represents an exchange of one customer between route i and j. The operators $(1,2)$, $(2,1)$ and $(2,2)$ represent exchange operation of the corresponding number of customers between vehicle lines.

Through this exchange process, the algorithm is able to realize the orderly and systematic search of customers on a given route and find improved solutions. The search order we implement is to use $(0,1)$, $(1,0)$, $(1,1)$, $(0,2)$, $(2,0)$, $(2,1)$, $(1,2)$ and $(2,2)$ on any given group of paths to generate neighborhoods. After building a solution, we need a criterion to accept or reject the result of this optimization. When choosing an acceptance criterion, we can consider many solutions as potential candidates. Based on this, researchers have proposed the following two selection strategies to select candidate solutions [22]:

(1) The First-Best (FB) strategy will select the first solution in the neighborhood $N_\lambda(S)$ of solution S that decreases in value with respect to the objective function.
(2) The Global-Best (GB) strategy is to search all solutions in the neighborhood $N_\lambda(S)$ of solution S and select the solution with the largest decrease in objective value under a given objective function.

In this paper we adopt the FB strategy.

4   Experimental Results

In order to verify the effectiveness of the proposed algorithm, RG-MMAS is compared with the most popular algorithms. In this paper, three representative algorithms, HRH [19], IACO [11] and VNS [20], are used as the comparison algorithms in this paper. HRH is a hybrid algorithm of genetic algorithm and decomposition framework, IACO is a modified ant colony algorithm, and VNS is a representative local search algorithm. In order to ensure the fairness of comparison, we use Java to simulate the three algorithms, taking the same number of generations as the stopping criterion, and refer to the respective literatures for detailed parameter settings. And the population size is 20–100, the number of iterations is 200–600. The pheromone factor α and heuristic factor β are set to 2 and 1, respectively. The pheromone volatilization factor is set to 0.85. And number of subproblem generation is set to 20. They were also tested 30 times for each instance. All simulation experiments were run on a Windows 10 system configured as Intel (R) Core (TM) i5-1135G7 2.60-GHz CPU and 16G RAM.

In this paper, the test problems of MDVRPTW proposed in literature [5] is selected as the data set of this experiment. And the data sets are available on the internet at http://www.hec.ca/chairedistributique/data.

The experiment in this paper is divided into two parts. Firstly, the maximum duration of the path is not considered, and the experimental results can be obtained as shown in Table 1. The best result for each instance in the table is shown in bold.

Table 1. Experimental results of IACO, VNS, HRH, and RG-MMAS(Du restriction is not considered).

No.	Best-known	IACO	VNS	HRH	RG-MMAS
P01	1083.98	1083.34	1083.98	**1081.98**	**1081.98**
P01	1763.07	1762.40	1780.27	1765.35	**1758.36**
P02	2408.42	**2395.91**	2408.42	2423.98	2411.79
P03	2958.23	2950.21	2936.85	**2890.64**	2901.61
P04	3134.04	3096.27	3038.56	3037.33	**3018.64**
P05	3904.07	3774.33	3800.31	3758.36	**3740.14**
P06	1423.35	**1419.22**	1420.39	1420.75	1421.62
P07	2150.22	2107.05	2121.97	2116.27	**2102.41**
P08	2833.80	2771.18	2788.76	2754.10	**2747.57**
P09	3717.22	3585.74	3581.35	3597.52	**3568.33**
P11	1031.49	1012.46	**1005.73**	1010.39	**1005.73**
P12	1500.48	**1489.31**	1496.04	1491.24	1499.12
P13	2020.58	2016.49	**2014.52**	2019.22	2023.53
P14	2247.72	2274.76	2252.83	**2238.79**	2256.39
P15	2509.75	2499.51	2600.21	2531.64	**2477.40**
P16	2943.90	2905.73	3211.79	2896.37	**2890.42**
P17	1250.09	1240.25	**1238.48**	1239.13	1240.63
P18	1809.35	1803.48	1796.21	1794.72	**1787.48**
P19	2310.92	2301.92	2320.95	2327.48	**2301.90**
P20	3131.90	3156.29	3094.66	**3085.36**	3106.75

It can be seen from the table that RG-MMAS dominates in 12 of the 20 cases, HRH dominates in 4 cases, VNS and IACO dominate in 3 cases respectively. Therefore, it can be roughly concluded that RG-MMAS can effectively solve most MDVRPTW problems and is very competitive.

Besides, for analyzing the advantages of our proposed algorithm in detail, we re-grouped all the data sets according to the size of customers. Small data sets contain less than 100 customers, medium data sets contain between 100 and 200 customers, and large data sets contain more than 200 customers. Table 2 shows the details of grouping Relative Percentage Deviation (RPD) values according to the above rules.

It can be observed that RG-MMAS algorithm performs best in small-scale data sets, but its advantage is not obvious. With the expansion of data sets, the gap between the RG-MMAS and other algorithms is gradually widened and it occupies a dominant position. This is because, with the increase of data size allocated to each depot and the number of customers on the each path are also gradually increase. The decomposition framework is used in RG-MMAS,

Table 2. Results of the algorithm on data sets of different sizes.

Number of customers	Number of customers per depot	Number of customers per route	IACO	VNS	HRH	RG-MMAS
1–100	15.4	8.16	−62.73%	−49.24%	−62.93%	**−65.30%**
100–200	34.2	10	−35.37%	−47.26%	−74.85%	**−77.91%**
200–300	51.6	10.75	−144.88%	12.50%	−179.53%	**−240.36%**

and the local search algorithm can find a more optimal solution between the different routes. Therefore, RG-MMAS is outstanding in large-scale data sets and has good scalability.

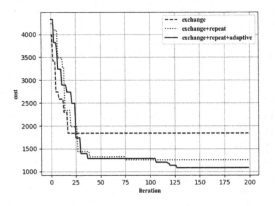

Fig. 1. Results of ablation experiments.

In the experimental phase, we also verify the influence of the three proposed innovations on the algorithm. Figure 1 shows the results of the ablation valida-tion. All three experiments are based on the decomposition framework. The short curve represents the experimental results after introducing λ-interchange oper-ator, the dotted line represents the experimental results after grouping border customers repeatedly on this basis, and the solid line represents the experimental results after introducing the technology of adaptively selecting customer size for repeated grouping. In the early iteration of the algorithm, because the repeated grouping of customers will lead to a period of pheromone confusion, the search direction is not clear at this time, so the short curve has the lowest target value. With the gradual development of search, the introduction of the definition of boundary customers and the appropriate scale of repeated grouping customers can have faster convergence speed.

Furthermore, it is proved that considering the constraint of path duration Du in the search process can improve the experimental results [20]. The prob-lem instances of MDVRPTW are characterized by the fact that the maximum

Table 3. Experimental results of IACO, VNS, HRH, and RG-MMAS(consider the Du limit).

No.	IACO	VNS	HRH	RG-MMAS
P01	1079.25	**1074.12**	1080.16	1076.33
P02	**1767.79**	1769.53	1783.60	1781.47
P03	2389.53	2373.65	2372.88	**2365.03**
P04	2871.90	2846.28	**2840.72**	2849.67
P05	2994.63	2987.29	2975.81	**2968.42**
P06	3629.72	3609.04	3614.25	**3598.40**
P07	**1415.84**	1423.79	1434.97	1421.62
P08	2121.38	2093.06	2090.73	**2085.34**
P09	2755.49	2741.39	2739.80	**2737.41**
P10	**3485.51**	3489.45	3495.37	3493.12
P11	1010.22	1007.24	**1001.96**	1005.39
P12	1478.45	1480.95	**1457.81**	1464.36
P13	2000.47	**1996.58**	2003.69	2007.85
P14	2235.89	2212.50	2216.32	**2208.49**
P15	2477.93	2467.24	2459.65	**2452.16**
P16	2896.48	2888.73	2876.30	**2872.36**
P17	1239.73	1226.49	**1220.41**	1228.05
P18	1814.52	1799.58	1783.89	**1774.28**
P19	2318.74	2325.39	**2292.45**	2300.43
P20	3053.19	3033.27	2973.44	**2947.29**

route duration Du is significantly shorter than the size of the depot opening time window. Therefore, ensuring that the vehicle has no idle time is critical to the viability of the solution. So consideration of path duration Du may make a solution previously considered infeasible acceptable. Table 3 shows the best solution for each problem instance after introducing forward relaxation for the four algorithms. The results show that after introducing the concept of forward relaxation, the optimal solution of the algorithm is improved, and the RG-MMAS algorithm still dominates 10 out of 20 examples in the final result.

5   Conclusion

In this paper, a repetitive grouping Max-Min Ant System is proposed to solve MDVRPTW, which is named RG-MMAS. All customers are added into the corresponding depot group according to certain rules in a decomposed framework, which makes the problem smaller and easier to solve. At the same time in order to improve the effect, three contribution are introduced into the algorithm. The first is to define boundary customers, which are repeatedly grouped into depot groups according to rules. The second is the adaptive selection range of boundary customers, which balance the convergence of the algorithm and resource consumption. Thirdly, λ-interchange is introduced to improve the local search capability of the algorithm. In the experimental stage, RG-MMAS is compared with the state-of-the-art algorithms on the benchmark data set. The experiment proves that RG-MMAS can effectively solve the MDVRPTW and has a good performance on the medium and large scale data set. It is a competitive heuristic algorithm.

In this paper, an adaptive change function is used to determine the scale of repeat customers according to the iteration stage of the algorithm. In the future, we can consider other ways to determine the scope of repeat customers, so as to better balance the solution quality and solution speed of the algorithm. And this paper is devoted to solving the multi-depot vehicle routing problem, which is widely used in real life. However, in real life, the problem is often not static, but dynamic changes. Therefore, in the future work, the dynamic multi-depot vehicle routing problem can be further studied.

Acknowledgements. This work was supported by the Provincial Natural Science Foundation of Shaanxi of China (No. 2019JZ-26).

References

1. Bae, H., Moon, I.: Multi-depot vehicle routing problem with time windows considering delivery and installation vehicles. Appl. Math. Model. **40**(13–14), 6536–6549 (2016)
2. Bettinelli, A., Ceselli, A., Righini, G.: A branch-and-cut-and-price algorithm for the multi-depot heterogeneous vehicle routing problem with time windows. Transport. Res. Part C: Emerg. Technol. **19**(5), 723–740 (2011)
3. Brandão, J.: A memory-based iterated local search algorithm for the multi-depot open vehicle routing problem. Eur. J. Oper. Res. **284**(2), 559–571 (2020)
4. Cordeau, J.F., Laporte, G., Mercier, A.: A unified tabu search heuristic for vehicle routing problems with time windows. J. Oper. Res. Society **52**(8), 928–936 (2001)
5. Cordeau, J.F., Laporte, G., Mercier, A.: Improved tabu search algorithm for the handling of route duration constraints in vehicle routing problems with time windows. J. Oper. Res. Society **55**(5), 542–546 (2004)
6. Dantzig, G.B., Ramser, J.H.: The truck dispatching problem. Manage. Sci. **6**(1), 80–91 (1959)
7. Desaulniers, G., Lavigne, J., Soumis, F.: Multi-depot vehicle scheduling problems with time windows and waiting costs. Eur. J. Oper. Res. **111**(3), 479–494 (1998)
8. Dorigo, M., Maniezzo, V., Colorni, A.: Ant system: optimization by a colony of cooperating agents. IEEE Trans. Syst., Man, Cybern., Part B (Cybernetics) **26**(1), 29–41 (1996)
9. Erdoğan, G.: An open source spreadsheet solver for vehicle routing problems. Comput. Oper. Res. **84**, 62–72 (2017)
10. Fan, H., Yang, X., Li, D., Li, Y., Liu, P., Wu, J.: Half-open multi-depot vehicle routing problem based on joint distribution mode of fresh food. Comput. Integ. Manufact. Syst **25**, 256–266 (2019)
11. Gao, J., Gu, F., Hu, P., Xie, Y., Yao, B.: Automobile chain maintenance parts delivery problem using an improved ant colony algorithm. Adv. Mech. Eng. **8**(9), 1687814016665297 (2016)

12. Kallehauge, B., Larsen, J., Madsen, O.B., Solomon, M.M.: Vehicle routing problem with time windows. In: Column generation, pp. 67–98. Springer (2005). https://doi.org/10.1007/0-387-25486-2_3
13. Koç, Ç., Bektaş, T., Jabali, O., Laporte, G.: The fleet size and mix location-routing problem with time windows: formulations and a heuristic algorithm. Eur. J. Oper. Res. **248**(1), 33–51 (2016)
14. Li, J., Li, Y., Pardalos, P.M.: Multi-depot vehicle routing problem with time windows under shared depot resources. J. Comb. Optim. **31**, 515–532 (2016)
15. Mancini, S.: A real-life multi depot multi period vehicle routing problem with a heterogeneous fleet: formulation and adaptive large neighborhood search based matheuristic. Transp. Res. Part C: Emerg. Technol. **70**, 100–112 (2016)
16. Martins, L.d.C., Bayliss, C., Juan, A.A., Panadero, J., Marmol, M.: A savings-based heuristic for solving the omnichannel vehicle routing problem with pick-up and delivery. Transport. Res. Proc. **47**, 83–90 (2020)
17. Nan, W., Shiqi, L., Junfeng, W.: Vehicle routing with time windows in material delivery for automobile general assembly line. Ind. Eng. J. **15**(2), 94 (2012)
18. Niu, M., Liu, R., Wang, H.: A max-min ant system based on decomposition for the multi-depot cumulative capacitated vehicle routing problem. In: 2021 IEEE Congress on Evolutionary Computation (CEC), pp. 620–627. IEEE (2021)
19. Noori, S., Ghannadpour, S.F.: High-level relay hybrid metaheuristic method for multi-depot vehicle routing problem with time windows. J. Math. Modell. Algorithms **11**(2), 159–179 (2012)
20. Polacek, M., Hartl, R.F., Doerner, K., Reimann, M.: A variable neighborhood search for the multi depot vehicle routing problem with time windows. J. Heuristics **10**(6), 613–627 (2004)
21. Rahimi-Vahed, A., Crainic, T.G., Gendreau, M., Rei, W.: Fleet-sizing for multi-depot and periodic vehicle routing problems using a modular heuristic algorithm. Comput. Oper. Res. **53**, 9–23 (2015)
22. Thangiah, S.R., Osman, I.H., Sun, T.: Hybrid genetic algorithm, simulated annealing and tabu search methods for vehicle routing problems with time windows. Computer Science Department, Slippery Rock University, Technical Report SRU CpSc-TR-94-27 69 (1994)
23. Toth, P., Vigo, D.: The vehicle routing problem. SIAM (2002)

Secure Access Method of Power Internet of Things Based on Zero Trust Architecture

Zhi-hua Wang[1]([⊠]), Ming-hui Jin[1], Lin Jiang[2], Chen-jia Feng[1], Jing-yi Cao[2], and Zhang Yun[1]

[1] State Grid Shanghai Electric Power Company, Shanghai 200122, China
wangzh@sh.sgcc.come.cn, zhang_yun@sh.sgcc.com.cn
[2] China Electric Power Research Institute, Beijing 100192, China
{jianglin,caojingyi}@epri.sgcc.com.cn

Abstract. With the development of the energy Internet, the terminals of the power Internet of Things are open and interactive, the grid protection boundary is ambiguous, and the traditional boundary based security system is difficult to ensure the secure access of the power Internet of things terminals. The program proposed a secure access method for the power system based on the zero-trust architecture. It takes the identity of the terminal as the center to carry out security authentication and dynamic access control. Based on the device fingerprint extraction and the Identification Public Key algorithm, the lightweight security authentication of the power terminal is realized. Through the trust calculation method based on terminal operating characteristics, continuous trust assessment and dynamic access control are carried out. The method is verified in the distributed power security access scenario. For common flood attacks, packet attacks and malicious code attacks, the security access method of the power Internet of Things based on the zero-trust architecture can effectively detect and block them in time, which can avoid attackers from conducting cyber attacks on the grid by controlling power IoT terminals (This work was supported by the State Grid Shanghai Technology Project, Research on Key Technologies for Network Security Access of Power Monitoring System Based on Zero Trust.).

Keywords: Power Internet of Things · Zero Trust · Secure Access · Identity Authentication · Access Control · Distributed Power

1 Introduction

With the rapid development of the application of new technologies such as power Internet of Things and secondary services such as spot market and co-control of load and storage of source network, network boundaries continue to extend, node devices are greatly increased, functional applications are deeper, and data interaction is more extensive, which greatly increases the exposure of power Internet of things and puts forward higher requirements for power grid security protection. The current network security protection system of power system is mainly based on boundary isolation protection, as shown in

© The Author(s), under exclusive license to Springer Nature Switzerland AG 2023
Y. Tan et al. (Eds.): ICSI 2023, LNCS 13969, pp. 386–399, 2023.
https://doi.org/10.1007/978-3-031-36625-3_31

Fig. 1. In accordance with the general principles of "security zone, network dedicated, horizontal isolation, vertical authentication", the production control area and information management area are divided, and the production control area is further divided into control area and non-control area. A horizontal isolation device is deployed between network areas, and an encryption authentication device is deployed along the vertical border.

The current defense system adopts the "One authentication, continuous trust" mechanism, which considers Intranet devices to be trusted. As shown in Fig. 1. However, the development of the power Internet of Things and the extensive connection of massive heterogeneous terminals have brought about the blurring of network boundaries and the diversification of access scenarios. The existing security protection system based on boundaries is faced with security risks, and there are deficiencies in security authentication and trust evaluation [1–4].

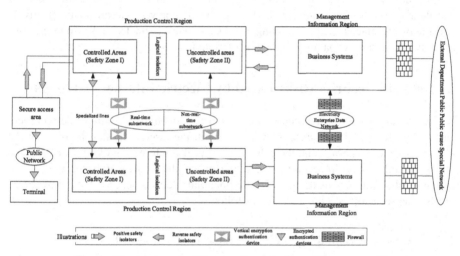

Fig. 1. The framework of the security protection of the power grid

It is necessary to change the protection thinking based on traditional boundary isolation, reconstruct the trust management and control system based on the concept of "zero trust", which takes the identity of system equipment as the center, continuous security assessment and dynamic access control, realize more effective security protection on the basis of boundary isolation protection, and provide guarantee and support for the safe operation of the existing power secondary system. In the electric Internet of Things environment, zero-trust technology should be introduced to carry out security authentication and continuous dynamic access control of Internet of Things terminals. Zero-trust concept carries out security control on any subject and any network request that accesses any resource, providing a new idea for active defense of electric Internet of Things [5–10].This paper firstly analyzes the network security risks faced by the electric Internet of Things, including the blurred boundary and internal and external attack risks faced by the electric Internet of Things [11].Secondly, the zero-trust framework is studied, including the framework and development process of zero-trust [12].Then, it studies the zero-trust based terminal security access framework of the electric Internet of Things. Based on the existing network security protection architecture based on isolation and encryption, it carries out research on key security issues of the electric Internet of Things, including continuous identity authentication, real-time trust assessment and dynamic access control technologies. In the process of identity authentication, a terminal security authentication method based on identification public key is proposed considering the characteristics of power Internet of Things such as limited computing power and narrowband transmission [13–15].The process of dynamic access control considers the behavior characteristics of terminals under malicious attacks, and proposes a trust degree calculation method based on the characteristics of terminals to build a continuous trust evaluation and dynamic access control mechanism to timely block abnormal behaviors of terminals and improve the security protection capability of the terminal side of the electric Internet of Things [15, 16].Finally, the application of the zero-trust architecture of the power Internet of Things in the typical scenario of secure access of distributed power supply is analyzed. Through the method verification, it can effectively resist network attacks such as flooding attacks and packet attacks, and help to improve the network security capability of the power system.

2 Network Security Risks Faced by Electric Internet of Things

The existing boundary based security protection system builds a relatively perfect security system for the power system. However, with the large-scale development of the electric Internet of Things and the enhancement of the automation of attack means and attack capability, it is difficult for the system to ensure the security of the terminal access of the electric Internet of things, and the potential security access of the power grid exists. The security risks are analyzed as follows.

2.1 The Security Boundary of Electric Internet of Things is Fuzzy

With the development of digital transformation and new business, the architecture of electric Internet of Things has become increasingly large and complex. As shown in Fig. 2, electric Internet of Things devices cover all aspects of power generation, transmission, transformation, distribution and use. The proportion of new energy and distributed energy is increasing, and the terminal security protection is not in place and the safety management problems are prominent. Smart home business is developing rapidly, and the scale of terminals has reached hundreds of millions. A large number of smart home terminals are at risk of being illegally controlled, and starting and stopping will have a serious impact on the safe operation of the power system. Electric power communication network is very complicated. Besides optical fiber network, mobile communication network, wireless LAN such as 5G and satellite communication together support the operation of power grid. Industrial control software and equipment have potential vulnerabilities, as well as the risk of deliberately implanted back doors. With the development of cloud computing, Internet of Things and other new technologies, the boundary of electric Internet of Things is constantly blurred, data interaction becomes more frequent, and protection becomes more difficult [16–18].

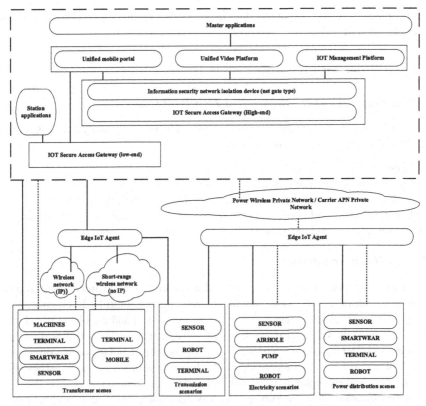

Fig. 2. Typical architecture of power Internet of Things

2.2 Electric Iot Faces Internal and External Attack Risks

Energy infrastructure is increasingly connected to the Internet, and the open and sharing attribute is enhanced. The power system faces severe risks of network attack. Power Internet of Things contains massive data assets, involving all electricity enterprises and residential users, with high confidentiality and high value, information leakage brings serious social impact and consequences. Cyber attacks are characterized by strong concealment, various attack modes and difficulty in prevention. A small breakthrough can develop into extensive damage. Malicious virus programs, APT attacks, zero-day vulnerabilities, ransomware and other forms of network attacks, traditional network security protection technology and equipment is difficult to fully detect and timely defense. The power system has the risk of blackout under large-scale network attack.

In view of the security risks faced by the power system under the electric Internet of Things environment and the high security requirements for the safe and stable operation of the key infrastructure of electric power, it is necessary to gradually change the traditional security protection concept based on boundary protection to the security authentication mechanism based on identity attributes. Based on the traditional power grid secondary security protection system, this paper introduces the concept of zero trust to study the security architecture of the electric Internet of Things based on zero trust, so as to enhance the security of the electric power system under the environment of the electric Internet of Things.

3   Zero-Trust Architecture

3.1   The Proposal and Development of Zero Trust Concept

Zero Trust Network Architecture was first proposed by Forrester analyst John Kindervag in 2010 and is based on five basic assumptions:

1. The network is in a dangerous environment all the time;
2. There are always internal and external threats in the network;
3. The degree of trust cannot be determined by the location in the network;
4. All devices, users and network traffic should be authenticated and authorized;
5. Dynamic adjustment calculation of security policy.

3.2   Zero-Trust Architecture

Zero-trust architecture follows the construction principles of "identity based, resource as the core, continuous trust assessment, dynamic access control". In essence, it constructs an access control system between the access subject and the object based on the authentication of the subject, trust assessment, and dynamic permission allocation. In the zero-trust architecture, as shown in Fig. 3, the access subject and the access object in the process of network access are considered. In the process of the access, trust agents are required to carry out dynamic access control and trust evaluation. Based on the digital identity information of access subjects such as personnel, equipment, applications and systems, combined with the state parameters of access subjects, access behavior characteristics, contextual environment risk level and other data, the trust assessment is carried out continuously through the trust assessment strategy, and the trust relationship between the access subject and the access object is dynamically established. Trusted access is endowed with the permission to access the object, carry out encrypted data communication, and realize the comprehensive protection of application, data, interface, service and other resources.

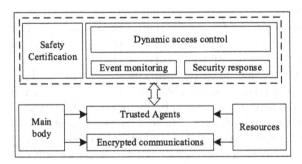

Fig. 3. Zero Trust Architecture Composition

3.3 Zero-Trust Architecture Solution

Zero-trust architecture is an emerging design concept of network security architecture. It is a network security protection framework formed by a series of logical components according to specific interactive relationships, and has a variety of technical solutions.

The SDP software-defined boundary is to establish a dynamic and fine-grained service access tunnel between visitors and resources, including the SDP controller, the SDP connection initiating host, and the SDP connection accepting host, as shown in Fig. 4a. The SDP controller is used to hide server addresses and ports to hide the network and implement access control. Users have access rights only at the application layer, but not at the network level.

The NIST architecture and core components are shown in Fig. 4b. When a user accesses a resource, access is granted through the policy engine (PE) and the corresponding policy enforcement point (PEP).PE defines decisions, such as rules that a role can access only a specified system. The PEP executes the decision to determine whether an access is legitimate or not based on the rules defined by the PE.

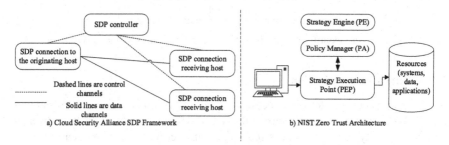

Fig. 4. Zero Trust Architecture Scheme

4 Security Framework of Electric Internet of Things Based on Zero Trust

4.1 Design of Security Protection Framework

According to the characteristics of electric Internet of Things, a zero-trust based terminal security access framework for electric Internet of Things is established, as shown in Fig. 5. When an access principal initiates an access to a resource, it first sends the request to a trusted proxy, and the trusted proxy performs security authentication on the identity of the access principal. Secondly, the access subject that passes the security authentication establishes the initial trust, and the access control module allocates the authority to the access subject, and establishes a two-way encrypted data channel between the access initiator and the authorized access resource for data communication. The access resources have been registered in the trusted agent, and the subject's access to the resources is realized through software-defined boundaries, while the real location of the accessed resources is hidden. Therefore, secure resource access can be realized to prevent illegal access subjects from launching malicious attacks on the terminals of the power Internet of things and affecting the safe operation of the power grid.

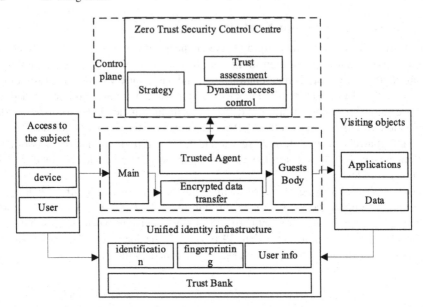

Fig. 5. The security framework of the Internet of Things based on zero trust

4.2 Security Authentication

To ensure the security of terminal access to the power grid, protect data transmission from unauthorized intrusions, and improve the active defense capability of the power grid, terminal security authentication is required. Considering the communication requirements such as narrowband communication and low power consumption between electric Internet of Things terminals, massive network endings greatly increase the difficulty of risk management and control of security authentication. Therefore, lightweight security authentication methods suitable for electric Internet of Things terminals should be studied.

This paper proposes a lightweight terminal security authentication method based on identification public key, which consists of three steps, as shown in Fig. 6. Firstly, the type of electric Internet of Things terminal device is analyzed, and the fingerprint of the device is extracted; Secondly, based on device fingerprint, the electric Internet of Things terminal identification key generation algorithm is designed to generate terminal device public and private keys. Finally, a terminal security authentication method based on identification public key is designed.

4.2.1 Device Fingerprint Extraction

According to the network security requirements of electric Internet of Things terminals, the types of control-related terminals are determined, and the fingerprint characteristics of devices are determined with the goal of ensuring uniqueness and low redundancy.

The extraction of fingerprint feature information should not only fully reflect the characteristics of a device, but also uniquely identify a device. In addition, the pressure such as computing power and energy consumption of the device should be considered.

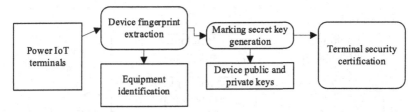

Fig. 6. Security authentication method based on identification public key

Therefore, through the multi-label feature selection algorithm based on multi-variable mutual information, the feature subsets filtered from the multidimensional feature information are as follows: the unique serial number of the device/module, the general parameter, the product detection serial number and the running state of the embedded module. This subset covers as much category information as possible with less redundancy, as described in Table 1.

Table 1. Terminal device fingerprinting

Fingerprint Feature	Description
Device serial number/ID	Unique identifier assigned to the device by the manufacturer
General parameters of the device	Type, name, model, function, etc.
Embedded module operation status	Security status, storage status, etc., high security, and not easy to copy

4.2.2 Electric Internet of Things Identification Public Key Generation Algorithm

The electric Internet of Things identification public key generation algorithm is based on IPK identification public key technology. SM9 algorithm is used to design the identification key pair generation method, and the existing public key system is transformed into the public key system combined with the identification of Internet of Things devices, so as to realize the relationship binding between identification and public key, combine the generation and distribution of keys, and realize the key management of mass terminals. As a lightweight key generation and management method, it directly simplifies the complexity and management difficulty of key generation, and reduces the cost of key system construction and operation and maintenance. The user private key and public key generation process of the Electric Internet of Things identification public key generation algorithm is shown in Fig. 7.

The process of generating electric Iot identification public key is as follows:

1. The terminal device extracts fingerprint information and forms fingerprint ID;

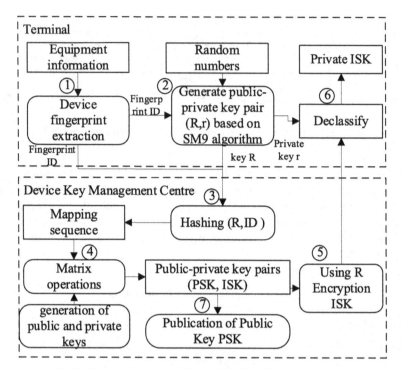

Fig. 7. The generation algorithm of the identification public key

2. The random number r generated by the terminal device generates the user's random public and private key pair (r, R) according to the SM9 algorithm, and transmits R and ID to the device key management center;
3. The device key management center hashes the R and ID sent by the device to get 32 groups of mapping sequences;
4. Public key PSK (Pre-SharedKey) and device private key ISK (IdentitySecureKey) were obtained by matrix calculation based on 32 groups of mapping sequences and randomly generated public key and private key;
5. The key management center of the device then uses the encryption algorithm with R as the public key to encrypt ISK and send the ciphertext back to the device;
6. The device uses the random private key r to decrypt the ciphertext and obtains its own private key ISK;
7. The device key management center will publish the public key PSK. Other users can decrypt the ciphertext sent by the device according to the public key of the device.

The preceding process prevents the device key management center from knowing the private key of the device and prevents other users from decrypting the ciphertext, ensuring the security of the private key and message.

In the process of the identification public key generation, the random public and private key pair generation of the terminal adopts the SM9 national secret algorithm, replacing foreign algorithms such as ECC, which has better performance and security, faster processing speed and less machine performance consumption. It gets rid of the dependence on foreign cryptography technology and realizes the core information security technology controlled from the level of cryptography algorithm.

4.2.3 Identity Authentication of Terminal Devices Based on Public Keys

The identity authentication of terminal devices is an important link of secure terminal access. Based on the lightweight identity key system, the digital signature of terminal devices is implemented for terminal identity authentication. During message data transmission between the Internet of Things terminal device and the regulatory system, in order to ensure that the message is not tampered with, the message needs to be digitally signed and checked. The lightweight SM2 algorithm is used from the edge side to the terminal side for signature verification, and the electric Internet of Things identification public key generation algorithm is used for key management. The digital signature and verification process is shown in Fig. 8.

Digital signature process: the terminal device of the sender first concatenates the message and device ID for Hash function encryption to obtain the encrypted message digest; Then, the encrypted message digest and device private key ISK are entered into the encryption algorithm for signature operation to obtain a digital signature. The obtained digital signature and the original message ID are put into the packet for transmission.

Check process: When receiving data packets, the receiver must verify the digital signature information to ensure the authenticity of data packets. The public key is used to decrypt the received signature information, and the decrypted message sequence is compared with the transmitted message sequence. If the result is consistent, the signature is valid, and the message is not tampered with; otherwise, the signature is invalid.

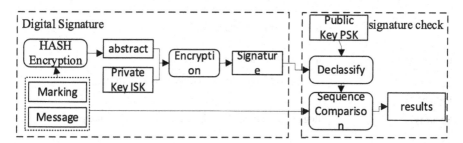

Fig. 8. Digital signature/verification process

In the process of secure access of terminal devices, the public and private keys are generated based on the electric Internet of Things identification public key generation algorithm, and the message is digitally signed based on the SM2 algorithm. The combination of the key management based on the electric Internet of Things identification public key generation algorithm and the SM2 digital signature is realized to ensure that the relevant data is the real data obtained from legitimate devices, and the security authentication of terminal devices is realized. Prevent impersonation attacks.

4.3   Dynamic Access Control

In the zero-trust-based access control mechanism, every time the electric Internet of Things terminal is connected to the grid, it needs to be re-authenticated and authorized. In other words, the single security authentication mechanism cannot continuously guarantee the legitimacy of the identity of the access subject. In the subsequent access, continuous trust assessment is also needed to calculate the trust degree of the access subject, and based on this, the subsequent dynamic permission allocation and access control of the access subject are carried out.

4.3.1   Continuous Trust Assessment

The continuous trust evaluation mechanism continuously collects, identifies and analyzes the network behavior of the access subject, and dynamically calculates the trust degree of the access subject. In this paper, a trust degree calculation method based on terminal operating characteristics is proposed, and the terminal operating characteristics measurement is analyzed to obtain the terminal trust degree.

Terminal operating characteristics include performance, reliability, and security, as shown in the following Table 2.

In the terminal operation characteristics metric, normal network communication is defined as a normal event and denoted by N, and malicious network attack traffic is defined as an abnormal event and denoted by A. The trustworthiness T_F of the current power IoT terminal operation is calculated based on the operational characteristics such as performance, reliability and security, where F denotes the terminal operational characteristics and the terminal trustworthiness T_F is denoted by $\{b_F, d_F, u_F\}$.

Table 2. Terminal operating characteristics

Feature	Specific classification
Performance	Processor, Memory, Disk usage, Network traffic
Reliability	Success rate, Loss rate, Mean time to failure
Security	Illegal connections, Port scans, Unauthorized attempts

$$\begin{cases} b_F = \dfrac{N}{N+A+1} \\ d_F = \dfrac{A}{N+A+1} \\ u_F = \dfrac{1}{N+A+1} \end{cases} \tag{1}$$

where b_F is the terminal credible probability; d_F is the terminal untrustworthy probability; and u_F is the degree of uncertainty of the terminal credible.

4.3.2 Dynamic Access Controls

Zero-trust network restricts the minimum access rights and limits the lateral expansion of attacks. In terms of dynamic access control, for power IoT terminal access scenarios, such as distributed power supply, load control, charging pile and other different types of services, the minimum access rights are assigned based on the trust degree of the access terminal, based on adaptive policy adaptation and optimized rights management. From small to large in order of complete access, monitoring access, restricted access and denied access.

Dynamically monitor the behavior status of power IoT terminals through continuous trust assessment. Reduce or cancel the authority when the terminal appears abnormal.

5 Application Scenario Validation

The distributed power terminal security access scenario was selected as the verification scenario. Multiple network attacks, such as forging inverters to attempt to establish a connection with the terminal information collection server and upload data, were simulated in the experiment. The goal was to verify the protection ability of the zero-trust architecture-based power IoT terminal security access method in power grid business.

5.1 Distributed Power Access Scenarios

Distributed power sources, such as distributed photovoltaics, can access the dispatching agency through the power scheduling data network or wireless network to send information such as telemetering and telecontrol data to the dispatching agency's main station. They can also receive telecontrol and teleadjustment commands issued by the dispatching agency's main station, as shown in Fig. 9. When the distributed power source terminal interacts with the power grid, it is necessary to perform security authentication and access control on the terminal equipment to prevent attackers from accessing the scheduling main station by forging or controlling the distributed power source entity and carrying out network attacks on the power grid.

5.2 Analysis of the Attack of Distributed Power Terminals

A simulation environment for distributed power generation system access was built using distributed power generation inverter terminals and data collection servers. The zero-trust-based security access method for power IoT terminals was prototyped, and the security access detection system was integrated with the distributed power generation system. The identity of distributed power generation terminals was authenticated using the identity-based public key security authentication method designed in this paper. The authenticated terminals established a connection with the data collection server of the power grid. The terminal trust measurement module based on terminal operation features evaluated simulated attacks such as flooding attacks, malicious code attacks, and message attacks, detected attacks in real-time, and restricted and blocked terminals that implemented malicious attacks by means of terminal access control, message concurrency control, and port control. This was done to prevent attackers from controlling

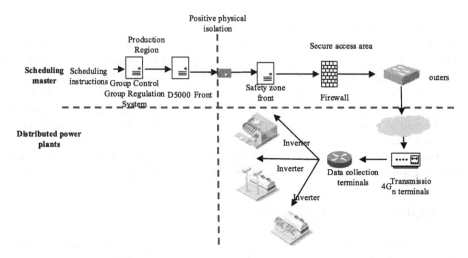

Fig. 9. Distributed power terminal access framework

distributed power generation terminals and conducting network invasion and destruction of the power grid. Through simulated attacks, the zero-trust-based security access method proposed in this paper can continuously identify abnormal terminal behavior, conduct dynamic terminal access control, and reduce network security risks caused by terminal invasion of the power grid.

6 Conclusion

This paper investigates a zero-trust framework for secure access of heterogeneous electric IoT terminals in order to address the security risks posed by a massive number of such terminals in the power grid. Considering the limited and widely dispersed resources of electric IoT terminals, the paper proposes a lightweight security authentication technology based on identity public keys. To mitigate the impact of potential network attacks on terminal behavior, the paper also introduces a dynamic terminal access control method based on terminal feature measurement. By conducting a technical verification in the distributed power access scenario, the proposed methods can effectively identify malicious traffic behaviors such as flooding attacks, and promptly block attacks, thereby reducing the risk of electric IoT terminal attacks on the power grid.

References

1. Bedi, G., Venayagamoorthy, G.K., Singh, R., et al.: Review of Internet of Things (IoT) in electric power and energy systems. IEEE Internet Things J. **5**(2), 847–870 (2018)
2. Xiaojian, Z., Liandong, C., Jie, F., et al.: Power IoT security protection architecture based on zero trust framework. In: 2021 IEEE 5th International Conference on Cryptography, Security and Privacy (CSP), pp. 166–170. IEEE (2021)
3. Zhang, H., Liu, B., Wu, H.: Smart grid cyber-physical attack and defense: a review. IEEE Access **9**, 29641–29659 (2021)

4. Tuballa, M.L., Abundo, M.L.: A review of the development of Smart Grid technologies. Renew. Sustain. Energy Rev. **59**, 710–725 (2016)
5. Sellitto, G.P., Aranha, H., Masi, M., Pavleska, T.: Enabling a zero trust architecture in smart grids through a digital twin. In: Dependable Computing - EDCC 2021 Workshops, EDCC 2021. Communications in Computer and Information Science, vol. 1462. Springer, Cham (2021). https://doi.org/10.1007/978-3-030-86507-8_7
6. Mir, A., Rashid, I., Kumar, K.R.: An augmented smart grid based SCADA security management system (SSMS) based on zero-trust architecture. In: Proceedings of the 2nd International Conference on ICT for Digital, Smart, and Sustainable Development, ICIDSSD 2020, 27–28 February 2020, Jamia Hamdard, New Delhi, India (2021)
7. Wu, Y.G., Yan, W.H., Wang, J.Z.: Real identity based access control technology under zero trust architecture. In: 2021 International Conference on Wireless Communications and Smart Grid (ICWCSG), pp. 18–22. IEEE (2021)
8. Sellitto, G.P., Aranha, H., Masi, M., Pavleska, T.: Enabling a zero trust architecture in smart grids through a digital twin. In: Dependable Computing - EDCC 2021 Workshops. EDCC 2021. Communications in Computer and Information Science, vol. 1462, pp. 73–81. Springer, Cham (2021). https://doi.org/10.1007/978-3-030-86507-8_7
9. Alipour, M.A., Ghasemshirazi, S., Shirvani, G.: Enabling a Zero Trust Architecture in a 5G-enabled Smart Grid. arXiv preprint arXiv:2210.01739 (2022)
10. He, Y., Huang, D., Chen, L., et al.: A survey on zero trust architecture: challenges and future trends. Wirel. Commun. Mob. Comput. **2022**, 1–13 (2022)
11. Sundmaeker, H., Guillemin, P., Friess, P., et al.: Vision and Challenges for Realising the Internet of Things. Cluster of European research projects on the internet of things, European Commission, vol. 3, no. 3, pp. 34–36 (2010)
12. Sultana, M., Hossain, A., Laila, F., et al.: Towards developing a secure medical image sharing system based on zero trust principles and blockchain technology. BMC Med. Inform. Decis. Mak. **20**(1), 1–10 (2020)
13. Lux, Z.A., Thatmann, D., Zickau, S., et al." Distributed-ledger-based authentication with decentralized identifiers and verifiable credentials. In: 2020 2nd Conference on Blockchain Research & Applications for Innovative Networks and Services (BRAINS), pp. 71–78. IEEE (2020)
14. Haqani, E.A., Baig, Z., Jiang, F.: A decentralised blockchain-based secure authentication scheme for IoT devices. In: Suma, V., Baig, Z., Kolandapalayam Shanmugam, S., Lorenz, P. (eds.) Inventive Systems and Control. Lecture Notes in Networks and Systems, vol. 436. Springer, Singapore (2022). https://doi.org/10.1007/978-981-19-1012-8_9
15. Jia, X., Hu, N., Yin, S., et al.: A2 chain: a blockchain-based decentralized authentication scheme for 5G-enabled IoT. Mob. Inf. Syst. **2020**, 1–19 (2020)
16. Mehta, M., Patel, K.: A review for IOT authentication–current research trends and open challenges. Mater. Today Proc. (2020)
17. Djenna, A., Harous, S., Saidouni, D.E.: Internet of things meet internet of threats: new concern cyber security issues of critical cyber infrastructure. Appl. Sci. **11**(10), 4580 (2021)
18. Pothumarti, R., Jain, K., Krishnan, P.: A lightweight authentication scheme for 5G mobile communications: a dynamic key approach. J. Ambient Intell. Humanized Comput., 1–19 (2021). https://doi.org/10.1007/s12652-020-02857-4

On the Complete Area Coverage Problem of Painting Robots

Gene Eu Jan[1], Kevin Fung[2], Chaomin Luo[3], and Hui-Ching Hsieh[4(✉)]

[1] President's Office, Tainan National University of the Arts, Tainan City, Taiwan
geneeujan@gmail.com
[2] Department of Industrial Engineering and Engineering Management, National Tsing Hua University, Hsinchu, Taiwan
s106034402@m106.nthu.edu.tw
[3] Department of Electrical and Computer Engineering, Mississippi State University, Starkville, MS, USA
Chaomin.Luo@ece.msstate.edu
[4] Postdoctoral Researcher, Tainan National University of the Arts, Tainan City, Taiwan
hsu.hsieh@msa.hinet.net

Abstract. Recently, we proposed a novel algorithm for painting problems based on the concepts of minimum spanning-tree based approach and triangle mesh to obtain an $O(n\log n)$ complete area coverage path planning algorithm on polygonal surfaces with minimum length, where n is the number of triangles. In this article, we reduced the time complexity to $O(n)$ by modifying the algorithm. Our proposed method adopts a mobile robot which navigates through an arrangement of areas to be covered without energy and time constraints. In the end, this robot will return to the original starting point. According to the performance analysis, our method is proven to be the fastest algorithm with minimum length to solve the complete area coverage planning for painting robots on polygonal surfaces. In addition, the number of turns has been significantly reduced by 30.26% using the scheme of triangle merge.

Keywords: Complete Area Coverage (CAC) · Painting Robots · Polygonal Surfaces · Triangle Mesh · Spanning Tree

1 Introduction

Mobile robots portray a crucial role for solving the complete area coverage (CAC) problem; it guides the robot to pass through all areas in the workspace. Agricultural crop harvesting equipment [32], automated harvesters [29], autonomous underwater covering vehicles, cleaning robots [25–27, 30, 31, 34], de-mining robots [15], lawn mowers, mine detectors [32], painting robots, vacuum robots [42], and window cleaners [9] and autonomous navigation of unmanned underwater vehicles [3, 6] are examples of the robotic applications that require CAC. Either single or multiple robot modes can complete these types of CAC applications.

© The Author(s), under exclusive license to Springer Nature Switzerland AG 2023
Y. Tan et al. (Eds.): ICSI 2023, LNCS 13969, pp. 400–413, 2023.
https://doi.org/10.1007/978-3-031-36625-3_32

In the literature, different models of the area coverage problem had been further addressed. Whether a map is required for the robot, the coverage models can be categorized as two types of algorithms, off-line and on-line [5]. Robots that requires a map of the workspace uses off-line algorithms (e.g. [15, 23, 24]) compared to on-line algorithms that do not need an environmental map (e.g. [1, 11, 15, 36]). Behavior based model (e.g. [1, 20–22, 39]), cell-decomposition based model (e.g. [23, 24]), Depth First Search approach (e.g. [14, 36]), graph-based model (e.g. [38]), and spanning-tree based approach (e.g. [14, 15]) are the five different classified categories we had previously researched related to area coverage. Ensuring the painting robot transverse each small triangle in the decomposition phase with three different fashions: exact, approximate, and semi- approximate [5], we can provably attain the complete area coverage planning.

Painting robots can be applied to spray painting products such as appliances [2], automobiles, furniture, and toys. These painting robots depend on the concepts of spray paint gun trajectory to achieve the uniformity of the paint's thickness [2].

Products can resemble their quality with suitable high- quality paint coating. Current researches have shown that there are two types of trajectory generation methods, automatic trajectory generation method [2] and typical teaching method. Therefore, there is a high demand of better solutions when manufacturing products in large quantities that requires high efficiency for manufacturers.

In [17], an algorithm with a runtime complexity of $O(n \log n)$ was proposed, where n is the number of triangles. In this article, we improved the previous research [17] to a runtime complexity of $O(n)$. Our new method for painting problems based on the concepts of spanning-tree based approach and triangle mesh was implemented to achieve an $O(n)$ complete area coverage path planning algorithm on polygonal surfaces with minimum length and less turns.

This article is organized as follows. Section 1 presents the introduction of complete area coverage and painting robots. The background of polygonal surfaces, spanning tress, and unfolding of an object is depicted in Sect. 2. Section 3 describes the proposed method's algorithm in further details. Performance analysis is shown in Sect. 4. The experimental results can be found in Sect. 5. Section 6 concludes this article.

2 Background

This section describes some of the well-known concepts such as polygonal surface [16], centroids, spanning tree problem [38], and the unfolding process in the following.

In the three-dimensional space, a polygonal surface defined as a set of triangles is considered to be a piecewise linear surface. Through the various types of computer vision and robot vision, these data can be captured by scanners and generated onto detailed surface models. The isosurfaces are exact due to the volume data with the concept of "marching cubes" algorithm in scientific visualization. It can contribute terrain data and polygonal models when compared to remote sensing, which is accomplished in Computer-Aided Design (CAD), computer games, computer graphics, and Electronic Design Automation (EDA) by subdivision of curved parametric surfaces. By using these methods described, we can then effortlessly acquire the surface model that is consisted from numerous numbers of polygons [16].

Theoretically, a spanning tree T of graph $G = (V, E)$ with shortest length among all spanning trees, is considered to be known as a minimum spanning tree for G. From any arbitrary node r, the tree roots will expand until the tree reaches all the nodes in V [34].

In addition, we can simply unfold the polygonal surface of a specific object into the 2-D construction and then apply the concepts of the complete area coverage to the path planning of painting robots, as depicted in Fig. 1.

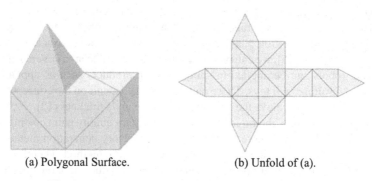

(a) Polygonal Surface.   (b) Unfold of (a).

Fig. 1. The process of a polygonal surface being unfolded.

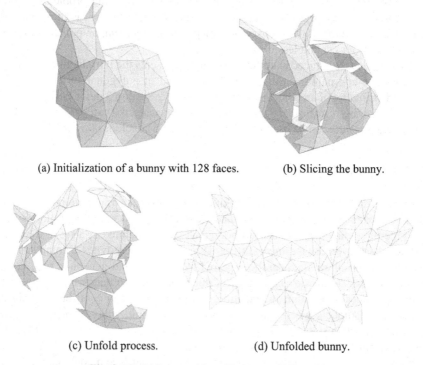

(a) Initialization of a bunny with 128 faces.   (b) Slicing the bunny.

(c) Unfold process.   (d) Unfolded bunny.

Fig. 2. Unfolding an object (Courtesy of Xi *et al.*).

Unfolding known as flattening or unwrapping can be implemented based on the method proposed by Xi et al. [41]. The concepts used to unfold a 3-D mesh object require a developed genetic based algorithm to unfold the 3-D mesh into a single connected patch. The object can be printed onto a piece of paper and refolded back to its original shape. In our research, we will be implementing Xi et al. method to assist us with the unfolding of a 3-D mesh object for painting robots. The process of initializing and unfolding the object is illustrated in Fig. 2. First, initialization of a bunny with 128 faces are shown in Fig. 2(a). The slicing of the bunny is presented in Fig. 2(b). Figure 2(c) represents the motion planning method to unfold. Lastly, the unfolded bunny with 128 faces is depicted in Fig. 2(d).

3   Algorithm and Its Illustration

Our proposed algorithm is demonstrated in this section based on the concepts of spanning-tree approach, combining small sub triangles into large triangles, and triangle mesh to acquire an $O(n)$ complete area coverage path planning algorithm on polygonal surfaces with minimum length.

Given a Delaunay triangulation by the set of vertices (expressed by S) [7, 18, 19, 39], first and foremost, connecting the centroids to their corresponding neighbor centroids, we then can develop the centroids edges (expressed by E_C). When the connected graph G with E_C is composed, we can then generate the spanning tree. Using the centroids, we can decompose the triangles and combine the small triangles into large triangles as shown in Fig. 3. Furthermore, the robot goes along the spanning tree in clockwise order and develops a complete area coverage path from the starting point.

Three different categories of all the divided triangles could be classified by their degrees of centroid in the spanning tree, degree = 1, 2, and 3. According to the degrees of the centroids in the spanning tree, the robot will then travel along a sub-triangles path that locally follows the directed routes. The illustration of three types with different degrees is depicted in Fig. 4.

3.1   Algorithm

Algorithm: Complete area coverage planning for industrial painting robots

BEGIN
Step 1: Generate the polygonal surfaces of an object with triangle mesh.
Step 2: Call function 1 to generate a spanning tree of the connected graph G.
Step 3: Call function 2 to circumnavigate the spanning tree.
Step 4: Plan the complete area coverage path in details.
END

Function 1: Generate a spanning tree of the connected graph G

BEGIN
Step 1: Embed centroids to all the triangles.

Step 2: Connect centroids to their corresponding neighbor centroids and create the connected graph G.
Step 3: Compose a spanning tree of graph G.
END

Function 2: Circumnavigate the spanning tree

BEGIN
Step 1: Decompose the triangles into six sub triangles.
Step 2: Combine the decomposed sub triangles into large triangles.
Step 3: Circumnavigate the spanning tree in a clockwise direction.
END

3.2   Illustration of Algorithm

The complete area coverage path planning algorithm for the painting robot on polygonal surfaces is illustrated in Fig. 5. Figure 5(a) presents the unfolded polygonal surfaces with inserted centroids. Figure 5(b) shows a constructed graph G. Figure 5(c) demonstrates the obtained spanning tree. The triangles are decomposed into six sub-triangles in Fig. 5(d). Figure 5(e) combines the six sub-triangles into large triangles. The circumnavigated spanning tree is presented in Fig. 5(f). The complete area coverage path planning of the polygonal surface is depicted in Fig. 5(g).

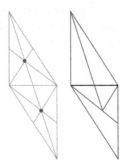

Fig. 3. Combining the small triangles to large triangles.

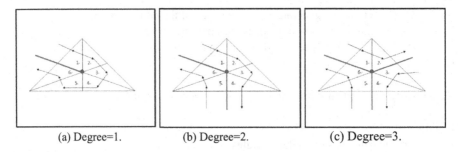

(a) Degree=1.   (b) Degree=2.   (c) Degree=3.

Fig. 4. Demonstration of the three different degrees.

(a) Embed centroids to all the triangles.

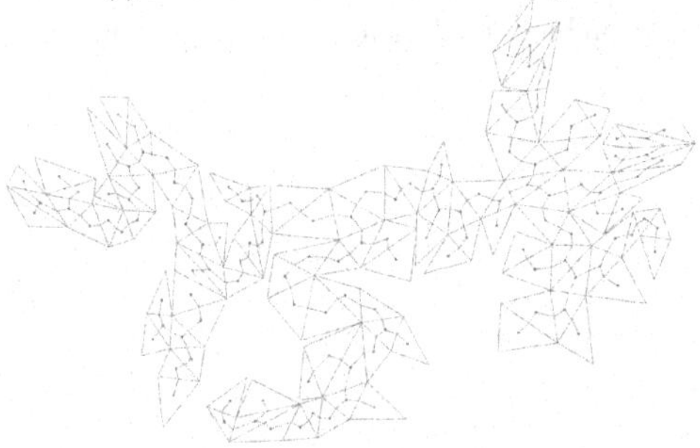

(b) Generate the connected graph G.

(c) Generate the spanning tree.

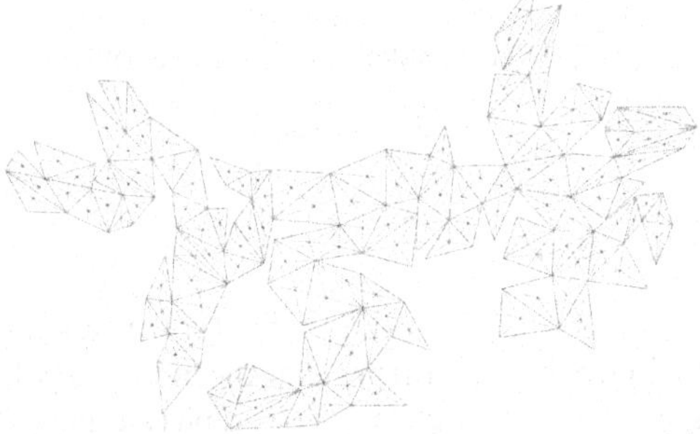

(d) Divide the triangles into six sub-triangles.

Fig. 5. Illustration of our proposed algorithm.

(e) Combine small triangles into large triangles.

(f) Circumnavigate the spanning tree in clockwise direction.

(g) Plan the complete area coverage path.

Fig. 5. (*continued*)

3.3   Illustration of Algorithm

The complete area coverage path planning algorithm for the painting robot on polygonal surfaces is illustrated in Fig. 5. Figure 5(a) presents the unfolded polygonal surfaces with inserted centroids. Figure 5(b) shows a constructed graph *G*. Figure 5(c) demonstrates the obtained spanning tree. The triangles are decomposed into six sub-triangles in Fig. 5(d). Figure 5(e) combines the six sub-triangles into large triangles. The circumnavigated spanning tree is presented in Fig. 5(f). The complete area coverage path planning of the polygonal surface is depicted in Fig. 5(g).

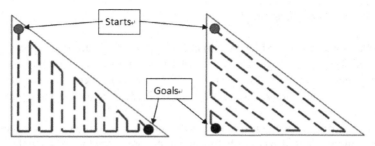

Fig. 6. The types of sweeping direction result in the difference of turn penalty.

The total length could be approximately obtained while adopting a different sweep direction to cover an area. In addition, the types of sweeping direction can result in the difference of turn penalty as demonstrated in Fig. 6.

4 Performance Analysis

Our route planning algorithm for the painting robot problems on polygonal surfaces is optimal in a known space.

Theorem. The time complexity of our coverage path algorithm with shortest path length for the painting robot problems on polygonal surfaces is $O(n)$, where n denotes the number of triangles.

Proof: Obviously, both the time complexities of steps 1 and 2 of function 1 are $O(n)$ since the numbers of triangles and Fermat points are bounded by $O(n)$.

We can generate a spanning tree with time of $O(n)$ [28] in step 3 of function 1.

Since that the numbers of Fermat points and triangles are bounded by $O(n)$, thus, the number of the sub-triangles and the time to obtain the covering path circumnavigates spanning tree in clockwise order are also bounded by $6 \times O(n)$ in steps 1 and 2 of function 2.

The time to sweep the sub-triangles is assumed to be $O(1)$ time. We need $6 \times O(n) \times O(1)$ to sweep all the sub-triangles of the complete area in step 4 of the algorithm.

In conclusion, from steps 1 to 4 of the algorithm the time complexity of our proposed algorithm is $O(n)$.

Comparison of different algorithms in coverage path problem is shown in Table 1.

5   Experimental Results

This research was conducted on a computer with the specifications equipping an Intel Core i5-5257U CPU running at 2.70 GHz with 8 GB of DRAM and an Intel Iris Graphics 6100 integrated graphics. The JavaScript code was experimented Macintosh OS Sierra (10.12.4) using Visual Studio Code, complied on Node.js, and executed using Google Chrome.

In Table 2, we have summarized the results for this research with the comparison of the execution time in milliseconds before and after combining the sub triangles. Table 3 provides the comparison of the number of turns before and after combining sub triangles. Figures 7 and 8 provides the column graph of the table's result respectively.

The combination of the decomposed sub triangles into large triangles increased the overall execution time by an average of 0.730 ms but successfully reduced the turn penalties by an average of 30.26% when compared to Jan [15]. Hence, this method is proven to be the most suitable for CAC as it is not energy and time consuming.

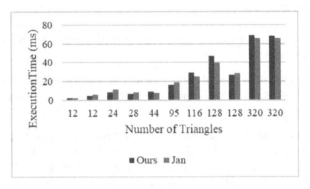

Fig. 7. Comparison between the number of triangles and the execution time.

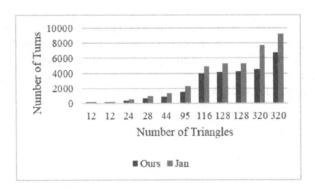

Fig. 8. Comparison between the number of triangles and the number of turns.

Table 1. Comparison of different algorithms in coverage path problem, where n denotes the number of obstacles and n is the number of cells comprising the area.

Algorithms	Gabriely [10]	Mannadiar [24]	Zuo [38]	Ryu [33]	Jan [15]	Ours
Time complexity	n	Postman problem (NP-hard)	Not available	n	$n \log n$	n
w/o concave polygons?	Without	With	Not available	Not available	Without	Without
Complete/ Approximate	Approximate	Complete	Complete	Approximate	Complete	Complete
Overlap/ Non-overlap	Non-overlap	Non-overlap	Overlap	Overlap	Non-overlap	Non-overlap
Returned the starting point at the end	Yes	Yes	No	No	Yes	Yes
Shortest length	Not available	Yes	No	No	Yes	Yes
Turn penalty	Under ave	Average	Above ave	Average	Above ave	Average

Table 2. Comparison of the execution time in milliseconds before and after combining the sub triangles.

Model	Number of triagles	Milliseconds (before)	Milliseconds (after)
Bunny	128	39.550	46.935
Bunny-body	116	24.730	29.290
MoneyBox	128	28.550	26.580
Periscope	28	8.165	6.235
Periscope2	44	7.830	9.135
Sphere1	320	65.985	68.650
Sphere2	320	65.630	68.135
Cube	12	2.045	2.155
Cube-ordered	12	6.020	4.705
Star-024	24	11.095	8.450
Star-4	95	19.030	16.395

Table 3. Comparison of the number of turns before and after combining sub triangles

Model	Number of triangles	Number of turns (before)	Number of turns (after)	Percentage Reduced (%)
Bunny	128	5281	4141	21.59
Bunny-body	116	4894	3944	19.41
MoneyBox	128	5238	4191	19.99
Periscope	28	1040	705	32.21
Periscope2	44	1363	924	32.21
Sphere1	320	7679	4477	41.70
Sphere2	320	9206	6705	27.17
Cube	12	283	174	38.52
Cube-ordered	12	283	174	38.52
Star-024	24	575	403	29.91
Star-4	95	2303	1573	31.70

6   Concluding Remarks and Future Works

In this research, a newly improved algorithm with minimum running time $O(n)$ for the painting robot path planning on a polygonal surface with the optimal approach is presented. In addition to reducing the time complexity from $O(n\log n)$ to $O(n)$, we also reduced the number of turns for the painting robot by an average of 30.26%. Our proposed algorithm adopts a robot which navigates through an arrangement of areas to be covered without energy and time consuming. At the end, this robot will return to the original starting point. Hence, our method is proven to address the complete area coverage problem with the fastest time complexity and minimum length for painting robots on polygonal surfaces.

This research is most suitable for painting robots used in industrial engineering such as the painting robots in the assembly lines of a factory. Future works of this research can be implemented for garbage collection, postman problems, and street cleaning.

References

1. Balch, T., Arkin, R.: Behavior-based formation control for multirobot teams. IEEE Trans. Robot. Autom. **14**(6), 926–939 (1998)
2. Chen, H., Xi, N., Sheng, W., Song, M., Chen, Y.: CAD-based automated robot trajectory planning for spray painting of free-form surfaces. Ind. Robot Int. J. **29**(5), 426–433 (2002)
3. Chen, J., Luo, C., Krishnan, M., Paulik, M., Tang, Y.: An enhanced dynamic Delaunay triangulation-based path planning algorithm for autonomous mobile robot navigation. In: Intelligent Robots and Computer Vision XXVII: Algorithms and Techniques (2010)
4. Chew, L.P.: Constrained delaunay triangulations. In: Proceedings of the Symposium on Computational Geometry, pp. 215–222 (1987)

5. Choset, H.: Coverage for robotics - a survey of recent results. Ann. Math. Artif. Intell. **31**, 113–126 (2001)
6. Chu, Z., Sun, B., Zhu, D., Zhang, M., Luo, C.: Motion control of unmanned underwater vehicles via deep imitation reinforcement learning algorithm. IET Intel. Transp. Syst. **14**(7), 764–774 (2020)
7. Delaunay, B.: Sur la sphère vide, Izvestia Akademii Nauk SSSR. Otdelenie Matematicheskikh i Estestvennykh Nauk **7**, 793–800 (1934)
8. Fang, G., Dissanayake, G., Lau, H.: A behaviour-based optimisation strategy for multi-robot exploration. In: Proceedings of the IEEE Conference on Robotics, Automation and Mechatronics, Singapore, pp. 875–879 (2004)
9. Farsi, M., Ratcliff, K., Johnson, P.J., Allen, C.R., Karam, K.Z., Pawson, R.: Robot control system for window cleaning. In: Proceedings of 11th International Symposium on Automation and Robotics in Construction, Brighton, UK, pp. 617–623 (1994)
10. Fazli, P., Davoodi, A., Pasquier, P., Mackworth, A.K.: Fault-tolerant multi-robot area coverage with limited visibility. In: Proceedings of the IEEE/RSJ International Conference on Intelligent Robots and Systems, Taipei, Taiwan, pp. 18–22 (2010)
11. Ferranti, E., Trigoni, N., Levene, M.: Brick & Mortar: an online multi-agent exploration algorithm. In: Proceedings of the IEEE International Conference on Robotics and Automation, Roma, Italy, pp. 761–767 (2007)
12. Gabriely, Y., Rimon, E.: Spanning-tree based coverage of continuous areas by a mobile robot. Ann. Math. Artif. Intell. **31**, 77–98 (2001)
13. Gage, D.W.: Randomized search strategies with imperfect sensors. In: Proceedings of SPIE, Mobile Robots VIII - The International Society for Optical Engineering, Boston, USA, pp. 270–279 (1994)
14. Hazon, M., Mieli, F., Kaminka, G.A.: Towards robust on-line multi-robot coverage. In: Proceedings of the IEEE International Conference on Robotics and Automation, Orlando, USA, pp. 1710–1715 (2006)
15. Hazon, N., Kaminka, G.A.: Redundancy, efficiency, and robustness in multi-robot coverage. In: Proceedings of the IEEE International Conference on Robotics and Automation, Barcelona, Spain, pp. 735–741 (2005)
16. Jan, G.E., Fung, K., Wu, P.Y., Leu, S.W.: Shortest path-planning on polygonal surfaces with $O(n\log n)$ time. In: Proceedings of 2016 IEEE International Conference on Control and Robotics Engineering, Singapore, April 2016, pp. 98–102 (2016)
17. Jan, G.E., Fung, K., Hung, T.H., Luo, C.: The optimal approach to the painting problems on polygonal surfaces. In: IEEE International Conference on CYBER Technology in Automation, Control, and Intelligent Systems, Waikiki Beach, Hawaii, USA, pp. 1172–1175 (2017)
18. Jan, G.E., Sun, C.C., Tsai, W.C., Lin, T.H.: An $O(n\log n)$ shortest path algorithm based on Delaunay triangulation. IEEE/ASME Trans. Mechatron. **19**(2), 660–666 (2014)
19. Jan, G.E., Tsai, W.C., Sun, C.C., Lin, B.S.: A Delaunay triangulation-based shortest path algorithm with $O(n\log n)$ time in the Euclidean plane. In: 2012 IEEE/ASME International Conference on Advanced Intelligent Mechatronics, pp. 186–189 (2012)
20. Jung, D., Cheng, G., Zelinsky, A.: Experimentals in realising cooperation between autonomous mobile robots. In: Fifth International Symposium on Experimental Robotics, Barcelona, Spain, pp. 609–620 (1997)
21. Jung, D., Cheng, G., Zelinsky, A.: Robot cleaning: an application of distributed planning and real-time vision. In: Zelinsky, A. (ed.) Field and Service Robotics, pp. 187–194. Springer, New York (1998)
22. Jung, D., Zelinsky, A.: An architecture for distributed cooperative-planning in a behaviour-based multi-robot system. Robot. Auton. Syst. **26**, 149–174 (1999)

23. Kurabayashi, D., et al.: Cooperative sweeping by multiple mobile robots with relocating portable obstacles. In: IEEE/RSJ International Conference on Intelligent Robots and Systems, Osaka, Japan, pp. 1472–1477 (1996)

24. Kurabayashi, D., Ota, J., Arai, T., Yoshida, E.: Cooperative sweeping by multiple mobile robots. In: IEEE International Conference on Robotics and Automation, Minneapolis, USA, pp. 1744–1749 (1996)

25. Lawitzky, G.: A navigation system for cleaning robots. Auton. Robot. **9**, 255–260 (2000)

26. Luo, C., Yang, S.X.: A bioinspired neural network for real-time concurrent map building and complete coverage robot navigation in unknown environments. IEEE Trans. Neural Netw. **19**(7), 1279–1298 (2008)

27. Luo, C., Yang, S.X., Li, X., Meng, M.Q.-H.: Neural-dynamics-driven complete area coverage navigation through cooperation of multiple mobile robots. IEEE Trans. Industr. Electron. **64**(1), 750–760 (2017). https://doi.org/10.1109/TIE.2016.2609838

28. Mannadiar, R., Rekleitis, A.I.: Optimal coverage of a known arbitrary environment. In: IEEE/RSJ International Conference on Intelligent Robots and Automation, pp. 5525–5530 (2010)

29. Najjaran, H., Kircanski, N.: Path planning for a terrain scanner robot. In: 31st International Symposium on Robotics, Montreal, Canada, pp. 132–137 (2000)

30. Oh, J.S., Choi, Y.H., Park, J.B., Zheng, Y.F.: Complete coverage navigation of cleaning robots using triangular-cell-based map. IEEE Trans. Industr. Electron. **51**(3), 718–726 (2004)

31. Oh, J.S., Park, J.B., Choi, Y.H.: Complete coverage navigation of clean robot based on triangular cell map. In: IEEE International Symposium on Industrial Electronics, Pusan, South Korea, pp. 2089–2093 (2001)

32. Ollis, M., Stentz, A.: First results in vision-based crop line tracking. In: IEEE International Conference on Robotics and Automation, Minneapolis, USA, pp. 951–956 (1996)

33. Ollis, M., Stentz, A.: Vision-based perception for an automated harvester. In: Proceedings of the IEEE/RSJ International Conference on Intelligent Robot and Systems, Grenoble, France, pp. 1838–1844 (1997)

34. Ortiz, F., et al.: Robots for hull ship cleaning. In: Proceedings of IEEE International Symposium on Industrial Electronics, pp. 2077–2082 (2007)

35. Prim, R.C.: Shortest connection networks and some generalizations. Bell Syst. Tech. J. **36**(6), 1389–1401 (1957)

36. Rekleitis, I.M., Dudek, D., Milios, E.E.: Multi-robot exploration of an unknown environment, efficiently reducing the odometry error. In: Proceedings of the 15th IEEE International Joint Conference on Artificial Intelligence, Nagoya, Japan, pp. 1340–1345 (1997)

37. Ryu, S.W., Lee, Y.H., Kuc, T.Y., Ji, S.H., Moon, Y.S.: A search and coverage algorithm for mobile robot. In: Proceedings of the 2nd International Conference on Ubiquitous Robots and Ambient Intelligence, Incheon, Korea, pp. 1–6 (2011)

38. Singh, A.: An artificial bee colony algorithm for the leaf-constrained minimum spanning tree problem. Appl. Soft Comput. **9**(2), 625–631 (2009)

39. Sun, C.-C., Jan, G.E., Leu, S.-W., Yang, K.-C., Chen, Y.-C.: Near-shortest path-planning on a quadratic surface with O(n log n) time. IEEE Sens. J. **15**(11), 6079–6080 (2015)

40. Wagner, I.A., Lindenbaum, M., Bruckstein, A.M.: Distributed covering by ant-robots using evaporating traces. IEEE Trans. Robot. Autom. **15**(5), 918–933 (1999)

41. Xi, Z., Lien, J.M.: Continuous unfolding of polyhedral-a motion planning approach. In: Proceedings of IEEE/RSJ International Conference on Intelligent Robots and Systems (IROS), Hamburg, Germany, pp. 3249–3254 (2015)

42. Yasutomi, F., Takaoka, D., Yamada, M., Tsukamoto, K.: Cleaning robot control. In: Proceedings of IEEE International Conference on Robotics and Automation, Philadelphia, USA, pp. 1839–1841 (1988)
43. Zuo, G., Zhang, P., Qiao, J.: Path planning algorithm based on sub-region for agricultural robot. In: Proceedings of the 2nd International Asia Conference on Informatics in Control, Automation and Robotics, vol. 2, pp. 197–200 (2010)

Reachability Map-Based Motion Planning for Robotic Excavation

Fan Li[1], Xiaotian Li[1(✉)], Feng Chi[2], and Yong Wang[2]

[1] Department of Mechanical Engineering, Tongji University, Shanghai, China
lixiaotian@tongji.edu.cn
[2] Shandong Lingong Construction Machinery Co., Ltd. (SDLG), Shandong, China

Abstract. The trend towards automation of excavation has necessitated the development of an efficient method for excavator motion planning. Traditional approaches typically model the excavator as a 4-DoF (degrees of freedom) tandem manipulator, despite its three closed kinematic chains and linear actuation via three hydraulic cylinders. To address this limitation, this study proposes a novel search-based motion planning approach that is tailored to the excavator's unique characteristics. Specifically, we employ the excavator's reachability to construct a graph that encodes the feasible poses and moving direction of the excavator bucket in an efficient manner. Moreover, we integrate the displacement of hydraulic cylinders into the cost function to ensure the smoothness of the excavator's movement. Experimental analysis based on a typical digging task demonstrates the planner's ability to generate full-bucket motion while providing guarantees on the smoothness of driving.

Keywords: Motion planning · Reachability map · Robotic excavation

1 Introduction

The excavator is a critical engineering machinery widely utilized in construction, mining, and dredging industries. With the increasing labor expenses and development of autonomous driving technology, fully automated excavating machines have become increasingly popular. Apart from the economic benefits, automated earth machines are indispensable in hazardous work conditions such as radiation zones [1,2]. To achieve these tasks, motion planning for excavator robots has become a research hotspot in the field of intelligent engineering machinery.

An excavator comprises a walking device and a working device with four degrees of freedom, including a rotating degree of freedom (cabin) and three translational degrees of freedom (boom, arm, and bucket). The working device is propelled linearly by three hydraulic cylinders and three closed kinematic chains. Depending on various factors, it can be described by three distinct mathematical expressions, which can be transformed into each other using geometric relationships:

This work is supported by the Shandong Provincial Key R&D Program (Major Technological Innovation Project) projects (No.2020CXGC011005).

- Task-space $\mathbf{q}_t = [x, y, z, \xi]$: it describes the pose of the bucket tip in Cartesian coordinate, where (x, y, z) and ξ represent the position and the attitude angle of the bucket, respectively.
- Joint-space $\mathbf{q}_j = [\theta_c, \theta_b, \theta_a, \theta_k]$: it is defined by four joint angles, including the angle between cabin and base θ_c, boom and cabin θ_b, arm and boom θ_a, and bucket and arm θ_k.
- Driving-space $\mathbf{q}_d = [\lambda_c, \lambda_b, \lambda_a, \lambda_k]$: it represents the displacement of the drivers. $\lambda_c = \theta_c$ denotes the rotation angle of the swing motor, and $\lambda_b, \lambda_a, \lambda_k$ indicate the displacements of the three hydraulic cylinders (boom, arm, and bucket).

In recent years, significant progress has been made in motion planning and trajectory optimization for automated excavators. Given the perceived state of the environment and the work task such as soil digging or rock moving, the motion planning problem involves locating the least-cost path in one workspace while satisfying certain constraints. Since the excavator's walking device is usually fixed during work, the planner typically focuses on the working device only. Most existing approaches [3–7] optimize variables either in task-space or joint-space, which are then transformed into driving-space and fed into the controller for excavator motion. The essence of these methods is to treat the excavator as a 4-DoF tandem manipulator, while ignoring the motion of the three hydraulic cylinders.

However, the relationship between the hydraulic drive positions and joint angles is non-linear, and smooth changes in joint-space trajectory may not correspond to smooth changes in driving-space variables, which is different from the rotary-driven manipulator. Thus, existing methods planning directly in joint-space or task-space cannot guarantee the stability of the hydraulic cylinder. Therefore, it is essential to develop a novel motion planning approach for the excavator that takes into account its operating characteristics and mechanical structure

In this study, a novel search-based motion planning method is proposed for the excavator. The core contributions of this work are as follows:

- The approach considers the unique operating characteristics and mechanical structure of the excavator by constructing a reachability map that incorporates drive-space during planning. It minimizes the traveled distance of the hydraulic cylinders, leading to smoother operation.
- A full reachability map-based motion planning approach for excavator is proposed. It is implemented in a typical excavator task, and experimental results demonstrate that the proposed planner can generate full-bucket motion for the excavator.

The proposed approach is expected to contribute significantly to the development of motion planning methods for excavators.

2   Related Works

2.1   Motion Planning Methods

In order to achieve the assigned task, an automated excavator employs sensor data, such as lidar or RGB-D camera, to perceive its environment. Subsequently, a feasible path that satisfies the higher-level task specifications while adhering to specific constraints can be derived. To this end, sampling-based methods, such as rapidly exploring random tree (RRT) [8] and Probabilistic roadmap (PRM) [9], have proven to be highly efficient in generating the path through random sampling. Nonetheless, due to their stochastic nature, these methods are ill-suited for our application, which necessitates frequent replanning as the excavator operates. Alternatively, search-based methods like Dijkstra [10] and A* [11] can solve this problem but are computationally demanding, particularly when utilized in conjunction with a large graph. In light of this, numerous variations have been developed based on these methods, and several studies [12,13] have investigated methods suitable for a dynamic environment.

In recent years, significant advances have been made in the field of motion planning and trajectory optimization for excavator. While most existing approaches focus on adhering to a desired or guided bucket position trajectory [3,4], some novel approaches have been proposed. Specifically, D. Jud et al. [5] utilize end-effector force-torque control to design a single dig cycle, which can be applied to various soil types. Similarly, E. Jelavic et al. [6] further consider the entire body of the excavator in their planning approach, enabling them to generate plans for locomotion over rough terrain. D. Lee et al. [7] develope a real-time optimization-based motion planning framework for hydraulic excavators, which can satisfy a range of constraints, including torque, power, and flow rate limits. It is worth noting that these methods treat the excavator as a 4-degree-of-freedom tandem manipulator, which is not entirely appropriate. Unlike ordinary manipulators, excavators have three closed kinematic chains and are driven linearly via three hydraulic cylinders. Therefore, planned paths based on the manipulator's structure may have a significant impact on the stability of hydraulic cylinder operation.

2.2   Reachability Map

As demonstrated in [14], researchers aim to employ internal maps or models to represent the reachability of their limbs for accomplishing diverse reaching tasks. Zacharias et al. [15] introduce a model known as the reachability map to characterize the reachability throughout the entire workspace. Building upon this work, Zacharias et al. [16,17] utilize the reachability map to position a mobile manipulator to execute linear constrained trajectories by matching a precomputed inverse kinematics database with discretized versions of the task path. However, since this kind of reachability map is based on positions, it fails to be transferred to a new frame when an extension frame is added to the end effector. To overcome this limitation, Dong et al. [18] proposed an orientation-based

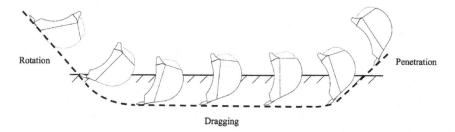

Fig. 1. Stage 2 of a typical excavation cycle. It consists of three progress including penetration, dragging, and rotation.

reachability map to enable reasonable placement of a robot's base. Furthermore, Makhal et al. [19] have leveraged the reachability map methodology to develop an open-source library named Reuleaux for robot reachability analyses and base placement. Similarly, Yang et al. [20] utilize the reachability map in the context of excavator motion planning and propose a novel compact reachability map representation based on the excavator's kinematic structure.

3   Proposed Method

3.1   Overview

The motion planning problems of excavators can be categorized into two types based on their application scenarios: planning in free space and planning during task execution. The former involves relatively few constraints, with the primary focus being on obstacle avoidance, while the latter requires the excavator to operate under a plethora of additional constraints, given that the bucket interacts directly with the soil. In this study, we employ trenching, a typical digging task, as an example to illustrate the planning process during task execution.

Trenching operations typically comprise three stages within a single digging cycle:

- In stage 1, the bucket tip moves towards the intended point of soil contact.
- In stage 2, the bucket tip cuts through the soil.
- In stage 3, the excavated soil is transported to the required area, such as above the truck, while ensuring that no soil is spilled.

During both stages 1 and 3 of the excavation process, the excavator's movement resembles that of free space motion, characterized by a minimal number of constraints, with obstacle avoidance being the primary goal. As illustrated in Fig. 1, stage 2 comprises the penetration of the bucket into the terrain, followed by dragging and rotation [21]. The constraints associated with this stage are discussed in Sect. 4.3.

Given that the excavator's walking device typically remains fixed during operation, our planning approach focuses solely on the working device. Moreover, as

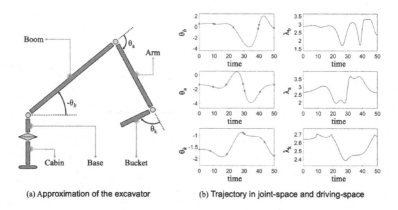

(a) Approximation of the excavator   (b) Trajectory in joint-space and driving-space

Fig. 2. Visualization of the traditional methods' approximation. (a) represents the 4-DoF tandem manipulator model used in these methods. (b) visualizes the change of joint angles and corresponding driver displacements. Planned trajectory by approximating the excavator to a manipulator can result in abrupt changes in drive-space.

the excavator's swing motion is often not involved in the excavation task, we plan the path in a 2-D task-space, specifically its sagittal plane. This approach can be extended to 3-D via a suitable transformation.

The excavator motion planning in 2-D task-space is as follows. Given an obstacle map which is the generated reachability map in our implementation, an initial pose of the bucket $\mathbf{p}_0 = (x_0, y_0, \theta_0)$, and a goal bucket pose $\mathbf{p}_g = (x_g, y_g, \theta_g)$, the motion planning problem is to find a safe, feasible, and near-minimal in both length and bucket trajectory represented using a sequence of poses $\mathbf{p}_0, \mathbf{p}_1, ..., \mathbf{p}_n$, with a certain deviation δ, where $\|\mathbf{p}_n - \mathbf{p}_g\| \leq \delta$. (x, y) and θ represent the location and orientation of the bucket tip, respectively. In this work, we do not model the bucker movement speed in our planning progress. Instead, we assume constant velocity and search for a feasible path at that pace.

3.2   Smoothness

In contrast to an ordinary manipulator, an excavator possesses three closed kinematic chains and is propelled linearly by three hydraulic cylinders. Despite the hydraulic drive position λ and joint angle θ having a 1-to-1 correspondence, the relationship between them is non-linear, particularly between θ_k and λ_k. As depicted in (b) of Fig. 2, we randomly selected five angles (blue points) for each joint-space parameter θ_a, θ_b, and θ_k, and interpolated cubic splines on these points (red curve). Then we individually transformed each angle to the drive-space using the geometric relationship and illustrated them. Notably, gradual changes in joint-space parameters can induce sudden alterations in drive-space.

A famous metric to evaluate the curve's curvature-variation is the integral of second order derivative squared as follows [22]:

$$\int \left[\frac{d^2}{dx^2} f(x) \right]^2 dx \ . \tag{1}$$

Inspired it, we develop a quantitative definition for the curve's smoothness, namely the length to curvature-variation ratio (LCR), which is defined as:

$$LCR = \frac{len(x)}{\int \left[\frac{d^2}{dx^2} f(x) \right]^2 dx} \ , \tag{2}$$

where $len(x)$ denotes the number of sampled points, and the integral is computed numerically. LCR can indicate the curvature changing level of the curves. A curve with a high LCR means high smoothness. In other words, its curvature varies gradually and without abruptness.

3.3   Reachablility Map

In the context of motion planning, it is imperative to account for the excavator's reachability and dexterity. To this end, the reachability map has emerged as a widely employed representation strategy for characterizing a manipulator's reachability across the entire workspace [15,18,19]. In this research, we propose an improved method for constructing a reachability map and apply it to the task of excavation.

Given that the excavator's working device largely operates within a two-dimensional plane, we construct a 2-D reachability map for the excavator, which is subsequently utilized to construct the graph. Although subtle differences may exist, the process of generating the fundamental reachability map typically entails three major steps, namely: workspace discretization, reachability assessment through inverse kinematics, and characterization of the workspace via the reachability map. Notably, we introduce a novel approach by encoding the hydraulic cylinder displacement within the reachability map, which diverges from traditional reachability mapping methodologies. The construction comprises the following steps:

- Discretizing: In this work, the excavation plane is discretized into tiny grids using 0.3 m × 0.3 m.
- Examining reachability: In each grid, N distributed points are generated equally on the inscribed circle. For each point obtained, we generate a pair sequence of bucket poses, in which the bucket tip direction point rotated in turn, as illustrated in (a) of Fig. 3. Then we use inverse kinematics (IK) to examine whether both poses are reachable.
- Characterizing the workspace: Once both valid IK solutions of a pair of poses are found, a line will be drawn in the visualization to mark it as reachable. Based on the parameters $[\gamma]$, the reachable directions of each grid can be finally represented by a sector area. Besides, we calculate the displacements

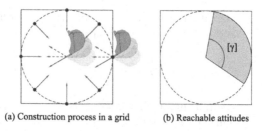

(a) Construction process in a grid   (b) Reachable attitudes

Fig. 3. Reachability map construction in a grid. The arrows in (a) denote bucket attitude angles. For each generated point on the inscribed circle, a pair of bucket poses are examined using IK. The grey area in (b) denotes reachable attitudes. It is parametrically represented as $[\gamma]$. (Color figure online)

of the three hydraulic cylinders between each pair of poses using geometric relationships, and record the average displacement d. The record displacements will be one of the cost functions which will be described in Sect. 3.5.

Eventually, the reachability map of the excavator can be represented as a four-dimensional array $\mathcal{M} = (x_i, y_i, \gamma_i, d_{ij})$, where (x_i, y_i) denotes the center position, and γ_i represents the all reachable heading angles of the i-th grid. d_{ij} denotes the average three hydraulic cylinders displacement of the j-th pair of poses in the i-th grid.

In practice, the resolution of the reachability map can be flexibly selected according to the situation, and the map only needs to be pre-generated for a particular excavator model.

3.4   Graph Construction

In this work, a variant of the well-known A* planner [11] is employed. To represent motion planning problems as graph search, a graph structure denoted as $G = (S, E)$ needs to be constructed, where S and E respectively represent the set of states of the graph and the transitions between them.

The conventional A* planner discretizes the search space, with nodes restricted to grid centers. The edges in E only permit the transfer of states between adjacent grid centers, resulting in a considerable amount of feasible space being overlooked. In contrast, we assign an approximately continuous state to each grid, defining a state s in S using the parameters $\mathbf{P}$ and γ_i in reachability map $\mathcal{M}$. Transitions in E are defined as a set of vectors representing the object's direction and distance, allowing the object to arrive anywhere in space rather than being limited to grid centers.

3.5   Cost Function

The cost of transition between pose $\mathbf{p}_{i-1}$ and $\mathbf{p}_i$ in graph can be represented as follows:

$$c(\mathbf{p}_{i-1}, \mathbf{p}_i) = c_{cell}(\mathbf{p}_{i-1}, \mathbf{p}_i) + w_{drive} * c_{drive}(\mathbf{p}_{i-1}, \mathbf{p}_i) , \tag{3}$$

where the latter term named driving cost is introduced to evaluate the movement of the hydraulic cylinders and is weighted with w_{drive}. Increasing the weight assigned to the actuation direction of the cylinder results in greater constraints on its changes, thereby effectively suppressing abrupt displacement changes in the hydraulic cylinder and promoting smoother motion. However, this comes at the cost of sacrificing optimality in path length.

$c_{cell}(\mathbf{p_{i-1}}, \mathbf{p_i})$ is computed by calculating the distance accumulated from the previous pose $\mathbf{p_{i-1}}$ to the current pose $\mathbf{p_i}$:

$$c_{cell}(\mathbf{p_{i-1}}, \mathbf{p_i}) = \sum_0^i ||x_i - x_{i-1}, y_i - y_{i-1}|| \, . \tag{4}$$

The driving cost $c_{drive}(\mathbf{p_{i-1}}, \mathbf{p_i})$ is used to assign different costs for each moving direction at a certain pose based on the average displacement of hydraulic cylinders calculated and stored in the reachability map, i.e.,

$$c_{drive}(\mathbf{p_{i-1}}, \mathbf{p_i}) = d \, . \tag{5}$$

Since the $2D$ map can be pre-generated and can be efficiently indexed due to its small size, $c_{drive}(\mathbf{p_{i-1}}, \mathbf{p_i})$ can not nearly increase the computational burden.

3.6   Heuristic

In this work, we define the heuristic $h(\mathbf{p_i})$ of a pose $\mathbf{p_i}$ as the sum of two components: $h_{p_i}(\mathbf{p_i})$ and $h_o(\mathbf{p_i})$, which estimate the predicted cost of reaching the desired end-effector position and orientation, respectively.

To compute $h_p(\mathbf{p_i})$, we ignore orientation to remove one dimensionality and reduce complexity. We use the A* planner to calculate the shortest distance from each grid center to the goal pose on the 2D obstacle map while avoiding obstacles. This distance is then used as $h_p(\mathbf{p_i})$, allowing for the introduction of obstacle information through a low-cost method that can guide the more expensive 3D search to avoid obstacles.

To achieve the desired bucket orientation at the goal pose, we introduce $h_o(\mathbf{p_i})$, which represents the difference between the current and desired orientations. One effective parameterization for describing this difference is through the axis-angle representation of a rotation [23]. Therefore, $h_o(\mathbf{p_i})$ is computed as the angle of rotation about a fixed axis specified by the axis-angle representation of the rotation between the bucket orientation of pose $\mathbf{p_i}$ and $\mathbf{p_g}$, i.e.,

$$h_o(\mathbf{p_i}) = ||\theta_g - \theta_i|| \, . \tag{6}$$

4   Experiment

In this section, we highlight the experimental results of the proposed planning method and its application in a typical single trenching cycle.

4.1 Implementation Details

In our experiment, we employed a model of an excavator with specific physical attributes, including a boom length of 5.70 m, an arm length of 2.93 m, a bucket length of 1.47 m, and a base height of 1.90 m. The excavator's hydraulic cylinders exhibit a range of motion extending from 1.8 m to 3.2 m, 2.1 m to 3.6 m, and 1.7 m to 2.8 m for the arm, boom, and bucket, respectively. We partitioned the workspace into discrete units with 0.3 m × 0.3 m resolution and specified a rotation resolution of 0.1π. The proposed planner was implemented using Matlab R2020a on a standard computer.

4.2 Motion Planning in Free Space

When planning tasks in free space, the initial and goal poses of the bucket, denoted by $\mathbf{p}0$ and $\mathbf{p}g$, are directly provided, which differs from planning during a task. Moreover, as the bucket traverses a longer distance with fewer constraints in free space than during trenching, the proposed heuristic and cost function can be evaluated more efficiently. Figure 4 compares different heuristics and cost functions. The planner that uses only Euclidean distance as the heuristic expands 436 nodes in (a). The incorporation of $h_p(\mathbf{p}_i)$, which considers the orientation of the bucket, is a significant improvement as it expands only 329 nodes (b). However, the use of $h_p(\mathbf{p}i)$ alone can make it challenging to achieve the desired bucket orientation at the goal pose. In addition, the introduction of the heuristic $h_0(\mathbf{p}i)$ results in an expanded planner of only 312 nodes, and it can efficiently obtain the proper bucket orientation together with the position. Further removal of the driving cost $cdrive(\mathbf{p}i-1, \mathbf{p}_i)$ leads to wasteful exploration of nodes, as evidenced by the expansion of 320 nodes in (c).

To evaluate the effect of the driving cost $c_{drive}(\mathbf{p}_{i-1}, \mathbf{p}i)$ on the smoothness of the hydraulic cylinder movement, we sampled 1000 groups of starting and ending poses and used two planners (one with the driving cost and one without) to plan paths, respectively. For each planned result, we calculated the hydraulic cylinder displacement and the LCR using Eq. 2. As shown in Table.1, the introduction of the driving cost $cdrive(\mathbf{p}_i)$ improved the smoothness of the arm, boom, and bucket hydraulic cylinders by 10.4%, 1.7%, and 2.3%, respectively.

Table 1. Comparison of planners without c_{drive} and with c_{drive}.

Planners	LCR_{λ_b}	LCR_{λ_a}	LCR_{λ_k}
without c_{drive}	7193.70	700.41	770.63
with c_{drive}	7942.33	712.26	753.06

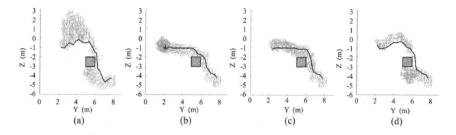

Fig. 4. Comparison of different heuristic and cost function. The black line and rectangle represent the planned path and obstacle, respectively, and the arraws represent the bucket's poses expanded by the planner.

4.3   Motion Planning During Trenching

As outlined in Sect. 3.1, a single digging operation typically comprises three stages. While the planning tasks in stages 1 and 3 resemble those in free space, our focus in this subsection is on addressing the motion planning challenge in stage 2.

To prevent the bucket's side plates from pressing against the soil, we impose constraints on the movement of the excavator's cabin section. The planning problem in stage 2 can be formulated as follows: given an approach position at which the bucket enters the earth's surface and a target terrain shape (assumed to be the horizontal plane without loss of generality), the proposed planner aims to identify a feasible bucket trajectory in the Cartesian coordinate system. Simultaneously, we prioritize achieving a large excavation volume that can be excavated by the volume of soil above the bucket edge [5]. The bucket tip velocity direction is denoted by β, while the bucket pose is indicated by α. It is worth noting that β differs from the bucket's attitude angle, ξ. The whole planning process operates on the reachability map-based graph and is influenced by the following constraints:

- Throughout the process, it is imperative that β increases in a monotonic fashion to prevent the displacement of debris from the bucket. Furthermore, a positive value for α is necessary to prevent the sections of the bucket aside from the tip from exerting pressure against the ground.
- During the penetration phase, the monotonic increase of α is necessary to prevent the bucket face from exerting pressure into the soil.
- During the dragging phase, a limit of π is imposed on β_{max}, which must be smaller than β. The bucket is then horizontally dragged towards the excavator until the excavated volume equals the bucket's capacity.
- In the rotation process, the bucket must be rotated upwards with a simultaneous translation until it reaches a nearly level position. The constraint $\alpha > 0$ ensures that the bottom and back of the bucket do not exert pressure against the soil.

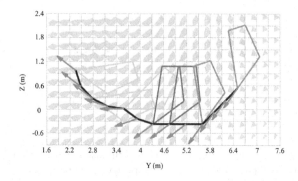

Fig. 5. The planned path in stage 2 including penetration, dragging and rotation, using proposed reachability map-based motion planner. The arrows indicate the direction of the bucket, which lies within the light blue reachable range. (Color figure online)

Figure 5 illustrates the planned trajectory for a single excavation operation incorporating the advancement of penetration, dragging, and rotation. As seen in the figure, the orientation of the bucket falls within the feasible range as defined by the reachability map.

5   Conclusions

In this work, a novel search-based motion planning methodology for excavators is proposed. The approach involves constructing a graph representation that encodes feasible excavator bucket poses and movement direction based on the excavator's reachability map. Moreover, the excavator's drive-space is accounted for through the introduction of drive cost, facilitating smoother operation of hydraulic cylinders. The proposed planner is evaluated in free space and during a typical digging operation, demonstrating its ability to generate full-bucket motion for excavators via experimental results.

Numerous avenues for future research are apparent. Presently, our approach has only been validated through simulation; next, we aim to validate our approach for practical implementation on real excavator hardware platforms. Furthermore, our current focus has been primarily on kinematic constraints; we intend to incorporate dynamic constraints into our method in future research.

References

1. Singh, S.: State of the art in automation of earthmoving. J. Aerosp. Eng. **10**(4), 179–188 (1997)
2. Yu, H., Liu, Y., Hasan, M.S.: Review of modelling and remote control for excavators. Int. J. Adv. Mech. Syst. **2**(1–2), 68–80 (2010)
3. Ha, Q., Nguyen, Q., Rye, D., Durrant-Whyte, H.: Impedance control of a hydraulically actuated robotic excavator. Autom. Constr. **9**(5), 421–435 (2000)

4. Tafazoli, S., Salcudean, S., Hashtrudi-Zaad, K., Lawrence, P.: Impedance control of a teleoperated excavator. IEEE Trans. Control Syst. Technol. **10**(3), 355–367 (2002)
5. Jud, D., Hottiger, G., Leemann, P., Hutter, M.: Planning and control for autonomous excavation. IEEE Robot. Autom. Lett. **2**(4), 2151–2158 (2017)
6. Jelavic, E., Hutter, M.: Whole-body motion planning for walking excavators. In: Proceedings of the 2019 IEEE/RSJ International Conference on Intelligent Robots and Systems (IROS), pp. 2292–2299 (2019)
7. Lee, D., Jang, I., Byun, J., Seo, H., Kim, H.J.: Real-time motion planning of a hydraulic excavator using trajectory optimization and model predictive control. In: Proceedings of the 2021 IEEE/RSJ International Conference on Intelligent Robots and Systems (IROS), pp. 2135–2142 (2021)
8. LaValle, S.M., James, J., Kuffner, J.: Randomized kinodynamic planning. Int. J. Robot. Res. **20**(5), 378–400 (2001)
9. Kavraki, L., Svestka, P., Latombe, J.C., Overmars, M.: Probabilistic roadmaps for path planning in high-dimensional configuration spaces. IEEE Trans. Robot. Autom. **12**(4), 566–580 (1996)
10. Dijkstra, E.W.: A note on two problems in connexion with graphs. Numer. Math. **1**(1), 269–271 (1959)
11. Hart, P.E., Nilsson, N.J., Raphael, B.: A formal basis for the heuristic determination of minimum cost paths. IEEE Trans. Syst. Sci. Cybern. **4**(2), 100–107 (1968)
12. Koenig, S., Likhachev, M.: Fast replanning for navigation in unknown terrain. IEEE Trans. Rob. **21**(3), 354–363 (2005)
13. Stentz, A.: Optimal and efficient path planning for partially known environments. In: Intelligent Unmanned Ground Vehicles: Autonomous Navigation Research at Carnegie Mellon, pp. 203–220. Springer, US (1997). https://doi.org/10.1007/978-1-4615-6325-9_11
14. Kawato, M.: Internal models for motor control and trajectory planning. Curr. Opin. Neurobiol. **9**(6), 718–727 (1999)
15. Zacharias, F., Borst, C., Hirzinger, G.: Capturing robot workspace structure: representing robot capabilities. In: Proceedings of the 2007 IEEE/RSJ International Conference on Intelligent Robots and Systems (IROS), pp. 3229–3236 (2007)
16. Zacharias, F., Borst, C., Beetz, M., Hirzinger, G.: Positioning mobile manipulators to perform constrained linear trajectories. In: Proceedings of the 2008 IEEE/RSJ International Conference on Intelligent Robots and Systems (IROS), pp. 2578–2584 (2008)
17. Zacharias, F., Sepp, W., Borst, C., Hirzinger, G.: Using a model of the reachable workspace to position mobile manipulators for 3-d trajectories. In: Proceedings of the 2009 9th IEEE-RAS International Conference on Humanoid Robots (Humanoid), pp. 55–61 (2009)
18. Dong, J., Trinkle, J.C.: Orientation-based reachability map for robot base placement. In: Proceedings of the 2015 IEEE/RSJ International Conference on Intelligent Robots and Systems (IROS), pp. 1488–1493 (2015)
19. Makhal, A., Goins, A.K.: Reuleaux: Robot base placement by reachability analysis. In: Proceedings of the 2018 Second IEEE International Conference on Robotic Computing (IRC), pp. 137–142 (2018)
20. Yang, Y., Zhang, L., Cheng, X., Pan, J., Yang, R.: Compact reachability map for excavator motion planning. In: Proceedings of the 2019 IEEE/RSJ International Conference on Intelligent Robots and Systems (IROS), pp. 2308–2313 (2019)
21. Singh, S.: Synthesis of Tactical Plans For Robotic Excavation. Ph.D. thesis, Carnegie Mellon University, Pittsburgh, PA (January 1995)

22. Moreton, H.P.: Minimum curvature variation curves, networks, and surfaces for fair free-form shape design. Ph.D. thesis, University of California, Berkeley (1992)
23. Murray, R.M., Li, Z., Sastry, S.S.: A mathematical introduction to robotic manipulation. CRC Press (2017)

Reinforced Vision-and-Language Navigation Based on Historical BERT

Zixuan Zhang[1], Shuhan Qi[1,2,3](✉) [iD], Zihao Zhou[1], Jiajia Zhang[1], Hao Yuan[1], Xuan Wang[1], Lei Wang[4], and Jing Xiao[4]

[1] Harbin Institute of Technology, Shenzhen, China
shuhanqi@cs.hitsz.edu.cn
[2] Peng Cheng Laboratory, Shenzhen, China
[3] Guangdong Provincial Key Laboratory of Novel Security Intelligence Technologies, Shenzhen, China
[4] Ping An Technology (Shenzhen) Co., Ltd., Shenzhen, China

Abstract. Vision-and-language navigation(VLN) task needs an agent navigate to a target location under the guidance of natural language instructions given by humans in novel 3D environments. We mainly study the vision-and-language navigation task in indoor scenes. It is difficult to obtain a large number of labeled training data, thus traditional multimodal methods perform poorly in the data-scarce environment. We introduce pre-trained BERT model into vision-and-language navigation task to extract environment features and reduce the computation consumption by fixing the instruction features during navigation. Meanwhile, we design Gated Recurrent Vision-and-Language BERT to effectively memorize and transmit historical information to alleviate memory loss problem during navigation. In addition, we conduct reinforcement learning and imitation learning simultaneously to address the exposure bias of imitation learning. Empirically, we conduct experiments in discrete environment and analyze experimental results to verify the effectiveness of our proposed method.

Keywords: Vision-and-language navigation · Pre-trained model · Reinforcement learning · Imitation learning

1 Introduction

It has always been a human dream to build a general robot that can complete the tasks assigned by humans in the process of communicating with humans in

This research was funded in part by Ministry of Science and Technology of China under Grant 2020AAA0104200, the Guangdong Provincial Key Laboratory of Novel Security Intelligence Technologies under Grant 2022B1212010005, Key Fields Research of Guangdong Province 2020B0101380001,the Shenzhen Foundational Research Funding under Grant JCYJ20200109113427092, JCYJ20220818102414030, in part by the PINGAN-HITsz Intelligence Finance Research Center, in part by the Ricoh-HITsz Joint Research Center, and in part by the GBase-HITsz Joint Research Center. The computing resources of Pengcheng Cloud Brain are used in this research.

Y. Tan et al. (Eds.): ICSI 2023, LNCS 13969, pp. 427–438, 2023.
https://doi.org/10.1007/978-3-031-36625-3_34

natural language. Now, with the great breakthroughs in the fields of computer vision and natural language processing, this dream has gradually become reality. Vision-and-language navigation is the basis to realize general robot.

Vision-and-language navigation is an decision-making problem under the navigation instructions given by humans in 3D scenes with incomplete environment information. The most critical and difficult part of the task is how an intelligent agent could better perceive the environment and make correct decisions. In the process of navigation, the agent needs to align textual instruction with visual observation with multi-modal fusion methods and then make action decision in terms of reasoning and grounding about fused information. The agent needs strong perceptual, memory and reasoning capabilities. In addition, there still exist a large performance gap between seen and unseen environments. Hence, it crucial to improve generalization ability of the model so that the navigation agent's performance is insensitive to different environments.

As an emerging and challenging research field, vision-and-language navigation has attracted interest of many researchers. Some researchers focus on multimodal information fusion such as multi-head attention [12]. Meanwhile, some researchers utilize Reinforcement Learning (RL) and Imitation Learning (IL). RL allows the agent to explore the state-action space outside the demonstration path and is used to balance exploitation and exploration when learning to navigate. Some methods such as Reinforced cross-modal matching(RCM) [9] are proposed to enhance agent's generalization ability.

Previous methods use traditional encoder-decoder models in cross-modal fusion, which need a large-scale and high-quality annotated data. However, it's hard to collect adequate high-quality data in 3D environments for training in vision-and-language navigation field. Thus we introduce pre-trained BERT model to address the lack of data and revise self-attention calculation in BERT model according to the characteristics of VLN to further reduce calculation and speed up computation. Since vision-and-language navigation could be modeled as a partially observable Markov decision-making process, in which the future observations depend on the current state and action of the agent. Meanwhile, only partial sub-instructions are relevant to the current observation and decision-making at each navigation step. Thus the agent needs to localize the relevant partial sub-instructions according to navigation progress in trajectory history, which requires the agent to memorize historical information. Therefore, we append gated recurrent neural network with residual connection to BERT model to handle memory loss problem. Furthermore, we incorporate reinforcement learning with imitation learning to enhance navigation agent's generalization capacity.

2   Vision-and-Language Navigation Based on Pre-training

One of the major challenges of vision-and-language navigation tasks is the acquisition of large-scale and high-quality annotated datasets. Pre-trained BERT models can reduce the demand of annotated data. However, because BERT models have many parameters which consume a lot of training resources and hinder the transmission of historical information, the appliance of pre-trained BERT in vision-and-language navigation is limited.

In order to address the problem of large resource consumption, we revise the BERT structure according to the characteristics of vision-and-language navigation tasks to reduce the calculation and enhance the agent's performance. As for the difficulty of memorizing historical information, we propose to add gated recurrent neural networks with residual connection to BERT model to memorize and transmit historical information, which alleviates the problem of memory loss. At the same time, with regard to exposure bias in imitation learning, we simultaneously conduct imitation learning and reinforcement learning to enhance the generalization ability of the agent.

2.1 Pre-trained Multi-modal BERT Model for VLN

The quality and scale of the labeled dataset substantially determine the performance of trained model in artificial intelligence. However, labeled datasets are hard to obtain or annotate and need to be reconstructed for different tasks.

Pre-training refer to training on a large scale of in-domain data to learn the general knowledge in the domain, and then fine-tuning on the labeled data of specific downstream tasks in the domain to transfer and reuse the general knowledge, and thus enhance the performance of the model.

We preprocess the pictures and text respectively and input them into Transformer. Text is represented by one-hot vector, and then input into the word embedding network to obtain token embedding. We encode the position of each word as position embedding(PE) and then concatenate PE with token embedding to provide position information.

We initialize parameters with Prevalent pre-trained model [10], which is the first pre-trained model focusing on vision-and-language navigation task. Prevalent is a self-supervised pre-training model that uses language instructions and visual observations to extract features and then combines them to improve navigation agents.

2.2 Recurrent BERT Model for Vision-and-Language Navigation

Previous multi-modal BERT pre-trained models are mainly used in static environment. We propose a Gated Recurrent Vision-and-Language Navigation BERT model(GRVLN-BERT), which uses an extra gated recurrent neural network(GRU) to pass historical information. The model fixes the language features and only uses them as Keys and Values for attention calculation to reduce computation during navigation. The model structure is shown in Fig. 1, which consists of instruction feature initialization, image feature process, and multi-modal information fusion.

Navigation Instruction Initialization Based on BERT Model. In a vision-and-language navigation task, assuming the vocabulary set is D, the navigation instruction $U = w_0, w_1...w_n, w_i \in D$ is given to the agent at the beginning, and U remains unchanged during whole navigation episode. Therefore, at the

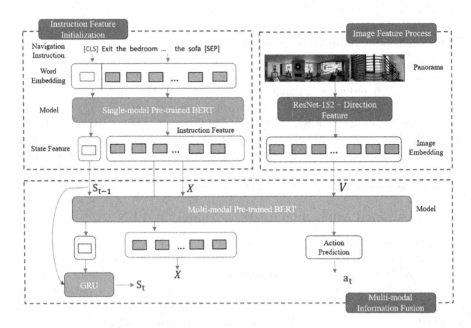

Fig. 1. The Structure of Gated Recurrent Vision-and-language Navigation BERT

initial state s_0, the start token [CLS] and the separation token [SEP] are added to the instructions: $U' = [CLS], w_0 w_1 ... w_n, [SEP]$. Then U' is input into the pre-trained BERT model.

$$s_0, X = GRVLN - BERT([U']) \quad . \tag{1}$$

As shown in Eq. 1, s_0 is the initial state. We regard the output embedding vector of [CLS] token as initial state representation, and X is the instruction feature encoded by GRVLN-BERT model. The navigation instruction U' will not change in the subsequent navigation steps. Therefore, X is directly used as the instruction feature input without update in the subsequent steps. We localize the relevant partial sub-instruction at the current step by calculating attention scores over the image features and historical information.

Image Feature Processing Based on BERT Model. The navigation agent receives local observations of current position continually as agent moves in the environment. In Room-to-Room environment, the visual input of the agent is a 360-degree panorama composed of 36 images, obtained at 12 horizontal directions and 3 vertical angles. In order to contain the visual feature and directional feature of each image, the feature of each image consists of 2048-dimensional visual feature s_v extracted by pre-trained ResNet network and 128-dimensional directional feature s_p composed of $[sin\psi; cos\psi; sin\omega; cos\omega]$. Therefore, the feature of each image is 2176-dimensional vector $s_i = [s_v^i, s_p^i]$, and we transform it into the dimension suitable for BERT input through a fully-connected layer.

BERT model extract features through self-attention mechanism, which calculates self-attention scores over all values. As shown in Eq. 2, if the vector of the current layer is h_{t-1}, the Query, Key and Value for self-attention calculation are all h_{t-1} in BERT's self-attention. In this way, each feature of h_t is extracted from all features of the previous layer h_{t-1} through self-attention mechanism.

$$h_t = softmax(\frac{QK^T}{\sqrt{(d_h)}})V, Q = K = V = h_{t-1} \quad . \tag{2}$$

In vision-and-language navigation tasks, $h_t = [s_t, X, V_t]$. The instruction will not change throughout the navigation period, and the model places emphasis on matching and understanding the language instruction in terms of current observation. Therefore, it is not necessary to use the instruction to extract the image features, which saves a lot of computation resources. The purpose of vision-and-language navigation is to make action decisions according to current observation panorama. As a long-horizon problem, current local observation only relates to partial sub-instruction, thus we refine features of relevant sub-instruction by calculate self-attention scores over instruction feature with image feature. Furthermore, state s_t is considered as the carrier of historical information, so its feature is also extracted from all features of the previous layer to memorize trajectory history.

$$X = X \quad . \tag{3}$$

$$V_{t+1} = softmax(\frac{QK^T}{\sqrt{(d_h)}})V, Q = V_t, K = h_t, V = h_t \quad . \tag{4}$$

$$s_{t+1} = softmax(\frac{QK^T}{\sqrt{(d_h)}})V, Q = s_t, K = h_t, V = h_t \quad . \tag{5}$$

Instruction feature X only serves as Key and Value for attention calculation to extract visual feature and state feature, while its own value will not change in the GRVLN-BERT model. The navigational action decision of the agent is not to make specific action, such as going ahead or turning left, but to select an image with the highest probability in the panorama comprised of 36 images and move towards it. This form of action decision-making is more more semantically informative and interpretable. And the attention layer of the multi-modal BERT model calculates the attention weights over each image composed of the panorama, which is consistent with the form of outputting the probabilities of selecting each image. Therefore, we directly use the attention weights on each image at the last multi-head attention layer as the decision probabilities.

In vision-and-language navigation tasks, the historical state refers to the summary of navigation instructions, historical navigation visual trajectory information, action decision sequences, etc. Although navigation is a partially observable Markov decision-making process, it is critical to record historical information for predicting the current navigation progress and correctly extracting key sub-instructions for decision-making. Utilizing only current image input to extract the related sub-instruction without historical state information degrades the inference ability of navigation agent. When encountering repetitive and similar scenes, the agent could not distinguish them so that make the same decisions at different stages of navigation. For instance, there are two similar rooms in the indoor scene, and the navigation instruction indicate the agent to turn left in the first room and turn right in the second room. However, without the assistance of historical information, the agent could not distinguish its current position, and has difficulty making correct navigation action.

We propose a method for recording and transferring historical state information in the BERT model. In the BERT model, the output corresponding to token [CLS] is generally used for specific tasks, such as classification. We consider the output corresponding to token [CLS] as state information s. Due to the self-attention mechanism of BERT, s carries the features of image input and navigation command at each navigation time step.

Previous methods directly utilize the output of BERT at last time step as new input of s_t, so that the model cannot distinguish important information to memorize and noise information to discard. Furthermore, when the historical state information is inputted into BERT, after the calculation of multiple self-attention layers, some historical information will be lost. Therefore, it is difficult for long-horizon sequential problem like vision-and-language to preserve historical information for a long time.

We add Gated Recurrent Neural Networks (GRU) [4] to the output state s in BERT model to address historical information loss problem. GRU utilize several control gate to control whether to preserve or discard some state information rather than directly transmit hidden state like recurrent neural network. The multiple gate mechanisms in GRU ensure early historical information can be preserved well in long-horizon sequential problem. Hence, GRU could solve the problem of long-term dependence well. We utilize skip connection to address the memory loss issue. We directly connect the state input to the hidden state input of GRU and preserve historical state information through skip connection.

In conclusion, the algorithm flow of our proposed Gated Recurrent Vision-and-Language Navigation BERT model is shown in Algorithm 1.

2.3   Training with Reinforcement Learning and Imitation Learning

Since vision-and-language navigation is long-horizon partially observable Markov decision-making process, there exists exposure bias problem. The root cause of exposure bias is inconsistency between sample spaces of expert demonstration and navigation agent policy. We propose We proposed a training method combining reinforcement learning and imitation learning, which simultaneously conducts both imitation learning and reinforcement learning. This combination makes full use of the advantages of faster convergence of imitation learning and better generalization of reinforcement learning.

Algorithm 1 Gated Recurrent Vision-and-Language Navigation BERT

Input: environment ENV, navigation agent GRVLN-BERT

  repeat

    Start a navigation episode and obtain navigation instruction

    Initialize time-step $t = 0$ and starting position pos_0

    Extract initial state feature and instruction feature $s_0, X = GRVLN - BERT([I])$

    repeat

      $O_t = ENV(pos_t)$ Receive environment observation at position pos_0

      $a_t = GRVLN - BERT([s_t, X, O_t])$ Predict action

      $s_{t+1}, pos_{t+1} = ENV(a_t)$ Perform action

      $t \leftarrow t + 1$

    until Navigation agent makes STOP action or t exceeds pre-defined threshold

  until Navigation agent GRVLN-BERT ends training

Output: navigation agent GRVLN-BERT

Imitation learning utilizes simple behavior cloning, which fits the expert's policy through minimizing the cross-entropy loss function:

$$L^{IL} = -\lambda \sum_t a_t^* log(p_t^a) \quad . \tag{6}$$

where a_t^* is expert's action and p_t^a is the probability that agent takes action a.

Reinforcement learning makes agent explore in an environment and self-update based on the reward given by the environment, which effectively increases the generalization of the agent. We use Advantage Actor-Critic (A2C) algorithm, whose objective function is to maximize the cumulative reward:

$$L^{RL} = -\sum_t a_t^s log(p_t^a) A_t \quad . \tag{7}$$

where A_t is advantage function, a_t^s is agent's action and p_t^a is the probability that agent takes action a.

Reinforcement learning and imitation learning share the network parameters of navigation agent. We sample expert demonstration and agent trajectory respectively during training. We calculate reinforcement learning loss and imitation learning loss individually, and the update agent's parameters with gradient descent over weighted total loss. This approach is equivalent using imitation learning to regularize reinforcement learning to accelerate the convergence of reinforcement learning algorithms.

$$L = L^{RL} + \lambda_{IL} L^{IL} \quad . \tag{8}$$

where λ_{IL} is the parameter that handles the weight of imitation learning.

3   Experiments

3.1   Dataset and Simulator for Indoor Scene Vision-and-Language Navigation

Matterport3D Simulator [2] is a software framework for various computer vision tasks, constructed using MatterPort3D panoramic RGB-D image dataset.

Room-to-Room is the first vision-and-language dataset consisting of navigation instructions and trajectories scene constructed by annotating navigation instruction-trajectory pairs in Matterport3D simulator. Each trajectory is a sequence of viewpoints in Matterport3D simulator and corresponds to 3 instructions.

3.2   Evaluation Metrics

We utilize some popular evaluation metrics for vision-and-language navigation tasks to measure the performance of navigation agent:

- Trajectory Length(TL): total length of trajectory in meters.
- Navigation Error(NE): the average distance between target location and agent's ultimate position in meters.
- Success Rate(SR): The ratio of the endpoint within a radius of 3 meters from the target location.
- Oracle Success Rate(OSR): The ratio of the nearest point among the agent's entire path within a radius of 3 meters from the target location.
- Success rate weighted by inverse Path Length(SPL):

$$SPL = \frac{1}{N} \sum_{i=1}^{N} S_i \frac{l_i}{max(p_i, l_i)} \quad . \tag{9}$$

As shown in Eq. 9, $S_i = 1$ if the navigation agent succeeds, and $S_i = 0$ if it fails. l_i is the shortest path length from the starting point to the terminus, and p_i is the path length of the actual trajectory that the navigation agent moves.

3.3   Experiment Analysis

The validation set of Room-to-Room is split into val-seen dataset and val-unseen dataset. Val-seen validation set refers to the indoor scenes that have appeared in the training set, and val-unseen validation set refers to the indoor scenes that the agent has never seen before. The test set of Room-to-Room only contains navigation instructions. The official of Room-to-Room dataset provides an evaluation server. After performing the navigation agent locally to collect the trajectories corresponding to the instructions in the test set, we upload these trajectories to the server for evaluation.

We initialize the parameters of GRVLN-BERT with Prevelent pre-trained model and simultaneously conduct both imitation learning and reinforcement

Table 1. Comparison among experiment results in val-seen validation environments

	TL	NE	SR	SPL
Random	9.58	9.45	0.16	–
Seq2Seq [1]	11.33	6.01	0.39	–
Speak-follower [3]	–	3.36	0.66	–
SMNA [7]	–	3.22	0.67	0.58
RCM+SIL [9]	10.65	3.53	0.67	–
EnvDrop [8]	11	3.99	0.62	0.59
AuxRN [11]	-	3.33	0.7	0.67
PREVALENT [10]	10.32	3.67	0.69	0.65
RelGraph [5]	10.13	3.47	0.67	0.65
VLNBERT [6]	11.17	3.16	0.70	0.65
GRVLN-BERT(ours)	**11.08**	**2.58**	**0.75**	**0.71**

learning.We set the learning rate to 10^{-5} and use the AdamW optimizer to train navigation agent for 300,000 epochs in total. Table 1 demonstrates the experimental results in val-seen validation set. The SPL indicator of GRVLN-BERT model increases from 0.65% to 0.71%, and the success rate SR improves from 0.70% to 0.75%. Under the condition of single-run without auxiliary data augmentation, the navigation agent achieves the state-of-the-art performance in NE, SR, and SPL metrics. Therefore, the memory and transfer of historical information greatly enhance the model's reasoning and grounding ability to help the agent make correct navigation action decisions.

As shown in Table 2, the navigational agent improves its success rate by 1% and reaches lower navigation error rate. Meanwhile, we find that the GRVLN-BERT model has a performance gap between val-seen and val-unseen environ-

Table 2. Comparison among experiment results in val-unseen validation environments

	TL	NE	SR	SPL
Random	9.77	9.23	0.16	–
Seq2Seq [1]	8.39	7.81	0.22	–
Speak-followe r[3]	–	6.62	0.35	–
SMNA [7]	-	5.52	0.45	0.32
RCM+SIL [9]	11.46	6.09	0.43	–
EnvDrop [8]	10.7	5.22	0.52	0.48
AuxRN [11]	–	5.28	0.55	0.50
PREVALENT [10]	10.19	4.71	0.58	0.53
RelGraph [5]	9.99	4.73	0.57	0.53
VLNBERT [6]	11.63	4.13	0.61	0.56
GRVLN-BERT(ours)	**12.49**	**3.81**	**0.62**	**0.56**

ments, and intend to enhance the generalization ability of GRVLN-BERT model in future research.

In the end, GRVLN-BERT agent performs navigation tasks in test environments with only given instructions. Meanwhile, we collect the agent's trajectories corresponding to give instructions and upload instruction-trajectory pairs to the test server we mentioned before to measure the agent's performance. As shown in Table 3, the GRVLN-BERT model performs better than previous models on each indicator. Especially, it reaches a better performance with an increase of 2% than the previous best model on success rate indicator.

Table 3. Comparison among experiment results in test environments

	TL	NE	SR	SPL
Random	9.89	9.79	0.13	0.12
Seq2Seq [1]	8.13	7.85	0.20	0.18
Speak-follower [3]	14.82	6.62	0.35	0.28
SMNA [7]	18.04	5.67	0.48	0.35
RCM+SIL [9]	11.97	6.12	0.43	0.38
EnvDrop [8]	11.66	5.23	0.51	0.47
AuxRN [11]	–	5.15	0.55	0.51
PREVALENT [10]	10.51	5.3	0.54	0.51
RelGraph [5]	10.29	4.75	0.55	0.52
VLNBERT [6]	11.68	4.35	0.61	0.57
GRVLN-BERT(ours)	**12.78**	**3.96**	**0.63**	**0.57**

3.4 Ablation Study

In order to verify the effects of different components of GRVLN-BERT model, we conduct detailed ablation studies. We study the impact of using instruction feature only as Key and Value in the cross-modal attention calculation(abbr.insKV) on navigation agent's performance and computational speed. As shown in Table 4, the instruction features extracted by the BERT module at the initial navigation stage has rich semantic information.

Table 4. Results about insKV module (val-seen,val-unseen)

	TL	NE	OSR	SR	SPL
GRVLN-BERT	11.08,12.49	2.58,3.81	0.79,0.70	0.75,0.62	0.71,0.56
GRVLN-BERT-noninsKV	14.57,15.27	3.98,4.80	0.66,0.60	0.60,0.52	0.55,0.45

We study the effect of historical information on the model's grounding ability. As shown in Table 5, GRVLN-BERT(normal) is the model using conventional state information transfer method without GRU module, and GRVLN-BERT(nonhistory) is the model without historical memory. We find that historical information can effectively increase the model's performance. Furthermore, our proposed historical information transfer based on RNN could enhance memory capacity and inference ability of the model.

Table 5. Results about memory module (val-seen,val-unseen)

	TL	NE	OS	SR	SPL
GRVLN-BERT	11.08,12.49	2.58,3.81	0.79,0.70	0.75,0.62	0.71,0.56
GRVLN-BERT (normal)	11.17,11.63	3.16,4.13	0.75,0.67	0.70,0.61	0.65,0.56
GRVLN-BERT (nonhistory)	10.76,10.05	3.64,4.91	0.73,0.60	0.68,0.57	0.63,0.53

We further analyze the effects of different training methods. As shown in Table 6, incorporating imitation learning and reinforcement learning combines the advantages of the both, so that it has excellent performance in val-seen scenarios and preferable generalization ability in val-unseen scenarios.

Table 6. Results about training method (val-seen,val-unseen)

	TL	NE	OSR	SR	SPL
GRVLN-BERT	11.08,12.49	2.58,3.81	0.79,0.70	0.75,0.62	0.71,0.56
GRVLN-BERT(IL)	9.62,9.27	2.92,4.88	0.77,0.60	0.72,0.54	0.70,0.52
GRVLN-BERT(RL)	14.21,14.01	3.80,4.62	0.71,0.64	0.63,0.56	0.57,0.49

4   Conclusion

In this paper, we introduce the pre-trained BERT model into vision-and-language navigation task. We revise the input and attention calculation of the BERT model according to the characteristics of vision-and-language navigation task to improve the agent's reasoning and grounding efficiency. In order to memorize and transmit historical trajectory information in the BERT model, we utilize gated recurrent neural network with residual connection to transmit hidden state and address the problem of memory loss. In order to solve the exposure bias issue of training navigation agent with imitation learning, a training method combining reinforcement learning and imitation learning is proposed to effectively enhance the agent's generalization ability. In the end, we find that GRVLN-BERT model achieve state-of-the-art result in some evaluation metrics through experiments and analysis, which proves the effectiveness of our proposed method.

References

1. Anderson, P., et al.: Vision-and-language navigation: 31st meeting of the ieee/cvf conference on computer vision and pattern recognition. In: Proceedings of the 2018 IEEE/CVF Conference on Computer Vision and Pattern Recognition (Dec. 2018), Piscataway, NJ, pp. 3674–3683. Institute of Electrical and Electronics Engineers (IEEE)
2. Chang, A., et al.: Matterport3D: Learning from RGB-D data in indoor environments. In: Proceedings of the 2017 International Conference on 3D Vision (United States, 2018), Institute of Electrical and Electronics Engineers Inc., pp. 667–676
3. Fried, D., et al.: Speaker-follower models for vision-and-language navigation. In In: Proceedings of the 32nd International Conference on Neural Information Processing Systems, Red Hook, NY, USA, pp. 3318–3329. Curran Associates Inc. (Dec 2018)
4. Hochreiter, S., Schmidhuber, J.: Long Short-Term Memory. In: Neural Computation, vol. 9(8), pp. 1735–1780 (1997)
5. Hong, Y., Rodriguez-Opazo, C., Qi, Y., Wu, Q., Gould, S.: Language and Visual Entity Relationship Graph for Agent Navigation. arXiv:2010.09304 [cs] (Dec 2020)
6. Hong, Y., Wu, Q., Qi, Y., Rodriguez-Opazo, C., Gould, S.: VLN BERT: a recurrent vision-and-language BERT for navigation. In: Proceedings of the 2021 IEEE/CVF Conference on Computer Vision and Pattern Recognition, Nashville, TN, USA, pp. 1643–1653. IEEE (June 2021)
7. Ma, C.-Y., et al.: Self-Monitoring Navigation Agent via Auxiliary Progress Estimation. arXiv:1901.03035 [cs] (Jan 2019)
8. Tan, H., Yu, L., Bansal, M.: Learning to navigate unseen environments: back translation with environmental dropout. In: Proceedings of the 2019 Conference of the North American Chapter of the Association for Computational Linguistics: Human Language Technologies, vol. 1 (Long and Short Papers), Minneapolis, Minnesota, pp. 2610–2621. Association for Computational Linguistics (June 2019)
9. Wang, X., et al.: Reinforced cross-modal matching and self-supervised imitation learning for vision-language navigation. In: Proceedings of the IEEE Conference on Computer Vision and Pattern Recognition, CA, USA, pp. 6629–6638. Computer Vision Foundation/IEEE (2019)
10. Hao, W., Chunyuan Li, X. L. L. C. J. G.: Towards learning a generic agent for vision-and-language navigation via pre-training. In: Computer Vision and Pattern Recognition (Feb 2020)
11. Zhu, F., Zhu, Y., Chang, X., Liang, X.: Vision-language navigation with self-supervised auxiliary reasoning tasks. In: Proceedings of the 2020 IEEE/CVF Conference on Computer Vision and Pattern Recognition, Los Alamitos, CA, USA, pp. 10009–10019. IEEE Computer Society (2020)
12. Wu, Z., Liu, Z., Wang, T., Wang, D.: Improved speaker and navigator for vision-and-language navigation. In: IEEE MultiMedia (Feb 2021)

Stock Prediction and Portfolio Optimization

Meta–heuristics for Portfolio Optimization: Part I — Review of Meta–heuristics

Kyle Erwin[1(✉)] and Andries Engelbrecht[1,2,3]

[1] Computer Science Division, Stellenbosh University, Stellenbosch, South Africa
kyle.erwin24@gmail.com, engel@sun.ac.za
[2] Department of Industrial Engineering, Stellenbosch University,
Stellenbosch, South Africa
[3] Center for Applied Mathematics and Bioinformatics,
Gulf University for Science and Technology, Mubarak Al-Abdullah, Kuwait

Abstract. Portfolio optimization is a popular application of meta–heuristic approaches, like evolutionary and swarm intelligence algorithms, due to their simplicity and ability to approximate optimal solutions efficiently. Portfolio optimization is process of selecting weights that balance opposing objectives, i.e. minimizing risk and maximizing return, for a given set of assets. This paper reviews several meta–heuristic approaches to the unconstrained portfolio optimization problem. This paper also serves as the background for a companion publication that empirically analyses the performance of these meta–heuristics for said problem.

Keywords: Set–based Particle Swarm Optimization · Particle Swarm Optimization · Artificial Bee Colony · Firefly Algorithm · Genetic Algorithm · Portfolio Optimization

1 Introduction

Portfolio optimization is the optimization of a financial model that predicts the behavior of a collection of assets given market relevant information. The perceived risk of an asset, the expected return of an asset, and the liquidity of an asset, are all examples of relevant market information. The model returns a value where, depending on the model, higher or lower values represent more worthwhile portfolios. Typically, portfolio models require a weight to be specified for each asset. These asset weight represent how much each asset contributes to the overall portfolio. Optimizing a portfolio model is to search for the set of asset weights that either maximize or minimize the value returned by the portfolio model. This optimization process is complicated when a portfolio model contains multiple objectives or has constraints that limit the feasible solution set. In fact, when constraints that introduce nonlinearity and non-convexity (such as boundary constraints and cardinality constraints) are added, the problem becomes NP-Hard [1,2,21]. Thus, approaches such as quadratic programming cannot be

Y. Tan et al. (Eds.): ICSI 2023, LNCS 13969, pp. 441–452, 2023.
https://doi.org/10.1007/978-3-031-36625-3_35

efficiently utilized to obtain solutions. Furthermore, there may exist multiple but equally good solutions to a portfolio optimization problem. Due to these characteristics that complicate the portfolio optimization problem, researchers in finance and computer science have turned toward meta–heuristics to approximate solutions because meta–heuristics are simple and efficient optimizers for complex optimization problems [16,21,25].

Popular meta–heuristic approaches to portfolio optimization include artificial bee colony (ABC), firefly algorithm (FA), genetic algorithms (GAs) and particle swarm optimization (PSO). These algorithms have been used repeatedly to solve interesting and different portfolio optimization problems [3–6,11,18,30,31,34]. A recently proposed approach to the portfolio optimization problem is set-based particle swarm optimization (SBPSO) that redefines the problem as a combinatorial optimization problem. The motivation for SBPSO for portfolio optimization is that continuous–based optimization approaches approaches, e.g. ABC, FA, GA and PSO, succumb to the curse of dimensionality [32]. That is, as the number of assets increases, the more difficult it is for these algorithms to find optimal solutions.

This paper, the first of two, details the portfolio optimization problem and the aforementioned algorithms. A companion publication [15] empirically assess the algorithms on a set of portfolio optimization problems. It is hypothesized that redefining portfolio optimization as a combinatorial optimization problem results in better solutions for larger portfolio optimization problems.

The remainder of this paper is organized as follows: Sect. 2 familiarizes the reader with portfolio optimization. The algorithms mentioned above are described in Sect. 3. Section 4 concludes the paper and outlines the work for the next paper.

2   Mean–variance Portfolio Optimization

An investor may use a portfolio model to assist in the investment decision process. For example, given information about the market and information about the assets, the portfolio model can be used to assess the quality of a portfolio. The best portfolios can be found by optimizing the portfolio model. Harry Markowitz's mean-variance model, which can be described as a two-objective nonlinear quadratic programming problem, is one such portfolio model [23,29]. The model, referred to as the mean–variance portfolio model, represents an optimal tradeoff between return and risk. Return is calculated using

$$R = \sum_{i=1}^{n} R_i w_i, \tag{1}$$

where R_i is the return of asset i and w_i is the weighting of asset i. The weight of an asset can be though of as how much that asset contributes to the portfolio. Risk is calculated as

$$\bar{\sigma} = \sum_{i=1}^{n} \sum_{j=1}^{n} w_i w_j \sigma_{ij}, \tag{2}$$

where n is the number of assets, and σ_{ij} is the covariance between assets i and j. Finally, the mean–variance portfolio model is formally described as the weighted difference of the R and $\bar{\sigma}$ terms defined above:

$$\lambda\bar{\sigma} - (1 - \lambda)R, \tag{3}$$

where λ represents an investor's risk tolerance in the range $[0, 1]$.

The mean–variance portfolio model is optimized (minimized) by searching for the set of weights thats returns the lowest value given a λ value. When λ is 0, return is maximized. Conversely, when λ is 1, risk is minimized. The λ parameter is a trade-off coefficient that balances these two objectives. Thus, portfolios with differing return and risk characteristics are found by minimizing equation (3) for various λ values. Lastly, the mean variance model has two constraints. The first requires that the sum of the asset weights must equal 1.0, and the second requires that the asset weights must be be non-negative.

3   Meta–heuristics for Portfolio Optimization

This section describes ABC, FA, a GA for portfolio optimization, PSO and SBPSO for portfolio optimization in Sects. 3.1 to 3.5, respectively. These meta-heuristics, with the exception of SBPSO, have been selected due to their popularity [7–10, 19, 24, 26, 27] and SBPSO is compared with these algorithms in [15].

3.1   Artificial Bee Colony

ABC is an optimization algorithm based on the foraging behavior of honey bees [22]. The colony is guided towards global optima based on the quantity of nectar found at various food sources located throughout the search space. Three types of artificial bees are used to perform the search process, each type having a particular role. Employed bees exploit the aforementioned food sources and share this information with the colony via a waggle dance. Onlooker bees watch this waggle dance and adjust their direction according to the quality of nectar. When all known food sources have been adequately exploited, scout bees search for new sources of nectar. The number of employed bees, onlooker bees and scout bees are equal to the number of food sources. Algorithm 1 contains pseudo code for ABC.

Each employed bee moves towards a food source, $\boldsymbol{x}_i$, and determines a neighboring food source, $\boldsymbol{o}_i$, using

$$\boldsymbol{o}_i = \boldsymbol{x}_i + \varphi_i(\boldsymbol{x}_i - \boldsymbol{x}_k), \tag{4}$$

where $\boldsymbol{\varphi}_i$ is a vector of random values in the range [-1,1], and $\boldsymbol{x}_k$ is a randomly chosen food source. Then, the employed bee selects the most abundant food source between $\boldsymbol{x}_i$ and $\boldsymbol{o}_i$, and dances. In the case of maximization problems, the amount of nectar available at each food source is determined by the objective

---

Algorithm 1: Artificial Bee Colony

$t = 0$;
Let f be the objective function;
Let *limit* be the food source trial limit;
Let $x_{i,t}$ be the trial counter for food source i, initially zero;
Create a colony of n_E employed bees, n_O onlooker bees, and n_S scout bees;
Create and initialize n food sources uniformly within a predefined hypercube of
   dimension d;
while *stopping condition is not true* **do**

    for *each employed bee $i = 1, \ldots, n_E$* **do**
        Select food source x_i;
        Create neighbor food source v_i using equation (4);
        if $f(x_i) < f(o_i)$ **then**
           $x_i = o_i$;
        else
           $x_{i,t} = x_{i,t} + 1$;

    for *each onlooker bee $i = 1, \ldots, n_O$* **do**
        Select food source x_i based on probabilities calculated using equation
        (6);
        Create neighbor food source o_i using equation (4);
        if $f(x_i) < f(o_i)$ **then**
           $x_i = o_i$;
        else
           $x_{i,t} = x_{i,t} + 1$;

    for *each scout bee $i = 1, \ldots, n_S$* **do**
        Select food source x_i;
        if $x_{i,t} ¿ limit$ **then**
           Create and initialize food source x_i' uniformly within the predefined
           hypercube;
           if $f(x_i) < f(x_i')$ **then**
               $x_i = x_i'$;
               $x_{i,t} = 0$;

  $t = t + 1$

---

function value. However, in the case of minimization problems, e.g. the mean–variance portfolio optimization problem, the nectar of x_i is determined by

$$\kappa_i = \begin{cases} \frac{1}{1+f(x_i)} & \text{if } f(x_i) \geq 0 \\ 1 + |f(x_i)| & \text{if } f(x_i) < 0 \end{cases}, \tag{5}$$

where κ_i is the nectar of food source i, and f is the objective function.

Next, each onlooker bee chooses a food source based on a probability determined by the food source's nectar relative to all other food sources. The probability, p_i, of selecting food source x_i is

$$p_i = \frac{\kappa_i}{\sum_{j=1}^{n} \kappa_j}, \tag{6}$$

where n is the number of food sources. The onlooker bees exploit the selected food source by visiting o_i.

If a food source does not improve over a fixed number of iterations, scout bees are dispatched to explore new food sources. The number of iterations is determined by the *limit* control parameter. If the new food source is better than the old food source, it replaces the old food source. The new food source is a randomly generated position within the bounds of the search space.

3.2 Firefly Algorithm

Fireflies generate flashes of light by a process of bioluminescence to facilitate communication. FA models these flashing–based interactions between fireflies to find global optima [33]. The following principles are used to guide the search process. Firstly, the brightness of a firefly represents the attractiveness of a solution. Secondly, a less bright firefly moves towards a brighter one. Lastly, brightness increases as the distance between fireflies decreases.

Algorithmically, the light intensity of a firefly is implemented as the objective function value. Like ABC, a transformation function can be used to accommodate minimization problems. That transformation function is

$$I(x_i) = \begin{cases} \frac{1}{1+f(\boldsymbol{x}_i)} & \text{if } f(\boldsymbol{x}_i) \geq 0 \\ 1 + |f(\boldsymbol{x}_i)| & \text{if } f(\boldsymbol{x}_i) < 0 \end{cases}, \tag{7}$$

where I is the light intensity of firefly $\boldsymbol{x}_i$ and f is the objective function.

The attractiveness, β, of a firefly is relative to the distance between it and the firefly looking at it. If firefly i is looking at the firefly j, the attractiveness can be calculated as

$$\beta(r) = \beta_0 e^{-\gamma r^2}, \tag{8}$$

where r is the Euclidean distance between firefly i and firefly j, β_0 is the attractiveness when $r = 0$ and γ is the light absorption coefficient that models light absorbed by the air. The distance between firefly i and firefly j is calculated as

$$r_{ij} = \sqrt{\sum_{k=1}^{d}(x_{i,k} - x_{j,k})}, \tag{9}$$

where d is the number of dimensions.

A firefly moves toward another brighter firefly if the move increases the brightness of the firefly; otherwise, the firefly stays where it is. A firefly updates its position using

$$\boldsymbol{x}_i(t + 1) = \boldsymbol{x}_i(t) + \beta_0 r^{-\gamma r_{i,j}^2}(\boldsymbol{x}_j - \boldsymbol{x}_i) + \alpha(g - 0.5), \tag{10}$$

where α controls the influence of g, a random number sampled from a uniform distribution in [0,1]. Algorithm 2 presents the FA for portfolio optimization.

Algorithm 2: Firefly Algorithm

$t = 0$;
Let f be the objective function;
Create and initialize n fireflies uniformly within a predefined hypercube of
  dimension d;
while *stopping condition is not true* **do**
  for *each firefly* $i = 1, \ldots, n$ **do**
    for *each firefly* $j = 1, \ldots, n$ **do**
      if $f(x_j) < f(x_i)$ **then**
        Move firefly i towards firefly j using equation (10) to create x_i';
        if $f(x_i') < f(x_i)$ **then**
          $\lfloor$ $x_i = x_i'$
  $t = t + 1$

3.3 Genetic Algorithm

GAs are inspired by Darwin's theory of natural selection [13, 20]. GAs simulate
genetic evolution by representing individuals as collections of genotypes, where
each genotype corresponds to a characteristic of the objective function. The survival strength of individuals within a population is determined by the objective
function to be optimized. Individuals are then selected based on their survival
strength for reproduction, which produces offspring for the next generation of
individuals.

One of the most popular GAs for portfolio optimization was that proposed
by Chang *et al* [5]. The GA, which was used to solve the mean–variance portfolio
optimization problem, is summarised in Algorithm 3. Each individual represents
a candidate solution to the mean–variance portfolio optimization problem. The
genes of an individual represent the asset weights. The following operators are
used to evolve the candidate solutions to the portfolio optimization problem:
Tournament selection is used to select parents that recombine to create new
individuals. The breeding process, referred to as uniform crossover, selects genes
from two parents, where each gene from either parent has an equal chance of
being selected. The result of the breeding process is a new individual, referred
to as an offspring. Lastly, a randomly chosen gene of the offspring is mutated.
The severity of the mutation is determined by the control parameter, S_m.

3.4 Particle Swarm Optimization

PSO is meta–heuristic inspired by the flocking behavior of birds [12]. Particles, which are a metaphor for birds, find solutions to an optimization problem
by iteratively updating their positions in the search space, and by exchanging
locally available information about their positions. The position of a particle is a
randomly initialized point within the bounds of an optimization problem. Each
particle also has a velocity, a cognitive guide and a social guide. The velocity

Algorithm 3: Genetic Algorithm for Portfolio Optimization

$t = 0$;
Let f be the objective function;
Let P_m represent the mutation probability of 0.5;
Let S_m represent the mutation severity;
Let τ represent the size of the tournament;
Create and initialize the population of n individuals uniformly within a
  predefined hypercube of dimension d;
while *stopping condition is not true* **do**
    for *each individual $i = 1, \ldots, n$* **do**
        Select parents $x_{i_1}, x_{i_2} \in P$ using τ-tournament selection;
        Apply uniform crossover on x_{i_1}, x_{i_2} to create x'_i;
        Randomly sample gene j from $\{1, \ldots, d\}$;
        if $U[0,1] \leq P_m$ **then**
            $x'_{i_j} = (1.0 - S_m) * x'_{i_j}$;
        else
            $x'_{i_j} = (1.0 + S_m) * x'_{i_j}$;
        Evaluate C using f;
    $t = t + 1$;

of a particle (initially a vector of zeroes) is the current trajectory of the parti-cle. The cognitive guide of a particle is the best position found thus far by the particle. The social guide of a particle is the best position found thus far in the neighborhood of that particle. A neighborhood is a network topology, e.g. a star topology or ring topology. The velocity of a particle is updated at each time step t using

$$v_i(t+1) = wv_i(t) + c_1 r_{1,i}(t)(y_i(t) - x_i(t)) + c_2 r_{2,i}(t)(\hat{y}_i(t) - x_i(t)), \quad (11)$$

where v_i is the velocity of particle i, w is the inertia weight used to regulate exploitation and exploration [28], c_1 and c_2 are positive constants referred to as the cognitive and social acceleration coefficients, respectively, r_1 and r_2 are vectors of random values sampled from a standard uniform distribution in [0,1], y_i is the personal best position of particle i and $\hat{y}_i$ is the best position within the particle's neighborhood, and x_i is the position of particle i. Finally, a particle's position is updated using

$$x_i(t+1) = x_i(t) + v_i(t+1). \quad (12)$$

3.5 Set–based Particle Swarm Optimization

SBPSO is a set–based implementation of PSO designed to solve combinatorial optimization problems. The position of a set–particle is a set which contains elements from U, where U is the universe of all elements in regard to a specific problem domain. The velocity of a set–particle is a set of operations that either

---

Algorithm 4: Particle Swarm Optimization

$t = 0$;

Let f be the objective function;

Create and initialize a swarm, S, of n particles uniformly within a predefined
hypercube of dimension d;

for *each particle* $i = 1, \ldots, n$ **do**

    Initialize x_i;

    Let y_i represent the personal best position of particle x_i, initialized to $x_i(t)$;

    Let $\hat{y}_i$ represent the neighbourhood best position of particle x_i, initialized
    to the best x_i in i's neighbourhood;

    Initialize $v_i(t)$ to $\mathbf{0}$;

while *stopping condition is not true* **do**

    for *each particle* $i = 1, \ldots, n$ **do**

        if $f(x_i(t)) < f(y_i(t))$ **then**

            $y_i(t+1) = x_i(t)$;

        for *particles* $\hat{i}$ *with particle* i *in their neighbourhood* **do**

            if $f(x_i(t)) < f(\hat{y}_i(t))$ **then**

                $\hat{y}_i(t+1) = x_i(t)$;

    for *each particle* $i = 1, \ldots, n$ **do**

        Update particle i's velocity using equation (11);

        Update particle i's position using equation (12);

    $t = t + 1$;

---

add or remove elements from the position. For example, $(+, e)$ will add element
e to the position and $(-, e)$ to remove element e from the position, where $e \in U$.
Algorithm 5 contains the pseudo–code for SBPSO.

This paper summarizes the version of SBPSO for portfolio optimization proposed by Erwin and Engelbrecht [14,17]. The authors proposed using SBPSO to select subsets of assets where the asset weights of these subsets was optimized by a traditional PSO. These subsets of assets were represented as the positions of the set–based particles. The asset weights of the assets in a set–particle's position were optimized at every time step. Thus, the dimensionality of the optimization problem that the PSO optimized was significantly reduced. The PSO optimized until there was no improvement in solution quality over three iterations. Afterwards, the objective function value of the global best position in the PSO was assigned to the corresponding set–particle. If the asset weights found by the PSO contained any zero–weighted assets, then those assets were removed from the position of the corresponding set–particle. If it were the case that a set–particle contained only one asset, then objective function was immediately calculated since the asset can only ever have a weight of 1.0.

Algorithm 5: Set–based Particle Swarm Optimization for Portfolio Optimization

---

$t = 0$;

Let f be the function described in Algorithm ??;

Create and initialize a swarm, S, of n particles uniformly from the set universe U;

for *each particle $i = 1, \cdots, n$* **do**
> Let Y_i represent the personal best position of particle X_i, initialized to $X_i(t)$;
> Let $\hat{Y}_i$ represent the neighborhood best position of particle X_i, initialized to the best X_i in i's neighborhood;
> Initialize $V_i(t)$ to the empty set;

while *stopping condition is not true* **do**
> **for** *each particle $i = 1, \cdots, n$* **do**
> > **if** $f(X_i(t)) < f(Y_i(t))$ **then**
> > > $Y_i(t+1) = X_i(t)$;
> >
> > **for** *particles $\hat{i}$ with particle i in their neighborhood* **do**
> > > **if** $f(X_i(t)) < f(\hat{Y}_i(t))$ **then**
> > > > $\hat{Y}_i(t+1) = X_i(t)$;
>
> **for** *each particle $i = 1, \cdots, n$* **do**
> > Update velocity using Equation (13);
> > Update position using Equation (14);
> $t = t + 1$;

Formally, the velocity update for SBPSO for portfolio optimization is

$$V_i(t+1) = \lambda_c(t)r_1 \otimes (Y_i(t) \ominus X_i(t))$$
$$\oplus \lambda_c(t)r_2 \otimes (\hat{Y}_i(t) \ominus X_i(t)), \qquad (13)$$
$$\oplus (1 - \lambda_c(t))r_3 \otimes A_i(t)$$

where V_i is the velocity of set–particle i; $X(t)$ is the position of set–particle i; $Y_i(t)$ and $\hat{Y}(t)$ are the cognitive and social guides of set–particle i; $A_i(t)$ is shorthand for $U \backslash (X_i(t) \cup Y_i(t) \cup \hat{Y}_i(t))$; r_1, r_2, and r_3 are random values, each sampled from a standard uniform distribution in the range $[0,2]$, and $\lambda_c(t)$ is an exploration balance coefficient equal to $\frac{t}{n_t}$, where n_t is the maximum number of iterations. The operators $\otimes$, $\ominus$, and $\oplus$ are defined in Appendix A. The positions of set–particles are updated by using

$$X_i(t+1) = X_i(t) \boxplus V_i(t+1), \qquad (14)$$

which simply applies the addition and removal functions in V_i to X_i. Refer to Appendix A for more details on the $\boxplus$ operator.

4   Conclusion

The portfolio optimization field includes a variety of algorithms which can be used to optimize a portfolio model. This paper described the mean–variance portfolio optimization problem. This paper also reviewed several popular meta–heuristics for portfolio optimization, namely artificial bee colony (ABC), fire-fly algorithm (FA), genetic algorithms (GAs) and particle swarm optimization (PSO). A new set–based approach, set-based particle swarm optimization (SBPSO), to the portfolio optimization was also discussed. SBPSO for portfolio optimization uses a bi-stage search process where SBPSO selects a subset of assets and then the asset weights are optimized by PSO. It is hypothesized that SBPSO performs better than the aforementioned algorithms for portfolio optimization problems as a result of redefining the problems as combinatorial optimization problems. A companion paper [15] explores this hypothesis by evaluating the algorithms on a set of portfolio optimization problems and analyzing the results.

Appendices

A   Set Operators

The following operators are used in equation (13) and equation (14):

$\oplus$: Given two velocities, e.g. V_1 and V_2, $\oplus$ is the union of V_1 and V_2:

$$\oplus : \mathcal{P}(\{+,-\} \times U)^2 \to \mathcal{P}(\{+,-\} \times U)$$
$$V_1 \oplus V_2 = V_1 \cup V_2 \tag{15}$$

$\ominus$: Given two positions, e.g. X_1 and X_2, $\ominus$ is the set of operations required to convert X_2 into X_1:

$$\ominus : \mathcal{P}(U)^2 \to \mathcal{P}(\{+,-\} \times U)$$
$$X_1 \ominus X_2 = (\{+\} \times (X_1 \backslash X_2)) \cup (\{-\} \times (X_2 \backslash X_1)) \tag{16}$$

$\otimes$: Given a scalar, η and a velocity, V, $\otimes$ is the multiplication of a velocity by a scalar:

$$\eta \otimes : [0,1] \times \mathcal{P}(\{+,-\} \times U) \to \mathcal{P}(\{+,-\} \times U)$$
$$\eta \otimes V = B \subseteq V \tag{17}$$

where B is a set of $\lfloor \eta \times |V| \rfloor$ elements randomly selected from V.

$\boxplus$: Given a position, X, and a velocity $\boxplus$ is the application of the functions in V to the position X:

$$X \boxplus V : \mathcal{P}(U) \times \mathcal{P}(\{+,-\} \times U) \to \mathcal{P}(U)$$
$$X \boxplus V = V(X) \tag{18}$$

References

1. A parallel variable neighborhood search algorithm with quadratic programming for cardinality constrained portfolio optimization. Knowledge-Based Systems **198**, 105944 (2020)
2. A rapidly converging artificial bee colony algorithm for portfolio optimization. Knowledge-Based Systems **233**, 107505 (2021)
3. Bacanin, N., Tuba, M.: Firefly algorithm for cardinality constrained mean-variance portfolio optimization problem with entropy diversity constraint. Sci. J. **2014** (2014)
4. Bacanin, N., Tuba, M., Pelevic, B.: Constrained portfolio selection using artificial bee colony (ABC) algorithm. Optimization **17**, 2831–2844 (2014)
5. Chang, T., Meade, N., Beasley, J., Sharaiha, Y.: Heuristics for cardinality constrained portfolio optimisation. Comput. Oper. Res. **27**(13), 1271–1302 (2000)
6. Chen, A.H.L., Liang, Y., Liu, C.: An artificial bee colony algorithm for the cardinality-constrained portfolio optimization problems. In: Proceedings of the IEEE Congress on Evolutionary Computation, pp. 2902–2909 (Jun 2012)
7. Chen, W., Zhang, H., Mehlawat, M.K., Jia, L.: Mean–variance portfolio optimization using machine learning-based stock price prediction. Appl. Soft Comput. **100**, 106943 (2021)
8. Chiadamrong, N., Suthamanondh, P.: Fuzzy multi-objective chance-constrained portfolio optimization under uncertainty considering investment return investment risk and sustainability. Int. J. Knowl. Syst. Sci. (IJKSS) **13**(1), 1–39 (2022)
9. Corazza, M., di Tollo, G., Fasano, G., Pesenti, R.: A novel hybrid PSO-based metaheuristic for costly portfolio selection problems. Ann. Oper. Res. **304**(1), 109–137 (2021)
10. Cura, T.: A rapidly converging artificial bee colony algorithm for portfolio optimization. Knowl.-Based Syst. **233**, 107505 (2021)
11. Deng, G.F., Lin, W.T., Lo, C.C.: Markowitz-based portfolio selection with cardinality constraints using improved particle swarm optimization. Expert Syst. Appl. **39**(4), 4558–4566 (2012)
12. Eberhart, R., Kennedy, J.: A new optimizer using particle swarm theory. In: Proceedings of the Sixth International Symposium on Micro Machine and Human Science, pp. 39–43 (Oct 1995)
13. Engelbrecht, A.: Computational Intelligence - An Introduction (2. ed.). Wiley (2007)
14. Erwin, K., Engelbrecht, A.P.: Improved set-based particle swarm optimization for portfolio optimization. In: Proceedings of the IEEE Swarm Intelligence Symposium. pp. 1573–1580 (2020)
15. Erwin, K., Engelbrecht, A.P.: Meta-heuristics for portfolio optimization: Part ii - empirical analysis. In: Proceedings of the 7th International Conference on Swarm Intelligence. ICSI '23, Springer International Publishing (2023)
16. Erwin, K., Engelbrecht, A.: Meta-heuristics for portfolio optimization. Soft Computing (Apr 2023)
17. Erwin, K., Engelbrecht, A.P.: Set-based particle swarm optimization for portfolio optimization. In: Dorigo, M., Stützle, T., Blesa, M.J., Blum, C., Hamann, H., Heinrich, M.K., Strobel, V. (eds.) ANTS 2020. LNCS, vol. 12421, pp. 333–339. Springer, Cham (2020). https://doi.org/10.1007/978-3-030-60376-2_28
18. Golmakani, H.R., Fazel, M.: Constrained portfolio selection using particle swarm optimization. Expert Syst. Appl. **38**(7), 8327–8335 (2011)

19. Jia, X., Cai, X.: A policy gradient based particle swarm optimizer for portfolio optimization problem. In: Proceedings of the Chinese Control Conference, pp. 1991–1996 (Jul 2022)
20. John, H.: Adaptation in Natural And Artificial Systems. MIT Press, Cambridge (1992)
21. Kalayci, C.B., Ertenlice, O., Akbay, M.A.: A comprehensive review of deterministic models and applications for mean-variance portfolio optimization. Expert Syst. Appl. **125**, 345–368 (2019)
22. Karaboga, D.: An idea based on honey bee swarm for numerical optimization. Tech. Rep. 6, Erciyes University, Engineering Faculty (2005)
23. Markowitz, H.: Portfolio selection. J. Finance **7**(1), 77–91 (1952)
24. Mba, J.C., Mai, M.M.: A particle swarm optimization copula-based approach with application to cryptocurrency portfolio optimisation. J. Risk Financial Manage. **15**(7), 285 (2022)
25. Okkes, E., C.B., K.: A survey of swarm intelligence for portfolio optimization: Algorithms and applications. Swarm Evol. Comput. **39**, 36–52 (2018)
26. Ramshe, M., Gharahkhani, M., Feiz, A., Sadjadi, S.J.: A firefly algorithm for portfolio optimization problem with cardinality constraint. Int. J. Indust. Eng. **8**(1) (2021)
27. Sahala, A.P., Hertono, G.F., Handari, B.D.: Implementation of improved quick artificial bee colony algorithm on portfolio optimization problems with constraints. AIP Conf. Proc. **2242**(1), 030008 (2020)
28. Shi, Y., Eberhart, R.: A modified particle swarm optimizer. In: Proceedings of the IEEE International Conference on Evolutionary Computation, pp. 69–73 (1998)
29. Stuart, A., Markowitz, H.M.: Portfolio selection: efficient diversification of investments. Quart. J. Oper. Res. **10**(4), 253 (1959)
30. Suthiwong, D., Sodanil, M.: Cardinality-constrained portfolio optimization using an improved quick artificial bee colony algorithm. In: Proceedings of the International Computer Science and Engineering Conference, pp. 1–4 (Dec 2016)
31. Tuba, M., Bacanin, N.: Upgraded firefly algorithm for portfolio optimization problem. In: Proceedings of the 16th International Conference on Computer Modelling and Simulation, pp. 113–118 (Mar 2014)
32. Woodside-Oriakhi, M., Lucas, C., Beasley, J.: Heuristic algorithms for the cardinality constrained efficient frontier. Eur. J. Oper. Res. **213**, 538–550 (2011)
33. Yang, X.S.: Firefly algorithms for multimodal optimization. In: Proceedings of the Stochastic Algorithms: Foundations and Applications, pp. 169–178. Springer, Berlin Heidelberg, Berlin, Heidelberg (2009)
34. Zhu, H., Chen, Y., Wang, K.: Swarm intelligence algorithms for portfolio optimization. In: Tan, Y., Shi, Y., Tan, K.C. (eds.) ICSI 2010. LNCS, vol. 6145, pp. 306–313. Springer, Heidelberg (2010). https://doi.org/10.1007/978-3-642-13495-1_38

Meta-heuristics for Portfolio Optimization: Part II—Empirical Analysis

Kyle Erwin[1]($\boxtimes$) (iD) and Andries Engelbrecht[1,2,3] (iD)

[1] Computer Science Division, Stellenbosh University, Stellenbosch, South Africa
kyle.erwin24@gmail.com, engel@sun.ac.za
[2] Department of Industrial Engineering, Stellenbosh University,
Stellenbosch, South Africa
[3] Center for Applied Mathematics and Bioinformatics, Gulf University for Science
and Technology, Mubarak Al-Abdullah, Kuwait

Abstract. A companion paper identified five meta-heuristic approaches for the unconstrained portfolio optimization problem. Four of which, artificial bee colony (ABC), firefly algorithm (FA), a genetic algorithm (GA) and particle swarm optimization (PSO), are very popular approaches to the problem. The fifth meta-heuristic identified, set-based particle swarm optimization (SBPSO), is a new set–based approach that redefines the portfolio optimization problem as a combinatorial optimization problem. This paper investigates the performance of SBPSO against the aforementioned popular approaches to portfolio optimization. It is shown that SBPSO is a highly competitive approach to portfolio optimization. Furthermore, SBPSO scaled to larger portfolio problems without a reduction in performance, while being significantly faster than the other algorithms.

Keywords: Set–based Particle Swarm Optimization · Particle Swarm Optimization · Artificial Bee Colony · Firefly Algorithm · Genetic Algorithm · Portfolio Optimization

1 Introduction

Portfolio optimization is a continuous optimization problem where the weights of assets in a portfolio are optimized according to a portfolio model — typically the mean–variance model. The mean–variance model specifies a tradeoff coefficient, λ, that balances risk and return. Further details on the mean–variance model are given in the companion paper [4]. The companion paper also identifies artificial bee colony (ABC) [5], firefly algorithm (FA) [8], a genetic algorithm (GA) [1] and particle swarm optimization (PSO) [2] as four popular approaches to portfolio optimization and describes the details of the algorithms. A fifth approach to portfolio optimization, set-based particle swarm optimization (SBPSO) [3], is also discussed in the companion paper. SBPSO is a bi-stage optimization algorithms where in the first stage subsets of assets are selected, and in the second stage, the assets weights are optimized.

Y. Tan et al. (Eds.): ICSI 2023, LNCS 13969, pp. 453–464, 2023.
https://doi.org/10.1007/978-3-031-36625-3_36

The companion paper hypothesized [4] that SBPSO performs better than the aforementioned algorithms for portfolio optimization problems as a result of redefining the problems as combinatorial optimization problems. This paper compares the performance of ABC, FA, the GA and PSO with that of SBPSO for portfolio optimization. The algorithms were evaluated on five common benchmark problems, where each problem increased in size. The control parameters of each algorithm were tuned to each benchmark problem in order to fairly compare the algorithms with one another. A variety of performance measures were used to access the quality of the solutions identified by the algorithms. It was found that SBPSO performed well across the benchmark problems. Furthermore, SBPSO was able to scale to larger portfolio problems while being significantly faster than the other algorithms. ABC and PSO performed well, especially for the smaller benchmark problems. The GA tested was able to find good quality solutions, but did not perform as well as ABC, PSO and SBPSO. FA was the worst performing algorithm across all benchmark problems.

The remainder of this paper is organized as follows: The empirical process used to assess the algorithms is explained in Sect. 2. The results are given in Sect. 3. Section 4 concludes the paper.

2   Empirical Process

This section details the empirical process used to assess the performance of SBPSO. The benchmark problems are discussed in Sect. 2.1, Sect. 2.2 discusses control parameter tuning, and the performance measures used are listed in Sect. 2.3.

2.1   Benchmark Problems

The OR Library[1] provides five portfolio optimization benchmark problems based on real world data. The benchmark problems are summarized in Table 1. The OR Library also provides a Pareto-optimal front (POF) for each benchmark problem that contains 2000 pairs of R and $\bar{\sigma}$ values. The algorithms minimized the mean–variance model (refer to equation (1) in [4]) for 50 evenly spaced λ values in the range [0,1] for each benchmark problem. The final R and $\bar{\sigma}$ values for each optimization problem were record. Thus, each algorithm produced a solution set of 50 pairs of R and $\bar{\sigma}$ values for each benchmark problem. These solution sets were treated as POFs and were compared with the true POFs provided by the OR Library. The mean–variance model requires that each asset weight is ≥ 0 and that the sum of the asset weights must equal 1.0. To satisfy these constraints, solution vectors were normalized after setting any negative values to 0. If all values in a solution vector were negative, a sufficiently large objective function value was given to deter the algorithm from that area of the search space. Lastly, the algorithms were given 5000 iterations to optimize mean–variance model for each problem instance.

---

[1] http://people.brunel.ac.uk/mastjjb/jeb/orlib/portinfo.html.

Table 1. Summary of the OR Library data sets for portfolio optimization

Stock Market	Region	Number of Assets
Hang Seng	Hong Kong	31
DAX 100	Germany	85
FTSE 100	UK	89
S&P 100	USA	98
Nikkei 225	Japan	225

2.2   Control Parameter Tuning

The control parameter values of each algorithm were optimized to maximize each algorithm's performance and ensure a fair comparison of the algorithms. This was done using 128 parameter sets (that were generated using sequences of Sobol pseudo–random numbers) that spanned the parameter space for each algorithm (refer to Table 2). Note that SBPSO did not require parameter tuning since it used a linearly increasing coefficient to balance time spent exploring the search space and refining solutions. To select the optimal parameter set, the following process was repeated for each benchmark problem and for each algorithm: The parameter sets were ranked according to their average objective function value for each of the λ values over 30 independent runs. The parameter set with the highest average ranking was selected as the optimal control parameter values.

Table 2. Parameter ranges used in the tuning process

Algorithm	Parameter	Range
ABC	$limit$	[0,100]
FA	β_0	[0.00,1.00]
	γ	[0.00,10.00]
	α	[0.00,1.00]
GA	S_m	[0.01,0.50]
	τ	[2,20]
PSO	w	[0.00,1.00]
	c_1	[0.00,2.00]
	c_2	[0.00,2.00]

2.3   Performance Measures

A variety of performance measures were used to evaluate the performance of the algorithms for portfolio optimization. The following portfolio–related measures were used: The number of assets in the portfolios (K), return (R), risk ($\bar{\sigma}$).

Pareto–optimality measures were also used, i.e. the number of optimal portfolios (N) found with in the true POF, generational distance (GD), inverted generational distance (IGD), and hypervolume (HV). GD, IGD, and HV are defined in Appendix A. The time in seconds to reach the stopping condition was also recorded. The performance measure values were recorded at after optimizing each λ value over 30 independent runs.

3   Results

This section presents and discusses the results of this paper. Mean and standard deviation values of the performance measures are given and the obtained POFs are visualized. The Hang Seng, DAX 100, FTSE, S&P and Nikkei 225 results are discussed in Sects. 3.1 to 3.5, respectively.

3.1   Hang Seng

Table 3 shows that, on average, ABC obtained higher return values than all other algorithms. However, ABC performed the worst with respect to risk. On the other hand, FA obtained the worst average return value, but the best average value for risk. The GA, PSO and SBPSO performed similarly to each other with regard to return and risk, and their standard deviations for return and risk were significantly smaller than those of the ABC and FA —indicating that the GA, PSO and SBPSO were more consistent in their performance. The similarity in performance of the GA, PSO and SBPSO can also be seen in the size of the portfolios they obtained. On average all three algorithms obtained portfolios that contained three assets. Conversely, the portfolios obtained by ABC were much larger with an average of 15 assets per portfolio. With regard to the number of optimal portfolios obtained, ABC and FA found slightly more solutions than the other algorithms. SBPSO was significantly faster than all the other algorithms in reaching the stopping condition of 5000 iterations. SBPSO was twice as fast as the the second fastest algorithm, ABC, and nearly 10 times as fast as the slowest algorithm, FA. PSO and GA were also quick to reach the stopping condition. Figure 1 shows that all algorithms were able to approximate the POF, except FA.

3.2   DAX 100

Like the previous results, ABC obtained the highest average value for return but also the highest average value for risk (refer to Table 4). The profitability of the portfolios obtained by ABC were on average twice that of the FA. Conversely, the average risk value for the FA was almost half that of ABC. SBPSO was slightly more profitable than PSO and GA with slightly less risk, thus making SBPSO a competitive approach to this problem. ABC found more optimal portfolios than the other algorithms, with an average of 49 portfolios — six more than the closest algorithm (i.e. SBPSO). Furthermore, the portfolios obtained by ABC contained

Table 3. Hang Seng results for each performance measure

		N	K	R	$\bar{\sigma}$	GD	IGD	HV	Time
ABC	$\bar{x}$	37	14	0.007848	0.002110	0.000176	0.000210	0.814630	11.04
	σ	2	2	0.000142	0.000125	0.000009	0.000000	0.000000	0.41
FA	$\bar{x}$	38	5	0.006084	0.001783	0.003959	0.000556	0.748081	38.95
	σ	4	1	0.000592	0.000268	0.000485	0.000085	0.028972	3.71
GA	$\bar{x}$	32	3	0.007576	0.001884	0.000279	0.000222	0.779399	18.10
	σ	1	0	0.000052	0.000042	0.000065	0.000005	0.025320	0.64
PSO	$\bar{x}$	33	3	0.007607	0.001897	0.000219	0.000212	0.786254	15.02
	σ	1	0	0.000003	0.000001	0.000049	0.000002	0.026135	0.69
SBPSO	$\bar{x}$	33	3	0.007607	0.001897	0.000218	0.000212	0.781949	4.52
	σ	0	0	0.000003	0.000000	0.000045	0.000002	0.026784	2.87

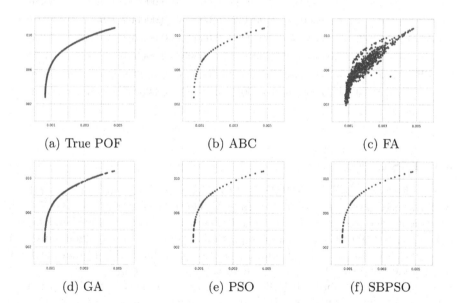

(a) True POF   (b) ABC   (c) FA

(d) GA   (e) PSO   (f) SBPSO

Fig. 1. Obtained Pareto-optimal fronts for Hang Seng

more assets than the other algorithms, but by a smaller margin than for the Hang Seng benchmark problem. Figure 2 shows that SBPSO was able to approximate the true POF as well as ABC and better than the PSO, FA, GA. The points on the POF obtained by PSO were more centered along the curve than the points obtained by ABC and the GA, which were more evenly distributed by comparison. Nonetheless, the PSO was as competitive as ABC, and better than the GA, with regard to GD. The FA failed to approximate the true POF as indicated by its scattered POF and poor GD and IGD results. On average, SBPSO reached the stopping condition in approximately nine seconds, which was significantly faster than the other algorithms. ABC reached the stopping condition in 43.68 s. The PSO and GA took around 50 s to reach the stopping condition. Lastly, the FA was the slowest algorithm with an average time of 166.26 s — at least three times slower than the other algorithms.

Table 4. DAX 100 results for each performance measure

		N	K	R	$\bar{\sigma}$	GD	IGD	HV	Time
ABC	$\bar{x}$	49	14	0.008633	0.001219	0.000228	0.000304	0.787923	43.68
	σ	1	1	0.000021	0.000029	0.000033	0.000000	0.000000	0.56
FA	$\bar{x}$	34	11	0.004485	0.000768	0.003702	0.000568	0.844798	166.26
	σ	7	2	0.000606	0.000143	0.000571	0.000103	0.033878	21.28
GA	$\bar{x}$	39	10	0.008387	0.000953	0.000295	0.000307	0.893540	50.86
	σ	1	0	0.000034	0.000025	0.000152	0.000004	0.057716	2.02
PSO	$\bar{x}$	42	5	0.008456	0.000967	0.000222	0.000304	0.816574	52.40
	σ	1	0	0.000017	0.000013	0.000053	0.000001	0.057681	1.73
SBPSO	$\bar{x}$	43	4	0.008469	0.000972	0.000250	0.000306	0.878138	8.99
	σ	0	0	0.000006	0.000000	0.000061	0.000008	0.065429	3.19

3.3   FTSE 100

The performance of SBPSO, PSO, and the GA were similar for portfolio–related performance measures, as seen in Table 5. The return and risk results for the aforementioned algorithms were much the same. Like the previous benchmarks, SBPSO and PSO obtained smaller portfolios than the other algorithms with fewer number of assets. The results show that ABC found 49 optimal portfolios on average, whereas the next best algorithm, the FA, found 13 fewer. Figure 3 shows that SBPSO obtained POFs that better resemble the true POF than the FA and GA, and slightly better than PSO. Further visual analysis shows that points of the POF obtained the FA were scattered and do not follow the trend of the true POF. The average GD and IGD scores for ABC were the best out of all algorithms, however, its average HV score was the worst. The poor HV score coincides with the concentration of points on the upper part of its obtained

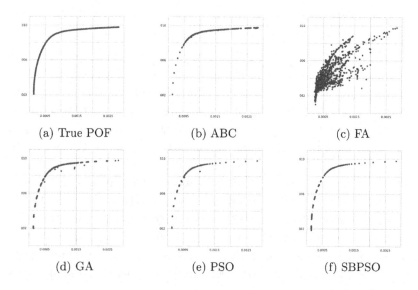

(a) True POF   (b) ABC   (c) FA

(d) GA   (e) PSO   (f) SBPSO

Fig. 2. Obtained Pareto-optimal fronts for DAX 100

POF. The fastest algorithm to reach the stopping condition was SBPSO, with an average time of 10.57 s. Furthermore, the search time of SBPSO was consistent with that of the previous benchmark problem, despite the increase in problem size. ABC was the second fastest algorithm with an average time to reach the stopping condition of 47.09 s, while PSO and the GA were approximately 10 s slower. FA was the slowest algorithm.

3.4   S and P 100

Table 6 shows that the return and risk values for SBPSO and PSO were very similar for the S&P benchmark problem. Once again, ABC was the most profitable algorithm but had the highest level of risk. The GA, although not as profitable as ABC, PSO and SBPSO, was a very risk averse approach. Both ABC and PSO performed well in the GD and IGD performance measures. The PSO, however, performed better than ABC, with regard to the HV measure. SBPSO was able to approximate the POF as well as or better than the other algorithms, as seen in Fig. 4. Furthermore, the good GD and IGD results for SBPSO quantify its ability to approximate the true POF for the S&P 100 benchmark problem. Yet again the FA produced a scattered POF that had little resemblance to the true POF. ABC, PSO and the GA took approximately a minute to reach the stopping condition, while the FA took almost four times that. SBPSO, the fastest algorithm thus far, reached the stopping condition in 16.45 s.

Table 5. FTSE 100 results for each performance measure

		N	K	R	$\bar{\sigma}$	GD	IGD	HV	Time
ABC	$\bar{x}$	49	16	0.007277	0.001073	0.000166	0.000232	0.679689	47.09
	σ	1	1	0.000011	0.000005	0.000020	0.000000	0.020228	1.02
FA	$\bar{x}$	36	12	0.004745	0.000779	0.003347	0.000467	0.714994	186.79
	σ	4	2	0.000445	0.000098	0.000356	0.000087	0.013245	25.15
GA	$\bar{x}$	29	12	0.006635	0.000778	0.000408	0.000238	0.766213	58.45
	σ	1	1	0.000163	0.000029	0.000321	0.000024	0.035444	1.29
PSO	$\bar{x}$	29	6	0.006667	0.000782	0.000190	0.000232	0.704657	59.86
	σ	0	0	0.000008	0.000001	0.000039	0.000001	0.049243	0.95
SBPSO	$\bar{x}$	30	5	0.006683	0.000785	0.000247	0.000235	0.741994	10.57
	σ	0	0	0.000007	0.000001	0.000038	0.000007	0.049170	3.55

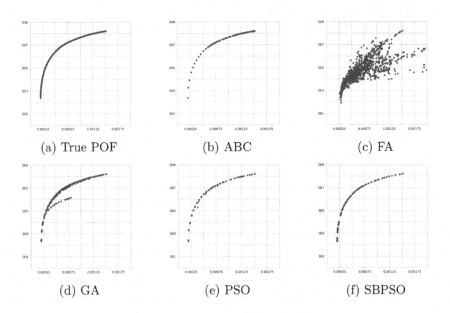

(a) True POF   (b) ABC   (c) FA

(d) GA   (e) PSO   (f) SBPSO

Fig. 3. Obtained Pareto-optimal fronts for FTSE 100

Table 6. S&P 100 results for each performance measure

		N	K	R	$\bar{\sigma}$	GD	IGD	HV	Time
ABC	$\bar{x}$	49	24	0.007843	0.001398	0.000236	0.000241	0.841527	56.54
	σ	1	3	0.000031	0.000035	0.000031	0.000000	0.018891	1.21
FA	$\bar{x}$	38	18	0.005203	0.000891	0.003536	0.000535	0.825091	230.75
	σ	5	2	0.000691	0.000246	0.000453	0.000093	0.016919	24.39
GA	$\bar{x}$	35	6	0.007430	0.001177	0.000810	0.000260	0.881914	54.52
	σ	2	0	0.000088	0.000052	0.000265	0.000007	0.042699	1.31
PSO	$\bar{x}$	44	8	0.007705	0.001245	0.000269	0.000241	0.892730	64.55
	σ	1	0	0.000010	0.000012	0.000096	0.000001	0.052169	1.26
SBPSO	$\bar{x}$	44	6	0.007716	0.001248	0.000275	0.000247	0.875057	16.45
	σ	0	0	0.000006	0.000001	0.000044	0.000003	0.044175	2.78

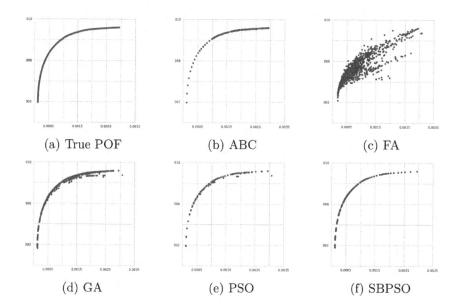

(a) True POF   (b) ABC   (c) FA

(d) GA   (e) PSO   (f) SBPSO

Fig. 4. Obtained Pareto-optimal fronts for S&P 100

Table 7. Nikkei 225 results for each performance measure

		N	K	R	$\bar{\sigma}$	GD	IGD	HV	Time
ABC	$\bar{x}$	49	24	0.003348	0.000858	0.000470	0.000210	0.924789	254.99
	σ	0	6	0.000022	0.000015	0.000221	0.000001	0.046806	2.91
FA	$\bar{x}$	28	12	0.001086	0.000962	0.004191	0.000490	0.734680	1087.28
	σ	6	7	0.000484	0.000114	0.000561	0.000056	0.064953	117.14
GA	$\bar{x}$	46	22	0.003185	0.000858	0.001702	0.000276	0.906150	248.82
	σ	4	3	0.000105	0.000068	0.000344	0.000019	0.043527	2.82
PSO	$\bar{x}$	44	7	0.003264	0.000782	0.000703	0.000218	0.957354	266.89
	σ	2	0	0.000023	0.000022	0.000126	0.000008	0.037560	4.40
SBPSO	$\bar{x}$	44	4	0.003310	0.000794	0.000260	0.000223	0.916014	63.19
	σ	0	0	0.000007	0.000001	0.000060	0.000005	0.047097	6.13

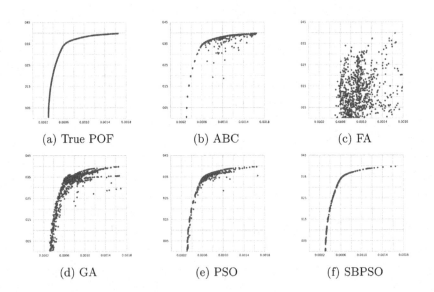

(a) True POF   (b) ABC   (c) FA

(d) GA   (e) PSO   (f) SBPSO

Fig. 5. Obtained Pareto–optimal fronts for Nikkei 225

3.5   Nikkei 225

Table 7 shows that the average return value of SBPSO was very close to that of ABC (the most profitable algorithm) while nearly being as risk averse as PSO (the least risky algorithm). All algorithms found a high number of optimal portfolios in comparison with previous benchmark problems, except for the FA. The portfolios obtained by the ABC and GA contained statistically significantly more assets than the FA, PSO and SBPSO. Figure 5 shows that all algorithms struggled to approximate the true POF, except for SBPSO. The Pareto optimality measures show that ABC and SBPSO performed well in comparison with the other algorithms. The FA took approximately 1000 s seconds to reach the stopping condition — an order of magnitude worse than ABC, PSO and the GA, and two orders of magnitude worse than SBPSO. In contrast, ABC, PSO and the GA took around 250 s to reach the stopping condition.

4   Conclusion

This paper performed a thorough investigation into the performance of the set-based particle swarm optimization (SBPSO) for portfolio optimization in comparison with four popular algorithms, namely artificial bee colony (ABC), firefly algorithm (FA), PSO, and a genetic algorithm (GA). SBPSO is a recently proposed algorithm for portfolio optimization where SBPSO is used for asset selection while a particle swarm optimization (PSO) algorithm is used to determine the asset weights. Results were collected across multiple indepdent runs on five benchmark problems. The control parameters were tuned to each benchmark problem in order to fairly compare the algorithms. It was found that SBPSO was performed similarly to or better than the other algorithms in many ways. SBPSO was able to scale to larger portfolio problems without a reduction in performance while being significantly faster than the other algorithms. Furthermore, SBPSO was the only algorithm to accurately obtain the true Pareto-optimal front (POF) for the largest benchmark problem. In comparison, the ability of the other algorithms to approximate the true POF deteriorated as the dimensionality of the portfolio problems increased. It was also found that ABC and PSO are good and competitive approaches to portfolio optimization. The GA was the fourth best performing algorithm overall, while the FA was the worst. In conclusion, SBPSO is a highly competitive algorithm to portfolio optimization and is suitable for problems that require solution quickly and efficiently.

Appendices

A   Pareto-optimality Measures

Generational distance (GD): Given an obtained POF, Q, and the true POF, Q_{true}, GD measures the average Euclidean distance of solutions Q to the nearest solution Q_{true} [7]:

$$GD = \frac{\sqrt{\sum_{i=1}^{|Q|} d_i^2}}{|Q|} \tag{1}$$

where d_i is the Euclidean distance between the i'th solution in Q and the nearest solution in Q_{true}. GD returns a minimum value, i.e., 0.0, when all solutions in Q are in Q_{true}.

Inverted generational distance (IGD): Given Q and Q_{true}, IGD measures the average Euclidean distance of the solutions in Q_{true} to the nearest solutions in Q [6]:

$$IGD = \frac{\sqrt{\sum_{i=1}^{|Q_{true}|} d_i^2}}{|Q_{true}|} \tag{2}$$

Like GD, lower values indicate better performance.

Hypervolume (HV): Given Q and a reference point, HV measures the volume of the objective space dominated by Q [9]:

$$HV = volume(\cup V_k) \; \forall q_k \in Q \tag{3}$$

where, for each solution $q_k \in Q$, V_k is the hypercube constructed between q_k and the reference point.

References

1. Chang, T., Meade, N., Beasley, J., Sharaiha, Y.: Heuristics for cardinality constrained portfolio optimisation. Comput. Operat. Res. **27**(13), 1271–1302 (2000)
2. Eberhart, R., Kennedy, J.: A new optimizer using particle swarm theory. In: Proceedings of the Sixth International Symposium on Micro Machine and Human Science, pp. 39–43 (Oct 1995)
3. Erwin, K., Engelbrecht, A.P.: Improved set-based particle swarm optimization for portfolio optimization. In: Proceedings of the IEEE Swarm Intelligence Symposium, pp. 1573–1580 (2020)
4. Erwin, K., Engelbrecht, A.P.: Meta-heuristics for portfolio optimization: Part i - review of meta-heuristics. In: Proceedings of the 7th International Conference on Swarm Intelligence. ICSI '23, Springer International Publishing (2023)
5. Karaboga, D.: An idea based on honey bee swarm for numerical optimization. Tech. Rep. 6, Erciyes University, Engineering Faculty (2005)
6. Tsai, S.J., Sun, T.Y., Liu, C.C., Hsieh, S.T., Wu, W.C., Chiu, S.Y.: An improved multi-objective particle swarm optimizer for multi-objective problems. Expert Syst. Appl. **37**(8), 5872–5886 (2010)
7. Van Veldhuizen, D.A., Lamont, G.B.: On measuring multi-objective evolutionary algorithm performance. In: Proceedings of the Congress on Evolutionary Computation. vol. 1, pp. 204–211 vol 1 (Jul 2000)
8. Yang, X.-S.: Firefly algorithms for multimodal optimization. In: Watanabe, O., Zeugmann, T. (eds.) SAGA 2009. LNCS, vol. 5792, pp. 169–178. Springer, Heidelberg (2009). https://doi.org/10.1007/978-3-642-04944-6_14
9. Zitzler, E., Thiele, L., Laumanns, M., Fonseca, C.M., da Fonseca, V.G.: Performance assessment of multi-objective optimizers: an analysis and review. IEEE Trans. Evol. Comput. **7**(2), 117–132 (2003)

Hierarchical Node Representation Learning for Stock Prediction

Zhihan Yue[1] and Ying Tan[1,2(✉)]

[1] Key Laboratory of Machine Perception (MOE), School of Intelligence Science and Technology, Institute for Artificial Intelligence, Peking University, Beijing 100871, China
`{zhihan.yue,ytan}@pku.edu.cn`
[2] Nanjing Kangbo Intelligent Health Academy, Nanjing 211100, China

Abstract. The stock market is a highly complex and dynamic system, where relationships between stocks play a critical role in predicting price movements. To capture these relationships, we propose a novel approach called the hierarchical predictive representation (HPR). The pairwise attention network is first employed to identify effective relationships between stocks. Then, the hierarchical node matching identifies the most predictive relationship subset at various hierarchical levels. By concatenating representations from various levels, our method achieves a comprehensive representation that reflects local to global information. We further introduce a representation ensemble mechanism to leverage multiple relationships, enhancing the model's predictive performance. Extensive experiments on various datasets demonstrate the superiority of HPR compared to existing state-of-the-art methods.

Keywords: Stock Prediction · Representation Learning · Graph Neural Networks

1 Introduction

Stock markets are complex and constantly changing systems, with various relationships between stocks. Capturing these relationships is crucial for predicting stock trends. For example, when a leading stock in an industry rises, it often prompts other stocks in the same industry to follow. The challenge is to extract features from stock relationships that can enhance predictive performance.

Recently, researchers have started using Graph Neural Networks (GNNs) to extract features from stock relationships. Due to the problem of over-smoothing [2], current GNN models struggle to stack many layers, limiting their ability to extract coarse-grained features. In contrast, existing graph pooling methods, which combine global attention and other node selection mechanisms, offer improved global feature extraction capabilities by aggregating nodes within the entire graph. However, financial markets display complex hierarchical structures from markets to primary and secondary industries to themes. These structures are vital for modeling stock relationships, but current efforts still struggle to fully capture intermediate-level structures.

Y. Tan et al. (Eds.): ICSI 2023, LNCS 13969, pp. 465–477, 2023.
https://doi.org/10.1007/978-3-031-36625-3_37

Besides, current graph neural networks face limitations in selectively extracting features from relationships. For example, Zhu et al. [21] reveals that in some heterogeneous graphs, the performance of a multilayer perceptron (MLP) exceeds that of models such as GCN [11], GAT [16], and GraphSAGE [8]. This suggests that these models might have difficulties aggregating features that enhance predictive performance, resulting in weaker performance compared to an MLP model.

To address these challenges, we propose the hierarchical predictive representation (HPR) method. This method aims to minimize joint prediction errors by adaptively extracting relationship representations, allowing related stocks to pair up and generate predictive signals. The node matching process spans multiple levels, enabling the learning of predictive representations at any granularity. Furthermore, we introduce a representation ensemble mechanism that creates more comprehensive relationship features for prediction. The effectiveness of this method is demonstrated through experiments on various datasets.

The major contributions of this paper are summarized as follows:

- We propose a stock relationship selection method based on pairwise attention networks, which adaptively selects relationships that contribute to predictions, resulting in more accurate relationship representations.
- We present a hierarchical representation learning method that extends node matching across multiple levels, enabling the model to learn predictive representations at any granularity. This method addresses the limitations of existing work by capturing the complex hierarchical structures in financial markets, including intermediate-level information.
- Extensive experiments are conducted to evaluate our method. The proposed method outperforms existing state-of-the-art baselines on three real-world datasets. Qualitative analysis shows that without feeding the correlation coefficients of the point pairs into the model, our method can automatically learn to match stocks with higher correlations.

2   Related Work

The history of stock prediction methods has evolved significantly over the years. In the beginning, classical methods focused on time series analysis models, such as Auto-Regressive Moving Average (ARMA) [1], Autoregressive Integrated Moving Average [13], and Vector Auto-Regression (VAR) [15]. However, these methods relied on specific linear assumptions about stochastic processes, which limited their ability to effectively handle complex time series tasks, like stock prediction.

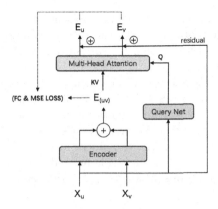

Fig. 1. Structure of the pairwise attention network.

To overcome the constraints of classical methods, researchers have turned to deep learning techniques for predicting stock trends, which have demonstrated higher precision compared to their predecessors. For example, Zhang et al. [20] developed a state-frequency memory-based recurrent neural network that decomposes hidden states of memory cells into multiple frequency components, each simulating latent trading patterns at specific frequencies, resulting in improved stock prediction performance. Qin et al. [14] incorporated input and temporal attention mechanisms into RNNs, enabling adaptive feature selection and extraction of long-term dependencies. Dai et al. [3] augmented Temporal Convolutional Networks (TCN) with attention mechanisms to handle time-varying distributions in stock data, achieving promising results on ultra-high-frequency market data. Deng et al. [4] proposed a knowledge-driven TCN model that extracts structured event embeddings from financial texts and combines them with stock market data to enhance stock trend prediction performance.

Furthermore, some studies have explored GNNs to model cross-sectional relationships. Hsu et al. [9] proposed a financial graph attention network that aims to recommend stocks based on stock price event sequences and sector information. Kim et al. [10] introduced a hierarchical attention network for leveraging relational data in stock market prediction. However, current efforts have not focused on learning market representations across multiple scales, making it difficult to fully capture the interactions among different levels within the market.

3   Method

3.1   Pairwise Attention Network

The key to select stock relationships lies in determining which relationships are beneficial for prediction. Traditional approaches involve statistical methods and empirical rules; however, these methods often require significant human effort and may result in lower accuracy. To address this issue, the pairwise attention network is proposed for extracting joint representations of any stock pairs and train these joint representations with the stock prediction task. Based on the prediction error of the stock pairs, we can automatically verify the predictive ability of the relationships.

The structure of the pairwise attention network is depicted in Fig. 1. The pairwise attention network takes the features of an adjacent stock pair X_u and X_v as input, computes the joint representation $E_{(uv)}$, and subsequently combines the joint representation with the original features to predict both stocks. The sum of the prediction errors of the two stocks at this point is referred to as the joint prediction error. By calculating the joint prediction error, we can determine the effectiveness of a relationship for the prediction task.

Joint Representation of Node Pairs. In the pairwise attention network, we obtain the joint representation of nodes u and v as follows:

$$E_{(uv)} = f_e(X_u) + f_e(X_v) \tag{1}$$

where $X_u, X_v \in \mathbb{R}^{T \times D}$ represent the features of stock nodes u and v, respectively, T denotes the time length, and D represents the feature dimension size. $(uv) \in \mathcal{E}$. $E_{(uv)}$ is the learned joint representation, which combines the features of stock nodes u and v. Note that we use the same encoder f_e for u and v to ensure $E_{(uv)} = E_{(vu)}$. In the experiments, the encoder f_e is a single-layer fully connected network with the ReLU activation function, which is designed to reduce the network parameters and training cost.

When predicting stock u or v, the joint representation $E_{(uv)}$ is introduced to incorporate the relationship information of the two stocks into the prediction. However, the joint representation is not always helpful for prediction, and in some cases, its contribution may be minimal or even detrimental to the prediction task. To address this issue, we introduce an attention mechanism, allowing the network to adaptively adjust its attention to relationship based on the input data. Specifically, the individual stock representations E_u and E_v for the prediction task can be obtained as follows:

$$Q = \text{Stack}(f_q(X_u), f_q(X_v)) \tag{2}$$

$$K = V = E_{(uv)} \tag{3}$$

$$W_t = \text{Softmax}(\frac{Q_{:,t}K_t}{\sqrt{D}}) \tag{4}$$

$$E_u = W_{:,1} \odot V + X_u \tag{5}$$

$$E_v = W_{:,2} \odot V + X_v \tag{6}$$

where f_q is a single-layer fully connected network. Stack$(\cdot)$ is used to concatenate matrices along a new dimension, and in this case, we set the concatenated dimension as the first dimension. $\odot$ represents element-wise multiplication.

The representation model is trained with the mean squared error objective, enabling it to have predictive capabilities. Specifically, we expect the joint representation $E_{(uv)}$ to not only predict the individual situations of stocks u and

v but also predict their common trends. In this paper, we consider the mean of the returns of u and v as their common trend. The prediction process and loss functions can be represented as follows:

$$\hat{Y}_u = f_{\text{pred}}(E_u) \tag{7}$$

$$\hat{Y}_v = f_{\text{pred}}(E_v) \tag{8}$$

$$\hat{Y}_{(uv)} = f'_{\text{pred}}(E_{(uv)}) \tag{9}$$

$$L_{u,v} = \text{MSE}(Y_u, \hat{Y}_u) + \text{MSE}(Y_v, \hat{Y}_v) \tag{10}$$

$$L_{(uv)} = \text{MSE}((Y_u + Y_v)/2, \hat{Y}_{(uv)}) \tag{11}$$

where $\text{MSE}(\cdot)$ is the mean squared error function. $\hat{Y}_u, \hat{Y}_v, \hat{Y}_{(uv)} \in \mathbb{R}^T$ are the predictions of the returns for u, v, and the mean returns for u and v, respectively, and $Y_u, Y_v \in \mathbb{R}^T$ are the true returns. $L_{u,v}$ represents the loss function for optimizing joint prediction errors, and $L_{(uv)}$ represents the loss function for common trend prediction.

The final loss function is the weighted sum of the two:

$$L_{\text{total}} = L_{u,v} + \alpha L_{(uv)} \tag{12}$$

Relationship Filtering. To comprehensively utilize and test the predictive capabilities of each relationship, we use all $\{(X_u, X_v)|(uv) \in \mathcal{E} \cup \{(vv)|v \in \mathcal{V}\}\}$ in the training set as input node pairs. Even if the original graph edge set $\mathcal{E}$ does not include self-loops, all self-loops will be included in the input during training. In this way, we can quantitatively evaluate whether the introduction of a certain relationship can improve predictive performance by referring to the model's performance under independent predictions.

For the prediction task, an effective stock relationship (uv) should result in a joint prediction error that is smaller than the sum of the individual prediction errors of stocks u and v. We use $\hat{Y}_{c|ab}$ to denote the prediction value of the model for $c \in \{a, b\}$ when the input node pair is (X_a, X_b). The condition can be represented as:

$$
\begin{aligned}
L_{u,v} &= \text{MSE}(Y_u, \hat{Y}_{u|uv}) + \text{MSE}(Y_v, \hat{Y}_{v|uv}) \\
&< \text{MSE}(Y_u, \hat{Y}_{u|uu}) + \text{MSE}(Y_v, \hat{Y}_{v|vv})
\end{aligned} \tag{13}
$$

In the financial field, the generalization ability of models are highly concerned. Overfitted models often perform well on the training set but have poor actual predictive performance out-of-sample. To avoid overfitting, we partition a validation set that does not intersect with the training set, calculate the joint prediction error on this validation set, and test whether the above condition is satisfied. If edge (uv) does not satisfy this condition, it proves that the introduction of the relationship between u and v will lead to a decline in out-of-sample predictive performance. In the effective relationship filtering stage, such edges will be removed from the edge set, resulting in the filtered edge set $\mathcal{E}'$.

3.2   Hierarchical Node Matching

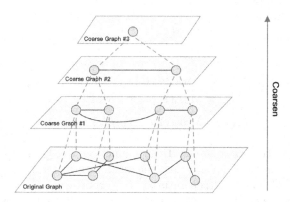

Fig. 2. An example of hierarchical node matching tree. Solid lines represent edges in the graph, while dashed lines represent the correspondence between nodes in the upper and lower graphs.

This section proposes a hierarchical node matching method, which can extract representations of any granularity. In the hierarchical node matching, the node aggregation process can be abstracted as a binary tree, as shown in Fig. 2. The bottom layer of the binary tree is the original stock graph, which is continuously aggregated to the top layer to obtain representations of any level. In each level of the tree, we need to decide how to pair nodes two by two to form nodes in the upper layer of the graph. Here, we adopt a layer-by-layer recursive approach, gradually building the entire tree from the leaf nodes upwards.

In Sect. 3.1, we obtained the joint representation $E_{(uv)}$ and joint prediction error $L_{u,v}$ for each adjacent node pair in the original graph at the bottom of the binary tree through the pairwise attention network, as well as the filtered edge set $\mathcal{E}'$. In the process of constructing the upper binary tree, we need to find a node matching scheme that minimizes the sum of the joint prediction errors of all node pairs. That is, selecting a subset $\mathcal{M}$ from the filtered edge set $\mathcal{E}'$, such that:

- Under the premise of maximizing the number of selected edges, the sum of the joint prediction errors should be minimized.
- There are no common nodes between any two edges.

For the edge (uv), its edge weight is marked as:

$$W_{(uv)} = \max_{(uv)\in\mathcal{E}'}\{L_{u,v}\} - L_{u,v} + \epsilon \tag{14}$$

where ϵ is an extremely small value greater than 0. This problem can be seen as the maximum weight matching for a general graph. The goal is to find several disjoint edges in the graph to maximize the sum of their edge weights. The optimal solution of this problem corresponds to the optimal matching set $\mathcal{M}$ we need. In this paper, we use the Blossom Algorithm [6] to solve the optimal matching scheme. The Blossom Algorithm is an efficient algorithm for solving the maximum matching problem in general graphs, with a time complexity of $O(|\mathcal{V}||\mathcal{E}|\log|\mathcal{V}|)$. Its core idea is to find augmenting paths by constructing a special blossom tree and continuously updating the matching.

Through the optimal node matching scheme, we can obtain the first coarsened graph $\mathcal{G}^{(1)} = (\mathcal{V}^{(1)}, \mathcal{E}^{(1)})$. In the lower-level graph, each (uv) corresponds to a node in the coarsened graph, and its joint representation $E_{(uv)}$ is used as the feature of the coarsened graph node. To construct the edges of the coarsened graph, we define the edge connection condition between nodes $a, b \in \mathcal{V}^{(1)}$ in the coarsened graph as: $(ab) \in \mathcal{E}^{(1)}$ if and only if there exists $(uv) \in \mathcal{E}'$ such that $u \in \text{child}(a)$ and $v \in \text{child}(b)$, where $\text{child}(x)$ represents the set of nodes in the lower-level graph corresponding to node x. In addition, the prediction target of the coarsened graph is defined as the common trend of the child nodes, i.e., $Y_x^{(1)} = (Y_u + Y_v)/2$ for $\text{child}(x) = u, v$.

By repeatedly coarsening the obtained coarsened graph, we can obtain coarsened graphs at various levels. The coarsening process starts from the original stock relationship graph and gradually coarsen upwards, obtaining representations of various granularities. Each level are aggregated based on the node matching results obtained from the previous level, forming the node features of the next level, thus obtaining higher-level representations. In this way, we can obtain a stock relationship graph with a multi-level structure, and as the number of layers increases, the receptive field of the learned representations for stocks also continuously expands, thereby learning representations at various granularities.

3.3   Multi-Granularity Representation Ensemble

In each level of the hierarchical matching, the pairwise attention network learns predictive representations based on the node relationships at that level. At the lowest level, the learned representations primarily utilize the relationships between individual stocks to enhance stock trend predictions. In other layers, each node predicts the average return of its descendant nodes (i.e., common trends), and the learned representations leverage relationships between disjoint sets of stocks (such as industries or sectors) to enhance the prediction of common trends in stock portfolios. This design ensures that the learned representations reflect information at various levels, from local to global. To apply the learned representations to stock prediction tasks, we concatenate the representations from the coarser upper layers to the original graph at the lowest level. After concatenation, each lowest-level node representation X_u' is formed by concatenating the representations of all its ancestor nodes and its own representation, providing a multi-granularity representation from local to global.

In node matching, each node will match with only one related node. However, in reality, a stock's rise or fall often relates to multiple stocks. Moreover, we observed that the matching results from training the network multiple times and executing the matching algorithm exhibit some differences. To fully exploit the contributions of various relationships to predictions, we introduce a representation ensemble method. In this method, we run the entire algorithm multiple times, obtaining a set of multi-granularity representations each time. We average multiple sets of multi-granularity representations to obtain the ensemble representation. The ensemble representation integrates information from various relationships, providing stronger predictive signals.

4   Experiments

4.1   Experimental Settings

The experiments in this article involve three datasets: ACL18 [17], KDD17 [20], and CH [19]. KDD17 and ACL18 are widely-used US stock datasets for stock prediction tasks, while CH is a dataset collected from the Chinese stock market.

- The ACL18 dataset comprises historical time series data for 88 stocks from NASDAQ and NYSE markets, spanning from September 2012 to September 2017. Data from September 2012 to February 2016 is utilized for training, data from March 2016 to August 2016 for validation, and data from September 2016 to September 2017 for testing.
- The KDD17 dataset includes 50 stocks from the US market, covering a time range from January 2007 to December 2016. Data from January 2007 to December 2014 is allocated for training, data from January 2015 to December 2015 for validation, and data from January 2016 to December 2016 for testing.
- The CH dataset consists of the constituent stocks of the SSE 50 Index in the Chinese market, incorporating 49 stocks and ranging from January 2013 to December 2020. Data from January 2013 to December 2018 is designated for training, data from January 2019 to December 2019 for validation, and data from January 2020 to December 2020 for testing.

In this paper we employs the same graph construction, evaluation metrics, and input features as Yue et al. [19]. To evaluate the performance of the learned representations, the GraphSAGE [8] is leveraged as the prediction model, replacing the original input features with the learned ensemble representations and feeding them into the model. This model comprises three SAGE convolution layers connected in series. The tensor obtained by concatenating the outputs of the three SAGE layers is input into a two-layer fully connected network to generate the final output. We trained the prediction model using the same stock ranking loss function as Yue et al. [19]. In this case, we avoided using overly complex prediction models to prevent introducing other variables and to verify the superiority of the HPR method concerning representation learning.

The experimental parameters are set as follows: The parameter α is selected from the set $0.5, 1, 2$. The coarsening depth d is chosen from $2, 3, 4, 5$, the feature dimension size D is selected from $16, 32, 64, 128$, and the number of runs used for representation ensemble is chosen from $5, 10, 20$. All parameters are selected based on optimal performance on the validation set. As for the prediction model, the hidden size of the SAGE convolutional layer is set to 64, while the hidden size of the output layer is set to 32. The HPR model is optimized using an AdamW optimizer [12] with a learning rate of 0.001. All experiments were conducted on an NVIDIA GeForce RTX 3090 GPU.

4.2 Performance Comparison

Table 1. Performance comparison results.

Dataset	Model	IC	Annual Return	Sharpe	Calmar
ACL18	GAT	3.50%	21.01%	0.97	1.05
	GCN	2.20%	19.50%	0.96	1.37
	GraphSAGE	1.80%	29.65%	1.40	2.91
	TGC	1.69%	28.51%	1.73	2.88
	DTML	-0.44%	-8.49%	-0.72	-0.40
	G-Transformer	0.26%	5.82%	0.46	0.40
	NLGA	2.49%	37.60%	**2.80**	5.14
	HPR	**3.54%**	**59.41%**	2.49	**9.05**
KDD17	GAT	2.16%	12.39%	0.67	0.93
	GCN	-0.78%	19.23%	0.80	1.35
	GraphSAGE	1.31%	22.63%	0.87	1.57
	TGC	-0.01%	16.16%	1.37	2.48
	DTML	2.14%	12.72%	0.87	1.61
	G-Transformer	0.95%	21.35%	1.70	2.71
	NLGA	2.63%	26.24%	2.04	5.20
	HPR	**5.71%**	**66.72%**	**2.14**	**6.32**
CH	GAT	4.18%	27.58%	0.96	1.59
	GCN	2.72%	36.59%	1.28	2.67
	GraphSAGE	2.71%	36.85%	1.35	2.90
	TGC	3.81%	34.28%	1.26	2.50
	DTML	2.81%	37.91%	1.52	1.31
	G-Transformer	4.18%	46.01%	1.88	2.77
	NLGA	4.32%	70.84%	1.93	**5.84**
	HPR	**5.60%**	**76.42%**	**2.02**	5.48

In this section, the HPR method is compared with other state-of-the-art methods, including NLGA [19], G-Transformer [5], DTML [18], TGC [7], GAT [16], GCN [11], and GraphSAGE [8]. The evaluation results are presented in Table 1. Since GAT, GCN, and other GNN models only focus on the local features of the graph, their performance is relatively poor. As the NLGA method incorporates non-local feature representation and diversified stock selection mechanisms, it significantly outperforms traditional GNN models. The HPR method introduces predictive representations of arbitrary granularity, achieving the best performance. Specifically, as shown in Table 1, compared to the best baseline methods, HPR improves the annualized returns by 21.8%, 40.5%, and 5.6% for the ACL18, KDD17, and CH datasets, respectively.

4.3   Ablation Study

Table 2. Performance comparison of the HPR method and its three variants on the CH dataset.

	IC	Annual Return	Sharpe	Calmar
Current Model (HPR)	5.60%	**76.42%**	**2.02**	**5.48**
Without Representation Ensemble	4.89%	47.96%	1.54	3.35
Single-Level Matching	**6.04%**	63.71%	1.62	3.62
Random Matching	4.41%	38.19%	1.42	2.97

To evaluate the impact of multi-granularity representation ensemble and hierarchical node matching on model performance, this section conducts an ablation study. We compared the proposed HPR method with its three variants on the CH dataset. The three variants are: without representation ensemble, single-level matching, and random matching. The variant without representation ensemble runs the algorithm only once, aiming to verify the effectiveness of the representation ensemble. The single-level matching variant performs node matching only at a single level, without multi-level node matching, aiming to verify the effectiveness of hierarchical representation learning. The random matching variant ignores the optimal matching results obtained by the algorithm and replaces them with random matching results, aiming to verify the effect of node matching.

The ablation study results are shown in Table 2. Among them, the current model achieves the best overall performance. The single-level matching HPR model has a worse annualized return and Sharpe ratio than the current model, but a better information coefficient. Note that the information coefficient is not an investment performance indicator; it only represents the correlation between the prediction signal and the return. Although the single-level matching variant has a higher information coefficient, its actual investment performance is not satisfactory. In contrast, the current model can take advantage of richer features from intermediate levels, thus having stronger profit potential.

4.4   Analysis

This section analyzes the node matching results. We run the algorithm ten times and record how many times each edge is matched in these runs. In this way, we

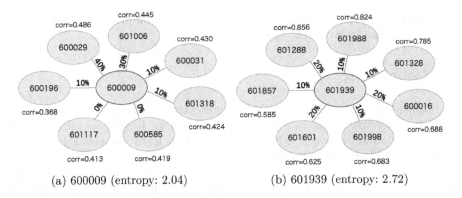

(a) 600009 (entropy: 2.04)   (b) 601939 (entropy: 2.72)

Fig. 3. Example of the matching results on two nodes.

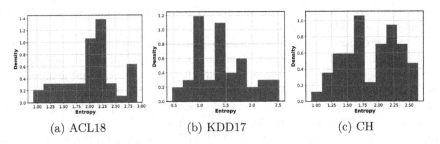

(a) ACL18   (b) KDD17   (c) CH

Fig. 4. The distribution of the entropy for each dataset.

can obtain the matching frequency for each edge and discover the pattern of node matching results. The node matching results for two randomly selected nodes are shown in Fig. 3. The marked percentages in the figure represent the matching frequency, and we also mark the Pearson correlation coefficients of the return rates between the node and its neighbors. In Fig. 3a, the matching is concentrated on the node pairs with larger correlation coefficients. Note that the correlation coefficients between node pairs were not input to the model, indicating that the model automatically learned the correlation between node pairs. In contrast, in Fig. 3b, most neighbor nodes belong to the same industry (banking). In this case, the matching results are more dispersed. This indicates that although some relationships have higher correlation coefficients, these relationships are highly homogeneous.

Next, the uncertainty of each node's matching result is measured using information entropy. Figure 4 shows the distribution of information entropy on each dataset. We found that for all datasets, information entropy presents a bimodal distribution. This indicates that for some stocks, certain relationships are particularly important for their prediction, making the probability of these relationships being selected significantly higher than other relationships, resulting in smaller entropy for these stocks. On the contrary, for another group of stocks, the differences in the impact of various relationships on prediction are not significant, leading to larger entropy.

5   Conclusion

In this paper, we propose a hierarchical representation learning method, designed to learn predictive representations at any granularity. Firstly, we introduce the pairwise attention network, which is employed to learn effective relationships between stocks. By jointly representing node pairs and their joint prediction errors, the pairwise attention network filters relationships that can enhance predictions. Next, we propose a hierarchical node matching method, which learns node representations at different levels and identify the most predictive relationship subset by matching nodes. Furthermore, we introduce a representation ensemble method that integrates multiple node matching results, thereby enhancing the model's generalization capabilities.

Through experiments and analyses conducted on multiple datasets, we verify the predictive performance of our method. Our approach demonstrates a significant performance advantage in predicting stock returns and achieves substantial excess returns. Additionally, we conduct ablation experiments to confirm that both representation ensembles and hierarchical node matching are essential components of the current model. We then analyze the node matching results, discovering that its information entropy exhibits a bimodal distribution phenomenon. The distribution of information entropy helps us better interpret the node matching results. Overall, the proposed method is generic and has the potential for applying in other graph representation tasks.

Acknowledgements. This work is supported by Science and Technology Innovation 2030 - 'New Generation Artificial Intelligence' Major Project (Grant Nos.: 2018AAA0100302) and partially supported by the National Natural Science Foundation of China (Grant No. 62250037, No. 62076010 and No. 62276008).

References

1. Brockwell, P.J., Davis, R.A.: Introduction to Time Series and Forecasting. STS, Springer, Cham (2016). https://doi.org/10.1007/978-3-319-29854-2
2. Chen, D., Lin, Y., Li, W., Li, P., Zhou, J., Sun, X.: Measuring and relieving the over-smoothing problem for graph neural networks from the topological view. In: Proceedings of the AAAI Conference On Artificial Intelligence, vol. 34, pp. 3438–3445 (2020)
3. Dai, W., An, Y., Long, W.: Price change prediction of ultra high frequency financial data based on temporal convolutional network. Proc. Comput. Sci. **199**, 1177–1183 (2022)
4. Deng, S., Zhang, N., Zhang, W., Chen, J., Pan, J.Z., Chen, H.: Knowledge-driven stock trend prediction and explanation via temporal convolutional network. In: Companion Proceedings of The 2019 World Wide Web Conference, pp. 678–685 (2019)

5. Ding, Q., Wu, S., Sun, H., Guo, J., Guo, J.: Hierarchical multi-scale gaussian transformer for stock movement prediction. In: IJCAI, pp. 4640–4646 (2020)
6. Edmonds, J.: Paths, trees, and flowers. Can. J. Math. **17**, 449–467 (1965)
7. Feng, F., He, X., Wang, X., Luo, C., Liu, Y., Chua, T.S.: Temporal relational ranking for stock prediction. ACM Trans. Inform. Syst. (TOIS) **37**(2), 1–30 (2019)
8. Hamilton, W., Ying, Z., Leskovec, J.: Inductive representation learning on large graphs. In: Advances in Neural Information Processing Systems 30 (2017)
9. Hsu, Y.L., Tsai, Y.C., Li, C.T.: Fingat: Financial graph attention networks for recommending top-k k profitable stocks. IEEE Trans. Knowl. Data Eng. **35**(1), 469–481 (2021)
10. Kim, R., So, C.H., Jeong, M., Lee, S., Kim, J., Kang, J.: Hats: A hierarchical graph attention network for stock movement prediction. arXiv preprint arXiv:1908.07999 (2019)
11. Kipf, T.N., Welling, M.: Semi-supervised classification with graph convolutional networks. In: International Conference on Learning Representations (2017). https://openreview.net/forum?id=SJU4ayYgl
12. Loshchilov, I., Hutter, F.: Decoupled weight decay regularization. arXiv preprint arXiv:1711.05101 (2017)
13. Nelson, B.K.: Time series analysis using autoregressive integrated moving average (arima) models. Acad. Emerg. Med. **5**(7), 739–744 (1998)
14. Qin, Y., Song, D., Chen, H., Cheng, W., Jiang, G., Cottrell, G.: A dual-stage attention-based recurrent neural network for time series prediction. arXiv preprint arXiv:1704.02971 (2017)
15. Sims, C.A.: Macroeconomics and reality. Econometrica: J. Economet. Soc., 1–48 (1980)
16. Veličković, P., Cucurull, G., Casanova, A., Romero, A., Liò, P., Bengio, Y.: Graph attention networks. In: International Conference on Learning Representations (2018). https://openreview.net/forum?id=rJXMpikCZ
17. Xu, Y., Cohen, S.B.: Stock movement prediction from tweets and historical prices. In: Proceedings of the 56th Annual Meeting of the Association for Computational Linguistics, vol. 1: Long Papers, pp. 1970–1979 (2018)
18. Yoo, J., Soun, Y., Park, Y.c., Kang, U.: Accurate multivariate stock movement prediction via data-axis transformer with multi-level contexts. In: Proceedings of the 27th ACM SIGKDD Conference on Knowledge Discovery & Data Mining, pp. 2037–2045 (2021)
19. Yue, Z., Tan, Y.: Non-local graph aggregation for diversified stock recommendation. In: Data Mining and Big Data: 7th International Conference, DMBD 2022, Beijing, China, 21–24 November 2022, Proceedings, Part II, pp. 147–159. Springer (2023). https://doi.org/10.1007/978-981-19-8991-9_12
20. Zhang, L., Aggarwal, C., Qi, G.J.: Stock price prediction via discovering multi-frequency trading patterns. In: Proceedings of the 23rd ACM SIGKDD International Conference on Knowledge Discovery and Data Mining, pp. 2141–2149 (2017)
21. Zhu, J., Yan, Y., Zhao, L., Heimann, M., Akoglu, L., Koutra, D.: Beyond homophily in graph neural networks: Current limitations and effective designs. Adv. Neural. Inf. Process. Syst. **33**, 7793–7804 (2020)

Application of APSO-BP Neural Network Algorithm in Stock Price Prediction

Ying Sun[1,2(✉)], Jingbo He[1], and Yuelin Gao[1,2]

[1] School of Mathematics and Information Sciences, North Minzu University, Yinchuan 750021, Ningxia, China
sunying@nun.edu.cn
[2] Ningxia Province Key Laboratory of Intelligent Information and Data Processing, North Minzu University, Yinchuan 750021, Ningxia, China

Abstract. In recent years, with the rapid development of the economy, more and more people have entered the stock market for investment. Due to the volatility characteristics of the stock market, stock price prediction is often a nonlinear time series prediction. And the fluctuation of stock prices will be affected by many factors, so it is difficult to predict through a simple model. For solving this problem, a hybrid adaptive particle swarm optimization and BP neural network algorithm (APSO-BP) is proposed. The APSO-BP algorithm effectively integrates the global search ability of the PSO algorithm and the local search ability of the BP algorithm and further improves the prediction accuracy. Two sets of real stock data of China's stock market are applied to empirical analysis, and the results show that the algorithm is more effective than the standard BP algorithm in solving this problem and can provide timely risk warning information for investors.

Keywords: PSO Algorithm · BP Neural Network Algorithm · APSO-BP Algorithm · Stock Price Prediction

1 Introduction

The stock market is an important part of the financial market, the foundation and core of the capital market, and an important mechanism for resource allocation under the condition of market economy. It plays a role in promoting economic development [1]. Since its establishment in 1990, China's stock market has been concerned by a large number of investors. In order to increase earnings more substantially, the fluctuation of stock price has always been the concern of many investors. However, due to the late establishment of China's stock market compared with foreign mature markets, there are shortcomings such as incomplete data, large fluctuations, unreasonable distribution and so on [2], how to use reasonable methods and models to predict the changes of stock prices is a hot research issue in the academic circles at present.

At present stage, the prediction of stock prices is mainly focused on two aspects: one is the traditional prediction model based on statistical principles, mainly autoregressive conditional heteroscedasticity (ARCH) model and stochastic volatility (SV) model.

Y. Tan et al. (Eds.): ICSI 2023, LNCS 13969, pp. 478–489, 2023.
https://doi.org/10.1007/978-3-031-36625-3_38

However, because these models are based on statistical analysis theory, the prediction process requires higher sample data. The larger the amount of data, the more complete and clear the data, the better the result. But in fact, in practice, even if there is a large sample size, it is not necessarily to find the statistical regularity, even if there is a statistical regularity, it is not necessarily typical [2]; The other is a new algorithm based on artificial intelligence methods, mainly including neural networks [3, 4], support vector machine [5, 6], Machine Learning [7, 8] and intelligent algorithms mixed with various algorithms [9–11].

In this paper, a BP neural network integrating adaptive PSO algorithm (APSO-BP) is proposed to predict the stock price. The algorithm organically combines the global search of PSO algorithm with the local optimization ability of BP neural network. Simulation experiments are carried out using the real data of two stocks in Shanghai and Shenzhen stock markets. The results show that the prediction accuracy of PSO-BP algorithm is greatly improved compared with BP neural network algorithm. It can be seen that the algorithm is effective in solving such problems, and it also provides a reference prediction means for investors in China's stock market.

2 BP Neural Network Algorithm

BP neural network algorithm is a multilayer feedforward neural network with error back propagation, which is proposed by Rumelhart et al. [12] in 1986. In BP network, the signal propagates forward. If the actual output of the output layer is different from the ideal output, it turns into the back propagation of the error signal. In the process of error back propagation, the error is transmitted back layer by layer through the connection weight of neurons, and the weight and threshold are continuously modified by error feedback, so that the actual output is closer to the expected output.

The specific training process of BP neural network algorithm is as follows [12]:

step 1: (Initialization) Creating *sample* set $\{(X_p, Y_p), \quad p = 1, 2, \cdots, S\}$, the dimension of the input signal X_p determines that there are n neurons in the input layer, the dimension of the ideal output Y_p determines that the output layer (H^{th} layer) has m_H neurons. The number of hidden layers is $H - 1$, there are m_{H-1} neurons in each layer. The activation function of the ith layer neurons is $F_i(*)$. Initialize the connection weight $W^{(i)}$ with a set of different small random numbers, where $W^{(i)}$ is the connection weight between the ith layer and its leading layer. Learning rate is α, the accuracy controlling parameter is ε.

step 2: For each sample (X_p, Y_p) in the sample set.

step 2.1: Calculating the actual output O_p corresponding to X_p:

$$O_p = F_H\left(\ldots\left(F_2\left(F_1\left(X_p W^{(1)}\right)W^{(2)}\right)\ldots\right)W^{(H)}\right) \tag{1}$$

step 2.2: Calculating the error between the actual output O_p and the ideal output of the pth sample:

$$E_p = \frac{1}{2}\sum_{j=1}^{m}\left(y_{pj} - o_{pj}\right)^2 \tag{2}$$

step 2.3: Calculating the error of all samples:

$$E = \sum E_p \tag{3}$$

step 3: Error back propagation is used to correct the connection weight and threshold. Because the output layer knows the ideal output and there is no ideal output for all hidden layers, the correction formula of hidden layer and output layer is different:

Output layer:

$$w_{pq} = w_{pq} + \Delta w_{pq}$$
$$\Delta w_{pq} = \alpha \delta_q o_p$$
$$= \alpha o_p (1 - o_q)(y_q - o_q) o_p \tag{4}$$

where w_{pq} is the connection weight between the qth neuron of the output layer and the pth neuron of the leading layer, O_p and O_q are actual outputs.

Hidden layer:

$$v_{hp} = v_{hp} + \Delta v_{hp} \tag{5}$$

$$\Delta v_{hp} = \alpha \delta_{pk-1} o_{hk-2} = \alpha o_{pk-1}(1 - o_{pk-1})$$
$$(w_{p1}\delta_{1k} + w_{p2}\delta_{2k} + \cdots + w_{pm_k}\delta_{m_k k}) o_{hk-2} \tag{6}$$

where v_{hp} is the connection weight between the hth neuron of the $k - 2^{th}$ hidden layer and the pth neuron of the $k - 1^{th}$ hidden layer. $w_{p1}, w_{p2} \ldots w_{pm_k}$ and $\delta_{1k}, \delta_{2k} \ldots \delta_{m_k k}$ are the connection weights and errors from the neuron of the $k - 1^{th}$ hidden layer to each neuron of the $k - 2^{th}$ hidden layer, respectively.

step 4: (Terminate validation) If the termination criteria is met, the training will be terminated, otherwise set $t := t + 1$ and turn to **Step 2**.

3   Improvement Strategy of Hybrid Algorithm

3.1   Coding Strategy and Calculation of Fitness Value

In this section, the position vector of population is initialized by real number coding, that is, the position vector of each particle corresponds to the connection weight and threshold of a group of BP neural network, and its dimension is: $D = (n \times m_1 + 1), +(m_1 \times m_2 + 1) + \cdots, +(m_{H-1} \times m_H + 1)$, the specific correspondence is as follows:

$$x = (x_1, \cdots, x_n, x_{n+1}, x_{n+2}, \cdots, x_{2n+1}, x_{2n+2}, \cdots, x_{D-m_{H-1}-1}, \cdots, x_{D-1}, x_D)$$
$$\downarrow$$
$$\left(W_{11}^{(1)}, \cdots, W_{1n}^{(1)}, \theta_1^{(1)}, W_{21}^{(1)}, \cdots, W_{2n}^{(1)}, \theta_2^{(1)}, \cdots\cdots, W_{m_H 1}^{(H)}, \cdots, W_{m_H m_{H-1}}^{(H)} \theta_{m_H}^{(H)}\right)$$

It can be seen that this coding method can ensure that all connection weights and thresholds of BP neural network can be optimized in the subsequent updating process of PSO algorithm. At the same time, in order to ensure that PSO algorithm can really optimize BP neural network, the fitness value of particles is calculated as follows: first, the position vector of each particle is decoded into the connection weight and threshold of BP neural network; Secondly, each sample is brought into the network in turn to calculate the error between its actual output and ideal output; Finally, the mean value of all sample errors is taken as the fitness value of the particle.

3.2 Particle Learning Strategy Based on BP Neural Network

The back error propagation mechanism of BP neural network comes from the steepest descent principle. Theoretically, its training should approach downward along the slope of the error surface, but in some complex networks, it is difficult to escape once it falls into the local extremum. In order to help it jump out of the local extremum and continue to approach the global optimal value, this section gives a particle learning strategy based on BP neural network.

First, set an evolutionary strategy selection probability:

$$p_r = \frac{t}{T_{\max}} \tag{7}$$

where $T_{\max}$ is the maximum number of iterations, t is the current number of iterations. Then we can see that p_r increases from $1/T_{\max}$ to 1 with the number of iterations.

Secondly, *rand* between [0, 1] is randomly generated. When *rand* $< p_r$, the error back propagation mechanism in Sect. 2 is used to modify the connection weight and threshold. After the correction is completed, a new particle position is recoded to form a new particle position, and this position is directly used to compare and update the global optimal position. Otherwise, the speed update Formula (8) and position update Formula (9) with linearly decreasing inertia weight (10) are used to update the current individual [13, 14].

$$v_{i,d}^{T+1} = wv_{i,d}^T + c_1r_1\left(p_{ibest,d} - x_{i,d}^T\right) + c_2r_2\left(g_{best,d} - x_{i,d}^T\right) \tag{8}$$

$$x_{i,d}^{T+1} = x_{i,d}^T + v_{i,d}^{T+1} \tag{9}$$

$$w = w_{\max} - \frac{T(w_{\max} - w_{\min})}{T_{\max}} \tag{10}$$

Finally, when all individuals are updated, a new particle population is formed.

At the early stage of iteration, the selection probability p_r is very small, so all individuals basically use the update strategy of PSO algorithm to find the optimal position in the global scope, which overcomes the shortcoming that BP algorithm is easy to fall into the local extremum. However, with the increase of the number of iterations, the p_r gradually increases, that is, the possibility of implementing the error back propagation mechanism of BP algorithm to update the particle position increases in the later stage of the algorithm, which ensures the efficient convergence of the algorithm in the local range.

4 A BP Neural Network Algorithm Integrating Adaptive PSO Algorithm (APSO-BP)

The specific process of the APSO-BP algorithm is shown below:

Step 1: (Initialization) Based on the basic principle of neural network, the topological structure of neural network is established according to the input and output sample sets. Assume that the population size is N, randomly initialize position vector and velocity vector, set the self-cognitive coefficient c_1 the social cognitive coefficient c_2, the maximum value w_{max} and the minimum value w_{min} of adaptive inertia weight, the learning rate α, the accuracy Controlling Parameter ε, the error parameter E, the maximum number of iterations T_{max} and the generation number $t := 1$;

Step 2: Calculate the fitness value of particles by the strategy in Sect. 3.1;

Step 3: Update every particle's optimal position P_{ibest} and the global best position g_{best};

Step 4: The particle learning strategy of hybrid BP neural network is used to update the current population;

Step 5: (Terminate validation) If the termination criteria is met, the connection weight and threshold of the global best position are output to the BP neural network, otherwise set $t := t + 1$ and turn to **Step 2**;

Step 6: The optimized BP neural network is used to continue training until the termination condition is met, and the trained network is output.

5 Empirical and Simulation

5.1 Data Selection

Figure 1 shows the trend of Shanghai Composite Index and Shenzhen composite index from May 22, 2017 to May 17, 2019. It can be found that during this period, the overall volatility of both Shanghai and Shenzhen stock markets is relatively large. During this period, after the Shanghai stock index hit a new high of 3587.03 points in the year on January 29, 2018, it began to fluctuate downward. As of the close on December 28, the Shanghai stock index closed at 2493.90 points, with a cumulative decline of 24.59% in the year. Therefore, the continuous rise after the opening of the market in 2019 has brought new hope to the majority of investors. In this long-term rising market, it will inevitably lead some investors to relax their awareness of risk prevention and control and pursue higher investment. Until May 6, the Shanghai and Shenzhen stock markets jumped sharply and opened low, and the market risk aversion increased, resulting in the Shanghai index closing at 2906.46 points, down 5.58%. Shenzhen composite index closed at 8943.52 points, down 7.56%. The small-caps index closed at 5605.28 points, down 7.27%. Gme index reported 1494.89 points, down 7.94%.

The purpose of academic research on the prediction of stock prices or trends is to help investors effectively avoid risks. Therefore, this is the reason why this paper selects the relevant data during this period to test the APSO-BP algorithm. We hope to use this volatile data to train an effective prediction network, so as to help investors find risks in a seemingly beautiful and calm market in time and avoid losses. In this section, we selects the data of 485 trading days of "SH600030" and "000498" as the sample data, the first 465 trading days (2017/5/22–2019/4/16) are used as training samples, and the last 20 data (2019/4 /17–2019/5/17) are used as test samples. Since the collected data include different data such as stock price and turnover, in order to avoid the problem of network paralysis due to the large data gap, it is necessary to normalize the original data to standardize the data between [−1, 1] by Formula (11).

$$y = 2*\frac{x - x_{min}}{x_{max} - x_{min}} + (-1) \tag{11}$$

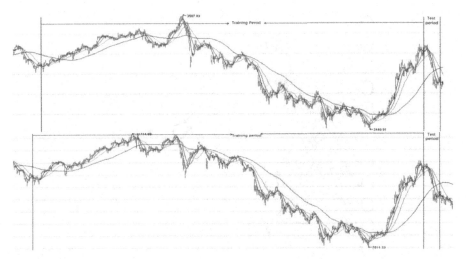

Fig. 1. Tendency of Shanghai index and Shenzhen Composite Index

5.2 Topological Structure of BP Neural Network

Through a lot of practice and theoretical research [15], it shows that the three-layer BP neural network can effectively approach the nonlinear continuous optimization problem. Therefore, this neural network is used to predict the stock price. The Purelin function is used as the activation function of the output layer, and the Tansig function is used as the activation function of the hidden layer.For the prediction of stock prices, it is obvious that the recent trading data have a greater impact on the current price fluctuations than the long-term ones. Therefore, this section uses 15 elements of the stock in the past three days, such as "opening price, highest price, lowest price, closing price and trading volume", as the input data. After three days, the "closing price" of the stock is used as the output data to establish a model. It can be seen that there are 15 neurons in the input layer and 1 neuron in the output layer. In the construction of the whole network, the most complex is the selection of the number of hidden layer neurons. Because the number is too large, the training time will be too long, and the network fault tolerance will also decline; If the number is too small, the network may not converge and the fitting effect may be poor. Generally, the number of neurons in the hidden layer is determined by Formula (12) [8], and then calculated and selected according to specific problems.

$$m_1 = \sqrt{n + m_2} + a \tag{12}$$

where m_1 is the number of hidden layer neurons, n is the number of input layer neurons, m_2 is the number of output layer, a is a constant between 1 and 10. According to the above construction method, the topological structure of BP neural network used in this paper is shown in Fig. 2.

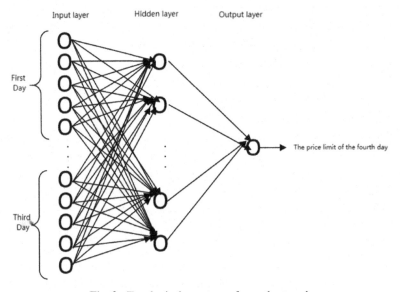

Fig. 2. Topological structure of neural network

5.3 Parameter Setting

In order to determine the number of hidden layer neurons of APSO-BP algorithm, this section uses the data of "SH600030" to conduct simulation experiments on each value in the appropriate range [5, 14] obtained from the empirical formula. The error values obtained are shown in Table 1.

Table 1. The error of APSO-BP algorithm with different hidden layer neurons

Num	5	6	7	8	9	10	11	12	13	14
Error Average	0.00301	0.00270	0.00281	0.00283	0.00275	0.00203	0.00289	0.00271	0.00304	0.00354

Therefore, it can be found that when the number of neurons in the hidden layer is 10, the minimum error average can be obtained. Therefore, there are 15 neurons in the input layer, 10 neurons in the hidden layer and 1 neuron in the output layer, the performance of BP neural network is the best.

Other parameters: the population size of the particles $N = 30, c_1 = 0.5, c_2 = 0.5$, the maximum value of adaptive inertia weight $w_{max} = 0.9$, the minimum value $w_{min} = 0.4$, the maximum number of iterations $T_{max} = 30$, the number of termination iterations of BP algorithm in the last stage is 10, learning rate $\alpha = 0.5$, accuracy Controlling Parameter $\varepsilon = 0.00001$. All algorithms in this paper are coded in MATLAB software and run in the environment of Intel (R) core (TM) i5-3570k CPU @ 3.40 GHz 3.40 GHz processor and 4.00 GB installed memory.

5.4 Analysis of Empirical Results

In order to illustrate the effectiveness of APSO-BP algorithm, it is compared with standard BP neural network algorithm. Because in each iteration process, each particle in APSO-BP algorithm is equivalent to optimizing the network once, and finally BP algorithm is iterated for up to 10 times, the training times of standard BP algorithm is selected as $30 * 30 + 10 = 910$, and other parameters are the same as APSO-BP algorithm.

For the stock "SH600030", from the Fig. 3 (a), we can be seen the fitting effect of the APSO-BP algorithm is better than the standard BP algorithm. The APSO-BP can basically keep the synchronous fluctuation with the real value and closer to the real value. The fitting data of the standard BP algorithm is too sensitive and unstable to the fluctuation, and the fitting values at the positions 1, 2 and 3 of the image are much larger than the real value. It shows that the fitting value of this algorithm fluctuates more violently when the market rises, resulting in excessive amplification of the simulation data. Therefore, in most cases, the fitting value of the standard BP algorithm is not as smooth and accurate as that of the APSO-BP algorithm. However, in some extreme positions, such as position 4, it can be found that the real stock price has a rapid rise. At this time, the sensitivity of the standard BP algorithm makes its fitting at this position better than that of the APSO-BP algorithm.

Figure 3 (b) shows the prediction results of the two algorithms for the closing price of the next 17 days. It is not difficult to see from the trend of the image that the prediction results of the APSO-BP algorithm are closer to the real value than those of the standard BP algorithm before the 7th day. On the 8th day (May 6, 2019), which is the focus of the prediction process, both algorithms can predict the possibility of continued decline tomorrow on the 7th day, and the APSO-BP algorithm has a large decline. It can be used as early warning information for investors. However, in the prediction of the next few days, the APSO-BP algorithm is not as stable as the standard BP algorithm. The biggest reason is that the input data in the later period fluctuates too much due to the precipitous drop on the 8th day. The accuracy of the APSO-BP algorithm model in the early stage of simulation is high, which also leads to the decline of fault tolerance after extreme conditions.

Figure 4 (a) is the fitting diagram of "SZ000498" stock, and Fig. 4 (b) is the prediction diagram of the test sample data. In general, the performance of APSO-BP algorithm is better than that of standard BP algorithm. From the fitting diagram after training, the fitting of the APSO-BP algorithm to the training samples is closer to the real value, and the fitting accuracy is higher. At positions 1 and 2 marked in the figure, it is obvious

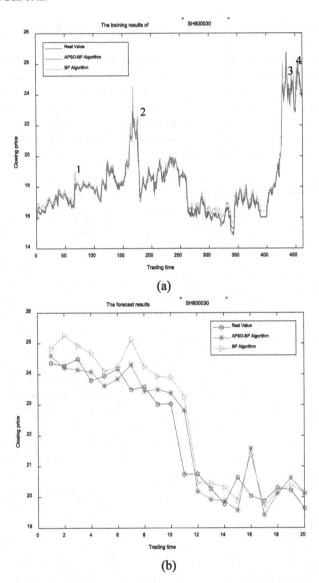

Fig. 3. Training and prediction results of stock "SH600030"

that the fitting value of the standard BP algorithm deviates from the real value, but the simulation effect of the stock price change trend of the BP algorithm at these two places can still be seen. Especially at position 1, the top characteristics of the real data can be basically fitted. Therefore, it can be seen that the BP neural network fused with the APSO algorithm not only retains the characteristics of the original algorithm, but also improves the fitting accuracy and accuracy of the algorithm.In Fig. 4 (b), it can be found that the 17 predicted closing prices obtained by the APSO-BP algorithm basically

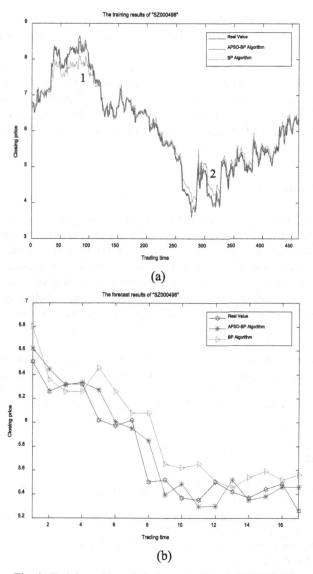

Fig. 4. Training and prediction results of stock "SZ000498"

fluctuate around the real value, with small deviation and high accuracy compared with the standard BP algorithm. Similarly, when observing the data of the eighth sample, the two stocks SZ000498 and SH600030 showed different shapes. Although the latter fell sharply in the eight samples, but SZ00049 also kept falling or trading sideways in the previous few days. Therefore, the two algorithms gave signals of continued decline in the eighth sample; However, by observing the price trend of SZ000498, we can find that it has risen slightly at the seventh sample, causing the false impression of bottoming and rebound. It is precisely because of this that the standard BP algorithm does not give any falling signal at the seventh sample, but the APSO-BP algorithm at this time gives three consecutive falling predictions, effectively reminding consumers of the huge risks hidden here.

Through the simulation results of the above two stocks, it can be seen that the APSO-BP algorithm has found the nonlinear relationship, which contained in each stock, after training a large number of sample data, so it shows better prediction accuracy in the test stage, especially at some important time, APSO-BP algorithm can give timely warning information. So BP neural network with APSO algorithm is more suitable for the prediction of such problems.

6 Conclusions

In this paper, a BP algorithm integrating adaptive PSO is proposed to predict the stock price. Due to the particularity of the stock market, it will be affected by national policies, macro-economy, external stock market, etc.. It cannot be 100% predicted accurately. However, from the simulation results of two stocks, it can be found that using APSO-BP algorithm to predict the stock price is effective, especially in extreme cases, The algorithm can give early warning signals in advance, so as to help investors find and pay attention to these potential risks in time, and provide reference for their trading of stocks.

Funding Statement. This research was funded by the Natural Science Foundation of NingXia Hui Autonomous Region (grant number 2021AAC03185), Research Startup Foundation of North Minzu University (grant number 2020KY QD23), First-Class Disciplines Foundation of NingXia (grant number NXYLXK2017B09) and Major Project of North Minzu University (grant number 2019MS003).

References

1. Huang, M.Y., Rojas, R.R., Convery, P.D.: Forecasting stock market movements using Google trend searches. Empirical Econ. **59**(6), 2821–2839 (2020)
2. Wang, H.Y., Wang, T.T.: Multifractal analysis of the Chinese stock, bond and fund markets. Physica A **512**, 280–292 (2018)
3. Yu, P.F., Yan, X.S.: Stock price prediction based on deep neural networks. Neural Comput. Appl. **32**, 1609–1628 (2020)
4. Liu, G., Ma, W.: A quantum artificial neural network for stock closing price prediction. Inf. Sci. **598**, 75–85 (2022)
5. Guo, Z.Q., Wang, H.Q., Liu, Q.: Financial time series forecasting using LPP and SVM optimized by PSO. Soft. Comput. **17**(5), 805–818 (2013)
6. Ali, M., Khan, D.M., Aamir, M., et al.: Predicting the direction movement of financial time series using artificial neural network and support vector machine. Complexity **2021**, 2906463 (2021)
7. Vara, P.V., Srinivas, G., Venkataramana, L.Y., et al.: Prediction of stock prices using statistical and machine learning models: a comparative analysis. Comput. J. **65**(5), 1338–1351 (2021)
8. Patel, J., Shah, S., Thakkar, P., Kotecha, K.: Predicting stock and stock price index movement using Trend Deterministic Data Preparation and machine learning techniques. Expert Syst. Appl. **42**(1), 259–268 (2015)

9. Adebiyi, A.A., Adewumi, A.O., Ayo, C.K.: Comparison of ARIMA and artificial neural networks models for stock price prediction. J. Appl. Math. **614342**, 1–7 (2014)
10. Tung, K.T., Loan, N.T.B., Chanh, L.Q., Hanh, L.T.M.: Applying artificial neural network optimized by fireworks algorithm for stock price estimation. ICTACT J. Soft Comput. **6**(3), 1183–1190 (2016)
11. Kky, A., Sang, W., Dw, A.: Prediction of stock price direction using a hybrid GA-XGBoost algorithm with a three-stage feature engineering process. Expert Syst. Appl. **186**, 115786 (2021)
12. Rumelhart, D.E., Hinton, G.E., Williams, R.J.: Learning internal representations by error propagation. Parallel Distrib. Process. Explor. Microstruct. Cogn. **1**, 318–362 (1996)
13. Kennedy, J., Eberhart, R.: Particle swarm optimization. In: Proceedings of ICNN 1995 - International Conference on Neural Networks (1995)
14. Shi, Y.H., Eberhart, R.C.: A modified particle swarm optimizer. In: Proceedings of IEEE ICEC Conference, pp. 69–73 (1998)
15. Course, A.P.: Artificial Neural Networks[M]. Springer International Publishing, Switzerland (2017)

The Research in Credit Risk of Micro and Small Companies with Linear Regression Model

Ying Yan and Bo Li[✉]

Chengdu Neusoft University, Chengdu 61000, China
{yanying,li-bo}@nsu.edu.cn

Abstract. The paper proposes a credit risk quantification and prediction model for Micro, Small, and Medium Enterprises by establishing a decision tree for training data sets and employing a linear programming approach. This method combines the results of decision trees with other constraints and objectives determined by banks, reducing the tendency of decision trees to over-fit the data model. The proposed model aims to achieve accurate prediction of quantitative credit risk scores for MSMEs, providing valuable decision-making support for banks.

Keywords: linear regression model · Micro Small and Medium Enterprises · credit risk

1 Introduction

The credit risk faced by banks, especially in China, has become one of the most important risks in the banking industry. Commercial banks rely on credit to achieve capital flow and efficient reporting, but this exposes them to market risks arising from credit business. The ability of banks to control and manage credit risk is crucial for the sustainable development of the banking industry and the overall economy [1].

One of the challenges in credit risk management is the lack of long-term cooperation between banks and companies, especially micro, small and medium-sized enterprises (MSMEs). While there has been increasing attention towards establishing long-term partnerships with enterprises in recent years, the main banking systems in China still focus on financial management in accordance with the requirements of commercial banks. The credit management of commercial banks in China is typically classified into five levels, namely normal, concern, subordinate, doubtful, and loss, with the latter three levels indicating micro non-performing loans.

MSMEs, due to their relatively small scale and lack of mortgageable assets, are not as resilient as large enterprises in terms of risk tolerance. Banks usually

Supported by the Ministry of Education's Industry School Cooperation Collaborative Education Project.

Y. Tan et al. (Eds.): ICSI 2023, LNCS 13969, pp. 490–500, 2023.
https://doi.org/10.1007/978-3-031-36625-3_39

provide loans to stronger enterprises with stable supply and demand relationships based on credit policies, taking into consideration the information of enterprise transaction notes and the influence of upstream and downstream enterprises. Banks assess the credit risk of SMEs based on their strength and reputation, and then determine whether to grant loans and credit strategies such as loan amount, interest rate, and maturity based on credit risk and other factors.

Credit risk in commercial banks is the potential loss that occurs when borrowers fail to repay their loans in full and on time as per the agreed contract, leading to financial losses for the bank. Despite efforts to build sound techniques, credit risk cannot be completely eliminated and is inherent in the banking system.

Traditional approaches used by banks to assess credit risk include internal ratings that consider quantitative and subjective factors such as leverage, earnings, and reputation through a scoring system. Many studies have focused on extracting features from various factors to improve credit risk assessment performance. Additionally, non-economic factors such as education, family environment, and financial education can also influence credit risk in MSMEs.

Quantitative methods, such as linear models, quadratic discriminant analysis, and generalized linear models like logistic regression, are commonly used in bank credit risk assessment. Some companies in the United States have used logistic regression techniques to construct regression models for predicting corporate credit default risk, which have shown higher predictive power compared to generic models.

Researchers have proposed credit risk optimization strategies by addressing internal management problems of MSMEs, market positioning issues of banks, and technical challenges in credit investigation [2]. Some scholars have also provided an overview of credit risk problems from the perspective of big data, suggesting practical strategies such as optimizing business processes, increasing data processing centers, and mining from various aspects like management system and data reserves to enhance operating income for the banking industry [3]. The credit risk is a significant aspect of commercial banking, and various methods, including quantitative techniques and optimization strategies, are employed to assess and manage credit risk in MSMEs and other borrowers. However, complete elimination of credit risk is not possible, and continuous efforts are needed to improve credit risk management in the banking system.

Qu Lei proposes credit risk optimization strategies by addressing internal management problems of micro and small enterprises, market positioning problems of banks, and technical problems of credit investigation. This could involve improving the internal management practices of micro and small enterprises, enhancing the market positioning of banks, and addressing technical challenges in credit investigation. Other scholars also highlight the importance of big data in addressing credit risk problems and increasing operating income for the banking industry. They suggest practical strategies such as mining data from various sources, optimizing business processes, and increasing the construction of data processing centers. These strategies may involve improving the management systems and business processes related to credit risk assessment, as well as enhancing data reserves and processing capabilities [4–6].

The credit evaluation methods for Micro, Small, and Medium Enterprises (MSMEs) discussed by Cheng Yang involve utilizing deep mining of incoming and outgoing invoice data to construct a multi-level credit risk evaluation model. This approach involves analyzing invoices data to assess the credit risk of MSMEs at multiple levels, taking into account various factors such as invoice amounts, payment patterns, and transaction history. To quantify the credit risk, Cheng Yang proposes using the weighted average method, where different factors are assigned different weights based on their importance in determining the credit risk. This allows for an effective quantification of credit risk, which can be used to predict the creditworthiness of MSMEs.

Shao utilizes logistic regression models to process input and output invoices of MSMEs, which are then used to develop an effective credit decision model. This approach involves analyzing invoice data to identify patterns and trends that can help assess the creditworthiness of MSMEs and make informed credit decisions. Mengling Cui quantifies risk evaluation indexes using a logistic regression model and develops corresponding credit strategies based on the production chain and related policies in which the enterprises are located. This approach takes into account the specific industry and policy context in which the MSMEs operate, and develops credit strategies accordingly. Ma Mingyu addresses the problem of credit risk rating of enterprises to allocate credit lines and interest rates. This involves classifying enterprises by industry and reconstructing the rating system. A linear programming model is then built to solve for the optimal credit strategy, taking into account the credit risk ratings of the enterprises. Fatao Wang et al., on the other hand, used a nonlinear Least Squares Support Vector Machine (LS-SVM) model to empirically analyze online credit risk in supply chains caused by corporate defaults. They established a supply chain online credit risk assessment index system to comprehensively identify the credit risk of SMEs.

These credit evaluation methods for MSMEs involve utilizing invoice data, logistic regression models, and linear programming techniques to assess credit risk, develop credit strategies, and make informed credit decisions. These approaches take into account various factors such as invoice data, industry context, and policy environment to provide effective credit risk assessment and management for MSMEs [7].

The paper further studies the current credit risk evaluation mechanism and analyzes the causes of credit risk in MSMEs. It proposes targeted measures based on a linear regression model to minimize credit risk in MSMEs.

2   Linear Regression Model

Linear regression is a statistical analysis method used to determine the quantitative relationship between two or more variables. In the context of credit risk analysis of MSMEs, a multiple linear regression model is utilized.

2.1 Linear Normalization

Linear normalization, also known as min-max normalization, is used for ease and speed of data processing. It is a linear transformation of the original data that maps the data values to a range between 0 and 1. Normalization is a dimensionless processing technique used to establish relative value relationships among the absolute values of physical system values. Linear normalization will be used for ease and speed of data processing [8, 9].

2.2 Least Squares

Least squares fitting is a mathematical optimization technique that finds the best functional match of the data by minimizing the sum of squares of the errors. It is commonly used to estimate unknown data and minimize the errors between the estimated data and the actual data.

2.3 Decision Tree

Decision tree is a popular data mining technique known for its fast learning speed and short classification time consumption. It is a decision analysis method that evaluates project risk and determines its feasibility based on the known probabilities of various scenarios through a large amount of data.

2.4 Confusion Matrix

Confusion matrix is a table used in machine learning for evaluating the prediction results of a classification model. It summarizes the classification performance by comparing the true category and the predicted category of the records in the dataset. It provides information on the number of predicted values that match the true values, allowing for an assessment of the model's accuracy.

It consists of four key metrics: True Positive (TP), False Positive (FP), True Negative (TN), and False Negative (FN). These metrics are calculated as follows:

True Positive (TP): The number of credit ratings that were predicted as positive (e.g., creditworthy) by both the model and the bank.

False Positive (FP): The number of credit ratings that were predicted as positive by the model, but negative (e.g., not creditworthy) by the bank.

True Negative (TN): The number of credit ratings that were predicted as negative by both the model and the bank.

False Negative (FN): The number of credit ratings that were predicted as negative by the model, but positive by the bank.

3 Model Empirical Analysis

3.1 Data Processing

The data processing in this approach is divided into two stages: data pre-processing and data processing. In the data pre-processing stage, irrelevant data is screened out, and profitability, single sales, and the number of stable coopera- tive enterprises are calculated. The same irrelevant data is eliminated to obtain the default situation of enterprises and calculate the percentage of successful transactions for evaluating enterprise reputation. Fuzzification is then applied to the data collected from enterprises with credit records to represent uncertainty or imprecision in the data. Attribute selection is performed using the information gain measure to select splitting attributes for decision tree construction.

The selected attributes, such as profitability, stable supply and demand, aver- age amount of single sales, percentage of successful transactions, and default, are used as indicators of creditworthiness and credit security. These indicators are combined to assess the creditworthiness and credit security of each enterprise, taking into consideration their repayment ability, successful transaction percent- age, and default percentage. The assessment results can be used for evaluating the reputation of enterprises and making credit decisions in the context of credit risk management.

The enterprise credit security scoring model takes into consideration sev- eral factors, including profitability, number of stable cooperative units, average amount of single sale, proportion of successful transactions, and violations. These factors are weighted according to their influence on enterprise repayment ability and corporate reputation.

First, the profitability of each enterprise is quantified by using relevant finan- cial data, such as sales amount and average amount of sales. The higher the average amount of sales, the more profitable the company is considered to be.

Second, the number of stable cooperative units for each enterprise is calcu- lated by screening the transactions between the enterprise and the same unit. A higher number of stable cooperative units indicates a more stable relationship between the enterprise and its partners, which is considered favorable for credit security.

Third, the average amount of a single sale is obtained by dividing the sum of the sales amount of each company by the number of transactions. This factor reflects the transaction size of each enterprise and is included in the credit scoring model.

Fourth, the proportion of successful transactions is calculated by dividing the number of successful transactions of each enterprise by the total number of trans- actions. This factor represents the success rate of the enterprise's transactions, which is considered an important indicator of creditworthiness.

Lastly, violations are taken into consideration in the credit scoring model. The nature and severity of violations committed by an enterprise, such as defaulting on loans or breaching contracts, can negatively impact its credit security.

Once the factors are quantified and weighted, the credit security of each enterprise is calculated based on the established credit scoring model. The higher the credit score, the better the enterprise's repayment ability and corporate reputation are considered to be.

Based on the credit rating derived from the credit scoring model, the compa- nies identified for lending are classified into three categories, such as high credit rating, medium credit rating, and low credit rating. A linear programming model is then developed to determine the loan amount, annual interest rate, and customer churn rate for each category of companies. The solution of the model is used to derive the bank's lending revenue and formulate the credit strategy for companies with different credit ratings.

The credit strategy may involve setting appropriate loan amounts, annual interest rates, and customer churn rates for each category of companies. For high credit rating companies, the bank may offer larger loan amounts at lower interest rates to encourage borrowing and maintain a long-term relationship. For medium credit rating companies, the loan amounts and interest rates may be moderate to manage the risk. For low credit rating companies, the bank may impose stricter loan terms, such as higher interest rates

By eliminating the invalid invoice items from the data set, the profit amount obtained by each enterprise is calculated by using the difference between the sum of the sales amount of each enterprise's sales invoice and the sum of the input amount in the input invoice information sheet, and then the profit margin of each enterprise is obtained based on the ratio of the profit amount obtained by each enterprise to the input amount in the input invoice information sheet. The calculation formula is as follows.

$$y_i = \frac{\sum x_i - x_i}{\sum x_i} \qquad (1)$$

where y_i means profit margin, x_i means input amount.

The enterprise credit security scoring model takes into consideration various factors, including profitability, number of stable cooperative enterprises, average amount of single sales, proportion of successful transactions, and violations. These factors are given different influence weights to quantify their impact on the repayment ability and corporate reputation of each enterprise.

To establish the credit security of each enterprise, the following steps are taken:

Normalization of data: The data related to default situation, supply and demand, profitability, and proportion of successful transactions for each enterprise are sorted by enterprise code and normalized. This ensures that the data is on a consistent scale and can be compared effectively.

Quantification of repayment ability and corporate reputation: The influence weights of the different factors are set, and the relevant data is brought into the credit scoring model to calculate the repayment ability and corporate reputation of each enterprise. The higher the values of profitability, number of stable cooperative enterprises, average amount of single sales, and proportion of successful transactions, the better the repayment ability and corporate reputation of the enterprise.

Credit rating classification: Based on the credit scores obtained from the credit scoring model, the enterprises are classified into three categories, such as high credit rating, medium credit rating, and low credit rating. This classification helps in determining the risk level associated with each enterprise and guides the bank's lending decisions.

Linear programming model for loan amount, annual interest rate, and customer churn rate: A linear programming model is developed to determine the optimal loan amount, annual interest rate, and customer churn rate for each category of credit rating. The objective of the model is to maximize the bank's lending revenue while considering the risk associated with different credit ratings.

Derivation of bank's lending revenue and credit strategy: The solution of the linear programming model provides the optimal loan amount, annual interest rate, and customer churn rate for each category of credit rating. This information is used to derive the bank's lending revenue, which is the total revenue generated from lending to all the enterprises. The credit strategy for companies with different credit ratings is determined based on the optimal loan amounts, annual interest rates, and customer churn rates obtained from the model. This helps the bank in making informed decisions on lending to different enterprises based on their credit ratings.

Overall, the enterprise credit security scoring model and the linear programming model enable the bank to quantify the credit security of each enterprise, classify them into different credit rating categories, and develop an optimal credit strategy for lending to different enterprises. This approach helps the bank in managing credit risk effectively and maximizing lending revenue.

3.2   Enterprise Credit Scoring Model

The data of each factor obtained by pre-processing are normalized. The degree of influence of profitability, stability of supply and demand, and the average amount of a single sale on the repayment ability of the enterprise were set as W_{b_1} and W_{b_2}, W_{b_3} and W_{b_4}, and W_{b_5}. The influence of repayment ability and creditworthiness on credit security is set as W_{B_1}, and W_{B_2}. The following linear programming model is developed. Establish the following credit scoring model for enterprises: establish the following linear programming model.

Based on the profitability of each firm after normalization b_1 with its weights w_{b_1}, stable supply and demand b_2 and their weights w_{b_2}. The corporate repayment capacity of each firm is calculated B_1 and the formula is as follows.

$$B_1 = b_1 * w_{b_1} + b_2 * w_{b_2} + b_3 * w_{b_3} \tag{2}$$

Similarly, using the percentage of successful transactions for each firm b_2 with their weights w_{b_1}, default cases b_4 with their weights w_{b_4}. The corporate reputation of each enterprise is calculated and the formula is as follows.

$$B_2 = b_4 * w_{b_4} + b_5 * w_{b_5} \tag{3}$$

Reuse of corporate repayment capacity B_1 with its weighting w_{B_1}, corporate reputation B_2 and their weights w_{B_2}. The credit security score is calculated and the formula is as follows.

$$O = B_1 * w_{B_1} + B_2 * w_{B_2} \tag{4}$$

3.3   Bank-to-Business Credit Strategy Model

It's important for enterprises to consider credit risk when formulating lending strategies. Enterprises with the worst credit ratings are typically considered to be high-risk borrowers, as their poor credit history may indicate a higher likelihood of defaulting on loans. As a result, it may be prudent for lenders to exclude these enterprises from their lending strategies in order to mitigate credit risk.

Additionally, using a Gaussian distribution to fit scatter plot curves can be a useful statistical technique for understanding the relationship between the annual interest rate and customer churn rate. A Gaussian distribution, also known as a normal distribution, is a common probability distribution that is characterized by a bell-shaped curve. By fitting a Gaussian distribution to the scatter plot data, it is possible to model the relationship between the two variables and gain insights into their potential correlations. This can help lenders better understand the impact of interest rates on customer churn rates, and make informed decisions about pricing and lending strategies based on the fitted curve. However, it's important to note that the suitability of a Gaussian distribution for fitting the scatter plot data would depend on the underlying characteristics of the data and the specific business context, and other statistical techniques may also be considered depending on the situation.

$$M = a * I^2 + b * I + c \tag{5}$$

where a is the quadratic term coefficient of the fitted function, b is the primary term coefficient of the fitted function, and c is the constant term coefficient of the fitted function.

we can obtain an expression for the annual return on bank lending.

$$Q = n_a * S_A(1 - M_A) + n_B S_B I_B(1 - M_B) + n_C S_C I_C(1 - M_C) \tag{6}$$

Therefore, the following linear programming model is developed.

$$\begin{cases} Q = 28 S_A I_A (1 - M_A) + 269 S_B I_B (1 - M_B) + 4 S_C I_C (1 - M_C) \\ 28 S_A + 269 S_B + 4 S_C \leq 10000 \\ 10 \leq S_A \leq 100 \\ 10 \leq S_B \leq 100 \\ 10 \leq S_C \leq 100 \\ 0.04 \leq I_A \leq 0.15 \\ 0.04 \leq I_C \leq 0.15 \\ 0.04 \leq I_B \leq 0.15 \\ M_A = -71.0765 I_A{}^2 + 20.8395 I_A - 0.6397 \\ M_B = -63.5670 I_B{}^2 + 19.2700 I_B - 0.6035 \\ M_C = -60.2853 I_C{}^2 + 18.7843 I_C - 0.5999 \end{cases} \tag{7}$$

4   Model Evaluation

4.1   Model Validation

Scatter chart and fitting curve of the relationship between bank loan annual interest rate and customer churn rate under three different credit ratings are obtained. From the Fig. 1, we can find that the proposed linear model fits the data of different credit ratings well.

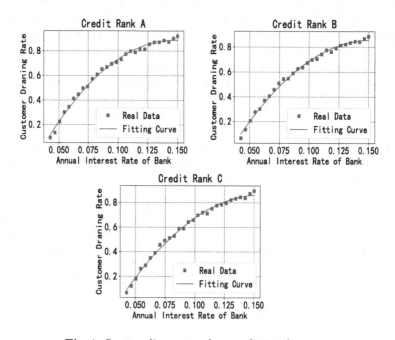

Fig. 1. Scatter diagram and curve fitting diagram.

4.2   Confusion Matrix

The confusion matrix is a table that is often used to describe the performance of a classification model on a set of data for which the true values are known. A high similarity between the results of the model's credit rating of enterprises and the results of the bank's credit rating of enterprises, as indicated in the confusion matrix, would mean that there are a high number of True Positives (TP) and True Negatives (TN), and low numbers of False Positives (FP) and False Negatives (FN). This would suggest that the model's evaluation results are accurate, with minimal errors and high accuracy.

The enterprise credit scoring model is tested by confusion matrix. The number of credit ratings judged by the bank is taken as the real value, and the number of credit ratings judged by the model is taken as the predicted value.

The confusion matrix is used for comparison, as shown in Table 1. From the result analysis, it can be seen that the evaluation result of the model on the enterprise credit rating is similar to that of the bank on the enterprise credit rating, which indicates that the evaluation result of the model is more accurate, the error is small and the accuracy is very high.

Table 1. Confusion matrix

	A	B	C	D
A^*	26	4	1	0
B^*	1	33	4	0
C^*	0	0	27	0
A^*	0	1	2	24

However, it's important to note that accuracy alone may not be the only metric to evaluate the performance of a credit scoring model. Depending on the specific requirements and objectives of the credit scoring model, other metrics such as precision, recall, F1-score, and area under the Receiver Operating Characteristic (ROC) curve may also be important to consider. These additional metrics can provide a more comprehensive assessment of the model's performance and help ensure that it meets the desired level of accuracy and reliability for its intended purpose.

5   Conclusion

The qualitative and quantitative analysis of data presented in this paper aims to understand the factors that influence credit risk for Micro, Small, and Medium Enterprises (MSMEs) using a linear regression model. The paper begins with data screening, which serves as a guide for predicting credit risk for MSMEs. This analysis has practical significance not only for credit risk prediction but also for other financial data analysis and modeling purposes.

The regression model developed in this paper for credit risk assessment of MSMEs without credit records is based on quantifying risk using a data set of enterprises with credit records. The results of the model not only accurately predict whether enterprises are likely to default, but also provide credit ratings through regression, which can better reflect the creditworthiness of MSMEs and offer guidance for the bank's credit strategy.

This paper contributes to the field of credit risk assessment for MSMEs by providing a comprehensive analysis of factors influencing credit risk and developing a regression model that can be used for MSMEs without credit records. The findings of this research have practical implications for financial institutions in managing credit risk for MSMEs, and the methodology employed can be extended to other data analysis and modeling tasks in the financial domain.

References

1. Umar, M., Ji, X., Mirza, N., Naqvi, B.: Carbon neutrality, bank lending, and credit risk: evidence from the eurozone. J. Environ. Manage. **296**, 113156 (2021)
2. Abdulnafea, A., Almasria, N.A., Alawaqleh, Q.: The effect of working capital management and credit management policy on jordanian banks' financial performance. Banks Bank Syst. **16**(4), 229–239 (2022)
3. Cheng, Z., Wang, J.: Research on financing models and countermeasures of small and micro enterprises under the background of "internet+". In: 2020 International Conference on Social Sciences and Big Data Application (ICSSBDA 2020), pp. 313–318. Atlantis Press (2020)
4. Xiao, P., Salleh, M.I.b., Cheng, J.: Research on factors affecting smes' credit risk based on blockchain-driven supply chain finance. Information. **13**(10), 455 (2022)
5. Liu, J., Liu, S.: Li, J: financial credit risk assessment of online supply chain in construction industry with a hybrid model chain. Int. J. Intell. Syst. **37**(11), 8790–8813 (2022)
6. Machado, M.R., Karray, S.: Assessing credit risk of commercial customers using hybrid machine learning algorithms. Expert Syst. Appl. **200**, 116889 (2022)
7. Long, J., Jiang, C., Dimitrov, S.: Clues from networks: quantifying relational risk for credit risk evaluation of smes. Financial Innov. **8**(1), 91 (2022)
8. Wang, Y., Jia, Y., Tian, Y.: Deep reinforcement learning with the confusion-matrix-based dynamic reward function for customer credit scoring. Expert Syst. Appl. **200**, 117013 (2022)
9. Wei, L., Han, C., Yao, Y.: The bias analysis of oil and gas companies' credit ratings based on textual risk disclosures. Energies **15**(7), 2390 (2022)

ICSI-Optimization Competition

Deep-Layered Differential Evolution

Kaiyu Wang, Zhenyu Lei, Ziqian Wang, Zhiming Zhang, and Shangce Gao[✉]

Faculty of Engineering, University of Toyama, Toyama, Japan
gaosc@eng.u-toyama.ac.jp

Abstract. Single-objective bounded optimization problems are a type of complex problem that frequently arises in industry. These problems are often challenging due to the limited internal information available. In this paper, we propose a new algorithm, called DDE, which is a variant of LSHADE that employs a deep hierarchical structure to enhance its performance. We evaluate the proposed algorithm on the ICSI Optimization Competition 2022. The obtained experimental results are compared with two CEC champions, and our DDE algorithm outperforms both of these algorithms, demonstrating its effectiveness.

Keywords: differential evolution · deep hierarchical structure · single objective optimization problem

1 Introduction

The advancement and development of various industries has given rise to numbers of complex optimization problems, among which single-objective continuous optimization problems are particularly significant [1], neural network training [2], automation control system [3], and network expansion planning problem [4]. Traditional mathematical methods suffer from miss internal information of the problem, are not applicable to this type of problems [5]. Evolutionary Algorithms (EAs), has been attached much attention due to its powerful search ability and population-based solution in dealing with so many problems [5]. EAs are considered an important methodology for solving complex optimization problems since they can treat them as a black-box with only inputs and outputs. Nowadays, researchers have proposed many classical algorithms, like gravitational search algorithm [6], particle swarm optimizer [7], covariance matrix adaptation evolution strategy (CMAES) [8], and differential evolution (DE) [9].

Among these, DE, which is first proposed in 1997 [10], achieved great success due to its straightforward structure and excellent search efficiency [11,12]. By apply the difference vector between individuals, DE can find a high-quality optimal more quickly and better. In recent years, researchers have proposed many variants of DE and achieved significant success. For example, Tanabe and Fukunaga proposed SHADE [13], which has historical memories and can retain the information about parameters of successful individuals. SHADE had the forth rank in the CEC 2013 competition. The following year, they proposed LSHADE [14] based on SHADE, which introduces the strategy of linear population size reduction (LPSR). It successfully help LSHADE ranks 1st place in the

Y. Tan et al. (Eds.): ICSI 2023, LNCS 13969, pp. 503–515, 2023.
https://doi.org/10.1007/978-3-031-36625-3_40

CEC 2014 competition since it can make a better balance between exploration and exploitation. In 2017, Janez Brest et al. proposed jSO [15], which is an well-perform variant of LSHADE. jSO uses a novelty mutation strategy, which has a variable scaling factor to influence the whole search procedure. jSO places the 2nd in the CEC 2017 competition optimization problems. As another success-ful variant of LSHADE, LSHADE-RSP [16] uses a selective pressure solutions with rank-based to improve the convergence speed. It scored 2nd in the CEC 2018 competition, demonstrating its effectiveness in solving optimization prob-lems. Janez Brest et al. designed jDE100 [17], which uses two populations and migrates the best individual of them. jDE100 has the 1st rank in the CEC 2019 100-digit challenge competition. In 2020, IMODE [18] is proposed by Karam M. S. et al., which has a multi-operator and dynamically changes the size of sub-populations. IMODE ranks the best in the CEC 2020 competition.

In this paper, we propose a novel variant of DE, named Deep-Layered Dif-ferential Evolution (DDE). It has multiple hierarchical structures with different ways of interaction between each hierarchy. This approach facilitates an increase in the diversity of the population, while simultaneously allowing DDE to achieve a better balance between exploitation and exploration.

The rest of this paper is as follows. In Sects. 2, the basic concepts of DE and the famous variant (LSHADE) are presented. In Sect. 3, we introduce the proposed algorithm and the several strategies utilized. In Sect. 4, experimental results of DDE against some presented champion algorithms are reported and analyzed. The conclusions are shown in Sect. 5.

2   Background

2.1   DE

DE is an evolutionary algorithm based on its population. It has three important strategies: mutation, crossover, and selection. The mutation strategy creates new individuals by perturbing the existing ones, and the crossover strategy generates new individuals by combining two or more existing individuals. The selection strategy decides which individuals will be retained to the new population based on their fitness values. The population P in DE is denoted as:

$$P = \{X_1^t, X_2^t, ..., X_i^t, ..., X_{NP}^t\}, \quad i = 1, 2, ..., NP \tag{1}$$

where t is iteration of search process and NP is the population size.

In the population, each individual consists of D variables:

$$X_i^t = \{x_{i1}^t, x_{i2}^t, ..., x_{iD}^t\} \tag{2}$$

where D denotes the dimension of an optimization problem.

At the beginning of a search process, the algorithm generates a random pop-ulation, and then generates a new population through three strategies and con-tinues until the termination condition is met.

Mutation: A mutation vector V_i^t is generated by a mutation strategy. There are currently five general mutation strategies in the DE and its variants, as follows:

1) DE/best/1:

$$V_i^t = X_b^t + F(X_{r_1}^t - X_{r_2}^t) \qquad (3)$$

2) DE/rand/1:

$$V_i^t = X_{r_1}^t + F(X_{r_2}^t - X_{r_3}^t) \qquad (4)$$

3) DE/pbest/1:

$$V_i^t = X_i^t + F(X_b^t - X_i^t) + F(X_{r_1}^t - X_{r_2}^t) \qquad (5)$$

4) DE/best/2:

$$V_i^t = X_b^t + F(X_{r_1}^t - X_{r_2}^t) + F(X_{r_3}^t - X_{r_4}^t) \qquad (6)$$

5) DE/pbest/2:

$$V_i^t = X_i^t + F(X_p^t - X_i^t) + F(X_{r_1}^t - X_{r_2}^t) \qquad (7)$$

where F indicates a scaling factor in the range of $[0,1]$. X_i^t, X_b^t, and X_p^t denote the i-th individual, the current best individual, and one of the top p high-quality individuals of the t-th iteration, respectively. The indexes r_1-r_4 represent mutually different numbers of random individuals. Also them are different from the indexes i, best, and pbest.

Crossover: After the mutation strategy generates the mutation vectors, the candidate solutions are created by a crossover operation. It crosses individuals in the V_i^t with its parent in the paternal populations X_i^t based on the probability, which is expressed as follows:

$$u_{i,j}^t = \begin{cases} v_{i,j}^t, & j = j_r \text{ or } randi(0,1) < C_r \\ x_{i,j}^t, & \text{otherwise} \end{cases} \qquad (8)$$

where $u_{i,j}^t$ is the i-th candidate solution in the t-th iteration, $i = 1, 2, ..., NP$ and $j = 1, 2, ..., D$. The crossover rate C_r controls the crossover probability. j_r is a random integer of $[1,D]$.

Selection: After the crossover operation, all candidate solutions are calculated for the fitness values. For a optimization problem with minimization, a smaller value implies a better quality of the current solution.

$$X_i^{t+1} = \begin{cases} U_i^t, & f(U_i^t) < f(X_i^t) \\ X_i^t, & \text{otherwise} \end{cases} \qquad (9)$$

where $f(x)$ indicates the fitness function. DE only retains the better solution between the paternal populations X_i^t and the candidates U_i^t. This selection strategy, also named greedy selection, is first used in DE and widely applied in other EAs.

2.2 LSHADE

LSHADE [14], a classical DE variant, has achieved much due to attention. After it obtained the championship in the 2014 CEC competition on single objective optimization, researchers investigate it a lot and have achieve great success based on LSHADE. For instance: iLSHADE [19], the 3rd in CEC 2016 competition, jSO [15], the 2nd in CEC 2017 competition, and LSHADE-RSP [16], the 2nd rank in CEC 2018 competition. In this section, we give a brief introduction to LSHADE [14], which is the basis for our presented algorithm (DDE). There are four main features in LSHADE: external archive, parameter adaptation, success-history-based parameter adaptation and LPSR.

A. External Archive
In LSHADE, an external archive A is used to maintain the diversity of population. Individuals that might have been eliminated in Eq. (9) are also preserved. The individual X_{r_2} in Eq. (7) is selected from $P \cup A$.

B. Parameter Adaptation
Different from original DE, each individual X_i in LSHADE, with its own parameters F_i and C_{ri} to mutation and crossover the candidate solution u_i. At the beginning of each generation, these parameters are determined based on the adaptive control parameters μ_F and μ_{C_r} as follows:

$$F_i = randc_i(\mu_F, 0.1) \tag{10}$$

$$C_{ri} = randg_i(\mu_{C_r}, 0.1) \tag{11}$$

where $randc_i$ and $randg_i$ denote random selection using the Cauchy and normal distributions, respectively. F_i and C_{ri} are bounded in [0,1].

After each generation of populations is generated, the following Eq. (12) and Eq. (13) are utilized to update the adaptive parameters μF and μC_r, respectively.

$$\mu_{F_{new}} = \frac{\sum_F \cdot F^2}{\sum_F \cdot F} \tag{12}$$

$$\mu_{C_{r new}} = \frac{\sum_{C_r} \cdot C_r{}^2}{\sum_{C_r} \cdot C_r} \tag{13}$$

C. Success Historical Memory Based Parameter Adaptation
As shown in Table 1, LSHADE stores a historical memory with H entries for the adaptive parameters μ_F and μ_{C_r}, then randomly selects them at the time of F_i and C_{ri} generation for each individual. When initialized, all parameters are set to an initial value.

Table 1. The historical memory M_F and M_{C_r}

Index	1	2	...	$H-1$	H
μ_F	μ_{F_1}	μ_{F_2}	...	$\mu_{F_{H-1}}$	μ_{F_H}
μ_{C_r}	$\mu_{C_{r1}}$	$\mu_{C_{r2}}$	...	$\mu_{C_{r_{H-1}}}$	$\mu_{C_{r_H}}$

D. LPSR

Unlike the fixed population size in the traditional EA, LSHADE proposes a linear decrease for the population size. As shown in Eq. (14), the LPSR strategy sets an initial population size N^{init} for the algorithm, and afterward decreases it according to the current number of evaluation times and finally reduces it to a minimum size N^{min}.

$$NP = round\,[(\frac{N^{min} - N^{init}}{MFET}) \cdot nFET + N^{init}] \qquad (14)$$

where round() is a function that returns a rounded integer. $MFET$ means the maximum function evaluation times, $nFET$ means the current function evaluation times.

3   Deep-Layered Differential Evolution (DDE)

In this section, we presented a Deep-Layered Differential Evolution (named DDE) based on LSHADE. It is a population-based algorithm with multiple hierarchical layers, and the different layers are communicated with each other in different ways. The pseudo-code and structure of DDE is given to Alg. 1 and Fig. 1, respectively.

DDE is divided into four different layers according to the characteristics of individuals in the population: population layer, personal-best layer, archive layer, and rank-based selective pressure layer. Four layers in DDE are elaborated as follows:

a) Population Layer: All individuals are placed in this layer. Each individual in the population layer is involved in the mutation and crossover operations. Other layers provide the search guidance for the population layer.

b) Personal-best Layer: The top p high-quality individuals of the population constitute the personal-best layer. During the mutation operation, an high-quality individual is randomly selected by the personal-best layer to guide the individuals in the population layer.

c) Archive Layer: The additional external archive A becomes the archive layer. Individuals in this layer provide more possibilities for the mutation operation of individuals. Each individual has the probability of being selected to an individual in this layer when a random individual X_{r_2} is selected. This provides the algorithm with a higher population diversity.

d) Rank-based Selective Pressure Layer: All individuals are rank-ordered and randomly selected for X_{r_1} and X_{r_2} as individuals in the rank selection pressure layer. After the rank-ordered, individuals with better fitness values would have

---

Algorithm 1: DDE

Input: Parameters N^{init}, N^{min}, D, H, $|A|$, P_j, p, F, C_r, $MFET$
Output: The optimal solution

1 **Initialization**: Randomly generate a population $\{X_1, X_2, ..., X_{NP}\}$
2 **while** *the termination criatera are not met* **do**
3   **for** $i = 1$ to NP **do**
4   Calculate the F_i and C_{ri} by Eqs. (15) and (16), respectively;
5   Cheak the constraint of F_i and C_{ri} with Eqs. (19) and (20), respectively;
6   Update the mutation vector V_i by Eq. (7);
7   Update the candidate solution $u_{i,j}$ uses Eq. (23);
8   Adjust the boundary of individual U_i;
9   Evaluate the $f(U_i)$;
10   Selection the individual X_i by Eq. (9);
11   **end**
12   Update A;
13   Update $\mu_{F_{new}}$ and $\mu_{C_{rnew}}$ by Eqs. (17) and (18), respectively;
14   Calculate the population size uses Eq. (14);
15 **end**

---

higher probabilities of being selected by the mutation operation, thus retaining more information about these individuals. The algorithm uses this method to obtain a faster convergence rate.

The other strategies used in DDE are briefly described as follows:

1. Maximum Search Operator: In the history memory H, an additional set of search operators is set to (0.9, 0.9) and remains constant during the evolution process [19]. This ensures that a fraction of individuals have high mutation and crossover rates, improving the population's diversity and the ability of algorithm to jump out of the local optimum region.

$$F_i = \begin{cases} randc_i(\mu_F, 0.1), & i < H \\ randc_i(0.9, 0.1), & \text{otherwise} \end{cases} \tag{15}$$

$$C_{ri} = \begin{cases} randn_i(\mu_{C_r}, 0.1), & i < H \\ randn_i(0.9, 0.1), & \text{otherwise} \end{cases} \tag{16}$$

2. Search Operator Update Strategy: History memory preserves some information of the original parameters by means of averaging with the original parameters when updated [16].

$$\mu_{F_{new}} = (\frac{\sum_F \cdot F^2}{\sum_F \cdot F} + \mu_{F_{old}}) / 2 \tag{17}$$

$$\mu_{C_{rnew}} = (\frac{\sum_{C_r} \cdot C_r{}^2}{\sum_{C_r} \cdot C_r} + \mu_{C_{rold}}) / 2 \tag{18}$$

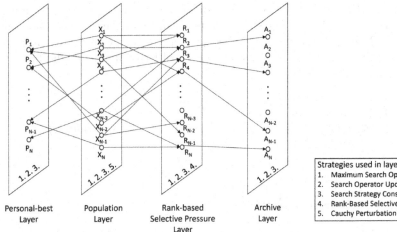

Personal-best Layer   Population Layer   Rank-based Selective Pressure Layer   Archive Layer

Strategies used in layer:
1. Maximum Search Operator
2. Search Operator Update Strategy
3. Search Strategy Constraint
4. Rank-Based Selective Pressure
5. Cauchy Perturbation

Fig. 1. Multiple hierarchical layers of DDE.

3. Search Strategy Constraint: The two search strategies generate different strategy constraints according to the current function evaluation times ($nFET$) during the evolution process. With different constraints, the algorithms have various focuses at different evolutionary stages [19].

$$
F_i = \begin{cases} 0.7, \; if \; F > 0.7 \; \text{and} \; nFET < 0.6 \cdot MFET \\ F_i, \; \text{otherwise} \end{cases}
\tag{19}
$$

$$
C_{ri} = \begin{cases} 0.7, \; if \; C_{ri} < 0.7 \; \text{and} \; nFET < 0.25 \cdot MFET \\ 0.6, \; if \; C_{ri} < 0.6 \; \text{and} \; nFET < 0.5 \cdot MFET \\ C_{ri}, \; \text{otherwise} \end{cases}
\tag{20}
$$

4. Rank-based Selective Pressure Strategy: By the rank-based selective pressure [16], each individual gets a score according to the rank of fitness, and the selected probability of an individual is calculated by a score, as shown in Eq. (22).

$$
R_i = k \cdot (NP - i) + 1
\tag{21}
$$

$$
pr_i = R_i / (R_1 + R_2 + \dots + R_{NP})
\tag{22}
$$

where k is the rank greediness factor, which is responsible for the rank selection. R_i is the rank score for individual i.

5. Cauchy Perturbation: The Cauchy perturbation strategy [20] determines whether an individual changes through the Cauchy distribution by the jump rate P_j. By employing the long-tailed property of the Cauchy distribution, the algorithm can create a very different set of individual, enough ability to explore the current region.

$$u_{i,j}^t = \begin{cases} v_{i,j}^t \,, \ j \ = \ j_r \text{ or } rand(0,1) \ < \ C_r \\ randc_i^t(x_{i,j}^t, \ 0.1) \,, \ \ j \ \neq \ j_r \ \& \ rand(0,1) \geq C_r \ \& \ rand(0,1) \ < \ P_j \\ x_{i,j}^t \,, \ \text{otherwise} \end{cases}$$

$$(23)$$

4   Experiments and Results

We tested DDE in the ICSI optimization competition' 2022 on single objective bounded optimization problems. The dimensions of benchmark functions in this test are D = 10, 20 and 50. Each function runs 50 times independently. The maximum function evaluation times ($MFET$) are set to 10000, 30000 and 70000 when D = 10, 20 and 50, respectively.

The PC configuration we used to test the program is: g++ (GNU Compiler) on a PC with an AMD Ryzen 7 1700X @ 3.4 GHz, Eight-Core processor and 16 GB RAM, using Ubuntu 18.04.6 LTS OS.

To show the performance of DDE, we used two champion EAs from the CEC single objective optimization competitions as the comparison algorithms, including:

1) The winner in the CEC 2017 competition, EBOwithCMAR [21], which uses the CMAES to improve the local search of the effective butterfly optimizer algorithm.
2) The winner in the CEC 2018 competition, HSES [22], which samples the search space with univariate sampled method, and later uses the CMAES to improve the search performance.

The parameters of DDE are set as shown in Table 2. The parameters of DDE are described with their meanings and the values used in the actual experiments. For EBOwithCMAR and HSES, we set the same parameters as in [21] and [22], respectively.

Table 2. Parameter Settings

Symbol	Description	Value
N^{init}	The initial population size	$5D$
N^{min}	The minimum population size	4
D	The dimension of the optimization problem to be solved	10/ 20/ 50
F	The initial scaling factor of the mutation strategy	0.3
C_r	The initial crossover rate of the crossover strategy	0.8
$|A|$	The size of external archive	1.0
H	The historical memory size	5
p	The p value for current-to-pbest/1 mutation	0.11
k	The rank greediness factor of rank-based selective pressure	3
P_j	The jump rate of the Cauchy distribution with perturbation	0.2

The obtained results are presented in Table 3, 4, and 5. In these tables, the error values of DDE are run 50 times independently for each problem. The best, worst, median, mean, and standard deviation of the error values are presented.

Next, we compare the performance between DDE with EBOwithCMAR and HSES. The results of error and standard are given in Tables 6, 7, and 8 for 10, 20 and 50 dimensions. And the totally result is shown in Table 9. We utilize the Wilcoxon rank-sum test with a significance level ($\alpha = 0.05$) is employed to determine whether there exist statistically significant differences between the performance of DDE and other competitors. $+$, $\approx$, and $-$ indicate that DDE is significantly better,tied or significantly worse than the comparison algorithm. The results in these tables clearly show that DDE is better than its competitors.

Table 3. DDE results for 10D problems

No.	Best	Worst	Median	Mean	Std
F01	2.85E−03	3.94E+04	2.26E+03	4.33E+03	6.70E+03
F02	1.79E−03	7.71E+03	4.41E+01	5.92E+02	1.51E+03
F03	0.00E+00	6.15E+00	6.06E+00	5.08E+00	2.24E+00
F04	2.98E+00	1.39E+01	5.97E+00	6.47E+00	2.59E+00
F05	3.13E−01	1.10E+03	4.38E+02	4.30E+02	2.28E+02
F06	0.00E+00	7.17E−01	5.66E−05	1.74E−02	1.01E−01
F07	3.08E−02	2.49E−01	1.24E−01	1.23E−01	4.75E−02
F08	0.00E+00	4.55E+05	4.37E+05	3.27E+05	1.96E+05
F09	2.85E−02	6.18E+05	3.04E+04	6.29E+04	1.12E+05
F10	4.21E−01	7.43E+01	1.37E+01	1.68E+01	1.44E+01

Table 4. DDE results for 20D problems

No.	Best	Worst	Median	Mean	Std
F01	3.98E+03	1.75E+05	3.88E+04	5.08E+04	4.41E+04
F02	1.81E−01	2.75E+03	5.49E+02	8.06E+02	8.57E+02
F03	5.37E−04	1.35E+01	9.49E+00	9.54E+00	1.69E+00
F04	3.98E+00	2.19E+01	1.19E+01	1.21E+01	4.57E+00
F05	7.94E+02	2.12E+03	1.57E+03	1.54E+03	3.78E+02
F06	0.00E+00	7.53E−02	5.01E−04	4.85E−02	3.10E−01
F07	4.29E−02	2.67E−01	1.74E−01	1.67E−01	5.29E−02
F08	4.23E+05	4.73E+05	4.35E+05	4.39E+05	1.44E+04
F09	1.39E+05	3.37E+07	9.55E+06	1.09E+07	9.12E+06
F10	5.04E+00	7.27E+01	2.52E+01	3.08E+01	1.80E+01

Table 5. DDE results for 50D problems.

No.	Best	Worst	Median	Mean	Std
F01	3.23E+05	1.60E+06	8.07E+05	8.48E+05	2.82E+05
F02	1.46E−01	4.44E+03	3.10E+02	8.09E+02	1.14E+03
F03	2.88E−02	8.51E+01	7.07E+01	5.28E+01	3.14E+01
F04	2.29E+01	6.44E+01	3.64E+01	3.85E+01	9.63E+00
F05	3.37E+03	7.36E+03	6.02E+03	6.00E+03	8.53E+02
F06	5.82E−03	1.33E+01	1.59E−01	7.17E−01	2.24E+00
F07	1.89E−01	5.09E−01	3.20E−01	3.22E−01	6.00E−02
F08	4.50E+05	5.52E+05	5.06E+05	5.08E+05	2.36E+04
F09	8.69E+07	9.07E+08	4.42E+08	4.85E+08	2.15E+08
F10	3.18E+02	8.57E+02	5.16E+02	5.16E+02	1.02E+02

Table 6. Experimental results of DDE and other comparison algorithms for D = 10.

Alg.	DDE	EBOwithCMAR	HSES
F01	**4.33E+03 ± 6.70E+03**	8.07E+06 ± 1.13E+07 +	6.87E+06 ± 7.51E+06 +
F02	**5.92E+02 ± 1.51E+03**	1.89E+07 ± 1.24E+08 +	7.80E+08 ± 8.53E+08 +
F03	**5.08E+00 ± 2.24E+00**	8.98E+00 ± 9.44E+00 +	1.61E+02 ± 6.57E+01 +
F04	**6.47E+00 ± 2.59E+00**	7.30E+01 ± 1.00E+01 +	6.75E+01 ± 1.45E+01 +
F05	**4.30E+02 ± 2.28E+02**	1.43E+03 ± 3.08E+02 +	1.36E+03 ± 3.59E+02 +
F06	**1.74E−02 ± 1.01E−01**	5.65E+01 ± 3.02E+01 +	1.16E+01 ± 5.25E+00 +
F07	1.23E−01 ± 4.75E−02	**1.20E−01 ± 4.44E−02** ≈	1.20E+00 ± 4.00E−01 +
F08	3.27E+05 ± 1.96E+05	**3.75E+05 ± 1.58E+04** −	4.10E+05 ± 2.33E+04 −
F09	**6.29E+04 ± 1.12E+05**	6.50E+06 ± 4.37E+06 +	4.11E+06 ± 4.91E+06 +
F10	**1.68E+01 ± 1.44E+01**	1.53E+02 ± 4.01E+02 +	1.00E+02 ± 5.13E+01 +

Table 7. Experimental results of DDE and other comparison algorithms for D = 20.

Alg.	DDE	EBOwithCMAR	HSES
F01	5.08E+04 ± 4.41E+04	**1.66E+03 ± 1.12E+03** −	3.84E+07 ± 5.65E+07 +
F02	**8.06E+02 ± 8.57E+02**	7.90E+03 ± 1.10E+04 +	9.09E+09 ± 1.45E+09 +
F03	**9.54E+00 ± 1.69E+00**	9.59E+00 ± 2.73E+00 +	5.85E+02 ± 1.47E+02 +
F04	**1.21E+01 ± 4.57E+00**	1.93E+02 ± 2.62E+01 +	1.60E+02 ± 2.31E+01 +
F05	**1.54E+03 ± 3.78E+02**	3.01E+03 ± 5.72E+02 +	2.31E+03 ± 5.88E+02 +
F06	**4.85E−02 ± 3.10E−01**	1.10E+02 ± 1.12E+02 +	6.19E+00 ± 6.48E+00 +
F07	**1.67E−01 ± 5.29E−02**	1.78E−01 ± 7.68E−02 ≈	2.37E+00 ± 4.14E−01 +
F08	**4.39E+05 ± 1.44E+04**	5.19E+05 ± 1.94E+04 +	5.13E+05 ± 1.55E+04 +
F09	1.09E+07 ± 9.12E+06	2.75E+08 ± 5.01E+08 ≈	**2.81E+07 ± 1.26E+08** −
F10	**3.08E+01 ± 1.80E+01**	7.57E+02 ± 2.45E+03 +	1.04E+04 ± 2.56E+03 +

Table 8. Experimental results of DDE and other comparison algorithms for D = 50.

Alg.	DDE	EBOwithCMAR	HSES
F01	8.48E+05 ± 2.82E+05	**5.49E+05 ± 1.52E+05** −	1.02E+09 ± 3.20E+08 +
F02	8.09E+02 ± 1.14E+03	**1.70E+02 ± 3.26E+02** −	4.13E+08 ± 2.91E+07 +
F03	5.28E+01 ± 3.14E+01	**1.79E+01 ± 2.86E+01** −	5.62E+03 ± 8.70E+02 +
F04	**3.85E+01 ± 9.63E+00**	4.13E+02 ± 1.44E+02 +	4.69E+02 ± 9.43E+01 +
F05	6.00E+03 ± 8.53E+02	7.45E+03 ± 1.00E+03 +	**6.00E+03 ± 1.46E+03** ≈
F06	**7.17E−01 ± 2.24E+00**	1.56E+02 ± 1.95E+02 +	2.43E+01 ± 7.27E+01 +
F07	3.22E−01 ± 6.00E−02	**2.21E−01 ± 6.32E−02** −	3.99E+00 ± 1.91E−01 +
F08	**5.08E+05 ± 2.36E+04**	6.55E+05 ± 4.46E+04 +	5.97E+05 ± 7.19E+04 +
F09	**4.85E+08 ± 2.15E+08**	9.27E+10 ± 2.29E+11 +	5.78E+09 ± 3.10E+10 +
F10	**5.16E+02 ± 1.02E+02**	6.81E+02 ± 3.10E+02 +	7.95E+04 ± 1.32E+04 +

Table 9. Statistical results for Wilcoxon rank-sum

DDE vs.	EBOwithCMAR	HSES
w/t/l	+/≈/−	
D = 10	8/1/1	9/0/1
D = 20	7/2/1	9/0/1
D = 50	6/0/4	9/1/0

5   Conclusions

Differential evolution algorithms and its variants have excellent performance on optimization problems. In this paper, we present the results of our proposed algorithm (DDE) in the ICSI optimization competition' 2022 on single objective bounded optimization problems.

In this work, we compared our proposed algorithm (DDE) with the EBOwith-CMAR and HSES. DDE has better overall results than the EBOwithCMAR and HSES algorithms in all dimensions. It shows that DDE has the excellent performance.

References

1. Gao, S., Wang, K., Tao, S., Jin, T., Dai, H., Cheng, J.: A state-of-the-art differential evolution algorithm for parameter estimation of solar photovoltaic models. Energy Convers. Manage. **230**, 113784 (2021)
2. Gao, S., Zhou, M., Wang, Y., Cheng, J., Yachi, H., Wang, J.: Dendritic neuron model with effective learning algorithms for classification, approximation, and prediction. IEEE Trans. Neural Network. Learn. Syst. **30**(2), 601–614 (2019)
3. Slowik, A., Kwasnicka, H.: Evolutionary algorithms and their applications to engineering problems. Neural Comput. Appl. **32**(16), 12363–12379 (2020). https://doi.org/10.1007/s00521-020-04832-8
4. Yar, M.H., Rahmati, V., Oskouei, H.R.D.: A survey on evolutionary computation: methods and their applications in engineering. Mod. Appl. Sci. **10**(11), 131139 (2016)
5. Eiben, A.E., Smith, J.: From evolutionary computation to the evolution of things. Nature **521**(7553), 476–482 (2015)
6. Wang, Y., Gao, S., Yu, Y., Cai, Z., Wang, Z.: A gravitational search algorithm with hierarchy and distributed framework. Knowl.-Based Syst. **218**, 106877 (2021)
7. Xia, X., et al.: Triple archives particle swarm optimization. IEEE Trans. Cybern. **50**(12), 4862–4875 (2019)
8. Arabas, J., Jagodziński, D.: Toward a matrix-free covariance matrix adaptation evolution strategy. IEEE Trans. Evol. Comput. **24**(1), 84–98 (2019)
9. Zhan, Z.H., Wang, Z.J., Jin, H., Zhang, J.: Adaptive distributed differential evolution. IEEE Trans. Cybern. **50**(11), 4633–4647 (2019)
10. Storn, R., Price, K.: Differential Evolution-a simple and efficient heuristic for global optimization over continuous spaces. J. Global Optim. **11**(4), 341–359 (1997)
11. Pant, M., Zaheer, H., Garcia-Hernandez, L., Abraham, A., et al.: Differential evolution: a review of more than two decades of research. Eng. Appl. Artif. Intell. **90**, 103479 (2020)
12. Opara, K.R., Arabas, J.: Differential evolution: a survey of theoretical analyses. Swarm Evol. Comput. **44**, 546–558 (2019)
13. Tanabe, R., Fukunaga, A.: Success-history based parameter adaptation for differential evolution. In: IEEE Congress on Evolutionary Computation (CEC). IEEE 2013, pp. 71–78 (2013)
14. Tanabe, R., Fukunaga, A.S.: Improving the search performance of SHADE using linear population size reduction. In: IEEE Congress on Evolutionary Computation (CEC). IEEE, pp. 1658–1665 (2014)
15. Brest, J., Maučec, M.S., Bošković, B.: Single objective real-parameter optimization: Algorithm jSO. In: IEEE Congress on Evolutionary Computation (CEC). IEEE 2017, pp. 1311–1318 (2017)
16. Stanovov, V., Akhmedova, S., Semenkin, E.: LSHADE algorithm with rank-based selective pressure strategy for solving CEC 2017 benchmark problems. In: 2018 IEEE Congress on Evolutionary Computation (CEC), IEEE pp. 1–8 (2018)

17. Brest, J., Maučec, M.S., Bošković, B.: The 100-digit challenge: Algorithm jDE100. In: IEEE Congress on Evolutionary Computation (CEC). IEEE 2019, pp. 19–26 (2019)
18. Sallam, K.M., Elsayed, S.M., Chakrabortty, R.K., Ryan, M.J.: Improved multi-operator differential evolution algorithm for solving unconstrained problems. In: IEEE Congress on Evolutionary Computation (CEC). IEEE 2020, pp. 1–8 (2020)
19. Brest, J., Maučec, M.S., Bošković, B.: iL-SHADE: Improved L-SHADE algorithm for single objective real-parameter optimization. In,: IEEE Congress on Evolutionary Computation (CEC). IEEE 2016, pp. 1188–1195 (2016)
20. Choi, T.J., Ahn, C.W.: An improved LSHADE-RSP algorithm with the Cauchy perturbation: iLSHADE-RSP. Knowl.-Based Syst. **215**, 106628 (2021)
21. Kumar, A., Misra, R.K., Singh, D.: Improving the local search capability of effective butterfly optimizer using covariance matrix adapted retreat phase. In: IEEE Congress on Evolutionary Computation (CEC). IEEE 2017, pp. 1835–1842 (2017)
22. Zhang, G., Shi, Y.: Hybrid sampling evolution strategy for solving single objective bound constrained problems. In,: IEEE Congress on Evolutionary Computation (CEC). IEEE 2018, pp. 1–7 (2018)

Dual-Populatuion Differential Evolution L-NTADE for ICSI-OC'2023 Competition

Vladimir Stanovov$^{(\boxtimes)}$ and Eugene Semenkin

School of Space and Information Technologies, Siberian Federal University, 660074
Krasnoyarsk, Russia
{vladimirstanovov,eugenesemenkin}@yandex.ru

Abstract. In this study the performance of a dual-population differential evolution algorithm L-NTADE (Linear population size reduction Newest and Top Adaptive Differential Evolution) is evaluated on the set of benchmark functions of the ICSI-OC'2023 competition. The original L-NTADE algorithm is modified by repairing crossover rate values in parameter adaptation, allowing more precise tuning. The performed computational experiments on the ICSI-OC'2022 and ICSI-OC'2023 have shown that L-NTADE is superior compared to one of the top methods from previous year competition.

Keywords: Differential evolution · Parameter adaptation · Numerical optimization

1 Introduction

The evolutionary numerical optimization methods nowadays represent one of the important directions of research in the area of evolutionary algorithms (EAs). This is because these algorithms apply to many real-world problems, so developing more efficient approaches plays a vital role in developing modern technologies. Among the existing directions in numerical black-box optimization, the methods based on differential evolution (DE) have become widely used due to their high efficiency and simplicity of implementation [1].

Today different modifications of DE are prize-winning algorithms in most competitions on numerical optimization. The most efficient approaches are derived from the L-SHADE algorithm [2], which used parameter adaptation for scaling factor F, crossover rate Cr and a control strategy for population size N. The modifications of L-SHADE changed the adaptation schemes or added new features, and although there was a significant progress [3], the main scheme of the algorithm remained the same. Several attempts were made to change the main scheme, for example in jDE100 [4] two populations were used, and in [5] the DE with an unbounded population was proposed.

In this study we further develop the recently proposed Linear population size reduction Newest and Top Adaptive Differential Evolution (L-NTADE) algorithm [6], which maintains two populations, one containing the newly generated solutions i.e. newest population, and the other - top solutions, i.e. best found through the whole search process. The algorithm is modified by adding the mechanism of repairing crossover rates, proposed in [7], which allows more accurate parameter tuning. The resulting L-NTADE_rcr is tested on the ICSI Optimization Competition 2022 (ICSI-OC'2022) [8] and ICSI-OC'2023 [9] and compared to the NL-SHADE-LM approach [10], which took the third place in previous year competition. The participants of the ICSI-OC'22 included the Fireworks Algorithm with Search Space Partition (FWASSP) [13], which took first place, Composite Evolutionary Strategy and Differential Evolution [14] and Surrogate-Assisted Differential Evolution [15].

The rest of the paper is organized as follows: Sect. 2 gives a description of the modern DE methods, Sect. 3 describes the proposed approach, Sect. 4 contains the experimental setup and results, and Sect. 5 concludes the paper.

2   Related Work

2.1   Differential Evolution

The idea behind the differential evolution algorithm, proposed in [11], was to use vectors of difference between positions of individuals in the population to generate new solutions. The algorithm, same as most numerical methods, such as real-coded genetic algorithms, particle swarm optimization and others, starts by initializing a set of N points in D-dimensional search space: $x_{i,j}$, $i = 1...N$, $j = 1...D$. The population size N is the first parameter of the algorithm.

After calculating target function values $f(x_i)$, the main cycle begins, containing mutation, crossover and selection operations. Nowadays most DE algorithms use the current-to-pbest mutation strategy, proposed in the JADE algorithm [12]:

$$v_{i,j} = x_{i,j} + F(x_{pbest,j} - x_{i,j}) + F(x_{r1,j} - x_{r2,j}), \tag{1}$$

where $pbest$ is one of the $pb*100\%$ best solutions, $r1$ and $r2$ are random indexes, F is the scaling factor parameter. $pbest$, $r1$ and $r2$ are generated to be different from each other and i. The mutant vector v is then used in crossover, which combines it with target vector x_i to generate trial vector u_i using crossover rate parameter Cr:

$$u_{i,j} = \begin{cases} v_{i,j}, & \text{if } rand(0,1) < Cr \text{ or } j = jrand \\ x_{i,j}, & \text{otherwise} \end{cases}, \tag{2}$$

where $jrand \in [1, D]$. After applying bound-constraint handling method, the selection (replacement) step is performed:

$$x_{i,j} = \begin{cases} u_{i,j}, & \text{if } f(u_i) \le f(x_i) \\ x_{i,j}, & \text{if } f(u_i) > f(x_i) \end{cases}. \tag{3}$$

In selection the newly generated trial vector replaces target vector only if it is at least as good in terms of target function.

Most modern DE-based algorithm used parameter adaptation scheme, proposed in the L-SHADE algorithm [2] and called success-history adaptation (SHA). SHA is used to tune F and Cr values by storing pairs of parameters in memory cells and using Cauchy/normal distribution to generate new values. The F and Cr values which helped to generate better solutions are saved and further used to update memory cells. In addition to this, the population size N is linearly reduced, allowing better exploration at the beginning and better exploitation at the end of the search. L-SHADE algorithm also used an archive of inferior solutions to increase the diversity of generated solutions.

3   Proposed Approach

The L-NTADE algorithm [6] considered in this study was inspired by the ideas proposed in the UDE [5]. L-NTADE deflects from a mainstream idea in DE algorithms, where a population of solutions is updated by replacing individuals with selection step. In L-NTADE there are two populations maintained, one containing the newly generated solutions, and the other - top individuals, i.e. best found through the whole search process.

The algorithm starts by initializing a population x_i^{new}, $i = 1, ..., N_{max}$. After calculating target function values, all the individuals are copied to the top population x^{top}. Both populations have the same size. As for the mutation strategy this study, the r-new-to-ptop/n/t is used, described as follows:

$$v_{i,j} = x_{r1,j}^{new} + F \times (x_{pbest,j}^{top} - x_{i,j}^{new}) + F \times (x_{r2,j}^{new} - x_{r3,j}^{top}). \tag{4}$$

The crossover step is modified by introducing repaired crossover rate, proposed in [7]. The L-NTADE uses the success-history based parameter adaptation scheme from the L-SHADE algorithm, so that F_i and Cr_i, $i = 1, ..., N_{cur}$ values are generated before mutation and crossover. During crossover the repaired value Cr_i^{rep} is calculated as the ratio of number of actually taken components from the trial vector. This can be formally presented as follows:

$$b_{i,j} = \begin{cases} 1, & \text{if } rand(0,1) < Cr \text{ or } j = jrand \\ 0, & \text{otherwise} \end{cases}, \tag{5}$$

where $b_{i,j}$ is a binary vector indicating which components are taken. The repaired value is calculated as $Cr_i^{rep} = \frac{\sum_{j=1}^{D} b_{i,j}}{D}$, and the trial vector is calculated as follows:

$$u_{i,j} = b_{i,j}v_{i,j} + (1 - b_{i,j})x_{i,j}. \tag{6}$$

The main advantage of the repaired crossover rate is that the actually applied Cr_i^{rep} value will be further used in parameter adaptation instead of the Cr_i.

Another feature of the L-NTADE is the rank-based selective pressure, applied for the $r2$ index. The individuals in the newest population are ranked according to fitness, and probabilities to be chosen are calculated proportional to rank values:

$$rank_i = e^{\frac{-kp \cdot i}{N}}, \tag{7}$$

where kp is the parameter controlling the pressure, and i is the individual number.

The selection step in L-NTADE is changed, and it updates the newest population individual with index nc, iterated from 1 to current population size N^{cur}, but the trial vector's fitness is compared to the fitness of the $r1$-th individual, not nc-th:

$$x_{nc} = \begin{cases} u_i, & \text{if } f(u_i) \leq f(x_{r1}^{new}) \\ x_{nc}, & \text{if } f(u_i) > f(x_{r1}^{new}). \end{cases} \tag{8}$$

This means that it is possible in L-NTADE, with the crossover rate sorting, to generate two new successful solutions with the same Cr value; hence, the same value is used in updating memory cells in the success-history adaptation mechanism.

The L-NTADE with the described modification for crossover rate repair will be further referred to as L-NTADE_rcr. The pseudocode of the method is presented in Algorithm 1.

4 Experimental Setup and Results

The experiments in this study were performed on two sets of test problems, namely the ICSI-OC'2022 and ICSI-OC'2023 competition on numerical optimization. The two benchmarks are similar in characteristics, but the ICSI-OC'2023 has modified functions. The tests functions are defined for $D = 10$, 20 and 50, and the computational resource is set to 10000, 30000 and 70000 accordingly. The experiments with L-NTADE_rcr were performed in comparison with NL-SHADE-LM method, which participated in ICSI-OC'2022 and took third place out of six. The L-NTADE_rcr algorithm was implemented in Python 3.8. The parameters used for testing are presented in Algorithm 1.

Table 1 shows the Mann-Whitney statistical tests with normal approximation and tie-breaking for different dimensions of ICSI-OC'2022 and ICSI-OC'2023. The values in the table are the number of wins ($+$), ties ($=$) and losses ($-$).

Table 1. L-NTADE_rcr vs NL-SHADE-LM

Algorithms	$10D$	$20D$	$50D$
L-NTADE_rcr vs NL-SHADE-LM (ICSI-OC'2022)	5+/4=/1-	6+/3=/1-	7+/0=/3-
L-NTADE_rcr vs NL-SHADE-LM (ICSI-OC'2023)	4+/4=/2-	4+/5=/1-	6+/2=/2-

Table 1 shows that the L-NTADE_rcr is superior compared to NL-SHADE-LM in most cases, although it has worse performance sometimes. In particular, in case of ICSI-OC'2022, it performed worse than NL-SHADE-LM for the 10-th function (composition function 3), and in $50D$ it also performed worse on the first two unimodal functions. In case if ICSI-OC'2023, the losses were on functions 3 and 6 ($10D$), function 6 ($20D$), and functions 1 and 5 ($50D$).

Algorithm 1. L-NTADE_rcr

1: Input: D, NFE_{max}, objective function $f(x)$
2: Output: x_{best}^{top}, $f(x_{best}^{top})$
3: Set $N_{max} = 10D$, $N_{cur}^0 = N_{max}$, $N_{min} = 4$, $H = 5$, $M_{F,r} = 0.4$, $M_{Cr,r} = 1$
4: Set $pb = 0.3$, $k = 1$, $g = 0$, $nc = 1$, $kp = 3$, $pm_F = 4$
5: Initialize population $(x_{1,j}^{new}, ..., x_{N_{max},j}^{new})$ randomly, calculate $f(x^{new})$
6: Copy x^{new} to x^{top}, $f(x^{new})$ to $f(x^{top})$
7: **while** $NFE < NFE_{max}$ **do**
8:     $S_F = \emptyset$, $S_{Cr} = \emptyset$, $S_{\Delta f} = \emptyset$
9:     Rank either x^{new} by $f(x^{new})$
10:     **for** $i = 1$ to N_{cur}^g **do**
11:         $r1 = randInt(N_{cur}^g)$
12:         Current memory index $r = randInt[1, H + 1]$
13:         **repeat**
14:             $F_i = randc(M_{F,r}, 0.1)$
15:         **until** $F_i \geq 0$
16:         $F_i = min(1, F_i)$
17:         Crossover rates $Cr_i = randn(M_{Cr,r}, 0.1)$
18:         $Cr_i = min(1, max(0, Cr))$
19:         **repeat**
20:             $pbest = randInt(1, N_{cur}^g * pb)$
21:             $r2 = randInt(1, N_{cur}^g)$ or with rank-based selection
22:             $r3 = randInt(1, N_{cur}^g)$
23:         **until** indexes $r1$, $r2$, $r3$ and $pbest$ are different
24:         Apply mutation to produce v_i with F_i
25:         Apply binomial crossover to produce u_i with Cr_i
26:         Calculate repaired crossover rate Cr^{rep}
27:         Apply bound constraint handling method
28:         Calculate $f(u_i)$
29:         **if** $f(u_i) < f(x_{r1}^{new})$ **then**
30:             $u_i \rightarrow x^{temp}$
31:             $F_i \rightarrow S_F$, $Cr^{rep} \rightarrow S_{Cr}$
32:             $\Delta f = f(x_{r1}^{new}) - f(u_i)$
33:             $\Delta f \rightarrow S_{\Delta f}$
34:             $x_{nc}^{new} = u_i$
35:             $nc = mod(nc + 1, N_{cur}^g)$
36:         **end if**
37:     **end for**
38:     Get N_{cur}^{g+1} with LPSR
39:     Join together x^{top} and x^{temp}, sort and copy best N_{cur}^{g+1} to x^{top}
40:     **if** $N_{cur}^g > N_{cur}^{g+1}$ **then**
41:         Remove worst individuals from x^{new}
42:     **end if**
43:     Update $M_{F,k}$, $M_{Cr,k}$
44:     $k = mod(k + 1, H)$
45:     $g = g + 1$
46: **end while**
47: Return x_{best}^{top}, $f(x_{best}^{top})$

Tables 2, 3 and 4 contain the results of L-NTADE_rcr for 10D, 20D and 50D, these values are provided for comparison with alternative approaches.

Table 2. L-NTADE_rcr results, 10D

No.	Best	Worst	Median	Mean	Std
F1	8.950219e+00	9.730551e+04	5.641668e+03	9.141224e+03	1.486696e+04
F2	3.419422e+01	1.700770e+04	3.125872e+03	4.941415e+03	4.999184e+03
F3	3.015668e-08	1.038233e+02	1.979264e-05	2.078428e+00	1.453499e+01
F4	9.900222e-01	2.132555e+01	4.950093e+00	6.780944e+00	5.229516e+00
F5	3.543656e-06	5.944432e-01	7.127903e-02	1.262006e-01	1.274296e-01
F6	6.666667e-01	6.666673e-01	6.666667e-01	6.666667e-01	8.927146e-08
F7	7.234773e-02	6.041305e-01	2.674638e-01	2.846075e-01	1.404930e-01
F8	9.244426e+05	1.004560e+06	9.499269e+05	9.554760e+05	2.105212e+04
F9	1.797772e+01	1.564120e+07	4.802558e+04	7.994148e+05	2.511913e+06
F10	3.235995e+01	2.651768e+02	8.705848e+01	9.859064e+01	4.761331e+01

Table 3. L-NTADE_rcr results, 20D

No.	Best	Worst	Median	Mean	Std
F1	5.074279e+04	2.638248e+06	4.290633e+05	6.169174e+05	4.838373e+05
F2	7.636088e-01	2.231104e+03	3.868582e+02	5.810163e+02	5.995728e+02
F3	4.326259e-05	2.400555e-01	4.031716e-02	7.117680e-02	6.747136e-02
F4	2.970055e+00	5.407645e+01	9.900194e+00	1.515631e+01	1.064077e+01
F5	9.778628e-01	1.091504e+02	8.502374e+00	1.555814e+01	2.061865e+01
F6	6.666667e-01	6.666826e-01	6.666676e-01	6.666693e-01	4.119123e-06
F7	1.472481e-01	4.897621e-01	2.858563e-01	2.870333e-01	7.706635e-02
F8	1.744102e+06	1.798552e+06	1.762762e+06	1.763426e+06	1.100994e+04
F9	1.772161e+04	2.220592e+07	2.041319e+06	2.773406e+06	3.328847e+06
F10	3.128034e+04	3.128035e+04	3.128034e+04	3.128034e+04	4.189348e-03

Figures 1, 2 and 3 show the convergence graphs of NL-SHADE-LM in comparison to L-NTADE_rcr on the ICSI-OC'2023 benchmark.

As can be seen from Figs. 1–3, the difference in the convergence curves are observed closer to the end of the search process. In particular, the L-NTADE_rcr algorithm shows better convergence closer to the end of the search process, especially on functions 7 and 9. In the middle of the search process the L-NTADE_rcr has larger error values compared to NL-SHADE-LM, probably due to larger diversity and using the linear population size reduction instead of non-linear in NL-SHADE-LM, which leads to smaller population size earlier. This increased diversity allows better exploration and, as a result, better convergence as the

Table 4. L-NTADE_rcr results, 50D

No.	Best	Worst	Median	Mean	Std
F1	2.115444e+07	1.996227e+08	6.672688e+07	6.739280e+07	3.320352e+07
F2	3.614452e+00	8.013684e+03	1.757400e+03	2.266013e+03	2.101087e+03
F3	1.878650e-03	5.600301e+02	5.138177e-01	1.339274e+02	2.373809e+02
F4	7.928732e+00	4.959956e+01	1.929715e+01	2.102762e+01	9.724358e+00
F5	2.351836e+02	1.823910e+03	7.344459e+02	7.979645e+02	3.183504e+02
F6	6.668969e-01	1.607636e+00	6.752279e-01	6.975358e-01	1.304470e-01
F7	1.573942e-01	4.755842e-01	2.868277e-01	2.883228e-01	6.360548e-02
F8	1.579738e+06	1.697448e+06	1.646863e+06	1.644364e+06	2.904634e+04
F9	2.211338e+07	2.588228e+08	9.392970e+07	1.096210e+08	5.895507e+07
F10	6.618763e+04	6.618808e+04	6.618775e+04	6.618778e+04	1.087299e-01

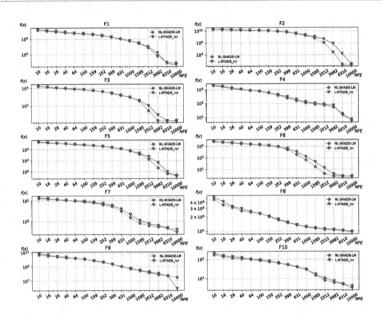

Fig. 1. NL-SHADE-LM vs L-NTADE_rcr, 10D.

worst individuals are removed from the population. The graphs for different dimensions in Figs. 1–3 are similar in structure, and both algorithms demonstrate the same convergence trends.

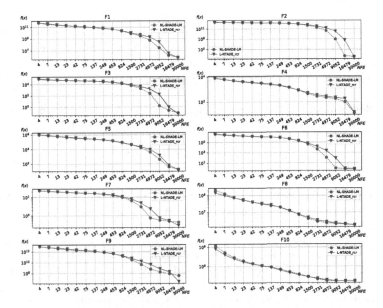

Fig. 2. NL-SHADE-LM vs L-NTADE_rcr, 20D.

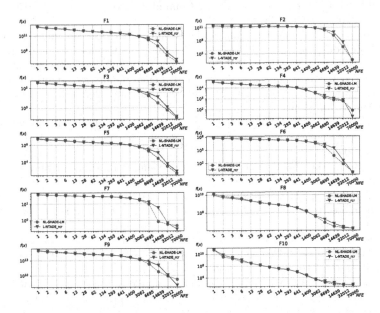

Fig. 3. NL-SHADE-LM vs L-NTADE_rcr, 50D.

5   Conclusion

In this study the L-NTADE_rcr algorithm with crossover rate repair was proposed. The experiments were performed on the ICSI-OC'2022 and ICSI-OC'2023 benchmarks, and the results have shown that the proposed algorithm is capable of outperforming one of the best algorithms from the previous year competition.

Acknowledgement. This research was funded by the Ministry of Science and Higher Education of the Russian Federation, Grant No. 075-15-2022-1121.

References

1. Ahmad, M.F., Isa, N.A., Lim, W.H., Ang, K.M.: Differential evolution: A recent review based on state-of-the-art works. Alex. Eng. J. **61**, 3831–3872 (2021). https://doi.org/10.1016/j.aej.2021.09.013
2. Tanabe, R., Fukunaga, A.S.: Improving the search performance of SHADE using linear population size reduction. In: Proceedings of the IEEE Congress on Evolutionary Computation, pp. 1658–1665 (2014) https://doi.org/10.1109/CEC.2014.6900380
3. Al-Dabbagh, R. D., Neri, F., Idris, N., Baba, M. S.: Algorithmic design issues in adaptive differential evolution schemes: Review and taxonomy. In: Swarm and Evolutionary Computation 43, pp. 284–311 (2018) https://doi.org/10.1016/j.swevo.2018.03.008
4. Brest, J., Maučec, M.S., Bošković, B. The 100-Digit Challenge: Algorithm jDE100. 2019 IEEE Congress on Evolutionary Computation (CEC), 19–26 (2019) https://doi.org/10.1109/CEC.2019.8789904
5. Kitamura, T.; Fukunaga, A. Differential Evolution with an Unbounded Population. In: Proceedings of the 2022 IEEE Congress on Evolutionary Computation (CEC), Padua, Italy, 18–23 July 2022. https://doi.org/10.1109/CEC55065.2022.9870363
6. Stanovov, V., Akhmedova, S., Semenkin, E.: Dual-population adaptive differential evolution algorithm L-NTADE. Mathematics **10**, 4666 (2022). https://doi.org/10.3390/math10244666
7. Gong, W., Cai, Z., Wang, Y.: Repairing the crossover rate in adaptive differential evolution. Appl. Soft Comput. **15**, 149–168 (2014). https://doi.org/10.1016/j.asoc.2013.11.005
8. Li, Y.: Definitions for the ICSI Optimization Competition'2022 on Single Objective Bounded Optimization Problems, Technical report, Peking University (2022)
9. Li, Y., Shipeng C.: Definitions for the ICSI Optimization Competition'2023 on Single Objective Bounded Optimization Problems, Technical report, Peking University (2023)
10. Stanovov, V., Akhmedova, S.: Differential evolution with biased parameter adaptation for ICSI-OC 2022 competition. Int. Conf. Swarm Intell. (2022). https://doi.org/10.1007/978-3-031-09726-3_38
11. Storn, R., Price, K.: Differential evolution - a simple and efficient heuristic for global optimization over continuous spaces. J. Global Optim. **11**(4), 341–359 (1997). https://doi.org/10.1023/A:1008202821328
12. Zhang, J., Sanderson, A.C.: JADE: adaptive differential evolution with optional external archive. IEEE Trans. Evol. Comput. **13**(5), 945–958 (2009)

13. Li, Y., Li, Y., Tan, Y.: Enhancing fireworks algorithm in local adaptation and global collaboration for solving icsi 2022 benchmark problems. Int. Conf. Swarm Intell. (2022). https://doi.org/10.1007/978-3-030-78743-1_41

14. Kudela, J., Holoubek, T., Nevoral, T.: Composite evolutionary strategy and differential evolution method for the ICSI'2022 competition. Int. Conf. Swarm Intell. (2022). https://doi.org/10.1007/978-3-031-09726-3_39

15. Kudela, J., Holoubek, T., Nevoral, T.: Surrogate-assisted differential evolution-based method for the ICSI'2022 competition. Int. Conf. Swarm Int. (2022). https://doi.org/10.1007/978-3-031-09726-3_40

Group Simulated Annealing Algorithm for ICSI-OC 2022

Mingyin Zou[1,2], Peichen Zhang[2], and Xiaomin Zhu[1(✉)]

[1] Strategic Assessments and Consultation Institute, Academy of Military Science,
Beijing 100097, China
zoumingyin20@alumni.nudt.edu.cn , xmzhu@nudt.edu.cn
[2] Laboratory for Big Data and Decision, National University of Defense Technology,
Changsha 410073, China

Abstract. Single-objective bounded optimization problems are widely used in real-world applications. Various evolutionary algorithms have been proposed to solve these problems. The Simulated Annealing Algorithm is one of the most popular algorithms with the advantage of fast convergence. When the problems have numerous local optimums, SA may fall into the first few local optimums. In this paper, to avoid falling into the local optimums, a Group Simulated Annealing Algorithm is proposed with a group strategy that separated each solution in the group for the annealing strategy. To assess the performance of the proposed algorithm, empirical experiments have been conducted on the test problems ICSI-OC 2022 with the comparison of the Simulated Annealing Algorithm.

Keywords: Simulated Annealing Algorithm · Single-objective bounded optimization · Group strategy

1 Introduction

Single-objective bounded optimization problems refer to the single-objective optimization problems including the boundary constraints, that are widely used in real-world applications, such as deep learning [8], machine learning [1], and complex network [11]. To solve this problem, various stochastic search algorithms such as evolutionary algorithms [5,9] and swarm intelligence algorithms [4,6] have been proposed. Simulated Annealing (SA) is one of the simplest and best-known single-objective evolutionary algorithms [3]. With the advantage of fast convergence, SA has a wide range of applications in various science and engineering fields. For example, In the field of Vehicle Routing Problem (VRP), SA with a restart strategy is proposed for an extension of the Green VRP [10].

This work was supported in part by the National Natural Science Foundation of China under Grants 61872378, 62002369, and 62102445 in part by the Scientific Research Project of National University of Defense Technology through grant ZK19-03.

In thermodynamics, SA is used to optimize the thermodynamic model for maximum exergy efficiency [2]. In bioinformatics, SA is used to optimize the protein structures [12]. However, when there are numerous local optimums in the fitness function as shown in Fig. 1, SA tends to fall into the first few local optimums as the temperature decreases. As a result, SA is unable to obtain a superior candidate solution.

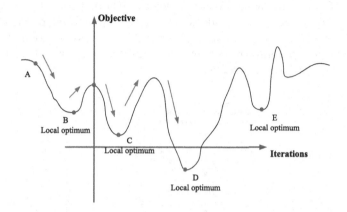

Fig. 1. Objective Function with Numerous Local Optimums.

In this paper, to solve the problem of premature convergence of SA, a Group Simulated Annealing Algorithm is proposed with a group strategy in which each solution in the group is separated for the annealing strategy. The rest sections of this paper are organized as follows. Section 2 describes the related work about SA. Section 3 describes the proposed Group Simulated Annealing Algorithm. Section 4 presents the notation descriptions, actual value, and the performance of the Group Simulated Annealing Algorithm. The conclusion and future work are drawn in Sect. 5.

2 Related Work

'Annealing' is a physics term for the process of heating up an object in cooling. SA applies the physical annealing of materials into an evolutionary algorithm, that can effectively solve the local optimal solution problem [7]. SA contains two parts: Metropolis Criterion and the annealing process. Metropolis Criterion provides a method to jump out in case of a local optimal solution by accepting new states with probability instead of using fully determined rules. In the annealing process, assuming that the previous decision variable is x_n. According to several variation criteria, the state of the system changes to x_{n+1}. Correspondingly, the objective function becomes as $f(x_{n+1})$, define the accept probability P for the system to change from x_n to x_{n+1} as:

$$P = \begin{cases} 1 & , \quad f(x_n) < f(x_{n+1}) \\ e^{-\frac{f(x_{n+1}) - f(x_n)}{T}} & , \quad f(x_{n+1}) \geq f(x_n) \end{cases} \tag{1}$$

The flowchart of SA is shown in Fig. 2. Firstly, SA randomly generates the initial solution and calculates the objective function. Then, the new solution is generated with perturbation. Next, the new solution is accepted according to the Metropolis Criterion. Then, SA judges whether the number of iterations n_{iter} reaches the preset value. If n_{iter} does not reach the preset value, a new solution is generated and accepted according to the Metropolis Criterion. If reached, SA reduces the temperature and returns to generate a new solution until the termination conditions are satisfied.

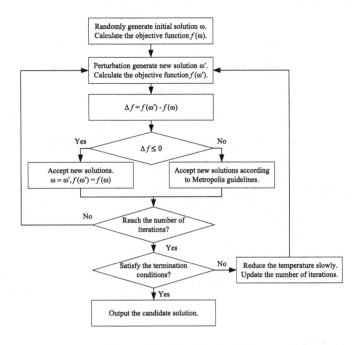

Fig. 2. The Flowchart of SA.

3   Detailed Description of the Algorithm

3.1   Group Strategy

The simulated annealing algorithm is greatly affected by the cooling coefficient and is prone to premature convergence. To avoid premature convergence, we design the grouping framework as shown in Fig. 3. In this framework, groups and the objective function of groups are initialized. Then separate the groups into several solutions. Next, a simulated annealing algorithm is performed for each solution. Then, all the solutions are combined back to a new group. Finally, judge the termination condition and output the candidate solutions.

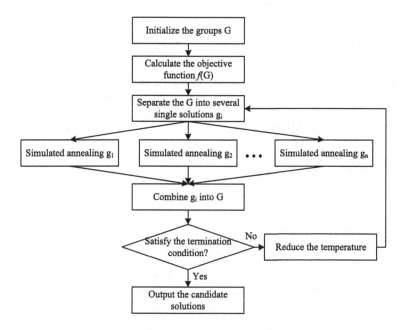

Fig. 3. Group Strategy Framework.

3.2 Group Simulated Annealing Algorithm

The detail of the Group Simulated Annealing Algorithm is shown in Alg.1. Firstly, the group of solutions is initialized with the objective function. Then, the current temperature is initialized according to the preset value. Next, when the times of fitness evaluation do not reach the MaxFE, keep running the algorithm. For each time in the annealing, each solution in the group mutates with probability p_{mu}. Then, the solutions that do not satisfy the boundary constraints are adjusted. After calculating the objective function of the new group, each solution in the new group is accepted according to the Metropolis criteria. Finally, the current temperature and solution variation range reduce via temperature decay coefficient λ.

4   Empirical Studies

4.1   Parameter Setting

The details of parameter sets are shown in Tab. 1.

Algorithm 1. Group Simulated Annealing Algorithm

Input: n_{Group}: number of solutions in the group; n_{Anneal}: times of annealing; $MaxFE$: max function evaluations; T_{init}: initial temperature; λ: temperature decay coefficient; σ: solution variation range; p_{mu}: posibility of mutation; D: dimensions of decision variables.

Output: Candatate solutions G

1: Initialize the group G;
2: Calculate the objective function $f(G)$;
3: Initialize the temperature T;
4: **while** $FE \leq MaxFE$ **do**
5:     **while** $n_{iter} \leq n_{Anneal}$ **do**
6:         **while** $i \leq n_{Group}$ **do**
7:             **for** $\mu \leq D$ **do**
8:                 **if** $Rand < p_{mu}$ **then**
9:                     $G_i'^{\mu} = G_i^{\mu} + \sigma \times Rand$
10:                     Adjust $G_i'^{\mu}$ to satisfy the boundary constraints;
11:                 **end if**
12:                 $\mu = \mu + 1$
13:             **end for**
14:             Calculate the objective function $f(G_i')$;
15:             **if** $Rand < \exp(-(f(G_i') - f(G_i))/(|f(G_i) + 10^{-6}| \times T))$ **then**
16:                 $G_i = G_i'$;
17:             **end if**
18:             $i = i + 1$;
19:         **end while**
20:         $T = T \times \lambda, \sigma = \sigma \times \lambda$;
21:         $n_{iter} = n_{iter} + 1$;
22:     **end while**
23: **end while**

Table 1. The Actual Value and Description of Experimental Parameter Setting.

Parameters	Actual value	Description
ID	1-10	the number of the function
D	10/20/50	the dimensions of the function
$MaxFE$	1E+5/3E+5/7E+5	the max function evaluation times
$runtime$	50	the number of runs
n_{anneal}	10	the number for each solution to perform SA
n_{Group}	10/30/70	the size of the group, $n_{Group} = MaxFE/n_{anneal}$
T	0.1	the initial temperature
p_{mu}	0.5	the posibility of mutation
σ	40	the initial variation interval
λ	0.99	the temperature decay coefficient

4.2   Experimental Environment

The testing platform for Group Simulated Annealing Algorithm is MATLAB R2020B. And the computer configuration information is i7-10750H CPU @ 2.59 GHz and 16 GB RAM, using win 10 OS.

4.3   Experimental Results

Fig. 4, 5, 6 shows the minimum value of ICSI-OC 2022 in 10D, 20D, and 50D. It can be seen that in different dimensions of ICSI-OC 2022, the GSA algorithm has better candidate solutions than the SA algorithm, and the solution is more stable.

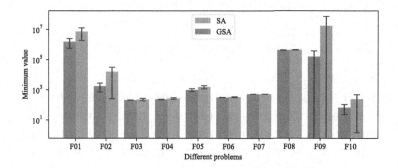

Fig. 4. The results for ICSI-OC 2022 with 10D in SA and GSA.

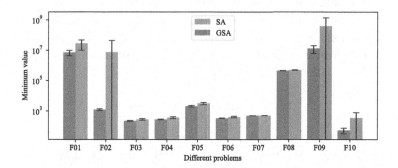

Fig. 5. The results for ICSI-OC 2022 with 20D in SA and GSA.

The recording table in different dimensions for Group Simulated Annealing is shown in Table 2, Table 3, Table 4.

4.4   Convergence Contrastive Analysis

Fig. 7 shows the convergence of SA and GSA during the evaluation. It can be clearly seen in the subfigure in Fig. 7 that the SA algorithm reaches convergence after 1000 iterations. In contrast, the optimized value of the GSA algorithm has been in a downward trend throughout the iterations.

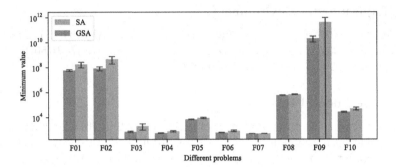

Fig. 6. The results for ICSI-OC 2022 with 50D in SA and GSA.

Table 2. Results for 10D problems.

No.	Best	Worst	Median	Mean	Std
F01	1.18E+05	4.08E + 06	1.43E + 06	1.55E + 06	9.73E + 05
F02	1.01E + 03	5.20E + 03	1.24E + 03	1.73E + 03	1.02E + 03
F03	2.00E + 02	2.10E + 02	2.03E + 02	2.04E + 02	3.01E + 00
F04	2.15E + 02	2.51E + 02	2.25E + 02	2.27E + 02	9.24E + 00
F05	4.41E + 02	1.39E + 03	9.82E + 02	9.57E + 02	2.07E + 02
F06	3.01E + 02	3.18E + 02	3.06E + 02	3.07E + 02	4.02E + 00
F07	5.03E + 02	5.09E + 02	5.06E + 02	5.06E + 02	1.36E + 00
F08	4.17E + 05	4.42E + 05	4.33E + 05	4.33E + 05	5.93E + 03
F09	8.34E + 02	1.28E + 06	7.74E + 04	1.59E + 05	2.26E + 05
F10	1.14E + 01	2.08E + 02	5.57E + 01	6.54E + 01	4.25E + 01

Table 3. Results for 20D problems.

No.	Best	Worst	Median	Mean	Std
F01	8.62E + 05	1.28E + 07	6.96E + 06	7.20E + 06	2.73E + 06
F02	1.05E + 03	1.93E + 03	1.20E + 03	1.24E + 03	1.65E + 02
F03	2.07E + 02	2.57E + 02	2.19E + 02	2.24E + 02	1.19E + 01
F04	2.46E + 02	3.09E + 02	2.82E + 02	2.80E + 02	1.40E + 01
F05	1.45E + 03	2.54E + 03	2.08E + 03	2.07E + 03	2.32E + 02
F06	3.16E + 02	3.62E + 02	3.36E + 02	3.37E + 02	1.13E + 01
F07	5.05E + 02	5.08E + 02	5.07E + 02	5.07E + 02	7.63E-01
F08	4.41E + 05	4.80E + 05	4.69E + 05	4.67E + 05	8.53E + 03
F09	9.83E + 05	3.35E + 07	1.24E + 07	1.34E + 07	6.99E + 06
F10	1.82E + 01	1.25E + 02	5.21E + 01	5.47E + 01	2.20E + 01

Table 4. Results for 50D problems.

No.	Best	Worst	Median	Mean	Std
F01	3.97E + 07	8.33E + 07	5.91E + 07	5.95E + 07	1.02E + 07
F02	3.11E + 07	1.46E + 08	8.47E + 07	8.43E + 07	3.03E + 07
F03	5.82E + 02	8.70E + 02	7.29E + 02	7.27E + 02	8.27E + 01
F04	4.86E + 02	6.29E + 02	5.75E + 02	5.69E + 02	3.26E + 01
F05	5.94E + 03	7.88E + 03	6.91E + 03	6.88E + 03	4.00E + 02
F06	5.59E + 02	6.59E + 02	6.02E + 02	6.06E + 02	2.70E + 01
F07	5.08E + 02	5.10E + 02	5.09E + 02	5.09E + 02	4.80E-01
F08	5.65E + 05	6.43E + 05	6.11E + 05	6.10E + 05	1.75E + 04
F09	5.69E + 09	5.23E + 10	1.96E + 10	2.08E + 10	1.01E + 10
F10	2.03E + 04	3.27E + 04	2.75E + 04	2.75E + 04	2.93E + 03

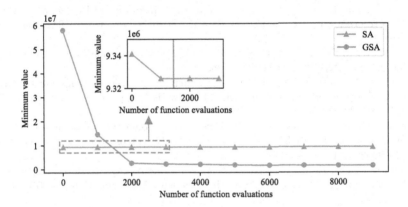

Fig. 7. The evolution of SA and GSA with ICSI-OC 2022 in 10D/F01.

5   Conclusion and Future Work

In this paper, a Group Simulated Annealing Algorithm is proposed for ICSI-OC 2022. To avoid premature convergence for the SA, a group strategy is developed in which each solution in the group is separated for the annealing strategy. Finally, the algorithm's performance is evaluated using ICSI-OC 2022 test issues with 10/20/50 dimensions. Compared with SA, GSA can get a better solution after convergence.

References

1. Alijla, B.O., Lim, C.P., Wong, L.P., Khader, A.T., Al-Betar, M.A.: An ensemble of intelligent water drop algorithm for feature selection optimization problem. Appl. Soft Comput. **65**, 531–541 (2018)
2. Çetin, G., Keçebaş, A.: Optimization of thermodynamic performance with simulated annealing algorithm: A geothermal power plant. Renewable Energy **172**, 968–982 (2021)
3. Delahaye, D., Chaimatanan, S., Mongeau, M.: Simulated annealing: from basics to applications. In: Gendreau, M., Potvin, J.-Y. (eds.) Handbook of Metaheuristics. ISORMS, vol. 272, pp. 1–35. Springer, Cham (2019). https://doi.org/10.1007/978-3-319-91086-4_1
4. Dorigo, M., Birattari, M., Stutzle, T.: Ant colony optimization. IEEE Comput. Intell. Mag. **1**(4), 28–39 (2006)
5. Glover, F., Laguna, M.: Tabu search. In: Handbook of Combinatorial Optimization, pp. 2093–2229. Springer (1998). https://doi.org/10.1007/978-1-4613-0303-9_33
6. Kennedy, J., Eberhart, R.: Particle swarm optimization. In: Proceedings of ICNN 1995-International Conference on Neural Networks, vol. 4, pp. 1942–1948. IEEE (1995)
7. Kirkpatrick, S., Gelatt, C.D., Jr., Vecchi, M.P.: Optimization by simulated annealing. Science **220**(4598), 671–680 (1983)
8. Liu, L., Yin, B., Zhang, S., Cao, X., Cheng, Y.: Deep learning meets wireless network optimization: Identify critical links. IEEE Trans. Netw. Sci. Eng. **7**(1), 167–180 (2018)
9. Mirjalili, S.: Genetic algorithm. In: Evolutionary Algorithms and Neural Networks. SCI, vol. 780, pp. 43–55. Springer, Cham (2019). https://doi.org/10.1007/978-3-319-93025-1_4
10. Vincent, F.Y., Redi, A.P., Hidayat, Y.A., Wibowo, O.J.: A simulated annealing heuristic for the hybrid vehicle routing problem. Appl. Soft Comput. **53**, 119–132 (2017)
11. Yeh, W.C., Lin, Y.C., Chung, Y.Y., Chih, M.: A particle swarm optimization approach based on monte carlo simulation for solving the complex network reliability problem. IEEE Trans. Reliab. **59**(1), 212–221 (2010)
12. Zhang, L., Ma, H., Qian, W., Li, H.: Protein structure optimization using improved simulated annealing algorithm on a three-dimensional ab off-lattice model. Comput. Biol. Chem. **85**, 107237 (2020)

Author Index

Printed in the United States
by Baker & Taylor Publisher Services